Meat Quality

Genetic and Environmental Factors

Chemical and Functional Properties of Food Components Series

SERIES EDITOR

Zdzisław E. Sikorski

Meat Quality: Genetic and Environmental Factors
Edited by Wiesław Przybylski and David Hopkins

Food Oxidants and Antioxidants: Chemical, Biological, and Functional Properties
Edited by Grzegorz Bartosz

Fermentation: Effects on Food Properties
Edited by Bhavbhuti M. Mehta, Afaf Kamal-Eldin and Robert Z. Iwanski

Methods of Analysis of Food Components and Additives, Second Edition
Edited by Semih Otles

Food Flavors: Chemical, Sensory and Technological Properties
Edited By Henryk Jelen

Environmental Effects on Seafood Availability, Safety, and Quality
Edited by E. Grazyna Daczkowska-Kozon and Bonnie Sun Pan

Chemical and Biological Properties of Food Allergens
Edited By Lucjan Jedrychowski and Harry J. Wichers

Chemical, Biological, and Functional Aspects of Food Lipids, Second Edition
Edited by Zdzisław E. Sikorski and Anna Kołakowska

Food Colorants: Chemical and Functional Properties
Edited by Carmen Socaciu

Mineral Components in Foods
Edited by Piotr Szefer and Jerome O. Nriagu

Chemical and Functional Properties of Food Components, Third Edition
Edited by Zdzisław E. Sikorski

Carcinogenic and Anticarcinogenic Food Components
Edited by Wanda Baer-Dubowska, Agnieszka Bartoszek and Danuta Malejka-Giganti

Toxins in Food
Edited by Waldemar M. Dąbrowski and Zdzisław E. Sikorski

Chemical and Functional Properties of Food Saccharides
Edited by Piotr Tomasik

Chemical and Functional Properties of Food Proteins
Edited by Zdzisław E. Sikorski

Meat Quality

Genetic and Environmental Factors

EDITED BY
Wiesław Przybylski, PhD
David Hopkins, PhD

CRC Press is an imprint of the
Taylor & Francis Group, an **informa** business

CRC Press
Taylor & Francis Group
6000 Broken Sound Parkway NW, Suite 300
Boca Raton, FL 33487-2742

First issued in paperback 2018

ISBN 13: 978-1-138-89407-5 (pbk)
ISBN 13: 978-1-4822-2031-5 (hbk)

Visit the Taylor & Francis Web site at
http://www.taylorandfrancis.com

and the CRC Press Web site at
http://www.crcpress.com

Contents

Preface

Domestication of animals has underpinned human development, and this has led to changes in lifestyle and diet. Meat is a source of highly nutritious and easily accessible protein, high energy fat, B vitamins, and other microelements (Higgs 2000; Jiménez-Colmenero et al. 2001; McAfee et al. 2010). Persistent and patient breeding work by breeders over generations has resulted in changes in animal genotypes leading to a diversity of breeds. Data suggest that globally there are about 790 cattle breeds, 920 sheep breeds, 410 goat breeds, 380 horse types, and about 440 pig types (Nowicki et al. 2001). These numbers indicate the significant genetic potential of domesticated farm animals and this resource has been central to the provision of food requirements for the human population given the high biological (nutritional) value of meat (Higgs 2000; Nowicki et al. 2001). By-products from animals have also aided human development (e.g., materials for clothes and manure for heat generation). There has been an increase in the number of domesticated farm animals in response to an increase in the global human population and the consequential growth in demand for food. For example, in the twentieth century the number of cattle has tripled and the number of pigs increased sixfold. In 2011, it was estimated that world's cattle population was around 1400 million animals, there were 1100 million sheep, and 980 million pigs (CSO 2013). The production of meat has increased significantly with a fivefold increase in the second half of the twentieth century. Statistics show that in 2011, global meat production reached 298 million tonnes. About 37% of it was pork, 34% poultry, 21% beef, and 5% sheep and goat meat (CSO 2013). During that period, intensive poultry production has shown a large increase.

Thanks to genetics and the development of computational methods, the growth rate of animals across species has shown a large increase (Roehe et al. 2003). For some species, fertility has improved as well. Meat is actually eaten quite commonly across the globe, but there is significant diversity in consumption patterns and amounts. In the poor countries of Africa and Asia, the meat intake amounts to about 4 kg per capita over the year while in the United States it reaches 120 kg/year (per capita) (Pisula 2011). It is predicted that along with the increase in the global human population, that meat production will keep rising as well. For example by 2050, human civilization is predicted to reach 9 billion people and meat production to reach 624 million tonnes, and this raises the question if such production is actually possible and whether the disproportionate consumption of meat between individual regions of the world will be reduced? (Pisula 2011). As research shows, it is clear that one of the important and deciding elements about the level of meat consumption will be its quality and price. Studies show that consumers expect meat to be characterized by high sensory quality, high nutritional value, and more and more frequently they take an interest in producer's guarantee of high quality, safety and care for the welfare of animals in the whole production chain (Pisula 2011). Moreover, literature increasingly mentions meat as a functional food because it contains n-3 fatty acids, bioactive peptides (carnosine, glutathione), choline, and some of the micro- and macro elements or vitamins (Jiménez-Colmenero et al. 2001).

One of the challenges faced by producers is the requirement to improve the quality of meat. Some of this has been stimulated by consumers increasingly demanding requirements of meat and breeding programs, which has lifted the daily gain or increased meat content in carcasses leading to lowered eating quality. For example, in pigs, over past 20 years, the fat content has been reduced by 30% (Higgs 2000). That selection pressure was a result of the market need for low-fat pork. Contemporary humans in highly industrialized countries need low-energy diets (excess of the energy—especially in the form of an animal fat, consisting of a high quantity of saturated fatty acids that leads to development of many diseases of the affluence). Thus, for example, in lean pigs it was noted that increased stress susceptibility and the formation of defective meat after slaughter like PSE (pale soft exudative) or too low an intramuscular fat content led to a decline in the sensory quality of the meat. Similarly, the intensification of animal breeding methods, that is, building of large-scale production poultry farms and introduction of mechanization and automation, resulted in the emergence of stressful living conditions (i.e., keeping animals on scaffolding floors). Progress and consolidation in the meat industry (lower number of slaughterhouses, but large-scale production) caused preslaughter handling and slaughter to become highly stressful as well. These issues have stimulated development of new approaches to breeding and preslaughter handling. These issues will be discussed in this book.

The topics covered in the book include animal welfare, nutrition, preslaughter marketing, slaughter technology, slaughter and carcass dressing as well as the impact of currently known genetic factors on the quality of meat. To present a comprehensive discussion of all issues related to meat quality, the book focuses on knowledge about meat definitions, meat quality, chemical composition, and the histological structure of meat. The book also describes the effects of postmortem changes on meat quality and covers the determinants of meat quality in a comprehensive way. Apart from the influence of genetic factors such as species, race, gender, or genes, the main issues to be considered in breeding programs to improve meat quality are also described across the species of cattle, sheep, and pigs. The book also outlines meat production systems to ensure high meat quality and the place of transgenic animal technology and meat quality is discussed.

The chapters have been written by authors from universities, research institutes, and breeding companies from different countries and continents: Australia, France, Ireland, Poland, Republic of China, Scotland, South Africa, the United States, and the United Kingdom. The chapters are based on the research and teaching experience of the contributors as well as on a critical evaluation of the current literature. The chapters in the book are new, prepared specially for this monograph. In the world literature there is no book that covers the issues of integrating production and meat quality. Previous books relate rather to processing technology or more individual species and they do not provide such an approach to the knowledge in this field. The target group of this book is those students who study food technology and nutrition, commodities sciences, animal sciences and engineering staff in the industry. The book is also addressed to food scientists in the industry and academia and to all those who are interested in meat science. It should also be of interest to the meat industry concerned with updating their theoretical knowledge about the determinants of meat quality.

Special thanks are due to the authors who prepared their chapters.

Wiesław Przybylski
David Hopkins

REFERENCES

Central Statistical Office. 2013. *Statistical Yearbook of Agriculture*. International Review, Warsaw.

Higgs, J.D. 2000. The changing nature of red meat: 20 years of improving nutritional quality. Review. *Trends Food Sci. Technol.* 11:85–95.

Jiménez-Colmenero, F., Carballo, J., and Cofrades, S. 2001. Healthier meat and meat products: Their role as functional foods. Review. *Meat Sci.* 59:5–13.

McAfee, A.J., McSorley, E.M., Cuskelly, G.J., Moss, B.W., Wallace, J.M.W., Bonham, M.P., and Fearon, A.M. 2010. Red meat consumption: An overview of the risks and benefits. *Meat Sci.* 84:1–13.

Nowicki, B., Jasek, S., Maciejowski, J., Nowakowski, P., and Pawlina, E. 2001. *Breeds of Farm Animals*. Scientific Publishing PWN, Warsaw, pp. 1–195.

Pisula, A. 2011. Introduction. In: Pospiech, E. and Pisula, A. (eds.), *Meat—Basis of Science and Technology*. SGGW-Warsaw Life Science University Press, Warsaw, pp. 15–21.

Roehe, R., Plastow, G.S., and Knap, P.W. 2003. Quantitative and molecular genetic determination of protein and fat deposition. *HOMO* 54:119–131.

Editors

Wiesław Przybylski is the professorial head of the Department of Food Gastronomy and Food Hygiene in the Faculty of Human Nutrition and Consumer Sciences at the Warsaw University of Life Sciences. He was vice dean of research for four years (2008–2012). He studied at the Pedagogical-Agricultural University in Siedlce and received a master's degree in animal science. He earned a PhD in animal science and then completed a 10-month postdoc scientific practice at the Station of Research of Meat in Theix and at the Station of Quantitative and Applied Genetics in Jouy en Josas (near Paris) of INRA (the Institut National de la Recherche Agronomique) in France. He prepared a habilitation thesis on the factors influencing the glycolytic potential in pig muscle in the Faculty of Animal Science in Warsaw University of Life Science. He received the title of professor of agricultural sciences—food technology from the president of the Republic of Poland. He worked in Siedlce Agricultural University, in Białystok Technical University and as director of the Primary and Secondary School in Brańszczyk. Professor Przybylski has supervised 75 graduate engineering students, 54 master's students, and 6 PhD students while tutoring a further 4 PhD dissertations. He lectures in the following courses: technology products of animal origin, meat technology, genetics, food technology. He has authored 265 publications and 5 books for secondary schools and universities. His areas of research interest are the effect of environmental and genetic factors on meat quality, traditional foods, and applications of multidimensional statistical methods in food technology. Professor Przybylski is a member of three scientific societies: Polish Society of Animal Sciences, Polish Society of Food Technology, and Polish Society of Nutritional Sciences. He was awarded the Medal of the National Education Commission and also the Medal of the Ministry of Agriculture and Rural Development. Professor Przybylski is married to Agnes and they have three daughters and three sons. He is a Christian grateful to God for everything.

David Hopkins has two degrees from the University of Melbourne and earned a PhD at the University of New England focusing on biochemical mechanisms responsible for tenderization of meat. He worked for the Tasmanian Department of Agriculture for six years before moving into a newly created position with NSW DPI (Centre for Red Meat and Sheep Development). As a senior principal research scientist with a focus on meat science he has built a team of people, and currently supervises three technical staff, two scientists and also supervises postgraduate students from around the world. He currently supervises three PhD students and a master's student.

He is the sole meat scientist in NSW DPI and has published extensively (more than 400 papers) on meat and carcass studies and currently sits on the editorial board of two journals, and is the editor in chief of the journal *Meat Science.* He is an adjunct professor at Charles Sturt University (CSU) and the University of New England (UNE). He has authored a number of seminal papers and has been invited to write 12 book chapters. He has led several projects in the Australian Sheep Industry Cooperative Research Centre and has conducted numerous studies with

Meat & Livestock Australia and Australian Meat Processor Corporation funding and works collaboratively with scientists internationally. He is currently studying the following issues: (1) factors impacting on the tenderization of meat, (2) growth, carcass composition, and meat quality in sheep, (3) application of electrical technologies to the processing industry, (4) application of Raman spectroscopy to the meat industry, (5) methods to improve the quality of lamb, beef, goat, and alpaca meat, and (6) assessment of technology for predicting meat yield and quality traits. David is a born-again Christian who grew up on a grazing property in the South West of Victoria, Australia. He is married to Kerry and they have two sons.

Contributors

Dominique Bauchart
Herbivore Research Unit
National Institute for Agricultural Research
Saint Genès Champanelle, France

and

Clermont Université, VetAgro Sup
Clermont-Ferrand, France

Lutz Bünger
Animal Breeding and Genetics
Scotland's Rural College
Edinburgh, United Kingdom

Heather A. Channon
Australian Pork Limited
Barton, Australia

Darryl N. D'Souza
Australian Pork Limited
Barton, Australia

Min Du
Department of Animal Sciences
Washington State University
Pullman, Washington

Temple Grandin
Department of Animal Sciences
Colorado State University
Fort Collins, Colorado

Jean-François Hocquette
Herbivore Research Unit
National Institute for Agricultural Research
Saint Genès Champanelle, France

and

Clermont Université, VetAgro Sup
Clermont-Ferrand, France

Benjamin W.B. Holman
Department of Primary Industries
Centre for Red Meat and Sheep Development
Cowra, Australia

David Hopkins
Department of Primary Industries
Centre for Red Meat and Sheep Development
Cowra, Australia

Danuta Jaworska
Department of Food Gastronomy and Food Hygiene
Warsaw University of Life Sciences—SGGW
Warsaw, Poland

Joseph P. Kerry
School of Food and Nutritional Sciences
University College Cork
Cork City, Ireland

Yuan H. Brad Kim
Department of Animal Sciences
Purdue University
West Lafayette, Indiana

Maria Koćwin-Podsiadła
Department of Bioengineering and Animal Husbandry
Siedlce University of Natural Sciences and Humanities
Siedlce, Poland

Roman Kołacz
Department of Environment Hygiene and Animal Welfare
Wrocław University of Environmental and Life Sciences
Wrocław, Poland

Danuta Kołożyn-Krajewska
Department of Food Gastronomy and Food Hygiene
Warsaw University of Life Sciences —SGGW
Warsaw, Poland

Elżbieta Krzęcio-Nieczyporuk
Department of Health Science
Siedlce University of Natural Sciences and Humanities
Siedlce, Poland

Robert Kupczyński
Department of Environment Hygiene and Animal Welfare
Wrocław University of Environmental and Life Sciences
Wrocław, Poland

Nicola R. Lambe
Kirkton Farm
Scotland's Rural CollegeCrianlarich, Scotland, United Kingdom

Didier Micol
Herbivore Research Unit
National Institute for Agricultural Research
Saint Genès Champanelle, France

and

Clermont Université, VetAgro Sup
Clermont-Ferrand, France

Suzanne I. Mortimer
NSW Department of Primary Industries
Agricultural Research Centre
Trangie, Australia

Paul E. Mozdziak
Prestage Department of Poultry Science
College of Agriculture and Life Sciences
North Carolina State University
Raleigh, North Carolina

James N. Petitte
Prestage Department of Poultry Science
College of Agriculture and Life Sciences
North Carolina State University
Raleigh, North Carolina

Brigitte Picard
Herbivore Research Unit
National Institute for Agricultural Research
Saint Genès Champanelle, France

and

Clermont Université, VetAgro Sup
Clermont-Ferrand, France

Rod Polkinghorne
Farmer/Consultant
Murrurundi, Australia

Eric N. Ponnampalam
Agriculture Research and Development
Department of Economic Development, Jobs, Transport and Resources
Victoria, Australia

Edward Pospiech
Institute of Meat Technology
Poznan University of Life Sciences
Poznań, Poland

Wiesław Przybylski
Department of Food Gastronomy and Food Hygiene
Warsaw University of Life Sciences —SGGW
Warsaw, Poland

Joe M. Regenstein
Department of Food Science
Cornell University
Ithaca, New York

Véronique Santé-Lhoutellier
Department of Quality of Animal Products
National Institute for Agricultural Research
Saint Genès Champanelle, France

Qingwu W. Shen
College of Food Science and Technology
Hunan Agricultural University
Changsha, China

Andrzej Sośnicki
Genus PIC
Department of Global Genetic Development & Technical Services
Hendersonville, Tennessee

Phillip E. Strydom
Animal Production Institute
Agricultural Research Council
Irene, South Africa

Claudia E.M. Terlouw
Herbivore Research Unit
National Institute for Agricultural Research
Saint Genès Champanelle, France

Andrzej Zybert
Department of Bioengineering and Animal Husbandry
Siedlce University of Natural Sciences and Humanities
Siedlce, Poland

1 Meat and Muscle Composition

Structure of Muscle, Chemical and Biochemical Constitution of Muscle, Nutritional Value. Species and Breed Characteristics

Véronique Santé-Lhoutellier and Edward Pospiech

CONTENTS

1.1 INTRODUCTION

The term meat most often describes all the parts of warm-blooded animals which are dedicated for consumption by people, that is, the meat of large animals for slaughter, game, and poultry. The ever broader access of humans to the water world makes the term practically encompass other living organisms such as fish and invertebrates. The definition of meat concerns not only the muscle system containing muscles, but also relates to bones, some viscera, and parts of animals which occur along with them and can be consumed by humans. Despite such a broad definition of meat, the present chapter will focus mainly on muscle structure and histochemical characteristics which determine the course of metabolic processes during an animal's life and postslaughter as well as denote the muscles' nutritional value, their suitability for culinary purposes and for processing.

1.2 STRUCTURE OF MUSCLE

The meat is the result of physical and biochemical changes of the muscle. The chemical composition and the structure of the muscles show a wide variability. This results from the influence of a number of factors among which particularly important ones include an animal's species, sex and age, meatiness and fatness, as well as muscle growth rate, type of muscle, and the location and function which the muscle serves *in vivo*. The effect of these factors is largely directed by the genome, however environmental effects are also significant and their role can even become dominant when breeding advancements have reached such a high level that further improvement is easier to achieve through changes in rearing and meat processing technology (Kwasiborski et al., 2008a,b; Gou et al. 2012; Hamill et al. 2012; Hocquette et al. 2012).

Muscle is attached to bones by tendons. Tendons transfer muscle contraction to the skeletal system and constitute fibrous and elastic connective tissue. Collagen or elastic fibers separate single muscle fibers, their groups in form of bundles, and at the end surround the whole muscle from the outside. Blood vessels and nerves are interwoven in connective tissue which stores fat tissue, shapes the nutritional value of meat, gives it certain physicochemical properties, and forms its marbling (Josza et al. 1993; Nishimura et al. 1999; Hamill et al. 2012; Dubost et al. 2013b).

Skeletal muscle consists of hundreds, or even thousands, of muscle fibers bundled together and wrapped in a connective tissue covering. Muscle fibers are grouped into clusters divided by connective tissue, ubiquitous in the whole muscle structure and which provides the mechanical stability to the muscle (Swartz et al. 2009). Single muscle fibers are surrounded by connective tissue called *endomysium*. Bundles of muscle fibers are surrounded by a connective tissue called *perimysium* and the muscle as a whole is covered by an external membrane called *epimysium* (Figure 2.1; Bailey and Light 1989; Torrescano et al. 2003; McCormick 2009). The membrane which surrounds thick clusters of muscle fibers is identified as primary. Further membranes which stem from the primary one are called secondary and tertiary membranes. They are distinct, smaller from the primary one, and comprise smaller and smaller units of the muscle. Muscle is also often surrounded by its own fascia. The direction of muscle fibers can be parallel to the long axis of the muscle, or oriented at an oblique angle to the muscle's longitudinal direction. Fiber orientation influences the eating properties of meat. This is important when preparing meat for consumption. A longitudinal muscle fiber layout in meat chops will lead to the perception of tendinous meat by the consumer, and a perpendicular layout to a perception of tender meat.

Muscle fibers are multinucleated cells surrounded by a cell membrane (sarcolemma). The cytoplasm, also called sarcoplasm for muscle cells, contains contractile elements constituted by myofibrils (Figure 1.1). The basic structural unit is the sarcomere corresponding to the anatomical and functional unit of a myofibril, bounded by Z lines (Figure 1.1). Muscle fibers are shaped like elongated prisms whose cross-section is usually a pentagon, quadrangle or, less often, a triangle. They are created as a result of the fusion of precursor cells during embryonic development (Lefaucheur and Gerrard 2000; Picard et al. 2002). Sometimes, with muscle defects so-called giant fibers are observed, whose cross-section is close to an oval (Fiedler et al. 1999; Lefaucheur and Gerrard 2000). Their existence was questioned and it has been proposed to treat them as muscular, pathological results of past stresses and not as an additional type of normal muscle cells (Sosnicki 1987). A muscle fiber contains approximately 1000 myofibrils, which run through the whole length of a cell and consist of sarcomeres. The number of fibers in each myofibril varies, depending on cell length, from a couple hundred to a few hundred thousand. At a standstill, a single sarcomere is usually 2.0 μm long, rarely above 3.0 μm (Millman 1998; Wheeler and Koohmaraie 1999). Regular layout of sarcomeres gives muscle fibers a characteristic striated appearance (alternating I and A bands). Striation is a consequence of varied refraction of light by those bands of the sarcomere seen under a light microscope. The variation results from the presence of two types of filaments: thin—actin and thick—myosin. The dark band (A band) is created by myosin filaments and partially present actin filaments which dominate in the light area (I), which is divided into two parts by the Z line. Striation differentiates muscles. Striated fibers are responsible for the movement of the body, which distinguishes them from smooth (nonstriated, innervated by the autonomic nervous system) and cardiac (involuntary striated muscle) fibers.

With longer sarcomeres, an M line can be observed in the middle of the A area, in the center of its lighter part defined as the H zone (Figure 1.1). Myomesin located in the H zone has been proposed as the main protein within the filament cross-linking the sarcomere, similar to α-actin in the Z-line. Actin filaments are anchored in Z lines and finish on the edge of the H area. This part of the sarcomere is rich in enzymes participating in glycolysis (see Chapter 3). Myosin filaments are located in the center of the sarcomere. These structures can only be observed using an electron microscope using specific staining (Bayraktaroglu and Kahraman 2011; Li et al. 2012).

Filaments as well as whole sarcomeres are well organized structures. The myosin filament, also known as the thick filament, encompasses two heavy chains and four light chains. They create the rod and head of the myosin filament. Using trypsin it is

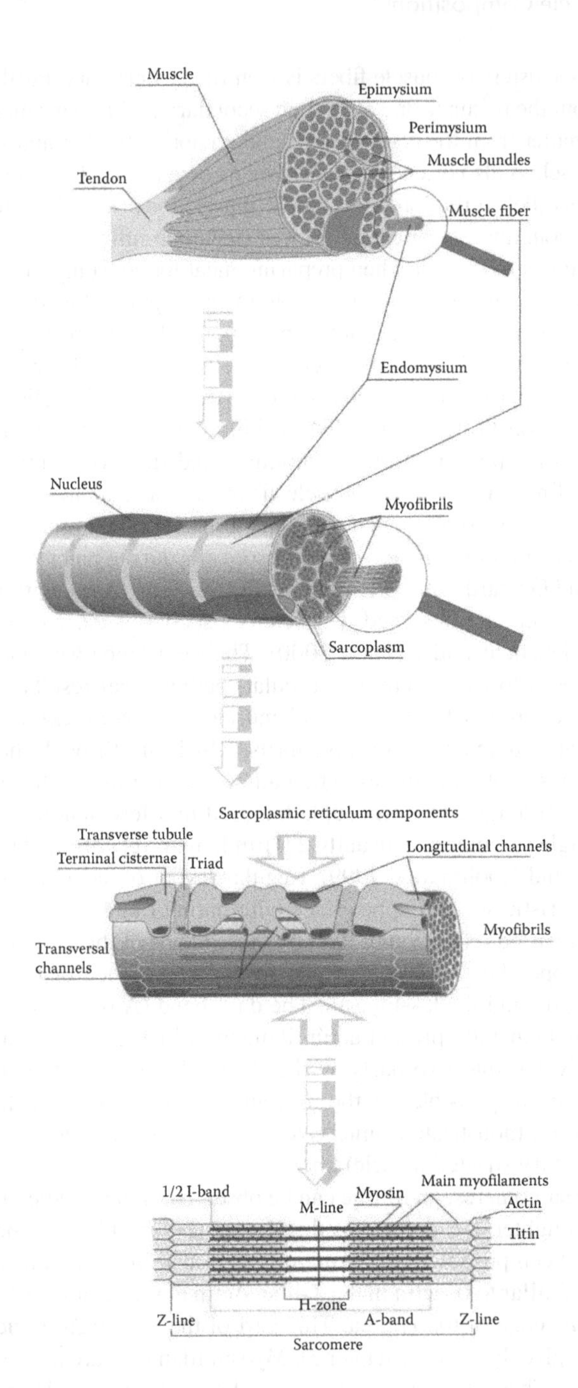

FIGURE 1.1 Muscle structure sketch.

possible to divide this filament into light and heavy meromyosin. The heavy portion comprises the myosin heads and shows ATPase activity, which means that it hydrolyses ATP and enables merging of myosin and actin filaments. Due to this reaction muscle contraction takes place. The actin filament (thin) has troponin proteins (C, T, and I) and tropomyosin (TPM) entwined in its structure alongside globular actin. The troponin complex together with TPM provide the calcium-sensitive switch that activates muscle contraction. Actin reacts with numerous proteins (e.g., tropomodulins and nebulette) and affects the organization of molecules as well as the structure of meat. Nebulette is a cardiac-specific isoform of nebulin and is important for the maintenance of TPM, troponin, and thin filament length (Pappas et al. 2011). Tropomodulins constitute a family of four proteins, which regulate actin function, stabilize its length and architecture (Yamashiro et al. 2012). For this reason actin is classed as a cytoskeletal protein. For the thin filament, the support element stabilizing its structure and position in the sarcomere is nebulin, a cytoskeletal protein. A similar role is played for the thick filament by titin, which is also one of the cytoskeletal proteins. Titin constitutes very thin filaments which run between Z and M lines. The portion of the thin filaments located in the I band is significantly folded (structures typical for immunoglobulin domains) which provides flexibility and elasticity of the muscle fiber and protects it from tearing. Protein C, which binds separate myofilaments into one also plays an important role in the forming of the thick filament. Aside from motor and structural functions, myosin and actin can also serve other tasks, for example, determining the ability for gel or emulsion formation, therefore investigating their part in shaping meat properties can be multifaceted (Acton et al. 1981; Asghar et al. 1985; Sakamoto et al. 2013).

Connective tissue has the specific function of strengthening the tissue structure, but its quantity and composition change with the age of animal (Bailey and Light 1989; McCormick 2009; Christensen et al. 2011). Fresh muscle tissue usually contains from about 2.25 mg/g muscle (*m. psoas major, m. gluteus medius*, and *m. longissimus dorsi*) to more than 9.0 mg/g (muscles from the chuck and round) of collagen (Prost et al. 1975; Rhee et al. 2004). This quantity depends on the age of the animal, type of muscle, and diet. Up to now 29 different types of collagen have been identified (Gelse et al. 2003; Lepetit 2007, 2008). Type I and III collagens are the most abundant in a mature skeletal muscle. The first type provides muscle stability and the other extensibility and elasticity to fibers.

A significant role in meat tenderness is played by the amount of collagen and the intra- and intermolecular cross-links. The animal's age has a large effect on the degree of collagen cross-linking and to some extent on the solubility with the former feature increasing and the latter decreasing as the animals get older (Bailey and Light 1989; Nishimura et al. 1999, 2008; Eggen et al. 2001). This change applies to the meat of all animal species, however the accretion rate varies according to species, for example, faster in pigs than in cattle (Wegner et al. 2000). Cross-links between collagen chains undergo degradation during meat aging (Nishimura et al. 1996, 1998, 2008). These changes are visible after extended postslaughter (in some cases of cattle muscles after more than 14 days) and depend on the type of muscle (Nishimura et al. 1999). Cracks appear both in *epimysium* and in *perimysium*. Because, however, over 90% of intramuscular collagen fibers are located in the *perimysium* membrane, its structure has a large impact on meat hardness and the differences between the tenderness of different muscles of the same carcass can be linked mainly to its thickness and content in muscles (Bailey and Light 1989; Brooks and Savell 2004). Proteoglycans and glycoproteins play a role in the weakening of collagen structure, and these are an integral component of connective tissue structure (Nishimura et al. 1996; Dubost et al. 2013a). After their degradation collagen can become more prone to proteolysis. The degradation of proteoglycans is also used to explain the postslaughter changes in collagen which lead to lowering the temperature required for its contraction. Muscles usually contain small amounts of elastin. The major difference between collagen and elastin arises from physicochemical properties mainly including denaturation and contraction temperatures. It is assumed that collagen denatures at about 65°C, while elastin at over 100°C, usually around 121°C (Bailey and Light 1989).

Denaturation and contraction temperatures are higher when the degree of collagen cross-linking and the content of hydroxyproline (hx), a dominant connective tissue amino acid, increases. The temperature of collagen contraction is higher in mammals' meat than in fish meat and it is largely affected by the degree of hydration (Bailey and Light 1989). The rate of heating and the content of water in the environment surrounding the meat significantly determines the changes which collagen undergoes during heat treatment (Pospiech et al. 1995). The more water and the slower the heating process, the more gelatin is created and the tenderer the meat becomes. Thus, "tendinous" meat is not suitable for short-lasting heat processes (frying, grilling), but is suited to stewing and moist cooking. Collagen belongs to the few proteins found in meat which do not contain all essential amino acids. Thus, sometimes the content of hx, the main amino acid of collagen, is used in industry, alongside the content of tryptophan (trp) (collagen does not contain it), to evaluate the trade value of meat products, especially sausages. The more hx and the less trp, the lower the product's value is (Brieskorn and Berg 1959). Muscle fibers are characterized by specific metabolic, morphological, and functional properties and furthermore, they show a dynamic nature. Thanks to nervous and hormonal signals, they adapt relatively easily and quickly to variable and desirable functions.

1.3 CHEMICAL AND BIOCHEMICAL CONSTITUTION OF MUSCLE

1.3.1 Fiber Type

By considering the morphological as well as functional characteristics, the fibers of skeletal muscle can be divided into a range of different types. Red and white fibers are the most different from each other. In between, however, there is a wide spectrum of different types of fibers, which possess intermediate metabolic and structural properties. Variation between fibers can be observed in most muscles (Ashmore and Doerr 1971; Klont et al. 1998; Wegner et al. 2000; Chang et al. 2003; Schiaffino and Reggiani 2011). Specific types of fibers can occur in muscle alongside each other even in the same muscle cluster and as a result muscle classification is based on the domination of particular types of muscle fibers.

A typical characteristic of red fibers is a large content of myoglobin, thus the red color. The process of their contraction is slow and based on aerobic metabolism. By contrast white fibers, besides having a smaller content of myoglobin, have a larger activity of glycolytic enzymes for generating energy rapidly through glycolysis. As a consequence, a higher level of carbohydrates is observed compared to red fibers. The diameter of white fibers is usually bigger than red fibers. Because this description is rather general and does not fully present the variation between fibers, the classification is constantly being improved. This is done using various cytochemical, immunohistochemical, as well as electrophoretic methods employing immunoblotting (Greaser et al. 2001; Chang et al. 2003; Choi et al. 2006; Ohlendieck 2010). Special attention is paid to isoforms of myosin, although the differences in muscle fiber types are also reflected by the activity of mitochondrial enzymes, phosphorylase, the content of glycogen and lipids as well as by the rate of the glycolysis and proteolysis processes which take place postslaughter (Doherty et al. 2004; Bouley et al. 2005; Okumura et al. 2005; Hamelin et al. 2006; Lametsch et al. 2006) (see Chapter 4). Red fibers are classified as type I fibers while white fibers constitute a bigger group including IIA, IIB, as well as IIX or IID fibers (Greaser et al. 2001; Greenwood et al. 2007; Schiaffino and Reggiani 2011). The most glycolytic character is shown by type IIB fibers while the remaining ones have the characteristics of intermediate fibers. Glycolytic and proteolytic processes are quicker in white fibers, which makes the tenderization of muscles dominated by them faster (see Chapter 3).

In the prenatal period of various animal species, the growth of their muscle mass is conditioned by simultaneous rise in the total amount of muscle fibers (hyperplasia) and growth in their thickness (hypertrophia). No increase in the number of muscle fibers is generally observed after birth, but their thickness increases (Hocquette et al. 1998; Wegner et al. 2000; Koohmaraie et al. 2002; Hemmings et al. 2009). Thicker fibers usually result in a greater hardness of muscles. A slightly different effect can be observed while comparing muscles from double muscled and normally muscled cattle. The fiber thickness of double muscled cattle is larger than that of the other group when comparing animals of the same age, and therefore the tenderness of their meat can be better (see Chapter 10). The reason for this effect is the smaller amount of connective tissue in the cross-section of the muscles of the same area (Ngapo et al. 2002a,b; Nishimura et al. 2002). The proportion of different types of muscle fibers has a significant effect on muscle mass (Carpenter et al. 1996; Greenwood et al. 2007; Ferguson and Gerrard 2014). In analogical muscles with the same total number of fibers obtained from different animals, a greater content of white fibers is going to result in a higher muscle mass, because white fibers are characterized by larger diameter.

1.3.2 Muscle Contraction

Muscle contraction properties influence meat qualities such as nutritional value and the culinary and processing aspects as well as the health characteristics. According to Huxley's "slide" theory, muscle contraction is a result of movement of actin filaments (c. 1 μm long) along myosin filaments (c. 1.6 μm long), while neither of the filaments changes its length in the process. The rate and size of muscle fiber contraction is varied despite the seemingly quite ordered structure of muscles (Ohlendieck 2010).

Muscle contraction is connected with energetic changes taking place in muscles. The main energy sources for an animal *in vivo* are fats, however energy can also come from carbohydrates or even proteins, although the latter are generally used for construction purposes. Cessation of blood circulation and very rapid exhaustion of oxygen resources in a muscle cell postslaughter causes anaerobic glycolysis to become the main energy source. Obtaining ATP as an energy source for many metabolic processes, including for muscles, is, however, not a simple reaction. The process is multistaged and can have an aerobic or anaerobic character (see: left and right side of Figure 1.2, respectively). The latter dominates postmortem.

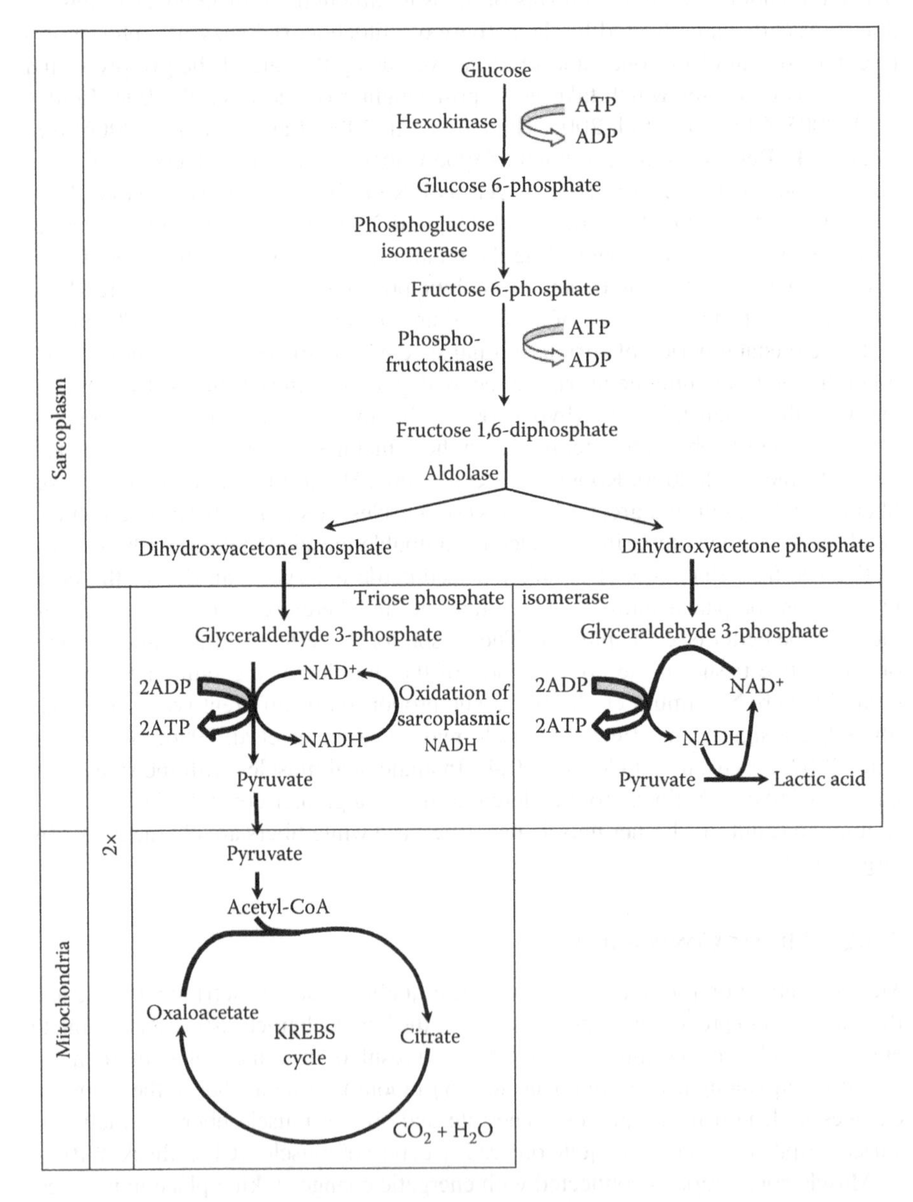

FIGURE 1.2 Schematic diagram of the main reaction in glycolysis.

Contraction *in vivo* is triggered by a nervous impulse, which results in a change of polarization of the cell membrane, sarcolemma. The process is simultaneous with the emission of acetylcholine as a neurotransmitter and is one of the important stages of the stimulation–contraction connection of a muscle. The stimulation is spread inside the muscle fiber through longitudinal channels, which are connected to the sarcoplasmic reticulum. The latter releases calcium ions, which trigger muscle contraction by binding regulatory proteins of muscle fibers, while inside myofibrils, calcium ions bind with troponin C (Tn-C). This increases the interaction between troponin I (Tn-I) and Tn-C and draws Tn-I away from the places on actin where TM, which prevents binding of actin and myosin, is located. The movement of TM on actin gives myosin access to places with which it binds. This process makes muscle contraction possible. As a result of this process the actin filament moves in relation to the myosin one, the sarcomere contracts and, as a consequence, the muscle fiber shortens.

The number of places released on actin by TM depends on its interaction with Tn-T and with the Tn-C–Tn-I complex, whose existence is dependent on Ca^{2+} ions. These processes are strongly dependent on the polymorphism of the proteins which take part in them. Interactions between these proteins are accompanied by dephosphorylation of ATP bound with myosin, in which a part is taken by myosin ATPase, activated by Ca^{2+} ions. The energy from this reaction is used in the creation of the actomyosin complex. ATP is needed for the sodium–potassium pump to work at transmitting the nervous impulse as well as for the calcium pump, which regulates the level of calcium ions in the sarcoplasmic reticulum. The sodium–potassium pump uses around 25% of the energy obtained from ATP degradation. The glycolysis process provides a large amount of energy, fulfilling the requirements when these reactions take place *in vivo* and shortly after slaughter, when for a very limited period they may still have an aerobic character.

1.3.3 Postmortem Muscle Contractile Activity and Meat Quality Implication

With anaerobic glycolysis, typical in the postmortem phase (Figure 1.2), only about 5% of energy is obtained from one unit of glucose in comparison to aerobic glycolysis. The result is very quick exhaustion of glycogen supplies and, as a consequence, a buildup of lactic acid, the final product of anaerobic activity. Due to this, the muscle undergoes acidification. Under these conditions glycolysis produces more NADH than can be regenerated by the aerobic way and oxidation of sarcoplasmic NADH takes place by way of pyruvate reduction to lactate. This means that anaerobic glucose degradation is still a process which delivers ATP, but in a substantially smaller quantity.

Despite the fact that the decline of pH in the muscle postmortem has been commonly attributed to the increase in the lactic acid level, some other patterns of H^+ ions concentration increase, related to cellular function and metabolism in which mitochondria may take part, are also considered and elucidated (Shen et al. 2008; Hudson 2012; England et al. 2013; Ferguson and Gerrard 2014).

Muscle contraction is one of the most important processes connected with protein transformation that takes place in meat accompanied by multiple chemical and physical reactions. The strength and course of rigor mortis depend on the directions of these reactions and the amount of energy obtained from them. Alongside glycolytic cycle enzymes, enzymes responsible for proteolysis are also important for meat quality, especially during the first hours after slaughter (England et al. 2013).

While observing the course of glycolysis, it has been noticed that if it is rapid, it denatures proteins, which most often causes a decline in the functional quality of meat, affecting mainly water holding capacity, color, and tenderness (Grześ et al. 2005, 2010; Mancini and Hunt 2005; Sayd et al. 2006; Thomson et al. 2008; Laville et al. 2009; Jacob and Hopkins 2014; Kim et al. 2014). These symptoms are typical of pale, soft, exudative (PSE) meat. An increased rate of glycolysis is mostly observed in muscles of pigs (see Chapter 4), in which white type IIB fibers dominate, however it has also been observed in cattle (Morzel et al. 2008). By contrast sometimes meat pH after slaughter stays at higher levels resulting in meat defined as dark, firm, dry (DFD) type, and the causative mechanism of this has been outlined in Chapter 3, while the consequences for meat quality are discussed in Chapter 2.

The contraction of actin and myosin which takes part during the first hours post-slaughter makes muscles lose flexibility and become stiff. Under normal cold storage conditions meat becomes more and more tender. A different phenomenon can be observed when the temperature of muscle tissue falls below 10–12°C and its pH is above 6.0–6.3, which may happen with very quick or shock cooling of carcasses or muscles immediately after slaughter and is characteristic of red meat observed in cattle and sheep carcasses. The above effect leads to the phenomenon of cold shortening (Locker and Hagyard 1963; Marsh et al. 1981). A so-called "supercontraction" of sarcomeres can be observed then. It is a profound muscle contraction, that is, up to 40% or even 60% of its length, while the bond between myosin and actin is very durable and makes the meat usually tougher than slowly chilled muscles. It is believed that the reason for this phenomenon, besides the outside factors mentioned above, is the disruption of the mechanism of releasing calcium ions from the sarcoplasmic reticulum which surrounds a muscle fibril. The excessive release of Ca^{2+} ions into the space surrounding myosin and actin's myofilaments results in acceleration of the contraction and, as a result the actin fibrils can meet or superimpose in the center of the A band (Smulders et al. 1990). Myosin filaments, on the other hand, will entirely fill the space between the sarcomeres Z disks. This effect can be prevented by using various technologies immediately after slaughter, mainly related to electrical stimulation or conditioning (Jacob et al. 2012). The toughness of cold-shortened muscles attributed to the shortening of sarcomeres through unusual overlapping between actin and myosin does not hinder the proteolysis of myofibrillar proteins. Some studies (Hwang et al. 2004; Weaver et al. 2008) revealed that proteolysis takes place in cold-shortened muscles without crucial changes in shear force. Tenderstretching performed in various forms on carcasses or single muscles and influencing the overlap between two main contractile proteins suggests that fiber density may have a real impact on shear force values alongside the degradation of any structural and regulatory proteins (Devine et al. 1999; Hopkins and Thompson 2001, 2002; England et al. 2012; Jacob and Hopkins 2014).

A phenomenon similar to cold shortening can take place while defrosting meat frozen immediately after slaughter while omitting the cooling process (Marsh and Thompson 1958). This is called thaw shortening and in this meat a relatively large amount of energy in the form of ATP is left. In such a case, muscle super contraction is caused by increased calcium ions released from the sarcoplasmic reticulum during meat defrosting.

The influence of sarcomere length on tenderness, one of the most important indicators of meat quality, is not always as obvious as in the case of cold-shortened muscles. Observations on pork muscles one day after slaughter showed a large variation in sarcomere length and tenderness ratings between muscles (Wheeler et al. 2000). However, if sarcomeres were extended to at least 2.0 μm, the muscle would be tender regardless of collagen content or proteolysis. Opposite results were obtained from earlier studies on beef muscles (Herring et al. 1965), where a limitation of 2.0 μm was not observed. In the case of lambs (Wheeler and Koohmaraie 1999) no relationship was observed between sarcomere length and proteolysis in the *Longissimus* and *Psoas major* muscles, but as the authors concluded, the size of sarcomeres might have an indirect effect on tenderization during aging due to its effect on initial tenderness.

The relationships between sarcomere length, proteolysis, and the final tenderness of meat are specific for animal species. Intrinsic factors such as the genotype, muscle structure, proximal composition and fiber type, and extrinsic factors such as pre- and postharvest treatment will impact the overall meat quality. All these factors have to be taken into consideration during the preparation of meat for consumption in any suitable form because they may influence not only sensory properties of meat, but also its nutritional value.

1.4 NUTRITIONAL VALUE

In most developed and European countries, animal proteins form a part of the daily diet, mostly combined with vegetable proteins, carbohydrates, lipids, and fiber. Meat production and consumption have increased over the last decades and have now leveled out at a steady total volume of meat production apart from China (USDA 2014). The nutritional quality of meat has become an increasingly important issue of interest to nutritionists, epidemiologists, technologists, and animal producers (Higgs 2000; Norat et al. 2002; Biesalski 2005). Meat is defined as any edible part of the striated muscle of an animal (a mammal, a fowl, a fish, etc.) and consists of proteins and heminic iron, and to a lesser extent, lipids with variable content of saturated and unsaturated fatty acids.

Meat is a heterogeneous food that varies according to the origin of the flesh (ruminants, monogastrics, and birds), the type of muscles, which are themselves complex in terms of structural and biochemical properties, and the preparations used to process them into various dishes. Yet, meats have some common nutritional features (Biesalski 2005). They are protein-rich (50%–80% of energy content) and (together with fish), are the fresh foods that present the highest protein content. In addition, these proteins are particularly rich in indispensable amino acids, notably lysine and histidine, which meets the supply requirements, from children to adults (FAO/WHO/UNU 2007). This means that proteins from meat products are used with great efficiency to increase or renew whole body proteins because it is not necessary to supply large amounts of proteins to cover the requirements of each indispensable amino acid. Furthermore, the high concentration of lysine in meat-proteins is useful to improve the quality of other proteins such as those from cereals. In addition, these proteins that are well digested in the small intestine (Silvester and Cummings 1995) and do not induce any notable reaction in the digestive tract which could increase endogenous losses, as might occur with high-fiber feedstuffs.

1.4.1 Macronutrients

Meat, like other animal products, is a source of macro- and micronutrients (Figure 1.3). Proteins are an essential component of our diet with a specific role to ensure amino acids supply, nitrogen, and energy; both of them required for the synthesis of proteins and nitrogen compounds. Meat proteins have good nutritional quality because of their balanced essential amino acids and high digestibility (Silvester and Cummings 1995; FAO/WHO/UNU 2007). Meat is also a source of lipids, mostly as phospholipids, triglycerides, and cholesterol. The proportion of lipids in meat lies between 2% and 6% according to species, muscle, and muscle type. For example, in white meat such as poultry meat, the amount of lipids barely reaches 2%. The hydrolysis of triglycerides releases three types of fatty acids: these can be either saturated or unsaturated (mono-unsaturated [MUFAs] or poly-unsaturated [PUFAs]).

1.4.1.1 Proteins

The muscle proteins represent one fifth of the muscle mass while the water content varies between 70% and 75%. As reported in Section 1.3.3, muscle proteins are either involved in contraction, regulation, signaling, or metabolism. Up to 1000 different proteins are found in muscle, however contractile proteins, namely actin and myosin, are the predominant ones. Myofibrils and cytoskeletal proteins represent more than 60% of total protein. Sarcoplasmic proteins represent about 30% of total protein. The sarcoplasmic proteins are mostly globular, and have molecular weights between 17 kDa (myoglobin) and 95 kDa (glycogen phosphorylase).

From a nutritional point of view, the amino acid composition is of importance because essential amino acids are required for cell renewal. Myoglobin provides heminic iron which is more bioavailable than nonheminic iron (Hurrell and Egli 2010). The muscle proteins except those of the connective tissue such as collagen, elastin, and reticulin have all the essential amino acids (Table 1.1). In humans, there are eight essential amino acids: trp, lysine, methionine, phenylalanine, threonine, valine, leucine, and isoleucine. Two others, histidine and arginine, are semiessential because only infants need an exogenous source (found in milk). Cysteine, glycine, and tyrosine are sometimes required for populations that are not able to synthesize these in sufficient quantities. For example, people suffering from phenylketonuria have a nonfunctional hepatic enzyme phenylalanine hydroxylase (PAH). They are not able to metabolize the amino acid phenylalanine (Phe) to the amino acid tyrosine, consequently tyrosine becomes an essential amino acid.

Then phenylalanine accumulates and is converted into phenylpyruvate (also known as phenylketone), which can be detected in the urine.

Meat is generally considered to be a source of saturated, mono-unsaturated, and poly-unsaturated fatty acids. Saturated and MUFAs are more involved in the

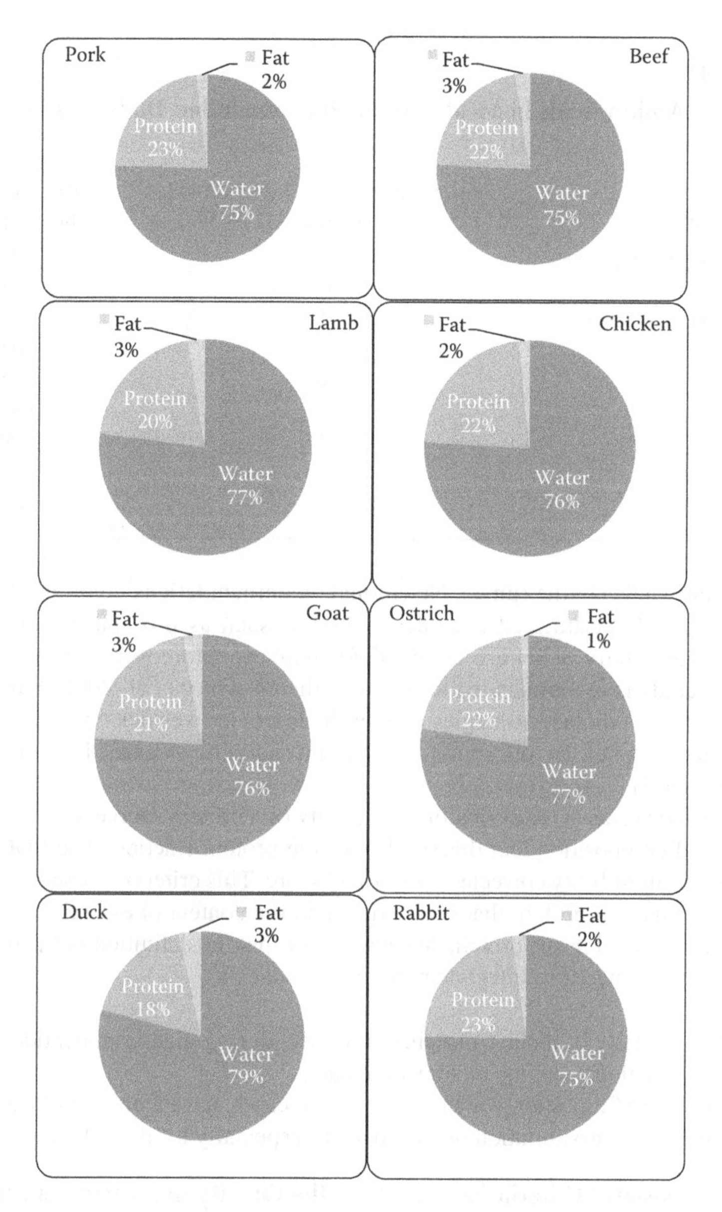

FIGURE 1.3 Proximate composition of meat from various animal species expressed in % (measurements in the *Longissimus lumborum* for pigs, cattle, and lamb; in breast muscles for poultry).

supply of cell energy while PUFAs have more functional properties. The composition and the content of the meat fatty acids are highly dependent on the ingested lipids for monogastrics such as pigs and poultry. The content varies according to muscle and species from 1% up to 10%. The relationship between composition and dietary fatty acids is less evident for ruminants because of the hydrogenation of

TABLE 1.1
Essential Amino Acids in Meat and the Recommended Daily Dose for Adults

Amino Acid	Recommended Daily Dose for Adults in mg/kg according to OMS, FAO	Values for an Adult of 70 kg (mg)
F phenylalanine (+ tyrosine)	25	1750
L leucine	39	2730
M methionine (+ cysteine)	15	1050
K lysine	30	2100
I isoleucine	20	1400
V valine	28	1960
T threonine	15	1050
W tryptophan	4	280
H histidine	10	700

dietary fatty acids in the rumen. Nutritional recommendations highlight the health benefits of polyunsaturated fatty acids (PUFA) such as n-6 and n-3 fatty acids, because these fatty acids are essential for humans. Moreover, the ratio of n-6/n-3 fatty acids is recommended to be lower than 4 (Garg et al. 2006; Simopoulos 2008). Different dietary strategies to alter the levels in livestock have been studied such as inclusion of fish oil, or n-3 PUFA plant rich oils, or linseed in animal diets (see Chapter 4).

The classic criteria for evaluating the quality of a protein source are based on the amino acid composition and digestibility of the protein fraction. The PDCAAS is the protein digestibility corrected amino acid score. This criteria is based on the proportion of ingested protein that is absorbed and the content of essential amino acid which is contained in the protein. We now know that the definition of the quality of dietary proteins needs to integrate new concepts such as

- The capacity of dietary proteins to release, during digestion, peptides having a local or systemic biological impact.
- The rate of digestion which may, in some cases, have a direct influence on whole body assimilation of amino acids, especially for the elder ones.

1.4.1.2 Classical Criteria for Evaluating the Quality of Protein Sources

Meat, as a source of proteins, provides amino acids and peptides. The amino acid composition of food protein affects its functional characteristics (solubility, emulsion, gelling, etc.). The essential amino acid composition also determines nutritional properties, which correspond to recommended needs. Meat proteins are particularly rich in lysine and histidine, but also in histidine-containing dipeptides such as anserine and carnosine, and tripeptides (glutathione). Naturally present in meat, these components can also be produced by processing or during digestion. Carnosine is a dipeptide (β-alanine-L-histidine) present only in meat and meat products (Crush 1970). Its content depends on muscle fiber type, with higher levels in type IIB muscle fibers (Aristoy and Toldra 1998; Cornet and Bousset 1999). In these muscles, the high speed of contraction requires rapid ATP production, which implies the recruitment of the glycolytic pathway. Therefore, when the degradation of glycogen exceeds the capacity of the Krebs cycle, carnosine, by its buffer capacity (up to 40% of buffering capacity of skeletal muscle), can neutralize the lactic acid formed (Abe 2000). Interestingly, carnosine is capable of inhibiting lipid oxidation catalyzed by iron, singlet oxygen, peroxyl, hydroxyl radicals, etc., by scavenging free radicals and chelating divalent metal ions (Rubtsov et al. 1991; Decker et al. 1992).

1.4.1.3 Bioactive Peptides

More and more data have proven the physiological effects of certain food-derived peptides on the activity of the digestive tract and other physiological functions (antihypertensive, opioid, immunomodulatory, antianxiolytic activities). For example, the sequence VAP coming from actin or the peptide IKP coming from myosin have an antihypertensive activity (Bauchart et al. 2007a). Among the peptides found in foods, some are already present, and abundant, in the original feedstuff and their synthesis does not follow the classical metabolic pathways of protein synthesis and degradation; these are, for example, carnosine and glutathione in meat products. Other peptides are present in proteins and can be only released during the various processes of meat transformation or during intestinal digestion.

1.4.1.3.1 Carnosine

Carnosine is only present in animal tissues. It is particularly abundant in brain and skeletal muscle from mammals. Its concentration is more elevated in muscles with glycolytic activity (Aristoy and Toldra 1998). Indeed chicken breast muscles contain a higher amount of carnosine compared to the thigh (Jung et al. 2013; Jayasena et al. 2014). It can also vary according to animal species (Crush 1970), age, and/or food intake (Purchas et al. 2004). The carnosine concentration in meat is not significantly affected by the type of aging or cooking processes (Bauchart et al. 2007a). The main biological activity of carnosine appears to be its buffering activity (Abe 2000) that allows, for instance, offsetting the decrease in intracellular pH linked to lactic acid production in muscles where anaerobic glycolysis is particularly active. Thus, in sportsmen, an increase in carnosine concentration in muscles seems able to attenuate fatigue after intensive exercise. Carnosine also has antioxidant properties by its ability to bind divalent metallic ions and its ability to trap free radicals (Guiotto et al. 2005). Moreover, it seems to be able to reduce aldehydes formed from unsaturated fatty acids during oxidative stress. It would also play a prominent role in protection against glycation and cross-linking of proteins (Hobart et al. 2004; Lee et al. 2005). Due to the specific position of the amino group of beta-alanine, carnosine is not degraded by dipeptide hydrolases, but by two specific enzymes called aminoacylhistidine dipeptidases (or carnosinases). Despite carnosinase activity in the small intestine and serum, ingestion of bovine meat induces a rapid increase in plasma concentration of carnosine in humans (Park et al. 2005). Yeum et al. (2010) reported in a human study that urinary concentrations of carnosine increased 13- to 14-fold after beef ingestion while carnosine was undetectable in plasma after ingesting pure carnosine or beef.

A study using an animal model (mini pigs) suggests that 1/5th of ingested carnosine (after a meal of meat) is actually absorbed and released in the blood (Bauchart et al. 2007b). This absorption is associated with an increased nonbicarbonate buffering capacity and antioxidant capacity of the serum (Antonini et al. 2002; Bauchart et al. 2007b). Moreover, even if the serum carnosinase hydrolyses an important part of the absorbed carnosine, the availability of its precursors (i.e., histidine, but more importantly β-alanine) could be sufficient to ensure an increase in carnosine synthesis and concentration within the skeletal muscle in humans (Harris et al. 2006). A recent study demonstrated the beneficial effects of dietary anserine and carnosine on cognitive functioning and physical activity of the elderly with chicken meat extract; corresponding to an intake of 300 mg of carnosine per day for three months (Szcześniak et al. 2014).

1.4.1.3.2 Glutathione

Glutathione is a tripeptide (GSH: L-γ-Glutamyl-L-cysteinyl-glycine) whose concentration is very high in the liver, but it is also found in skeletal muscle. Contrary to carnosine, GSH is not specific to animal products and is also present in significant quantities in vegetables such as broccoli and spinach (Wierzbicka et al. 1989). Because of the thiol function of cysteine, radical, glutathione exists in reduced (GSH) or oxidized (GSSG) forms. GSH is the major hydrosoluble antioxidant in animal cells and is an efficient free radical scavenger, protecting cells from reactive oxygen species (ROS) attacks, especially hydrogen peroxide. Any changes in GSH and GSSG concentrations directly reflect alterations of their redox state. Glutathione also plays a role in xenobiotics detoxification, metabolism of various molecules (leukotrienes, prostaglandins, formaldehyde, methylglyoxal, nitric oxide, etc.). Moreover, glutathione is involved in the regulation of expression and/or activation of transcriptional factors "oxidation-sensitive" necessary for the antioxidant response in the cell (Wu et al. 2004). Glutathione deficiencies contribute to oxidant stress which plays a key role in the aging process and the establishment of various pathologies (Alzheimer's, Parkinson's, inflammation of the digestive tract, etc.). Studies in humans and animals show that an adequate protein intake is essential to maintain glutathione homeostasis. It seems that a portion of the dietary glutathione can be absorbed intact and participates in the intracellular glutathione store in peripheral tissues (Favilli et al. 1997). Moreover, GSH plays an important role in maintaining the integrity of the intestinal mucosa. Finally, dietary GSH, in addition to biliary GSH, participate in the reduction of lipid peroxides present in the intestinal lumen.

1.4.1.3.3 Antihypertensive Peptides

Very few studies are available on the potential of meat proteins as sources of bioactive peptides. The most studied biological activity is the antihypertensive activity based on the inhibition of the angiotensin-converting enzyme (ACE). This activity seems to occur in normal feeding conditions and studies in humans have shown a significant decrease in arterial pressure compared to a control group following the repeated ingestion of fermented milk containing antihypertensive peptides (Jauhiainen and Korpela 2007). Several ACE-inhibiting peptides have been evidenced in controlled muscle proteins hydrolysates: from a skeletal muscle hydrolyzed with thermolysin (Arihara et al. 2001), from myosin hydrolyzed with thermolysin (Nakashima et al. 2002), from troponin C hydrolyzed with pepsin (Katayama et al. 2008), and from sarcoplasmic proteins hydrolyzed with a mixture of thermolysin, proteinase A, and protease type XIII (Jang and Lee 2005).

An *in vivo* study using an animal model showed that, following the ingestion of beef, numerous peptides are reproducibly released during the digestive process and that many of them contain an amino acid sequence known to provide an ACE-inhibiting potential (Bauchart et al. 2007a). To be active at the peripheral level, these sequences still need to be released intact by mucosal peptidases, to enter the blood circulation, and to be resistant to the peptidases present in plasma. The possibility of an absorption of antihypertensive peptides has been demonstrated in humans after an oral administration of the dipeptide Val-Tyr, but the occurrence (and significance) of such an absorption following the ingestion of dietary proteins containing bioactive sequences has to date not been reported. However, it was shown that a partial substitution of dietary carbohydrates by red meat can lower blood pressure in hypertensive patients (Hodgson et al. 2006).

1.4.1.4 Digestion Rate

It has been clearly shown that the kinetics of the digestion of dietary protein determines the effectiveness of their assimilation, however optimal kinetics are not necessarily the same for all subjects. For instance, for elderly people, it seems preferable to concentrate the daily protein supply on a single meal, or to ingest rapidly digested proteins, in order to accentuate the postprandial increase in plasma amino acids and stimulate protein synthesis (Mosoni and Patureau-Mirand 2003). From these observations, nutritional strategies have been developed to counteract aging-related muscle wasting (sarcopenia). Meat in this context could be a very important food because of its high concentration of highly digestible proteins providing the required protein in one meal. Comparing beef and soy protein in an isonitrogenous amount, Phillips (2012) demonstrated that beef protein intake elicited a greater response in muscle protein synthesis.

As in the field of bioactive peptides, many studies on protein digestion kinetics have been carried out with milk proteins (caseins, lactoserum proteins), and few data are available regarding the proteins of meat products. A recent study in old subjects has shown that meat can be considered as a source of rapidly digested proteins (Rémond et al. 2007). However, this property depends on the chewing capacity of elderly people, and more precisely on the level of bolus disruption before swallowing. The bolus disruption is less efficient in subjects with reduced chewing efficiency (e.g., with a full denture) compared to people of the same age normally toothed. Consequently, it induces a lesser increase in postprandial protein synthesis. The meat consumption can counteract muscle wasting in the elderly, especially when associated with physical activity (Campbell et al. 1999; Phillips 2012). But we must ensure that the textural properties of the meat product are appropriate to the chewing efficiency of elderly. In the future, the development of adapted forms of meat to meet the specific requirements of such a population is expected.

1.4.2 Micronutrients

1.4.2.1 Iron

Iron plays an important role as an oxygen carrier in hemoglobin in blood, or myoglobin in muscle, and it is also required for many metabolic processes. Dietary iron occurs in two forms, heme and nonheme iron, with heme iron being more readily absorbed and utilized by the body. The source of nonheme iron is mainly whole grains and legumes. These plant foods contain phytate, which has been found to affect the absorption of nonheme iron adversely. The same applies to coffee and tea due to the presence of tannins. These types of polyphenols have a strong inhibitory effect on the absorption of iron, mainly nonheme iron. Approximately 30% of heme iron is absorbed in the intestine, against only 7% of nonheme iron (Williamson et al. 2005).

Iron contained in meat is mostly in the heme form. The concentration of heme iron in meat depends on the animal species. For example, 100 g of red meat supplies approximately 3 mg of iron, that is, 30% of the daily recommended intake (Martin 2001), while the same quantity of white meat provides only 0.4 mg of iron. In modern societies, meat and meat products provide approximately 20% of the iron actually used by the body (Russel et al. 1999; Henderson et al. 2003). Iron is an essential trace element in human nutrition. Due to the high prevalence of anemia in developing and industrialized countries, it is necessary to maintain a suitable iron intake through the diet in order to achieve an appropriate status of this element in the body. For this reason, accurate knowledge of iron availability of foods is essential in order to plan intervention strategies that improve deficient situations of this nutrient.

Nonheme iron availability is conditioned by several dietary factors, such as classic factors (meat, ascorbic acid, fiber, phytic acid, and polyphenols) and new factors (caseinophosphopeptides and fructo-oligosaccharides with prebiotic characteristics) (Kibangou et al. 2005). Iron deficiency is a worldwide problem, even in developed countries. For example, in France, 29% of young children and 15% of teenagers are not achieving recommended iron intakes (Preziosi et al. 1994) and in the United Kingdom, 25% of women are not reaching the recommended intake (Henderson et al. 2003). Many studies have demonstrated the positive effect of including meat in the diet on iron incorporation, especially in population groups at risk such as children (Nathan et al. 1996), young women (Ball and Barlett 1999), and the elderly (Wells et al. 2003). In Chapter 13 some of the factors that impact on the level of iron and zinc in sheep meat are outlined and impact of animal nutrition on the level of these macronutrients is covered in Chapter 4.

1.4.2.2 Zinc

Zinc is associated with the reactivity of a wide variety of enzymes. By binding zinc, phytate decreases its bioavailability. Hence, meat is a better source of this element than vegetables (Ball and Ackland 2000; Hunt 2002). Thus 100 g of bovine meat supplies approximately 5 mg of zinc, that is, 40% of the daily recommended intake, 12 mg/day (Martin 2001). Meat contributes to 17% of the total zinc intake in France and 31%–34% in Denmark, UK, and New Zealand (Russel et al. 1999; Henderson et al. 2003). In the United States the contribution of meat to total zinc intake is 56% (Hunt 2002). The SUVIMAX study showed that 7% of the French population had serum levels of zinc below the deficiency threshold. Populations at risk for zinc deficiency are children (Hambidge and Krebs 2007) and the elderly (Lukito et al. 2004).

1.4.2.3 Selenium

Selenium plays a central role in human antioxidative defenses and it has been suggested that high selenium intakes reduce the risk of cancer and cardiovascular diseases (Rayman 2000). Meat and meat products are a good source of selenium (Daun and Akesson 2004; Demirezen and Uruç 2006). Thus, 100 g of meat supplies 6–8 μg of selenium, that is, approximately 10% of recommended intake, 60 μg/day (Martin 2001). Liver is particularly rich in selenium (40–100 μg for 100 g). Depending on the country, meat and meat products can supply 10%–25% of recommended intake (Russel et al. 1999; Williamson et al. 2005).

Other trace elements: Meat and meat products are also important sources of copper, cobalt, chromium, manganese, and nickel and greatly contribute to the recommended intakes of these elements (Demirezen and Uruç 2006).

1.4.2.4 Vitamins

Vitamin A is involved in bone growth, eye pigment synthesis, and prevention of respiratory diseases (Biesalski 2005). Meat and milk can be considered as a better source of vitamin A than vegetables like carrot. Indeed, they contain retinol or retinol esters, which are directly usable by the body, while vegetables contain β-carotene, a precursor of retinol, and the conversion rate of β-carotene into retinol is only 1:12. The concentration of vitamin A in meat ranges widely among species (2–100 μg per 100 g of meat). With 20 mg per 100 g, liver is the best source of vitamin A in the human diet, and its contribution to the recommended intake is approximately 75% (Biesalski 2005).

The B vitamins work as cofactors in different enzyme systems in the body and their deficiency is linked to many diseases. Meat and animal-derived foods are the only foods that naturally provide vitamin B12, a hydrosoluble component (Chan et al. 1995; Russel et al. 1999; Esteve et al. 2002; Biesalski 2005; Williamson et al. 2005). Meat contains 0.3–2 μg of vitamin B12, oxidative muscle having the highest content, while liver contains 65 μg (Ortigues-Marty et al. 2005). A daily consumption of 120 g of meat or a monthly consumption of 120 g of liver supplies recommended intakes. Vegetarians and vegans who exclude meat from their diet are at risk of vitamin B12 deficiency (Lloyd-Wright et al. 2000; Biesalski 2005). Due to reduced absorption, the elderly are also at risk of inadequate intakes (Miller 2002). Meat is also an important source of vitamin B2 (riboflavin), vitamin B3 (niacin), and vitamin B6 (pyridoxal) and greatly contributes to the daily intake of these vitamins (Chan et al. 1995; Esteve et al. 2002).

Vitamin D is essential for the development and maintenance of bone. Although most vitamin D comes from the action of sunlight on dehydrocholesterol in the skin, some human populations are particularly reliant on a dietary supply of vitamin D. With a level of 0.1 μg per 100 g, the contribution that meat makes to vitamin D dietary intake is important. In Europe, meat products contribute 20%–30% of daily intake of vitamin D (Williamson et al. 2005). Dunnigan and Henderson (1997) have suggested that meat contains a "magic meat factor" since the influence of meat on bone metabolism is greater than could be expected from the amount of vitamin D present alone. Reanalyses of the British National Diet and Nutrition surveys of adults and young children confirmed meat as a major contributor to natural dietary vitamin D intakes. Meat and meat products appear to provide around 20% of British intakes, compared to previous estimates of just 4% (Gibson and Ashwell 1997). There are small amounts of vitamin E in meat, but the recent trend to include seed oils or vitamin E in animal diet will have contributed to an increase of vitamin E content in meat (Sanders et al. 1997; Mercier et al. 1998).

1.5 SPECIES AND BREED CHARACTERISTICS

1.5.1 Ruminants: Saturated Fatty Acids

Because of lipid content and its high saturated/unsaturated fatty acid ratio, beef consumption suffers from a bad image in terms of nutritional quality. The major saturated fatty acids (SFA) in beef are myristic acid C14:0, palmitic acid C16:0, and stearic acid C18:0. The first two acids have been found to be significantly associated with CHD risk while for stearic acid (C18:0) some cholesterol-raising effects in humans have been reported (Derr et al. 1993). Higgs (2000) highlighted that the red meat produced today is leaner and lower in fat content than that produced 10 years ago. In order to increase the amounts of polyunsaturated fatty acids (PUFAs), and especially of n-3 PUFAs, in bovine and ovine meat several strategies including manipulation of the diet have been implemented (see Chapter 4) to alter the n-6:n-3 ratio (Raes et al. 2004) and achieve positive health effects. PUFA and MUFA are highly reactive to oxidation in meat storage or during processing, therefore addition of antioxidants in the diet of ruminants is recommended. The potential to genetically alter the levels of fatty acids in sheep meat is outlined in Chapter 13.

Recent studies have been conducted on the use of meat in the human diet to supply conjugated linoleic acid (CLA), a component with important pharmacological properties (anticarcinogenic and antiatherogenic activities). CLA is a term used to describe a group of positional and geometric isomers of octadecadienoic acid, of which ruminant meat and milk are the major dietary sources. CLA occurs naturally in higher concentrations in ruminant meat than in pork or poultry (Fogerty et al. 1988). CLA is formed both through the ruminal biohydrogenation of dietary linolenic acid and also through an endogenous synthesis pathway from transvaccenic acid. Indeed, concentrations of CLA in beef muscle have been found to range from 0.37 to 1.08 g/100 g. This is mainly due to the advantage of a pasture diet over concentrates in producing higher CLA levels in meat as has been demonstrated (Enser 2000).

1.5.2 Monogastric: Pig and Poultry—Lean Meat

Poultry has lower saturated fatty acid content than red meats. Pigs and poultry are excellent provider of proteins in high quantity and quality. Unlike ruminants, the fatty acid profile can be easily modified by the diet (Stewart et al 2001; Crespo and Esteve-Garcia 2002).

1.5.3 Exotic Meat: The Case of Ratites

Interestingly exotic animals production has demonstrated specific features common to ruminants and monogastrics. For example, ratites are flightless or "running" birds with different anatomical and physiological characteristics when compared to other major bird species. Ratites have no keel on their sternum (crista sterni) to anchor their wing muscles, and by consequence, they are unable to fly (Sick 1997). The ostrich is the most commercially exploited ratite specie. Due to a belief in the special healthy characteristics of their meats, ratites (ostriches, emus, and rheas) are receiving more and more attention by meat producers in developed markets (Sales and Horbańczuk 1998; Sales et al. 1999). Meat production is primarily under controlled farming and management schemes. Thus, the same factors of diet, age, sex, handling, stress, slaughter practices, postmortem aging, and processing that influence properties of meat from domesticated species can also influence ratite meat.

The main reference for amino acid and mineral composition of ostrich meat is Sales and Hayes (1996). The *Flexor cruris* and *Iliofibularis* muscles in the ostrich had slightly higher protein, less cholesterol, and much lower fat to protein ratio than turkey and beef (Paleari et al. 1998). The fatty acid profile of ostrich meat was generally closer to that of beef than turkey meat (Paleari et al. 1998). Oleic acid (C18:1) is the fatty acid with the highest concentration in ostrich meat, followed by palmitic acid (C16:0) and then linoleic acid (C18:2n-6). Ostrich fat might be a source of PUFA or essential fatty acids in human and livestock diets because there is a high PUFA to SFA ratio in breast fat from culled ostrich breeding females (Horbańczuk et al. 1998; Sales 1998) and a high content of C18:2, C18:3, and C20:4 from breast and back fat deposits in 14-month-old ostriches. According to McMillin and Hoffman (2009), differences in fatty acid content were found in meat from ostriches at 14 months of age compared with those 8 years of age (Hoffman and Fischer 2001). Within the normal age at slaughtering, the fatty acid profile of ostrich meat varied with bird age (10–11 compared with 14–15 months) and muscles (*Iliofibularis, Gastrocnemius*, and *Iliotibialis*). Muscles from older birds had increased total SFA and MUFA and decreased total PUFA (Girolami et al. 2003).

Analyzing minerals, heme and nonheme iron contents of rhea meat, Ramos et al. (2009) observed no differences for calcium, phosphorus, magnesium, and sodium in the *Obturatorius medialis* (*OM*), *Iliotibialis lateralis* (*IL*), and *Iliofibularis* (*I*) muscles. They did report however that there was more potassium, zinc, and copper in the *IL* muscle than in *OM* and *I* muscles. For manganese, *OM* and *IL* muscles showed a higher content in comparison with *I* muscles. For selenium, *IL* and *I* muscles showed the highest content compared to *OM* muscles. For total content of heme and nonheme iron, the *IL* muscle showed the highest content in respect to the other muscles. When compared to other meats, the mineral content of rhea meat showed an elevated level in phosphorus, selenium, and total and heme iron content.

Because of their favorable balance in indispensable amino acids, their very high digestibility, and the high bioavailability of their amino acids, meat proteins have a high biological value. Thus, as part of a well-balanced diet, meat and fish consumption does not need to exceed 120 g/d for healthy adults without any specific requirements. In these conditions, the part of the daily supply which is provided by meat is variable according to nutrients. It can exceed 60% for some indispensable amino acids, vitamin B12 and zinc, 40% for vitamin B3 and cholesterol, 20% for iron, selenium, riboflavin, vitamin B6, and pantothenic acid as well as saturated fatty acids.

1.6 FINAL REMARKS

This chapter gives an overview of the structure of the muscle and its composition, highlighting the fiber organization and the specific associated metabolic, morphological, and functional properties. Enzymes and fiber type play a major role in the metabolic response of the muscle cell postmortem, and consequently on the final quality of meat. In the future, new insights in the comprehension of muscle structure, organization, and metabolism linked with meat quality would be addressed in fetal muscle development programs or in the selection of breed adapted to climatic changes. The muscle and therefore meat as a source of macro- and micronutrients is at the beginning of exploration, especially taking into account the aging population and their specific need for proteins with high nutritional quality. As meat is a rich source of proteins, furthermore of good quality due to its balanced amino acids and high digestibility, its consumption is adapted to fulfill the requirements for elderly in food with high protein content to prevent sarcopenia.

REFERENCES

Abe, H. 2000. Role of histidine-related compounds as intracellular proton buffering constituents in vertebrate muscle. *Biochemistry-Moscow* 65:757–765.

Acton, J.C., Henna, M.A., and Satterless, L.D. 1981. Heat induced gelation and protein-protein interaction of actomyosin. *J. Food Biochem.* 5:101–103.

Antonini, F.M., Petruzzi, E., Pinzani, P., Orlando, C., Poggesi, M., Serio, M., Pazzagli, M., and Masotti, G. 2002. The meat in the diet of aged subjects and the antioxidant effects of carnosine. *Arch. Gerontol. Geriatr.* 8(Suppl):7–14.

Arihara, K., Nakashima, Y., Mukai, T., Ishikawa, S., and Itoh, M. 2001. Peptide inhibitors for angiotensin I-converting enzyme from enzymatic hydrolysates of porcine skeletal muscle proteins. *Meat Sci.* 57:319–324.

Aristoy, M.C., and Toldra, F. 1998. Concentration of free amino acids and dipeptides in porcine skeletal muscles with different oxidative patterns. *Meat. Sci.* 50:327–332.

Asghar, A., Samejima, K., and Yasui, T. 1985. Functionality of muscle proteins in gelation mechanisms of structure meat products. *Crit. Rev. Food. Sci. Nutr.* 22:27–106.

Ashmore, C.R., and Doerr, L. 1971. Comparative aspects of muscle fiber types in different species. *Exp. Neurol.* 31:408–418.

Bailey, A.J., and Light, N.D. 1989. *Connective Tissue in Meat and Meat Products*. Elsevier Applied Science, London and New York, pp. 170–194.

Ball, M.J., and Ackland, M.L. 2000. Zinc intake and status in Australian vegetarians. *Br. J. Nutr.* 83:27–33.

Ball, M.J., and Bartlett, M.A. 1999. Dietary intake and iron status of Australian vegetarian women. *Am. J. Clin. Nutr.* 70:353–358.

Bauchart, C., Chambon, C., Mirand, P.P., Savary-Auzeloux, I., Rémond, D., and Morzel, M. 2007a. Peptides in rainbow trout (*Oncorhynchus mykiss*) muscle subjected to ice storage and cooking. *Food Chem.* 100:1566–1572.

Bauchart, C., Morzel, M., Chambon, C., Mirand, P.P., Reynes, C., Buffière, C., and Rémond, D. 2007b. Peptides reproducibly released by *in vivo* digestion of beef meat and trout flesh in pigs. *Br. J. Nutr.* 98:1187–1195.

Bayraktaroglu, A.G., and Kahraman, T. 2011. Effect of muscle stretching on meat quality of biceps femoris from beef. *Meat Sci.* 88:580–583.

Biesalski, H.K. 2005. Meat as a component of a healthy diet-are there any risks or benefits if meat is avoided in the diet. *Meat Sci.* 70:509–524.

Bouley, J., Meunier, B., Chambon, C., De Smet, S., Hocquette, J.F., and Picard, B. 2005. Proteomic analysis of bovine skeletal muscle hypertrophy. *Proteomics* 5:490–500.

Brieskorn, C.H., and Berg, H-W. 1959. Ein vereinfachtes Verfahren zur Tryptophan-Bestimmung der Tryptophan-Peptid-Wert [A simplified method for of tryptophan determination for the tryptophan-peptide value]. *Zeitschrift Lebensmittel-Untersuchung Forschung* 109:302–306.

Brooks, J.C., and Savell, J.W. 2004. Perimysium thickness as an indicator of beef tenderness. *Meat Sci.* 67:329–334.

Campbell, W.W., Barton, M.L., Cyr-Campbell, D., Davey, S.L., Beard, J.L., Parise, G., and Evans, W.J. 1999. Effects of an omnivorous diet compared with a lactoovovegetarian diet on resistance-training-induced changes in body composition and skeletal muscle in older men. *Am. J. Clin. Nutr.* 70:1032–1039.

Carpenter, E., Rice, O.D., Cockett, N.E., and Snowder, G.D. 1996. Histology and composition of muscles from normal and callipyge lambs. *J. Anim. Sci.* 74:388–393.

Chan, W., Brown, J., Lee, S.M., and Buss, D.H. 1995. *Meat, Poultry and Game. Fifth Supplement to McCance and Widdowson's the Composition of Food.* Royal Society Chemistry Editions, Cambridge, UK, pp. 601.

Chang, K.C., de Costa, N., Blackley, R., Southwood, O., Evans, G., Plastow, G., Wood, J.D., and Richardson, R.I. 2003. Relationships of myosin heavy chain fiber types to meat quality traits in traditional and modern pigs. *Meat Sci.* 64:93–103.

Choi, Y.M., Ryu, Y.C., and Kim, B.C. 2006. Effect of myosin heavy chain isoforms on muscle fiber characteristics and meat quality in porcine *longissimus* muscle. *J. Muscle Foods* 17:413–427.

Christensen, M., Ertbjerg, P., Failla, S. et al. 2011. Relationship between collagen characteristics, lipid content and raw and cooked texture of meat from young bulls of fifteen European breeds. *Meat Sci.* 87:61–65.

Cornet, M., and Bousset, J. 1999. Free amino acids and dipeptides in porcine muscles: Differences between "red" and "white" muscles. *Meat Sci.* 51:215–219.

Crespo, N., and Esteve-Garcia, E. 2002. Nutrient and fatty acid deposition in broilers fed different dietary fatty acid profiles. *Poult. Sci.* 81:1533–1542.

Crush, K.G. 1970. Carnosine and related substances in animal tissues. *Comp. Biochem. Physiol.* 34(1):3–30.

Daun, C., and Akesson, B. 2004. Glutathione peroxydase activity, and content of total and soluble selenium in five bovine and porcine organs used in meat production. *Meat Sci.* 66:801–804.

Decker, E.A., Crum, A.D., and Calvert, J.T. 1992. Differences in the antioxidant metabolism of the carnosine in the presence of copper and iron. *J. Agric. Food Chem.* 40:756–759.

Demirezen, D., and Uruç, K. 2006. Comparative study of trace elements in certain fish, meat and meat products. *Meat Sci.* 74:255–260.

Derr, J., Kris-Etherton, P.M., Pearson, T.A., and Seligson, F.H. 1993. The role of fatty acid saturation on plasma lipids, lipoproteins and apolipoproteins. II. The plasma total and LDL-cholesterol response of individual fatty acids. *Metabolism* 42:130–134.

Devine, C.E., Wahlgren, N.M., and Tornberg, E. 1999. Effect of rigor temperature on muscle shortening and tenderisation of restrained and unrestrained *beef M. longissimus thoracicus et lumborum. Meat Sci.* 51:61–72.

Doherty, M.K., McLean, L., Hayter, J.R., Pratt, J.M., Robertson, D.H., El-Shafei, A., Gaskell, S.J., and Beynon, R.J., 2004. The proteome of chicken skeletal muscle: Changes in soluble protein expression during growth in a layer strain. *Proteomics* 4:2082–2093.

Dubost, A., Micol, D., Meunier, B., Lethias, C., and Listrat, A. 2013a. Relationships between structural characteristics of bovine intramuscular connective tissue assessed by image analysis and collagen and proteoglycan content. *Meat Sci.* 93:378–386.

Dubost, A., Micol, D., Picard, B., Lethias, C., Andueza, D., Bauchart, D., and Listrat, A. 2013b. Structural and biochemical characteristics of bovine intramuscular connective tissue and beef quality. *Meat Sci.* 95:555–561.

Dunnigan, M.G., and Henderson, J.B. 1997. An epidemiological model of privational rickets and osteomalacia. *Proc. Nutr. Soc.* 56:939–956.

Eggen, K.H., Pedersen, M.E., Lea, P., and Kolset, S.O. 2001. Structure and solubility of collagen and glycosaminoglycans in two bovine muscles with different textural properties. *J. Muscle Foods.* 12:245–261.

Enser, M. 2000. Producing meat for healthy eating. In *Proceedings of 46th International Congress of Meat Science and Technology* 2.II-L4, Buenos Aires, Argentina, pp. 124–127.

England, E.M., Fisher, K.D., Wells, S.J., Mohrhauser, D.A., Gerrard, D.E., and Weaver, A.D. 2012. Postmortem titin proteolysis is influenced by sarcomere length in bovine muscle. *J. Anim. Sci.* 90:989–995.

England, E.M., Scheffler, T.L., Kasten, S.C., Matarneh, S.K., and Gerrad, D.E. 2013. Exploring the unknowns involved in the transformation of muscle to meat. *Meat Sci.* 95:837–843.

Esteve, M.J., Farré, R., Frigola, A., and Pilamunga, C. 2002. Contents of vitamins B1, B2, B6, and B12 in pork and meat products. *Meat Sci.* 62:73–78.

FAO/WHO/UNU 2007. Protein and amino acid requirements in human nutrition. Report of a joint expert consultation (WHO Technical Report Series 935) p. 265.

Favilli, F., Marraccini, P., Iantomasi, T., and Vincenzini, M.T. 1997. Effect of orally administered glutathione on glutathione levels in some organs of rats: Role of specific transporters. *Br. J. Nutr.* 78:293–300.

Ferguson, D.M., and Gerrard, D.E. 2014. Regulation of post-mortem glycolysis in ruminant muscle. *Anim. Prod. Sci.* 54:464–481.

Fiedler, I., Ender, K., Wicke, M., Maak, S., Legerken, G., and Meyer, W. 1999. Structural and functional characteristic of muscle fibers in pigs with different malignant hypertermia susceptibility (MHS) and different meat quality. *Meat Sci.* 53:9–15.

Fogerty, A., Ford, G., and Svoronos, D. 1988. Octadeca-9,11-dienoic acid in foodstuffs and in the lipids of human-blood and breast-milk. *Nutr. Reports Int.* 38:937–943.

Garg, M.L., Wood, L.G., Singh, H., and Moughan, P.J. 2006. Means of delivering recommended levels of long chain n-3 polyunsaturated fatty acids in human diets. *J. Food Sci.* 71:R66–R71.

Gelse, K., Pöschl, E., and Aigner, T. 2003. Collagens—Structure, function, and biosynthesis. *Adv. Drug Delivery Rev.* 55:1531–1546.

Gibson, S., and Ashwell, M. 1997. New vitamin D values for meat and their implication for vitamin intake in British adults'. *Proc. Nutr. Soc.* 56:116A.

Girolami, A., Marsico, I., D'Andrea, G., Braghieri, A., Napolitano, F., and Cifuni, G.F. 2003. Fatty acid profile, cholesterol content and tenderness of ostrich meat as influenced by age at slaughter and muscle type. *Meat Sci.* 64:309–315.

Gou, P., Zhen, Z.Y., Hortós, M., Arnau, J., Diestre, A., Robert, N., Claret, A., Čandek-Potokar, M., and Santé-Lhoutellier, V. 2012. PRKAG3 and CAST genetic polymorphisms and quality traits of dry-cured hams. I. Associations in Spanish dry-cured ham Jamón Serrano. *Meat Sci.* 92:346–353.

Greaser, M.L., Okochi, H., and Sosnicki, A.A. 2001. Role of fiber types in meat quality. In *Proceedings of the 47th International Congress of Meat Science and Technology*, Kraków, Poland, pp. 34–37.

Greenwood, P.L., Harden, S., and Hopkins, D.L. 2007. Myofibre characteristics of ovine *longissimus* and *semitendinosus* muscles are influenced by sire-breed, gender, rearing type, age, and carcass weight. *Aust. J. Exp. Agric.* 47:1137–1146.

Grześ, B., Pospiech, E., Koćwin-Podsiadła, M., Kurył, J., Krzęcio, E., Łyczyński, A., Iwańska, E., and Mikołajczak, B. 2005. Evaluation of the effect of the RN gene on meat tenderness from the point of view of myosin heavy chains polymorphism. *Ann. Anim. Sci.* 2:51–54.

Grześ, B., Pospiech, E., Koćwin-Podsiadła, M., Łyczyński, A., Krzęcio, E., Mikołajczak, B., and Iwańska, E. 2010. Relationships between the polymorphism of myosin heavy chains and selected meat quality traits of pigs with different susceptibility to stress. *Arch. Tierz.* 53:64–71.

Guiotto, A., Calderan, A., Ruzza, P., and Borin, G. 2005. Carnosine analogues: Alpha, beta-unsaturated aldehyde scavengers based on the histidyl hydrazide moiety. *Curr. Med. Chem.* 12:2293–2315.

Hambidge, K.M., and Krebs, N.F. 2007. Zinc deficiency: A special challenge. *J. Nutr.* 137:1101–1105.

Hamelin, M., Sayd, T., Chambon, C. et al. 2006. Proteomic analysis of ovine muscle hypertrophy. *J. Anim. Sci.* 84:3266–3276.

Hamill, R.M., McBryan, J., McGee, Ch., Mullen, A.M., Sweeney, T., Talbot, A., Cairns, M.T., and Davey, G.C. 2012. Functional analysis of muscle gene expression profiles associated with tenderness and intramuscular fat content in pork. *Meat Sci.* 92:440–450.

Harris, R.C., Tallon, M.J., Dunnett, M., Boobis, L., Coakley, J., Kim, H.J., Fallowfield, J.L., Hill, C.A., Sale, C., and Wise, J.A. 2006. The absorption of orally supplied β-alanine and its effect on muscle carnosine synthesis in human *vastus lateralis*. *Amino Acids* 30:279–289.

Hemmings, K.M., Parr, T., Daniel, Z.C.T.R., Picard, B., Buttery, P.J., and Brameld, J.M. 2009. Examination of myosin heavy chain isoform expression in ovine skeletal muscles. *J. Anim. Sci.* 87:3915–3922.

Henderson, L., Irving, K., and Gregory, J. 2003. *The National Diet and Nutrition Survey: Adults Aged 19–64 Years. Vol. 3: Vitamin and mineral intake and urinary analyses.* HMSO, London.

Herring, H.K., Cassens, R.G., and Briskey, E.J. 1965. Further studies on bovine muscle tenderness as influenced by carcass position, sarcomere length, and fiber diameter. *J. Food Sci.* 30:1049–1054.

Higgs, J. 2000. The changing nature of red meat: 20 years of improving nutritional quality. *Trends Food Sci. Technol.* 11:85–95.

Hobart, L.J., Seibel, I., Yeargans, G.S., and Seidler, N.W. 2004. Anti-crosslinking properties of carnosine: Significance of histidine. *Life Sci.* 75:1379–1389.

Hocquette, J.-F., Botreau, R., Picard, B., Jacquet, A., Pethick, D.W., and Scollan, N.D. 2012. Opportunities for predicting and manipulating beef quality. *Meat Sci.* 92:197–209.

Hocquette, J.F., Ortigues-Marty, I., Pethick, D., Herpin, P., and Fernandez, X. 1998. Nutritional and hormonal regulation of energy metabolism in skeletal muscles of meat-producing animals. *Livest. Prod. Sci.* 56:115–143.

Hodgson, J.M., Burke, V., Beilin, L.J., and Puddey, I.B. 2006. Partial substitution of carbohydrate intake with protein intake from lean red meat lowers blood pressure in hypertensive persons. *Am. J. Clin. Nutr.* 83:780–787.

Hoffman, L.C., and Fisher, P. 2001. Comparison of meat quality characteristics between young and old ostriches. *Meat Sci.* 59:335–337.

Hopkins, D.L., and Thompson, J.M. 2001. The relationship between tenderness, proteolysis, muscle contraction and dissociation of actomyosin. *Meat Sci.* 57:1–12.

Hopkins, D.L., and Thompson, J.M. 2002. Factors contributing to proteolysis and disruption of myofibrillar proteins and the impact on tenderisation in beef and sheep meat. *Aust. J. Agric. Res.* 53:149–166.

Horbańczuk, J., Sales, J., Celeda, T., Konecka, A., Zieba, G., and Kawka, P. 1998. Cholesterol content and fatty acid composition of ostrich meat as influenced by subspecies. *Meat Sci.* 50:385–388.

Hudson, N.J. 2012. Mitochondrial treason: A driver of pH decline rate in post-mortem muscle? *Anim. Prod. Sci.* 52(12):1107–1110.

Hunt, J.R. 2002. Moving toward a plant-based diet: Are iron and zinc at risk. *Nutr. Rev.* 60:127–134.

Hurrell, R., and Egli, I. 2010. Iron bioavailability and dietary reference values. *Am. J. Clin. Nutr.* 91(Suppl.):1461S–1467S.

Hwang, I.H., Park, B.Y., Cho, S.H., and Lee, J.M. 2004. Effects of muscle shortening and proteolysis on Warner–Bratzler shear force in beef *longissimus* and *semitendinosus*. *Meat Sci.* 68:497–505.

Jacob, R.H., and Hopkins, D.L. 2014. Techniques to reduce the temperature of beef muscle early in the post mortem period—A review. *Anim. Prod. Sci.* 54:482–493.

Jacob, R.H., Rosenvold, K., North, M., Kemp, R., Warner, R.D., and Geesink, G.H. 2012. Rapid tenderisation of lamb *M. longissimus* with very fast chilling depends on rapidly achieving sub-zero temperatures. *Meat Sci.* 92:16–23.

Jang, A., and Lee, M. 2005. Purification and identification of angiotensin converting enzyme inhibitory peptides from beef hydrolysates. *Meat Sci.* 69:653–661.

Jauhiainen, T., and Korpela, T. 2007. Milk peptides and blood pressure. *J. Nutr.* 137:825–829.

Jayasena, D.D., Jung, S., Bae, Y.S., Kim, S.H., Lee, S.K., Lee, J.H., and Jo, C. 2014. Changes in endogenous bioactive compounds of Korean native chicken meat at different ages and during cooking. *Poult. Sci.* 93:1842–1849.

Josza, L., Lehto, M.U., Jarvinen, M., Kvist, M., Reffy, A., and Kannus, P. 1993. A comparative-study of methods for demonstration and quantification of capillaries in skeletal-muscle. *Acta Histochem.* 94:89–96.

Jung, S., Bae, Y.S., Kim, H.J., Jayasena, D.D., Lee, J.H., Park, H.B., Heo, K.N., and Jo, C. 2013. Carnosine, anserine, creatine, and inosine 5'-monophosphate contents in breast and thigh meats from 5 lines of Korean native chicken. *Poult. Sci.* 92:3275–3282.

Katayama, K., Anggraeni, H.E., Mori, T., Ahhmed, A.M., Kawahara, S., Sugiyama, M., Nakayama, T., Maruyama, M., and Muguruma, M. 2008. Porcine skeletal muscle troponin is a good source of peptides with angiotensin-I converting enzyme inhibitory activity and antihypertensive effects in spontaneously hypertensive rats. *J. Agric. Food Chem.* 56:355–360.

Kibangou, I.B., Bouhallab, S., Henry, G., Bureau, F., Allouche, S., Blais, A., Guérin, P., Arhan, P., and Bouglé, P. 2005. Milk proteins and iron absorption: Contrasting effects of different caseinophosphopeptides. *Pediatr. Res.* 58:731–734.

Kim, Y.H.B., Warner, R.D., and Rosenvold, K. 2014. Influence of high pre-rigor temperature and fast pH fall on muscle proteins and meat quality: A review. *Anim. Prod. Sci.* 54:375–395.

Klont, R.E., Brocks, L., and Eikelenboom, G. 1998. Muscle fiber type and meat quality. *Meat Sci.* 49(Suppl. 1): 219–229.

Koohmaraie, M., Kent, M.P., Shackelford, S.D., Veiseth, E., and Wheeler, T.L. 2002. Meat tenderness and muscle growth: Is there any relationship? *Meat Sci.* 62:345–352.

Kwasiborski, A., Sayd, T., Chambon, C., Santé-Lhoutellier, V., Rocha, D., and Terlouw, C. 2008a. Pig *Longissimus lumborum* proteome: Part I. Effects of genetic background, rearing environment and gender. *Meat Sci.* 80:968–981.

Kwasiborski, A., Sayd, T., Chambon, C., Santé-Lhoutellier, V., Rocha, D., and Terlouw, C. 2008b. Pig Longissimus lumborum proteome: Part II: Relationships between protein content and meat quality. *Meat Sci.* 80:982–996.

Lametsch, R., Kristensen, L., Larsen, M.R., Therkildsen, M., Oksbjerg, N., and Ertbjerg, P. 2006. Changes in the muscle proteome after compensatory growth in pigs. *J. Anim. Sci.* 84:918–924.

Laville, E., Sayd, T., Terlouw, C., Blinet, S., Pinguet, J., Fillaut, M., Glénisson, J., and Chérel, P. 2009. Differences in pig muscle proteome according to HAL genotype: Implications for meat quality defects. *J. Agric. Food Chem.* 57:4913–4923.

Lee, Y.T., Hsu, C.C., Lin, M.H., Liu, K.S., and Yin, M.C. 2005. Histidine and carnosine delay diabetic, deterioration in mice and protect human low density lipoprotein against oxidation and glycation. *Eur. J. Pharmacol.* 513:145–150.

Lefaucheur, L., and Gerrard, D. 2000. Muscle fiber plasticity in farm mammals. *J. Anim. Sci.* 77:1–19.

Lepetit, J. 2007. A theoretical approach of the relationships between collagen content, collagen cross-links and meat tenderness. *Meat Sci.* 76:147–159.

Lepetit, J. 2008. Collagen contribution to meat toughness: Theoretical aspects. *Meat Sci.* 80:960–967.

Li, K., Zhang, Y., Mao, Y., Cornforth, D., Dong, P., Wang, R., Zhu, H., and Luo, X. 2012. Effect of very fast chilling and aging time on ultra-structure and meat quality characteristics of Chinese Yellow cattle m. *longissimus lumborum. Meat Sci.* 92:795–804.

Lloyd-Wright, Z., Allen, N., Key, T., and Sanders, T. 2000. How prevalent is vitamin B12 deficiency among British vegetarians and vegans? *Proc. Nutr. Soc.* 60:1–16.

Locker, R.H., and Hagyard, C.J. 1963. A cold-shortening effect in beef muscles. *J. Sci. Food Agric.* 14:787–793.

Lukito, W., Wattanapenpaiboon, N., Savige, G.S., Hutchinson, P., and Wahlqvist, M.L. 2004. Nutritional indicators, peripheral blood lymphocyte subsets and survival in an institutionalised elderly population. *Asia Pac. J. Clin. Nutr.* 13:107–112.

Mancini, R.A., and Hunt, M.C. 2005. Current research in meat color. *Meat Sci.* 71:100–121.

Marsh, B.B., and Thompson, J.F. 1958. Rigor mortis and thaw rigor in lamb. *J. Sci. Food Agric.* 9:417–424.

Marsh, B.B., Lochner, J.V., Takahashi, G., and Kragness, D.D. 1981. Effects of early postmortem pH and temperature on beef tenderness. *Meat Sci.* 5:479–483.

Marsh, B.B., Ringkob, T.P., Russell, R.L., Swartz, D.R., and Pagel, L.A. 1988. Mechanisms and strategies for improving meat tenderness. *Reciprocal Meat Conf. Proc.* 41:113–121.

Martin, A. 2001. Apports nutritionnels conseillés pour la population française. *Tec & Doc Lavoisier*, 3rd edition, 605.

McCormick, R.J. 2009. Collagen. In Du, M., and McCornmick, R.J. (eds.), *Applied Muscle Biology and Meat Sci.* CRC Press, Taylor & Francis Group LLC, Boca Raton, FL, Chapter 7, pp. 129–148.

McMillin, K.W., and Hoffman, L.C. 2009. Improving the quality of meat from ratites. In: Kerry, J.P., and Ledward, D. (eds.), *Improving the Sensory and Nutritional Quality of Fresh Meat.* Woodhead Publishing Ltd., Cambridge, pp. 419–446.

Mercier, Y., Gatellier, P., Viau, M., Remignon, H., and Renerre, M. 1998. Effect of dietary fat and vitamin E on colour stability and on lipid and protein oxidation in turkey meat during storage. *Meat Sci.* 48:301–318.

Miller, J.W. 2002. Vitamin B12 deficiency, tumor necrosis factor-alpha, and epidermal growth factor: A novel function of vitamin B12. *Nutr. Rev.* 60:142–144.

Millman, B.M. 1998. The filament lattice of striated muscle. *Physiol. Rev.* 78:359–391.

Morzel, M., Terlouw, C., Chambon, C., Micol, D., and Picard, B. 2008. Muscle proteome and meat eating qualities of longissimus thoracis of "Blonde d'Aquitaine" young bulls: A central role of HSP27 isoform. *Meat Sci.* 78:297–304.

Mosoni, L., and Patureau Mirand, P. 2003. Type and timing of protein feeding to optimize anabolism. *Curr. Opin. Clin. Nutr. Metab. Care* 6:301–306.

Nakashima, Y., Arihara, K., Sasaki, A., Mio, H., Ishikawa, S., and Itho, M. 2002. Antihypertensive activities of peptides derived from porcine skeletal muscle myosin in spontaneously hypertensive rats. *J. Food Sci.* 67:434–437.

Nathan, I., Hackett, A.F., and Kirby, S.P. 1996. The dietary Intake of a group of vegetarian children aged 7–11 years compared with matched omnivores. *Br. J. Nutr.* 75:533–544.

Ngapo, T.M., Berge, P., Culioli, J., and De Smet, S. 2002a. Perimysial collagen crosslinking in Belgian Blue double-muscled cattle. *Food Chem.* 77:15–26.

Ngapo, T., Berge, P., Culioli, J., Dramsfield, E., De Smet, S., and Clayes, E. 2002b. Perimysial collagen crosslinking and meat tenderness in Belgian Blue double-musculed cattle. *Meat Sci.* 61:91–102.

Nishimura, T., Fang, S., Ito, T., Wakamatsu, J., and Takahashi, K. 2008. Structural weakening of intramuscular connective tissue during postmortem aging of pork. *Anim. Sci. J.* 79(6):716–721.

Nishimura, T., Hattori, A., and Takahashi, K. 1996. Relationship between degradation of proteoglycans and weakening of the intramuscular connective tissue during post-mortem ageing of beef. *Meat Sci.* 42:251–260.

Nishimura, T., Hattori, A., and Takahashi, K. 1999. Structural changes in intramuscular connective tissue during the fattening of Japanese Black cattle: Effect of marbling on beef tenderization. *J. Anim. Sci.* 77:93–104.

Nishimura, T., Liu, A. Hattori, A., and Takahashi, K. 1998. Changes in mechanical strength of intramuscular connective tissue during postmortem aging of beef. *J. Anim. Sci.* 76:528–532.

Nishimura, T., Nakashima, O., Listrat, A., Picard, B., Hocquette, J-F., and Hattori, A. 2002. Characteristics of intramuscular connective tissues from double-muscled young bulls. In *Proceedings of the 48th International Congress of Meat Science and Technology*, 25–30 August 2002, Rome, Italy, 2:582–583.

Norat, T., Lukanova, A., Ferrari, P., and Riboli, E. 2002. Meat consumption and colorectal cancer risk: Dose response meat analysis of epidemiological studies. *Int. J. Cancer* 98:241–256.

Ohlendieck, K. 2010. Proteomics of skeletal muscle glycolysis. *Biochimica et Biophysica Acta (BBA)—Proteins and Proteomics* 1804:2089–2101.

Okumura, N., Hashida-Okumura, A., Kita, K., Matsubae, M., Matsubara, T., Takao, T., and Nagai, K. 2005. Proteomic analysis of slow- and fast-twitch skeletal muscles. *Proteomics* 5:2896–2906.

Ortigues-Marty, I., Micol, D., Prache, S., Dozias, D., and Girard, C.L. 2005. Nutritional value of meat: Influence of nutrition and physical activity on vitamin B12 concentrations in ruminant tissues. *Reprod. Nutr. Dev.* 45:453–467.

Paleari, M.A., Camisasca, S., Beretta, G., Renon, P., Corsico, P., Bertolo, G., and Crivelli, G. 1998. Ostrich meat: Physico-chemical characteristics and comparison with turkey and bovine meat. *Meat Sci.* 48:205–210.

Pappas, Ch.T., Bliss, K.T., Zieseniss, A., and Gregorio, C.C. 2011. The nebulin family: An actin support group. *Trends Cell Biol.* 21(1):29–37.

Park, Y.J., Volpe, S.L., and Decker, E.A. 2005. Quantitation of carnosine in humans plasma after dietary consumption of beef. *J. Agric. Food Chem.* 53:4736–4739.

Phillips, S.M. 2012. Nutrient rich meat proteins in offsetting age related muscle loss. *Meat Sci.* 92:174–178.

Picard, B., Lefaucheur, L., Berri, C., and Duclos, M.J. 2002. Muscle fiber ontogenesis in farm animal species. *Reprod. Nutr. Dev.* 42:415–431.

Pisulewski, P.M., Franczyk, M., and Kostogrys, R.B. 2005. Health related effects of nutritionally modified foods of animal origin. *J. Anim. Feed Sci.* 14:71–85.

Pospiech, E., Domagała, A., Kałuża, E., and Stefańska, D. 1995. Influence of heating on the solubility of collagen in meat. [in Polish: Wpływ ogrzewania na rozgotowalność kolagenu mięsa]. *Roczniki Akademii Rolniczej w Poznaniu. Technol. Żywn.* CCLXX, 19(1):87–92.

Preziosi, P., Hercberg, S., Galan, P., Devanlay, M., Cherouvrier, F., and Dupin, H. 1994. Iron status of a healthy French population: Factors determining biochemical markers. *Ann. Nutr. Metab.* 38:192–202.

Prost, E., Pełczyńska, E., and Kotula, A.W. 1975. Quality characteristics of bovine meat. I. Content of connective tissues in relation to individual muscles, age and sex of animal and carcass quality grade. *J. Anim. Sci.* 41:534–540.

Purchas, R., Rutherfurd, S.M., Pearce, P.D., Vather, R., and Wilkinson, B.H.P. 2004. Cooking temperature effects on the forms of iron and levels of several other compounds in beef semitendinosus muscle. *Meat Sci.* 68:201–207.

Raes, K., Fievez, V., Chow, T.T., Ansorena, D., Demeyer, D., and De Smet, S. 2004. Effect of diet and dietary fatty acids on the transformation and incorporation of C18 fatty acids in double-muscled Belgian Blue young bulls. *J. Agric. Food Chem.* 52:6035–6041.

Ramos, A., Cabrera, M.C., del Puerto, M., and Saadoun, A. 2009. Minerals, haem and non-haem iron contents of rhea meat. *Meat Sci.* 81:116–119.

Rayman, M.P. 2000. The importance of selenium to human health. *Lancet* 356:233–241.

Rhee, M.S., Wheeler, T.L., Shackelford, S.D., and Koohmaraie, M. 2004. Variation in palatability and biochemical traits within and among eleven beef muscles. *J. Anim. Sci.* 82:534–550.

Rémond, D., Macheboeuf, M., Yven, C., Buffière, C., Mioche, L., Mosoni, L., and Patureau Mirand, P. 2007. Postprandial whole-body protein metabolism after a meat meal is influenced by chewing efficiency in elderly subjects. *Am. J. Clin. Nutr.* 85:1286–1292.

Rubtsov, A.M., Schara, M., Sentiurc, M., and Boldyrev, A.A. 1991. Hydroxyl radical scavenging activity of carnosine: A spin trapping study. *Acta Pharm. Jugosl.* 41:401–404.

Russel, D., Parnell, W., and Wilson, N. 1999. National nutrition survey. New Zealand food. New Zealand people. Key Results of the 1997. National Nutrition Survey. Ministry of Health/Wellington.

Sakamoto, T., Sasaki, S., Nakade, K., Ichinoseki, S., Tanabe, M., and Miyaguchi, Y. 2013. Molecular interaction studies evaluating the gelation of myosin B with glyceraldehyde 3-phosphate dehydrogenase after succinylation. *Food Sci. Technol. Res.* 19:229–235.

Sales, J. 1998. Fatty acid composition and cholesterol content of different ostrich muscles. *Meat Sci.* 49:489–492.

Sales, J., and Hayes, J.P. 1996. Proximate, amino acid and mineral composition of ostrich meat. *Food Chem.* 56:167–170.

Sales, J., and Horbańczuk, J. 1998. Ratite meat. *World's Poult. Sci. J.* 54:59–67.

Sales, J., Navarro, J.L., Martella, M.B., Lizurume, M.E., Manero, A., Bellis, L., and Garcia, P.T. 1999. Cholesterol content and fatty acid composition of rhea meat. *Meat Sci.* 53:73–75.

Sanders, S.K., Morgan, J.B., Wulf, D.M., Tatum, J.D., Williams, S.N., and Smith, G.C. 1997. Vitamin, E supplementation of cattle and shelf-life of beef for the Japanese market. *J. Anim. Sci.* 75:2634–2640.

Sayd, T., Morzel, M., Chambon, Ch. et al. 2006. Proteome analysis of the sarcoplasmic fraction of pig semimembranosus muscle: Implications on meat color development. *J. Agric. Food Chem.* 54:2732–2737.

Schiaffino, S., and Reggiani, C. 2011. Fiber types in mammalian skeletal muscles. *Physiol. Rev.* 91:1447–1531.

Shen, Q.W.W., Gerrard, D.E., and Du, M. 2008. Compound C, an inhibitor of AMP-activated protein kinase, inhibits glycolysis in mouse longissimus dorsi postmortem. *Meat Sci.* 78:323–330.

Sick, H. 1997. Ordem Rheiformes—Emas: Família Rheidae. In Sick, H. (ed.), *Ornitologia Brasileira*. Nova Fronteira, Rio de Janeiro, Brazil, pp. 168–171.

Simopoulos, A.P. 2008. The importance of the omega-6/omega-3 fatty acid ratio in cardiovascular disease and other chronic diseases. *Exp. Biol. Med. (Maywood)* 233:674–688.

Silvester, K.R. and Cummings, J.H. 1995. Does digestibility of meat protein help explain large-bowel cancer risk. *Nutr. Cancer- Int. J.* 24:279–288.

Smulders, F.J.M., Marsh, B.B., Swartz, D.R., Russell, R.L., and Hoenecke, M.E. 1990. Beef tenderness and sarcomere length. *Meat Sci.* 28:349–363.

Sosnicki, A. 1987. Histopathological observation of stress myopathy in m. longissimus in the pig and relationship with meat quality, fattening and slaughter traits. *J. Anim. Sci.* 65:584–596.

Stewart, J.W., Kaplan, M.L., and Beitz, D.C. 2001. Pork with a high content of polyunsaturated fatty acids lowers LDL cholesterol in women. *Am. J. Clin. Nutr.* 74:179–187.

Swartz, D.R., Greaser, M.L., and Cantino, M.E. 2009. Muscle structure and function. In Du, M., and McCornmick, R.J. (eds.), *Applied Muscle Biology and Meat Sci.* CRC Press, Taylor & Francis Group LLC, Boca Raton, FL, Chapter 1, pp. 1–45.

Szcześniak, D., Budzeń, S., Kopeć, W., and Rymaszewska, J. 2014. Anserine and carnosine supplementation in the elderly: Effects on cognitive functioning and physical capacity. *Arch. Gerontol. Geriatr.* 59:485–490.

Thomson, K.L., Gardner, G.E., Simmons, N., and Thompson, J.M. 2008. Length of exposure to high post-rigor temperatures affects the tenderisation of the beef *M. longissmus dorsi*. *Aust. J. Exp. Agric.* 48:1442–1450.

Torrescano, G., Sanchez-Escalante, A., Gimenez, B., Roncales, P., and Beltran, J.A. 2003. Shear values of raw samples of 14 bovine muscles and their relation to muscle collagen characteristics. *Meat Sci.* 64:85–91.

United States Department of Agriculture. 2014. Livestock and poultry: World markets and trade. Available at: http://www.thefarmsite.com/

Weaver, A.D., Bowker, B.C., and Gerrard, D.E. 2008. Sarcomere length influences postmortem proteolysis of excised bovine semitendinosus muscle. *J. Anim. Sci.* 86:1925–1932.

Wegner, J., Albrecht, E., Fiedler, I., Teuscher, F., Papstein, H.J., and Ender, K. 2000. Growth- and breed-related changes of muscle fiber characteristics in cattle. *J. Anim. Sci.* 78:1485–1496.

Wells, A.M., Haub, M.D., Fluckey, J., Williams, D.K., Chernoff, R., and Campbell, W.W. 2003. Comparison of vegetarians and beef-containing diet on haematological indexes and iron stores during a period of resistive training in older men. *J. Am. Diet. Assoc.* 103:594–601.

Wheeler, T.L., and Koohmaraie, M. 1999. The extent of proteolysis is independent of sarcomere length in lamb *longissimus* and *psoas major*. *J. Anim. Sci.* 77:2444–2451.

Wheeler, T.L., Shackelford, S.D., and Koohmaraie, M. 2000. Variation in proteolysis, sarcomere length, collagen content, and tenderness among major pork muscles. *J. Anim. Sci.* 78:958–965.

Wierzbicka, G.T., Hagen, T.M., and Jones, D.P. 1989. Glutathione in food. *J. Food Comp. Anal.* 2:327–337.

Williamson, C.S., Foster, R.K., Stanner, S.A., and Buttriss, J.L. 2005. Red meat in the diet. *Br. Nutr. Found. Nutr. Bull.* 30:323–355.

Wu, G., Fang, Y.Z., Yang, S., Lupton, J.R., and Turner, N.D. 2004. Glutathione metabolism and its implications for health. *J. Nutr.* 134:489–492.

Yamashiro, S., Gokhin, D.S., Kimura, S., Nowak, R.B., and Fowler, V.M. 2012. Tropomodulins: Pointed-end capping proteins that regulate actin filament architecture in diverse cell types. *Cytoskeleton* 69:337–370.

Yeum, K.I., Orioli, M., Regazzoni, L., Carini, M., Rasmussen, H., Russell, R.M., and Aldini, G. 2010. Profiling histidine dipeptides in plasma and urine after ingesting beef, chicken or chicken broth in humans. *Amino Acids* 38:847–858.

2 Meat Quality of Slaughter Animals

Phillip E. Strydom, Danuta Jaworska, and Danuta Kołożyn-Krajewska

CONTENTS

2.1 INTRODUCTION

Meat quality can be described according to different quality categories consisting of various attributes as listed in Table 2.1. Ultimately, meat quality is defined as the degree to which a specific product satisfies the needs and expectations of a particular buyer.

The Total Food Quality Model (TFQM) of Grunert et al. (1996) distinguishes between two concepts of quality perceptions or evaluations, namely expected quality (at the buying stage) and experienced quality (after consumption). Darby and Karni (1973) add a third dimension to quality and categorize the quality characteristics of products into search, experience and credence characteristics. "Search" characteristics refers to attributes like color or drip loss that the consumer observes before purchase, while "experienced" quality involves attributes like tenderness, flavor, juiciness evaluated after or during consumption and preparation. A "credence" quality cannot normally be evaluated and includes healthiness, nutritional value, or safety of a product and is a question of faith, trust, and reliable information. The relationship between quality expectation and quality experience (i.e., before and after purchase) is commonly believed to determine product satisfaction and consequently the probability of repurchasing the product (Issanchou 1996; Grunert et al. 2004; Banovic et al. 2009; Verbeke et al. 2010).

The various quality categories in Table 2.1 are affected by several intrinsic and extrinsic factors. Intrinsic factors include breed, genotype, age, and sex. Rearing system, feeding regime, preslaughter conditions (feed withdrawal and transport), stunning and slaughter procedures, chilling, and postslaughter handling of the meat, for example, aging, storage, and cooking parameters are important extrinsic factors affecting meat quality. The effects of these factors on meat quality are discussed in other chapters of this book.

TABLE 2.1
Groups of Meat Quality Characteristics

Quality Categories	Individual Attributes
Sensory quality	Raw meat: visual texture, color, visible fat, natural drip
	Heated meat: aroma, flavor, texture
Technological quality	WHC, pH value, protein, lipid and connective tissue properties, antioxidative status, emulsifying capacity, gel formation capacity
Nutritional quality	Protein, moisture and lipid content, vitamins, minerals, digestibility
Product safety	Microbiological quality, pesticides, heavy metal ions, antibiotics, hormones
Ethical considerations	General welfare: handling, farming system, transport, slaughter practices
	Farming system: feedlot, free range, organic farming, and outdoor rearing
	Slaughter procedure: general handling and killing method, for example, Kosher

2.2 SENSORY QUALITY OF MEAT

The sensory quality of meat can be divided into visual and eating quality. Visual quality relates to both fresh and cooked products. At the point of purchase the color of meat and fat, the ratios of meat and fat (intra-muscular, inter-muscular, and subcutaneous fat [SCF]), and the amount of purge are evaluated by the consumer before purchase decisions are made (Grunert et al. 2004; Ngapo and Dransfield 2006; Cho et al. 2007). Consumers expect that visual quality will be reflected in acceptable flavor, juiciness, texture of meat and absence of off-flavors (palatability) even though these visual traits are often unrelated to palatability traits.

2.2.1 Visual Quality

A bright red meat color (muscle part) is associated with the freshness of raw meat while the gray or tan color of cooked meat is associated with the typical color of cooked meat. Cured meat has a characteristic pink color (Suman and Joseph 2014). Myoglobin (Mb) is the main color pigment in meat that causes color variation. In fresh meat three forms of Mb exist, namely bright red oxymyoglobin (OxyMb), brown metmyoglobin (MetMb) that is the result of Mb oxidation and the deterioration of color, and purple–red deoxymyoglobin (DeoxyMb) that exists under anoxic conditions (Mancini and Hunt 2005). In addition, the amount of light reflected from the surface of the meat, which is influenced by the surface characteristics, will also influence how color is perceived by the consumer. Denatured globin hemichrome is the gray or tan-oxidized pigment in cooked meat (Suman and Joseph 2014). This pigment may also undergo chemical reduction under anaerobic conditions to form the pink-colored denatured globinhemochrome. Apart from premature browning, green pigments may also form on the meat surface during microbial spoilage and these pigments are metsulfmyoglobin and cholemyoglobin in fresh meat and nitrimetMb in cured meats. Various other factors can influence the formation of MetMb and browning of fresh meat that will be discussed in other chapters. While packaging of meat in anoxic conditions prevents MetMb formation and extends microbiological shelf life, consumers often discriminate against the purple–red color (Carpenter et al. 2001).

Excessive fat is considered by consumers as unhealthy (Brewer et al. 2001; Ngapo et al. 2007) even though it influences the "experienced" quality positively within certain limits (Miller 2002). According to various studies on fat color, mostly negative opinions of yellow fat are cited. However, Priolo et al. (2002) and Dunne et al. (2004) showed that different markets have different opinions concerning the color of fat in beef and sheep meat. Spanish consumers (southern Europe) are willing to pay premiums for lamb and beef with white carcass fat, while consumers from Ireland and the United Kingdom seem to be used to creamier colored fat from pasture animals. Walker et al. (1990) reported that yellow fat in beef carcasses is less acceptable for both the domestic (Australian) and export markets (Japan and Korea) than whiter fat and generally such carcasses are sold into the lower priced, manufacturing beef market. The Japanese market in particular specifies a requirement for white fat in

Australian beef. According to Forrest (1981), consumers in North America (United States and Canada) have become accustomed to the white fat of feedlot finished cattle, which has become an established practice since the 1970s. The yellow coloration of body fat in cattle is normally due to the presence of fat-soluble carotenoid pigments absorbed from the diet (Hill 1968) and is commonly attributed to pasture feeding. Although β-carotene is not the most abundant carotenoid in green plants, it is selectively absorbed, accounting for more than 80% of the yellow pigments in beef fat (Yang et al. 1992).

2.2.2 Eating Quality

The palatability of meat is determined by a combination of tenderness, juiciness, and meat flavor (Koohmaraie et al. 2002). Of these, tenderness is the most variable quality characteristic and is also rated by consumers as the most important sensory attribute and for these two reasons inconsistency in tenderness is regarded as a major problem for the beef industry (Destefanis et al. 2008). In the recently developed MSA (Meat Standards Australia) grading system used in Australia to categorize various cuts of the beef carcass according to the expected eating experience by the consumer, "tenderness" was indicated as the most important attribute, closely followed by "overall liking," while "flavor" and "juiciness" contributed less to the final score based on extensive testing (see Chapter 15) for best combinations and options (Watson et al. 2008a,b). However, based very much on the same attributes, Thompson et al. (2005a,b) recorded slightly higher correlation values between "overall liking" and "like flavor" than between "overall liking" and "tenderness" for lambs using a consumer panel. "Juiciness" recorded lower correlation values with "overall liking." In addition, Huffman et al. (1996) reported that most of the variation in overall palatability of steaks consumed at home could be explained by flavor perceptions ($R^2 = 0.67$).

Smith et al. (1983) reported that the basic "meaty" flavor of meat is of nonlipid origin but it is believed that some quantity of fat is necessary to produce a typical beef taste. Although the development of meat flavor is a complex process (Calkins and Hodgen 2007; Elmore and Mottram 2009), in simple terms meat flavor forms during cooking as a result of the Maillard reaction and lipid oxidation. Amino acids, peptides, and carbohydrates, in particular ribose, are important meat flavor precursors, which react together during heating in the Maillard reaction. The nature and composition of Maillard precursors (sugars and amino acids) and lipids can be influenced by several factors and processes originating pre- and postslaughter. The Maillard reaction is responsible for the typical meaty flavor and roast character of meat while lipid degradation provides compounds that add to fatty aromas and aroma differences between meats of different species. The challenge in meat flavor research is to gain a better understanding of how changes in water-soluble flavor precursors as a result of production and processing influence the flavor of cooked meat (Elmore and Mottram 2009). Recent development in analytical techniques such as capillary electrophoresis, ion chromatography and liquid chromatography–mass spectrometry, and gas chromatography–mass spectrometry will advance such efforts. The complexity describing the origin of flavor variation is highlighted by the list of 65 compounds and related characteristics of flavors and aromas identified in meat (Calkins and Hodgen 2007). Young et al. (1997) extracted 244 volatiles related to odor from the fat of sheep meat of which 10 were specifically responsible for typical sheep meat odor. Parallel to the complexity of the origin of flavors and odors in meat, is the different ways in which consumers of different backgrounds, culture, and eating habits respond to specific meat flavors (Killinger et al. 2004; Sañudo et al. 2007).

The sensory experience of tenderness by the consumer depends on the resistance of the meat structure to chewing combined with the release of fat and moisture that further assist in chewing and swallowing of the meat. The resistance of the meat structure during chewing will therefore depend on the sensory-detectable effects of connective tissue, the extent of muscle fiber contraction, and the degree of degradation of myofibrillar and cytoskeletal proteins as a result of aging (Rhee et al. 2004). The contribution of each of the three components to final eating quality, in particular tenderness, is muscle dependent and hence each muscle will respond differently to the various extrinsic and extrinsic factors that may affect quality (Bouton and Harris 1972; Bouton et al. 1976). For this reason Australia developed the MSA grading system based on numerous muscles cooked according to different cooking methods, while including all preharvest, harvest, and postharvest factors that can affect eating quality into the prediction models (Watson et al. 2008a,b).

It is expected that the amount of intramuscular fat (IMF) will also contribute to general eating quality and satisfaction, although the magnitude of the effect is often disputed. For example, early studies by Campion et al. (1975) reported that no more than 10% of the variation in the organoleptic properties of beef can be attributed to marbling, while Carpenter (1974) suggested that practically all organoleptic attributes are related in a positive manner to muscle fat, but muscle fat contributed only 15% of the variation in these attributes. More recently, Thompson (2004) showed that amount of IMF explained less variation recorded in the tenderness score than variation measured in the juiciness and flavor scores. In addition, the relationship between tenderness, flavor, and juiciness with IMF was curvilinear, meaning that as IMF increased, the increment in the three sensory attributes decreased and reached an inflexion point around 14%–17% IMF. Platter et al. (2003) showed slightly higher correlations between USDA (United States Department of Agriculture) marbling score and juiciness score (0.34) followed by tenderness (0.27) and flavor (0.22), while Tatum et al. (1982) showed that marbling accounted for 5%, 5%, and 15% of variation in sensory panel ratings for juiciness, tenderness, and flavor, respectively. Therefore, the relationships for all three attributes were relatively poor although this could have been attributed to a small range of fat levels (3.4%–7.3%; Platter et al. 2003), as opposed to the study of Thompson (2004). In addition, aged samples of young grain-fed animals were used suggesting that other components probably contributed to tenderness acceptability more than marbling did. Rated against overall acceptability the results of Platter et al. (2003) did show a weak but significant linear relationship with marbling score, suggesting that acceptability will increase by 10% for each increase in marbling score (1%–1.5% IMF). Smith et al. (2008) suggested that this "acceptability" mainly refers to flavor and juiciness when marbling is considered.

2.3 TECHNOLOGICAL QUALITY

The technological quality of meat includes parameters such as pH, color, water-holding capacity (WHC), the ratios and properties of fat, connective tissue, and protein which ultimately affect the potential quality of fresh meat and the ability of fresh meat to be converted to processed meat (Xiong 2014). In particular, the physicochemical functionality of muscle proteins makes an important contribution to the quality of fresh and processed meat. The functional properties of muscle proteins include texture forming and related properties, namely, water holding, fiber swelling, solubility/extractability, gelation, emulsification, and adhesion/binding. The three groups of proteins in muscle, namely stromal, myofibrillar, and sarcoplasmic proteins possess different technological qualities and have different functions in the production of fresh and processed meat (Rhee et al. 2004; Xiong 2014).

WHC is of paramount importance in fresh meat and meat product manufacturing and is defined as the ability of meat to hold on to its own or added water. Lean muscle contains 75% water held within muscle cells, between muscle cells (inter-fascicular) and between muscle bundles (extra-fascicular; Offer and Cousins 1992; Schäfer et al. 2000). Within muscle cells water is found within myofibrils (85%; intra-myofibrillar); or between myofibrils themselves or myofibrils and the sarcolemma (inter-myofibrillar; 15%). Because water is dipolar it is bound mainly by specific charges of surrounding proteins, but also through capillary forces. Three fractions of water exist and these are defined as bound (also termed protein-associated water), entrapped or immobilized (85%), and free water (Huff-Lonergan and Lonergan 2005; Pearce et al. 2011). Bound water is closely bonded to muscle proteins and does not move to other compartments, but only represents a very small fraction of total water in muscle. Entrapped or immobilized water is held by either steric (space) effects and/or attraction to bound water. It does not flow freely out of meat, but is rather affected by the rigor process (Honikel et al. 1986; Offer and Knight 1988) through changes to the cell structure integrity and pH. Finally, free water exists in the sarcoplasmic area in the muscle cell in narrow passages (called capillaries), is only held by weak surface forces and flows easily from tissue through minor physical forces such as lateral shrinkage of myofibrils. It mainly originates from entrapped water in unfavorable conditions. The focus of preserving water within meat is on "entrapped water" and is mainly determined by charges on myofibrillar proteins, the structure of muscle cells and components (myofibrils, membrane permeability, cytoskeletal protein linkages) and extracellular space created by various processes.

In the normal rigor process the muscle pH declines to the isoelectric point of most proteins (pH = 5.4). A lack of electrical charges on muscle proteins and myosin denaturation lead to myofibrillar lattice shrinkage causing a reduction in space between myofibrils that forces water out of this space (Honikel et al. 1986; Pearce et al. 2011). During the same process, disintegration of cellular membranes takes place that also contributes to leakage of water out of the intracellular spaces (Tornberg et al. 2000). Furthermore, cross-bridges form between the major filaments, actin and myosin, when muscle pH reaches the isoelectric point of myosin reducing the space within myofibrils (lateral shrinkage) (Offer and Trinick 1983). Additional decrease in intra-myofibrillar space is caused by sarcomere shortening during rigor development

(Honikel et al. 1986). These two processes, lattice shrinkage and contraction, force water out of the myofibril and into the extra-myofibrillar space and will contribute to purge. Myofibril shrinkage also causes entire muscle cell shrinkage creating channels between bundles of cells that function as channels for moisture to flow out of the muscle (Offer and Knight 1988; Offer and Cousins 1992). This happens if intermediate filament structures and costameres remain connected to the sarcolemma (no proteolysis) during conversion of muscle to meat (Kristensen and Purslow 2001; Melody et al. 2004). Peri- and endomysial breakages and shrinking also cause gaps and moisture accumulation in the perimysal network and between the muscle fibers and endomysium (Pearce et al. 2011). The environment for proteolytic enzymes, such as calpain, becomes increasingly less favorable for action on the intermediate structures as time progresses postmortem due to lowering of the pH, an increase in ionic strength, and an increasing inability to maintain reducing ability (Huff-Lonergan et al. 1996; Maddock et al. 2005). However, Melody et al. (2004) and Bee et al. (2007) showed that the proteins of intermediate structures can degrade within the first 45 min to 6 h after death which will be to the advantage of water preservation within the muscle. Davis et al. (2004) confirmed that less drip loss occurred in pork when desmin degradation was increased. It could therefore be expected that factors inhibiting proteolysis will contribute to drip loss. Since calpastatin is an effective regulator of calpain (Koohmaraie 1992; see Chapter 3) the genetically related increase in the expression of calpastatin (Ciobanu et al. 2004) has been found to reduce juiciness in pork. Likewise, the utilization of β agonists increases calpastatin and impairs proteolysis (Strydom et al. 2009) and could contribute to higher drip loss in steaks produced with β agonist supplements (Hope-Jones et al. 2012).

Pressure differences between intra- and extra-myofibrillar spaces will equalize if sufficient degradation of cytoskeletal proteins takes place and water between the two compartments is in equilibrium. During prolonged aging, water flow to extra-myofibrillar compartments will be reduced or even returned due to released contraction, but also by reverse osmosis created by proteolysis (Kristensen and Purslow 2001).

In cases where postmortem pH decreases very rapidly and muscle temperature is therefore high the myosin heads denature and shrink (Offer 1991). Denatured myosin will cause more significant lateral shrinkage of myofibrils and also exhibit lower binding capacity to water, consequently leading to excessive water loss, typically demonstrated in pale, soft, and exudative (PSE) pork. This condition is typically caused by genetic conditions in pigs such as the Halothane gene causing rapid decline in pH (but no change in final pH) and the mutation of the PRKAG3 gene (denoted RN^-) causing increased muscle glycogen at slaughter (Rosenvold and Andersen 2003; see Chapter 12). It is well-known that stress related preslaughter conditions can also cause PSE in normal pigs.

While the causes and consequences of PSE in pigs is well covered in the literature and addressed in industry, a similar condition in beef has only recently been identified (Warner et al. 2014a,b) in Australian grain-fed beef due to low cooling rates and a fast rate of pH decline (Strydom and Rosenvold 2014). Extensive knowledge gaps still exist regarding the effects of different pre- and postslaughter factors on the kinetics of enzymes (e.g., AMPK) involved in carbohydrate metabolism. Ferguson and Gerrard (2014) contribute this to the fact that the understanding of the anaerobic metabolism (postmortem) of energy is often molded on the knowledge of aerobic energy metabolism in the live animal.

The oxidative stability of meat not only affects the technological quality, but also influences sensory quality, like tenderness, color stability, and flavor (Aalhus and Dugan 2014). The fatty acid composition of meat can have major effects on the technological properties of meat such as color stability, oxidative stability, and firmness of fat (Wood et al. 2003). These properties in turn affect the shelf life and physical properties (oiliness or hardness of fat) of fresh and processed meats. The hardness of fat and cohesiveness between fat and muscle is a function of fatty acid (FA) type (saturated vs. unsaturated) due to different melting points (Wood et al. 2003). Wood et al. (2008) reported that fatter pigs had firmer SCFs due to higher 18:0 saturated fatty acid (SFA) (melting point 69°C) and lower 18:2n-6 polyunsaturated fat (PUFA) (melting point –5°C). While harder fat favors the technological quality of meat, the opposite is true with regard to the healthiness (Mann et al. 2006). The opposite also applies for ruminants as reported by Enser and Wood (1993) who found softer fat with less SFA for grain-fed lambs of higher fat levels. Despite the health advantage of higher levels of PUFA, in particular those that are limited in meat, through feed manipulation, higher PUFAs also contribute to lower oxidative stability that lead to rancid odors and poor flavor in fresh and processed meats. The thiobarbituric acid reacting substances (TBARS) test is used to measure oxidative stability of fat and values higher than 0.5 are regarded as critical for pork quality (Tarladgis et al. 1960), although Campo et al. (2006) reported a value of 2.3 for beef before rancid flavors over powered the typical beef flavor. According to Wood et al. (2008) values above 0.5 are seldom found in fresh pork even when PUFA proportions have been manipulated. However, minced and commuted products may show higher levels of oxidation than intact fresh meat due to exposure to pro-oxidants like iron that are released from cells (Sheard et al. 2000). Nute et al. (2007) reported TBARS as high as 6.2 in meat of lamb when lambs were supplemented with high PUFA feed components (combinations of fish oil, protected lipid supplement, and marine algae) as displayed in high oxygen packaging. In cattle Warren et al. (2008) also reported unfavorable TBARS for meat from grain-fed vs. grass-fed steers when packaged in high oxygen due to high PUFAs. Higher vitamin E levels in meat of grass-fed samples contributed to buffering capacity (Arnold et al. 1993) and were sufficient for keeping TBARS below critical levels, which was not the case for meat from grain-fed cattle. Color is also affected by lipid stability and Warren et al. (2008) demonstrated the faster deterioration of red color in high PUFA grain-fed beef.

Oxidative stability is not only limited to fat since any postmortem changes in muscle are also accompanied by an increase in oxidation (Harris et al. 2001). Oxidation of myofibrillar proteins causes a reduction in protein functionality (Xiong and Decker, 1995) through formation of protein disulfide linkages and conversion of certain amino acids such as histidine to carbonyl derivatives (Martinaud et al. 1997). Since proteins such as calpain have histidine components, protein oxidation can lead to inactivation of this enzyme and impairment of aging (Lund et al. 2007; Clausen et al. 2009; Kim et al. 2010) and even contribute to drip loss (Huff-Lonergan and Lonergan 2005).

2.4 MICROBIOLOGICAL QUALITY OF MEAT

2.4.1 Microbiological Spoilage of Meat

Meats contain all the nutrients required for the growth of microorganisms. In addition, the pH range in fresh and processed meats provides favorable growth conditions for most microorganisms.

Spoilage of meat and meat products is caused by the practically unavoidable contamination and subsequent decomposition of meat by bacteria and fungi, which are borne by the animal itself, by the people handling the meat, and by their implements. The organisms spoiling meat may be carried by the animal while still alive ("endogenous contamination") or may contaminate the meat after its slaughter ("exogenous contamination") (Lawrie and Ledward 2006).

While the oxidation–reduction potential of whole meats is low, redox conditions at the surfaces tend to be higher so that strict aerobes and facultative anaerobes generally find conditions suitable for growth. A facultative anaerobe can produce ATP (adenosine triphosphate) by aerobic respiration if oxygen is present, but is capable of switching to fermentation or anaerobic respiration if oxygen is absent. Aerobic spoilage microorganisms are usually dominated by pseudomonas, while anaerobic microflora are usually dominated by lactobacilli which produce spoilage by the slow accumulation of volatile organic acids. Meat of high ultimate pH, which is packaged anaerobically, spoils rapidly because the high pH allows anaerobic growth of bacterial species of higher spoilage potential than lactobacilli (Borch et al. 1996). Before overt spoilage develops, the spoilage status of meat can be accurately assessed from the bacterial numbers on meat only when there is knowledge of meat composition, storage conditions, and the types of bacteria present.

The types and genera of microorganisms most often found on fresh and spoiled meats and their symptoms of spoilage are listed in Table 2.2. *Brochothrix thermosphacta, Carnobacterium* spp., *Enterobacteriaceae, Lactobacillus* spp., *Leuconostoc* spp., *Pseudomonas* spp., *Weissella* spp., and *Shewanella putrefaciens* are predominant bacteria associated with the spoilage of refrigerated beef and pork (Jay 2000). The main defects caused by microorganisms in meat are off-odors and flavors, discoloration, slime formation, and gas production (Borch et al. 1996). Molds apparently do not grow on meats if the storage temperature is below –5°C (Lawrie and Ledward 2006). The following genera of molds have been recovered from various spoilage conditions of carcasses: *Mucor, Rhizopus*, which produce "whiskers" on beef, *Cladosporium*—a common cause of "black spot," *Penicillium* which produces green patches, and *Sporotrichum* and *Chrysosporium*, both of which produce "white spot" (Jay 2000).

Surface spoilage and the type of spoilage organisms are determined by the available moisture on fresh meat or meat products. Molds tend to predominate spoilage when the surface is too dry for bacterial growth. But molds never develop on meats when bacteria grow abundantly because the bacteria grow faster than molds, consuming available surface oxygen, which molds require for their metabolism (Jay 2000). During long-term refrigeration storage in vacuum packaging the predominant microorganisms are lactobacilli or *Bacillus thermosphacta* or both (Jay 2000).

TABLE 2.2
Spoilage of Meat and Meat Products by Microorganisms

Type of Microorganisms	Genera (Examples)	Symptoms of Spoilage
Aerobic bacteria	*Acinetobacter* *Bacillus* *Moraxella* *Pseudomonas*	Surface slime, discoloration, gas production, change in odor, and fat decomposition
Anaerobic and facultative anaerobic bacteria	*Lactobacillus* *Clostridium*	Putrefaction and foul odors, gas production, and souring
Molds	*Alternaria* *Aspergillus* *Cladosporium* *Chrysosporium* *Fusarium* *Geotrichum* *Penicillium* *Mucor* *Rhizopus* *Sporotrichum*	Sticky and "whiskery" surface, discoloration, change in odor, and fat decomposition
Yeast	*Candida* *Debaromyces* *Rhodotorula* *Torulopsis*	Surface slime, discoloration and change in odor and taste, and fat decomposition

2.4.2 Microbial Pathogens

Limiting the contamination and subsequent inactivation of pathogenic bacteria is very important to the safety of meat and meat products (Borch and Arinder 2002). There are numerous diseases that humans may contract from endogenously infected meat, such as anthrax, bovine tuberculosis, brucellosis, salmonellosis, listeriosis, trichinosis, or taeniasis. Meat safety concerns, challenges, and related issues that will continue being of concern in the twenty-first century may be divided into those associated with microbial pathogens and into other meat safety issues (Sofos 2008).

Contamination of meat from exogenous sources is probably a more likely threat to consumer well-being than endogenously infected meat. Two major sources of pathogenic bacteria causing foodborne disease in meat and meat products have been identified. Firstly, there are bacteria that reside in the intestinal tract of the living animal and secondly, bacteria from the environment that contaminate meat during slaughter and processing. The large intestine of animals contains some 3.3×10^{13} viable bacteria, which may contaminate meat and fat surfaces if the carcass is improperly dressed (Borch and Arinder 2002). Soiled hides are also a source of contamination in addition to improperly cleaned slaughter or dressing implements, contaminated water used for washing and cleaning and human handling. For example, the bolt of a captive bolt pistol may carry about 400,000 bacteria per square centimeter (Borch and Arinder 2002).

After slaughter, care must be taken not to infect the meat through contact with any of the various sources of infection in the abattoir, notably the hides and the soil adhering to them, water used for washing and cleaning, the dressing implements, and the slaughterhouse personnel. Bacterial genera which commonly infect meat while it is being processed, cut, packaged, transported, sold, and handled include *Salmonella* spp., *Shigella* spp., *Escherichia coli, Bacillus proteus, Staphylococcus epidermidis, Staphylococcus aureus, Clostridium. welchii, Bacillus cereus*, and fecal streptococci. These bacteria are all commonly carried by humans while infectious bacteria from the soil include *Clostridium botulinum* (Borch and Arinder 2002). Meat workers are important sources of pathogenic bacteria, most frequently indirectly by cross contamination.

Meat safety challenges associated with microbial pathogens may be divided into those dealing with problems caused by pathogens of current concern, pathogens of potential future concern, for example, those undergoing changes and adaptations, and the involvement of the environment in microbial pathogen concerns (Sofos 2008).

Microbial pathogens include *Salmonella, Campylobacter*, enterohemorrhagic *E. coli* including serotype O157:H7, *S. aureus, B. cereus*, and *Cl. botulinum*. A number of pathogens of current concern (e.g., *E. coli* O157:H7, *Listeria monocytogenes, Campylobacter jejuni, Yersinia enterocolitica*, etc.) were unknown or not suspected as causes of foodborne illness until recently. Additional foodborne pathogens recognized since 1970 include *Vibrio cholerae* serogroup non-O1, *Vibrio vulnificus*, Norovirus, *Cryptosporidium parvum, Cyclospora cayetanensis, Enterobacter sakazakii*, and prions. Some of these and other unsuspected pathogens may be of greater concern in the future or may become associated with animal health pandemics, for example, the avian influenza (AI) and foot-and-mouth disease (FMD) viruses. Other potentially important pathogens may be *Mycobacterium avium sub*sp. *Paratuberculosis, Escherichia albertii*, and *Clostridium difficile* (Sofos 2008).

Since any intestinal bacterial pathogen may be found in foods under unsanitary conditions, the same may be presumed for intestinal viruses, even though they may not proliferate in foods. Other than hepatitis A and rotaviruses, those of primary concern are Norwalk virus, calicivirus, and astrovirus. It has been shown that enteroviruses persisted in ground beef up to 8 days at 23°C or 24°C and were not affected by the growth of spoilage bacteria (Jay 2000).

2.4.3 Nonmicrobial Pathogens

Prions are unique proteins in that they can convert other proteins into damaging ones by causing them to alter their shape (Jay 2000). The normal cell prion protein (PrP) exists in the brain cell membrane where it carries out some vital functions and is then degraded by proteases. However, the pathogenic form is distorted and is resistant to proteases, and thus it accumulates in brain tissue and gives rise to diseases generically known as transmissible spongiform encephalopathies (TSEs). It has been postulated that the distorted prion molecule, acting as a template, converts normal protein to a distorted form. Bovine spongiform encephalopathy (BSE) is a prion disease of cattle and sheep that has been referred to as "mad cow disease" in the past. A TSE type prion disease of humans is Creutzfeldt–Jakob disease (CJD) and since humans are susceptible to the prions that cause CJD, the concern is whether humans can contract BSE from cattle. In March 1996, a new variant of CJD (nvCJD, V-CJD) was reported in the United Kingdom in a small group of people, all of whom were much younger than previous cases of CJD. This prompted speculation that nvCJD had been contracted from cattle (Jay 2000).

Trichinella spiralis is the etiological agent of trichinosis (trichinellosis), the roundworm disease of greatest concern from the standpoint of food transmission (Jay 2000). The global distribution of the various species of *Trichinella*, coupled with the wide range of hosts and various cultural eating habits involving raw or undercooked meat, are the main factors which favor human infections in both industrialized and nonindustrialized countries. *Trichinella* infection has been documented in domestic animals (mainly pigs) and in wildlife of 43 (21.9%) and 66 (33.3%) countries, respectively (Jay 2000). Although information on the occurrence of *Trichinella* in animals has not been reported in 92 countries, it is likely that the parasite is present in at least wild animals in many of these areas (Gajadhar et al. 2009). This disease can be prevented by the thorough cooking of meats such as pork or wild boar meat. Freezing will destroy the encysted forms, but freezing times and temperatures depend on the thickness of the product and the specific strain of *T. spiralis*. The lower the temperature of freezing, the more destructive it is to *T. spiralis* (Gajadhar et al. 2009).

2.4.4 Safety of Meats

Microorganisms do not survive thorough cooking of the meat, but several of their toxins and microbial spores do. For example, ochratoxin A (OTA) is a mycotoxin usually produced as a secondary metabolite by *Aspergillus* and *Penicillium* spp. fungi during the storage of foods. Many farm animal feedstuffs are based on cereals and cereal by-products, which are the preferred substrate for growth of *Penicillium* and *Aspergillus* spp. Animal-derived products for human consumption may be contaminated with OTA even if the animal has been fed only low levels of contaminated feedstuffs. Duarte et al. (2012) attributed the occurrence of OTA in pig-derived products to either direct fungal contamination, due to growth of toxigenic fungi in the outer layers of meat products, or indirect transmission via the ingestion of OTA-contaminated feed. In contrast, Bryden (2012) reported that animal products (e.g., fresh meat) are an unlikely source of mycotoxins or their metabolites with the possible exception of milk. However, of major concern is the impact of mycotoxins on animal health and production and strict control in the form of HACCP (Hazard Analysis Critical Control Points) programs are implemented throughout the world to control mycotoxins in feed and their possible carry-over to products.

In recent years, evidence of continuous adaptation and development of resistance by pathogenic microorganisms to antibiotics and potentially to traditional food preservation barriers such as low pH, heat, cold temperatures, dryness or low water activity, and chemical additives has emerged. Resistance of pathogens to antibiotics used in animal production or human medicine is of major concern in clinical settings, and will continue being important in the future (Sofos 2008).

The presence of infectious agents can be detected with rapid microbiological tests during the production and processing of meat, but testing by itself is not sufficient to ensure adequate food safety. The industry-standard HACCP system provides a comprehensive quality management framework as a part of which such tests can be conducted. Such advances may help in finding pathogen sources, evaluating control interventions, and developing, verifying critical control points and critical limits for pathogen control through HACCP programs.

2.5 DEFECTIVE MEAT

2.5.1 Introduction and Definition of Defective Meat

Defective meat has a limited functionality that will lead to unacceptable appearance (color), taste, texture (tenderness and juiciness), and processing ability or a combination of these. A meat defect can be caused as a result of intrinsic factors, such as genetics, or extrinsic factors such as environmental effects, and production and processing systems, or by a combination of these factors. Defective meat has a very broad definition, but includes any property of meat that will dissatisfy end-users for example, processors, butchers, or consumers.

The origin of defective meat can roughly be considered as occurring in either the pre- or postslaughter phase of meat harvesting. Preslaughter processes can further be divided into on-farm and slaughter plant conditions, while the postslaughter processes includes the early part when muscle changes to meat (see Chapter 3) and the later part when meat is further processed or handled.

2.5.2 Examples of Defective Meat and Their Characteristics

2.5.2.1 Meat Toughness

Meat toughness can be caused by any factor related to connective tissue composition and properties, the amount of muscle contraction during rigor mortis and the aging ability of the meat that is facilitated by proteolytic enzymes. Different levels of tenderness will be acceptable for different consumers. Factors that could induce meat toughness include genetics, metabolic modifiers, feeding regimes, age, gender, preslaughter stress, the slaughter process, and conditions (i.e., the conversion of muscle to meat and the aging duration and conditions; Koohmaraie 1996; Ferguson et al. 2001). Many of these factors also affect other meat quality attributes like color and WHC. Most of these factors, like conversion of muscle to meat, genetics, age, sex, and nutrition are addressed in other chapters. A selection of other factors contributing to tenderness and other quality defects will be discussed here.

2.5.2.2 Dark, Firm, and Dry Meat

Dark, firm, and dry (DFD) meat is characterized as having an abnormally dark color, a firm texture, and an increased WHC associated with a dry, sticky surface (Tarrant 1981). DFD meat originates from animals with deficient muscle glycogen levels at slaughter so that the conversion of muscle to meat through glycolysis (rigor) is impaired (Warriss et al. 1989). Such meat is characterized by a high ultimate pH (>6.0) and deficiencies in glucose and glycolytic metabolites (Newton and Gill, 1981). While normal muscle glycogen concentration in cattle and sheep ranges from

75 to 120 mmol/kg (Monin 1981; Lambert et al. 1998; Immonen et al. 2000; Pighin et al. 2014), the critical threshold lies between 45 and 57 mmol/kg below which normal final pH in meat (5.5–5.6) will not be attained. Wulf et al. (2002) reported an almost linear increase in ultimate pH (above unfavorable values) when beef loin muscle recorded a glycolytic potential (GP measured as amount lactate = 2 × [glycogen + glucose + glucose-6-phosphate] + lactate) <100 μmol/g tissue. Interestingly, values above 100 μmol/g tissue did not correlate with pH and it appears that final pH in such cases is determined by pH inhibition of glycolytic enzymes. This has practical impact as it suggests that animals with high GP will be able to stand more stress before slaughter before DFD is exhibited compared to those with lower GP.

The dark color and poor keeping quality of DFD meat are mostly rated as the main objection by consumers (Tarrant 1981) as it appears unattractive to consumers who often associate the condition with meat of old animals at least with beef. However, Cho et al. (2007) and Ngapo et al. (2004) showed that Korean and French consumers, respectively, mostly preferred darker colored pork chops with L^* values (56) close to DFD values ($L^* = 54$, Warriss and Brown 1993) as opposed to Danish and British consumers (Dransfield et al. 2005) preferring lighter colored pork suggesting that sensitivity to DFD may be consumer specific. Other objections mentioned in surveys included abnormal or bland taste, difficulty to brown during cooking, unacceptable curing or drying properties, and stickiness of the meat surface (Tarrant 1981). Wulf et al. (2002) and Viljoen et al. (2002) also reported off-flavors associated with dark cutting beef steak and a larger variation in tenderness. Interestingly, female consumers indicated more off-flavors in DFD meat than male consumers. High ultimate pH (>6.0) and deficiencies in glucose and glycolytic intermediates cause bacterial spoilage of DFD meat when bacteria utilize amino acids due to a lack of glucose. Consequently, amino acids are utilized sooner and spoilage becomes evident at lower microbial counts than in normal meat (Newton and Gill 1981).

2.5.2.3 Pale, Soft, and Exudative and Related Meat Defects

Pork is classified into four categories according to color, texture or firmness, and exudation or drip loss as summarized by Warner et al. (1993). Normal pork (red, firm, and nonexudative, RFN) is reddish pink, has a firm texture and is nonexudative, that is, has a desirable WHC and the muscle has followed a moderate pH decline during rigor. Pale, soft, and exudative (PSE) pork has an undesirable pale to gray color, has a very low WHC, and poor soft texture. For the same reason it yields very poor manufactured products. In addition, brined PSE meat absorbs more salt than normal meat and when cooked hams manufactured with PSE meat are sliced the meat will crumble easily in the PSE zones. Red, soft, and exudative (RSE) pork has a normal color, but is soft and exudative. Other related categories of pork have also been identified, namely pale, firm, and nonexudative (PFN), pale, firm, and exudative (PFE), and pale firm and dry (PFD) (Kanda and Kancchika 1992; Roserriro et al. 1993; Warner et al. 1993; see Chapters 3 and 12). The PSE condition (and some of the other related conditions of pale or exudative conditions) in pork mostly result from acute stress conditions immediately prior to slaughter which causes a rapid rate of glycolysis leading to low pH values at high muscle temperature (Sayre and Briskey 1963; Offer, 1991; Joo et al. 1999). Denaturation of myofibrillar (in particular myosin) and sarcoplasmic proteins under these unfavorable conditions causes drip loss, loss of structural integrity, and paleness in color.

While the condition of low muscle pH and high muscle temperature may also occur in beef, the pathways leading to this condition are different from that of pork, where preslaughter stress is the main cause. In beef, the condition is termed "high rigor temperature carcasses" and is defined as beef carcasses with pH dropping below 6 when muscle temperature is above 35°C (Thompson 2002). In contrast to PSE in pork, this condition in beef and sheep was not studied intensively or included in meat quality schemes until recently. Warner et al. (2014a) reported that in seven major Australian abattoirs up to 75% of carcasses fell within the criteria for high rigor temperature conditions (in 2006). As with PSE in pork rapid metabolism (glycolysis) postmortem is the key mechanism causing high rigor temperature meat. In addition, the condition causes changes in protein functionality through denaturation of contractile, proteolytic, and other proteins with consequent negative effects on WHC, color and color stability, and tenderization (Kim et al. (2014). In some, but not all cases meat may toughen due to muscle shortening under these conditions. Recent changes in the beef and sheep industry is most probably the reason for the increased awareness of high rigor temperature carcasses and their defects. Increased electrical inputs at slaughter and increased carcass weights and carcass fat levels are the major causative factors for high rigor temperatures according to Jacob and Hopkins (2014) and Warner et al. (2014a). Larger and fatter carcasses are the result of an increased demand for meat worldwide and therefore improved efficiency. As a result larger beef carcasses and higher numbers of grain-fed carcasses are produced that pose new challenges to efficient chilling and control of electrical inputs to meat optimum pH and temperature criteria (Thompson 2002). Incidentally, controlling carcass temperatures of grain-fed animals is further challenged by insulin resistance and associated factors leading to decreased heat tolerance and subsequent increased carcass temperatures at slaughter (DiGiacomo et al. 2014; Warner et al. (2014a).

2.5.2.4 Bruises

Bruises on carcasses are the result of direct trauma to soft tissue that causes rupturing of local capillaries, arterioles, or venules allowing blood to escape into the surrounding tissues and interstitium (Rezac 2013). In practical terms bruises can occur up until the point of exsanguination. Tissue damage due to injections can also be regarded as bruises. Although it is often suspected that bruised meat is more susceptible to spoilage (Hamdy and Carpenter 1974), no differences in microbiological loads or rates of bacterial growth have been found between normal and bruised meat (Rogers et al. 1992) provided they are treated identically within the abattoir (Gill and Newton 1978; Gill and Harrison 1982). Furthermore, apart from slightly higher salt contents in bruised meat no differences were found in composition or processing properties of bruised meat (Rogers et al. 1992). Rogers et al. (1992) could not find any microbiological or technological reason for the condemnation of bruised beef and therefore it can be concluded that bruised meat is condemned only due to its aesthetically unattractive appearance.

The 1991 US National Beef Quality Audit (NBQA) reported a 39.2% occurrence of bruises on beef carcasses produced in the United States with 14% registering more than one bruise and 5% requiring extensive trimming (Lorenzen et al. 1993). Most bruises occurred in the loin area of the carcass (23.4%) followed by the rib and chuck area (17% and 14%). In a 2000 NBQA on fed cattle, just under 50% of carcases had no bruises and 16 % had more than one bruise (McKenna et al. 2002). Most bruises occurred in the chuck and loin (>25% each) and <15% in the round. The most recent NBQA showed improvement with only 33% of carcasses reported as having bruises, of which only 4% had more than one bruise. More than 50% of bruises were found in the loin and much less in the ribs (14%) and chuck (21%) (McKeith et al. 2012). Strappini et al. (2010) report bruise prevalence figures of 8.6% and as high as 20.8% at two Chilean slaughterhouses. According to Gallo et al. (1999) 7.7% of carcasses in 22 Chilean abattoirs had bruises of which 4.8% were only subcutaneous, in 2.1% muscles were affected and in 0.8% of the carcasses subcutaneous, muscle and bone tissues were affected.

The Australian carcass bruising system classifies bruises by severity and location and can thereby give an indication of the seriousness of the problem and the origin of the incident or factor leading to the bruising thereby enabling preventative measures (Wythes et al. 1985). Bruises on carcasses from different producers or lots indicate an external problem, while random bruises on any carcass, point to an internal or slaughter plant problem. McCausland and Dougherty (1978) reported that between 43% and 90% of bruises occur after Australian cattle arrived at the abattoir. Although there is probably no single control measure to minimize bruising, improving the skills of handlers and drivers before and at the slaughter plant seems to be the most important action to lower bruise counts (Wythes et al. 1985; Romero et al. 2013). Various other measures and precautions could be added to proper handling. Weeks et al. (2002) and Strappini et al. (2010) found that bruising can be reduced by minimizing the transfer of cattle to slaughter houses through stock markets, especially older animals. This will minimize loading and handling of animals, mixing of unfamiliar groups, and duration of transport which are factors contributing to bruises (Knowles 1999). Extra care should be taken with the handling and transport of older animals and female animals (Wythes and Shorthose 1991; Strappini et al. 2010; Hoffman and Luhl 2012) probably because these animals are often leaner and therefore more prone to bruising than male and younger animals (Weeks et al. 2002; Hoffman and Luhl 2012). Jago et al. (1996) and Knowles et al. (1994) also reported that leaner deer and sheep are more prone to bruises. Due to higher activity in the breeding season of deer, leaner animals coincide with a specific time of the year when the slaughter of male animals should then be prevented. In contrast to studies on cattle, Knowles et al. (1994) found a higher incidence of bruises in lambs compared to ewes and attributed this result to a higher incidence of riding while animals were moved in handling facilities.

While shorter distances are propagated by certain studies to reduce bruising (Yeh et al. 1978; Hoffman et al. 1998) others found no effect of travel distance (Wythes et al. 1985; Hoffman and Luhl 2012; Romero et al. 2013). Other factors such as type of road, driving skills, design of the truck, type of animal, and loading density seem to interact with travel distance. For example, Strappini et al. (2010) could not explain lower bruising counts with longer distances but these could have been attributed to other factors, such as animals calming down during longer trips with less likeliness of animals going down. Proper truck design and design of holding and processing facilities could reduce the number and type of bruises (Wythes et al. 1985; Rezac 2013; see Chapter 6).

Romero et al. (2013) found that the reduction in lairage time from between 18 and 24 h to between 12 and 18 h will decrease bruise counts on beef carcasses by 50%. Reduction in lairage time also reduced the incidence of bruises in sheep (Cockram and Lee 1991), pigs (Fraqueza et al. 1998), and in deer (Jago et al. 1996). Aggressive interaction due to mixing of unfamiliar animals in a single group will contribute to bruising especially when these are male animals or mixed sexes and will further be exacerbated by the presence of horns or a mixed group with horns and no horns (McVeigh and Tarrant 1983; Minka and Ayo 2007; Hoffman and Luhl 2012).

While deep tissue bruises are more common with cattle, bruises on pig carcasses are most often only skin damage or blemishes although haematomas of the underlying tissue can also occur in serious cases. Most factors contributing to bruises in cattle, deer, and sheep also apply to causing skin blemishes in pigs, such as on farm handling, mixing of animals, fasting, handling at loading and offloading, transport and lairage time, actions, and conditions (Faucitano 2001). Dalla Costa et al. (2007) reported that two thirds of skin bruises originated during loading transport and offloading while lairage was responsible for the remaining one third. Therefore, procedures to prevent skin damage should be focussed accordingly. Proper handling of pigs on the farm will reduce aggressiveness and occurrence of skin bruises during preslaughter events. Avoiding mixing of unfamiliar groups during transport or in lairage by compartments in trucks and smaller groups in lairage will reduce fighting and skin blemishes (Faucitano 2001). While fasting has benefits like preventing PSE and deaths during transport, excessive fasting could lead to aggression and increased skin damage. More bruises were recorded on pigs transported in winter probably due to the animals huddling together and climbing over each other to cope with the cold (Correa et al. 2013). On and off-loading could be improved by hydraulic tail gates that could adjust the height of the ramps, by improved handling facilities and correct and limited use of goads, sticks, and boards to reduce the excitement of pigs. Results on stocking densities are contradictive as some studies suggest that low stock densities leave too much space for movement leading to aggressive behavior, while other report higher counts of skin damage when stocking densities are too high and pigs battle to assume a resting position (review: Faucitano 2001; see Chapter 6). Male animals were more prone to skin blemishes due to aggressive interactions (Warriss 2003) and increasing the lairage time from <1 h to 18 h also increased skin blemishes from 5 to 18 per carcass. Handling facilities and the procedures followed to herd pigs into holding pens or to the stun area could determine the way the animals respond and interact. Automization should reduce aggressive behavior, while herding of smaller groups and proper handling practices will also contribute to less skin damage. Design of facilities in particular races leading to the stun box should provide for easy handling and least interaction of animals and humans and among animals (Faucitano 2001).

2.5.2.5 Ecchymosis or Blood Splash

Ecchymosis or blood splash is a muscle condition mostly associated with stunning where localized hemorrhages of small blood vessels within muscle tissue occur that form spots on the surface of various muscles and organs. In severe cases, ecchymosis can extend deep into the muscle. Hemorrhages can range from microscopic to several centimeters in diameter and the severity of the case will determine the amount of downgrading of a cut. Petechial hemorrhages or speckle should not be confused with ecchymosis, but is an effusion of blood into fat or connective tissue layers of muscle and has the appearance of a rash (Petersen et al. 1986). Speckle is often regarded of low economic significance, since it is easy to remove from affected tissue (Gilbert and Devine 1982; Velarde et al. 2000).

Classic work by Tweed et al. (1931) determined that the two main causes of ecchymosis or blood splash during the slaughter process are increased blood pressure combined with severe muscular spasm, normally induced by stunning or other electrical inputs. Leet et al. (1977) has confirmed by electron microscopy that "super contracture" of small groups of muscle fibers lead to the rupturing of small blood vessels in the vicinity and the formation of the hemorrhages in the muscle. Although both ecchymosis and speckle are associated with the stunning of the animal, Kirton et al. (1978) showed that increased blood pressure following stunning alone is not responsible for blood splash as the condition still occurred in lambs after the jugular vein has been severed and an electrical current (such as stunning) was applied. Smulders et al. (1989) attributed blood splash to contraction of antagonistic muscles when animals are restrained at stunning. It is further postulated that vigorous contraction of muscle in the early stage of bleeding in nonelectric stunning may have the same aggravating effect on blood splash as tonic spasms caused by electrical stunning. Petechial hemorrhages or speckle occur as a result of muscle spasm during electrical stunning, but blood pressure was not described as a contributing factor in sheep (Gilbert and Devine 1982).

Ecchymosis and speckle can occur in pork, beef, and sheep meat (Gregory 1985), but is also of great significance in the slaughtering of deer (Mulley and Falepau 1999; Barnes and Barnes 2000). In a collection of studies reported by Mulley and Falepau (1999) incidences as high as 23% were recorded in the carcases sampled. Apart from type of stunning (see Chapter 8 for discussion on other species), factors involved in prevalence of ecchymosis in deer are: time between stunning and exsanguination, exsanguination method, carcass condition, stress levels of animals preslaughter, and distance travelled prior to slaughter (Barnes and Barnes 2000; Mulley et al. 2010), which mostly agree with the factors involved with ecchymosis in other farm species. Limiting of the interval between stunning and sticking is regarded as one of the most important measures to reduce ecchymosis in deer, which warrants the use of purpose built abattoirs to accommodate different sized animals of different species. Barnes and Barnes (2000) reported a reduction in blood splash from 20% to 3% when shorter times between stun to stick and hanging of the carcass were introduced in abattoirs. Thoracic stick further reduced blood splash. Although electrical stunning is mostly used in deer abattoirs some of the studies by Mulley and Falepau (1999) showed that stunning by captive bolt reduced ecchymosis. Mulley et al. (2010) reported that captive bolt stunning gave higher numbers of ecchymosis cases when stun to stick was longer than 10 s, followed by electrical stunning, but no differences were recorded between the two stunning methods when stun to stick time was <10 s. Mulley and Falepau (1999) recorded less hemorrhages in *masculus vastus lateralis* and *masculus rectus femoris* of deer when cranial extension of the hind leg was restricted subsequent to the grand mal seizure induced by electrical stunning. This finding contradicts the results of Smulders et al. (1989) reporting higher incidences of blood splash due to contraction of antagonistic muscles when animals were restrained.

Mulley and Falepau (1999) and Mulley et al. (2010) found the highest amount of blood splash with castrates followed by female deer despite intact male animals recording higher preslaughter activity, bruises, and higher ultimate pH, especially during breeding time. Since physical stress measured as elevated final pH values did not lead to increased blood splash, they speculated that the higher susceptibility of castrates was probably associated with emotional stressors, but no physiological explanation was offered for the relationship with the mechanisms of blood splash. Pearson et al. (1977) also did not find any association between prestun circulating cortisol levels of sheep as an indicator of physical stress and blood splash in sheep.

Smulders et al. (1989) showed that blood splash occurred more commonly in both the perimysium (speckle) and muscle tissue of the *Mm. longissimus, infraspinatus, semimembranosus, rectus femoris, gastrocnemius*, and *gracillus* of beef (veal), while Channon et al. (2000) and Velarde et al. (2000) specifically recorded defects in the shoulder blade subprimal of pigs and recorded very low incidences of ecchymosis in loins and hams of pork carcasses. Hind quarter muscles are also more frequently affected in deer than fore quarter muscles according to Mulley and Falepau (1999).

2.6 EVALUATION OF DIFFERENT ASPECTS OF MEAT QUALITY

2.6.1 Instrumental Methods to Measure Meat Quality

Quality as perceived and experienced by consumers is a highly subjective issue and may vary according to culture and experience. In terms of meat many of these traits can be measured by means of instruments online or at least shortly after slaughter to enable the sorting or description of carcasses according to certain quality characteristics. Other tests require time and sophisticated equipment and are mostly used for research.

2.6.1.1 Tenderness Measurements

Various mechanical instruments have been developed to measure the tenderness or texture of meat. Mechanical methods to measure tenderness of meat samples in a laboratory work on the principle of applying and measuring force to disrupt meat texture in some way and a full description of these is given by Purchas (2014). Equipment that can accurately predict tenderness or tenderness development with aging noninvasively at the point of slaughter will be of great value to the industry. George et al. (1997) compared the Australian Tendertec instrument to conventional USDA quality grading for segmenting carcasses into tenderness groups. The Tendertec operates on the principle of driving a blunt object into the muscle and measuring the distance travelled as well as the force, and therefore the amount of work/energy needed to push the rod against the muscle. The study reported poor correlations between variables of the Tendertec instrument and Warner–Bratzler shear force (WBSF) values and sensory scores for overall tenderness and amount of connective tissue of rib steaks aged for 14 and 28 days. Furthermore, the ability of the instrument to segregate carcasses in tenderness groups was poorer than that of the USDA grading system. However, the general poor relationship between raw beef muscle tenderness and that of cooked meat should be mentioned as a reason why instrumental measurements on raw samples often poorly predict the tenderness of cooked samples (Carpenter et al. 1965; Alsmeyer et al. 1966).

It is important to know how different mechanical methods relate to one another, how consistent their results are and how well they predict consumer scores for tenderness. The best known instrument is the Warner–Bratzler (WB) shear device that can be attached to various operating systems like the Instron materials testing machine, WB shear machine, or Lloyd texture analyser (Bourne 2002). Another popular instrument especially used in Australia and New Zealand is the tenderometer developed by MIRINZ (Meat Industry Research Institute of New Zealand). The main difference from the WB device is that it has a blunt wedge shaped "tooth" simulating a bite action as opposed to the WB device's shear action. Bouton and Harris (1972) showed a 0.94 correlation between WB and the tenderometer values and further studies by Graafhuis et al. (1991), Hopkins et al. (2011, 2013) developed models to relate the data of the two machines. The G2 is a recent upgrade of the tenderometer where the same testing action was retained, but the G2 provides a more cost effective and portable machine to industry than the previous tenderometer model or WB devices. Hopkins et al. (2011, 2013) reported a 1.2–1.3 times conversion of WB to tenderometer results. The difference is probably due to the sharper blade of the WB device, therefore registering lower values. Repeatability within samples was the same for both devices, but when less tender samples were used the tenderometer was more variable and the differences between the results of the two devices increased. In addition, more replicates of the same sample were needed for the tenderometer to obtain equal levels of precision than with the WB device. Hopkins et al. (2013, 2006) reported poor predictions of sensory tenderness and overall liking scores by sensory panels for both devices. It was suggested that reference data should have a higher variation in sensory scores, in particular on the lower end of tenderness scale, in order to improve the prediction models. Perry et al. (2001) concluded that shear force was a useful indicator of sensory tenderness, but it did not account for all the improvement in sensory scores when meat was aged. When data were adjusted for shear force (same shear force), samples aged for 14 days were scored higher by a trained panel than those aged for one day indicating that the relative contribution of connective tissue and myofibrillar components are not analyzed effectively by mechanical measurements.

Slice shear force (SSF) was developed by Shackelford et al. (1999) as a simplified method for measuring shear force. Instead of using a WB shear device that has a beveled edge V-shaped cutting blade, the SSF method uses a solid blade of the same thickness with a beveled shearing edge. The latter cuts through a single 1-cm-thick, 5-cm-long slice instead of six or more cores of meat. Shackelford et al. (1999) reported better accuracy in relation to trained sensory panel (TSP) scores with warm steaks directly after cooking than with cooled down steaks, a prerequisite for WBSF. This means the SSF was more practical and less time consuming than the

WBSF method. In addition, the SSF was also more accurate in the prediction of TSP scores than the WBSF in this study and showed high repeatability. Wheeler et al. (2002) confirmed that SSF performed at 72 h postmortem can accurately categorize carcasses into tender and tough groups (based on the *M. longissimus*), while other noninvasive methods such as color meters (BeefCam) and Minolta color values combined with marbling score and hump height could not achieve this.

Wheeler et al. (1997) demonstrated the importance of standardized protocols to firstly ensure repeatable measurements with WB machines (but obviously all instrumentation) so as to be able to compare values among laboratories. It was evident that different laboratories deviated from standard protocol for shear force measurement (Wheeler et al. 2005). Differences in protocols were numerous. Initial temperature of samples varied due to variation in thawing temperature, while cooking time differed due to variation in temperature control and cooking facilities. Higher initial and lower final temperatures resulted in lower shear values. Fiber orientation and consistent diameter of the cores were also emphasized for consistency. Shearing the cores twice instead of once through the center resulted in higher shear values due to shearing through harder and drier outer parts of the cores. Higher crosshead speeds increased shear values and Bratzler (1932) indicated an optimal speed of 229 mm/min for best results. Finally, some variation among machines measuring the shear force (e.g., Instron vs. WB machine) was also suspected as a reason for variation in results where protocols were standardized. The study concluded that high repeatability within laboratories and accurate comparisons among laboratories were possible if strict adherence to a standardized protocol was followed. Hopkins et al. (2010) emphasized that extreme care should be taken when large research programs use different laboratories for meat tenderness evaluations. Despite using similar protocols for shear force measurements, results differed between two laboratories that were mainly accounted to differences in cooking loss between the laboratories. Considering that every phase of a study where animals and processing are involved may contribute to variation, multiphase experimental designs should be in place to accommodate such sources of variation so that biases in estimates of parameters can be prevented.

When considering the different factors involved in meat tenderness (Koohmaraie 1996; Rhee et al. 2004), namely IMF content, collagen content and properties (cross-linking and heat stability), the degree of contraction of the muscle at the end of rigor mortis, and the extent of specific protein degradation, indirect methods to determine or describe variation in tenderness can also be used. Sarcomere length by laser defraction or microscopy aided by video image analyses measures the amount of contraction at rigor mortis. Collagen stability and content can be measured by heat treatment and assessment of hydroxyproline content, respectively. IMF content is always positively, but not closely related to tenderness and can be measured by solvent extraction, visual appraisal with the help of marbling charts or by near-infrared (NIR) spectroscopy, for example.

2.6.1.2 Video Image Analyses

Video image analysis uses hardware such as sophisticated video cameras and microscopes in combination with dedicated software to evaluate different physical and histological characteristics captured by the camera or microscope. This may include color (Chmiel et al. 2011) which can be more accurately measured since the whole surface of a steak is converted to pixels as opposed to smaller areas measured by standard chromameters and spectrophotometers. Marbling can be accurately assessed as percentage area occupied by fat on a steak surface. Histological tests include, sarcomere lengths that measure the degree of contraction of muscle at rigor mortis, myofibrillar length (MFL) that measures the amount of fragmentation of muscle fibers due to the action of proteolytic enzymes and myofibrillar fractures and detachments, that quantifies the number of Z-line breakages and the extent to which muscle fibers have detached during aging (Koohmaraie et al. 1987; Snyman et al. 2008). All three histological methods measure an aspect of tenderness.

2.6.1.3 Near-Infrared Spectrophotometry

Various instruments have been designed to use light in the NIR wavelength spectrum to evaluate certain meat properties pre- and postrigor. Chemical composition, technological parameters, sensory attributes, and a means of classification of carcasses into quality grades can be summarized as the different categories for which NIR has applications (review by Prieto et al. 2009a).

2.6.1.3.1 Chemical Composition

Various studies as summarized by Prieto et al. (2009a) have shown that chemical composition of fat, protein, and moisture in ground meat can be determined by NIR spectrophotometry with high accuracy; coefficients of prediction between (R^2) 0.87 and 0.99 and ratio performance deviation (RPD) between 2.56 and 28.46 (RPD higher than 2 are suitable for screening purposes). Meat consists of up to 75% water and the specific absorbance of O–H bonds at 1450–1940 nm in the NIR spectra explains its ability to predict water content. Likewise, the predictive ability for fat and protein is due to absorption of N–H bonds at 1460–1570 and 2000–2180 nm for protein and absorption of C–H bonds of fatty acids at 1100–1400, 1700, and 2200–2400 nm.

2.6.1.3.2 Technological Parameters

The ability to predict DFD prerigor could be of importance in hot boning of beef. Andrés et al. (2008), Cozzolino and Murray (2002), and Reis and Rosenvold (2014) demonstrated accurate predictions of DFD by means of NIR ($R^2 = 0.97$, RPD = 3.17) since the components in prerigor meat related to the occurrence of DFD, namely glycogen and lactic acid, are active in the NIR spectral range 1100–2500 nm. Other studies as summarized by Prieto et al. (2009a,b) record poorer results (as low as $R^2 = 0.07$ and RPD = 0.90) in this regard. In most cases poor results were related to the nature of the test sample (ground) and variation of the reference data, while Andrés et al. (2008) and Cozzolino and Murray (2002) using samples with a wide range of pH values as reference data and intact meat samples, showed good repeatability of the reference method.

Prieto et al. (2008) showed good predictions of L^* and b^* color values using absorbance in the regions of 1230–1400 nm and 1600–1710 nm, respectively, which relates to absorbance in C–H bonds of IMF. Redness (a^*) normally showed poorer prediction values, but Prieto et al. (2009b) reported an $R^2 = 0.86$ and RPD = 2.0

when light in the visible range was included. Andrés et al. (2008), Cozzolino et al. (2003), and Prieto et al. (2008) generally reported high predictability values for *L**, *a**, and/or *b** ($R^2 > 0.85$ and RSD > 2.17) although many other studies as reviewed by Prieto et al. (2009a) recorded predictability values of $R^2 < 0.35$ and RSD < 1.5.

NIR spectroscopy has a limited ability to predict WHC (Prieto et al. 2009a). WHC is actually predicted by its indirect association with protein, IMF, and water. Changes of a physicochemical nature related to these components determine various related aspects of WHC namely cooking loss, drip loss, and pressed out water and this complex relationship is poorly interpreted by NIR spectroscopy, although Forrest et al. (2000) and Ripoll et al. (2008) did report moderate accuracies of prediction ($R^2 = 0.86$, 0.86, and RPD = 1.76, 1.75). In addition, Geesink et al. (2003) showed that NIR reflectance spectra between 1000 and 2500 nm could predict drip loss with 55% accuracy (1.1% standard error) and could distinguish between carcasses with superior (<5%) and inferior (>7%) WHC.

The literature also presents contradictory reports on the ability of NIR spectroscopy to predict shear force tenderness and sensory qualities as reviewed by Prieto et al. (2009a). WBSF is closely related to the chemical composition of meat samples, especially, collagen, moisture, and IMF, but also various other histological and biochemical factors play a role which also applies to sensory tenderness (Koohmaraie 1996; Rhee et al. 2004). Prieto et al. (2008) did report favorable correlations between WBSF and absorbance in the IMF region of 1300–1400 nm which they contributed to the diluting effect of IMF on the effect of collagen on tenderness. In most cases, poor relationships have been found that relate to the reference methods and the poor homogeneity of the samples as discussed later on in this section.

2.6.1.3.3 Sensory Attributes of Meat

Sensory attributes of meat are poorly predicted by NIR ($R^2 = 0.003$–0.69; RPD = 0.57–1.67; Prieto et al. 2009a,b) mostly due to the complex nature of sensory attributes and the robustness of sensory scores as discussed below. Interestingly, Prieto et al. (2009b) predicted flavor with an accuracy of $R^2 = 0.59$ and related the predictability to the spectra range of fatty acids (1100–1400 nm) suggesting that these would provide sufficient variation in flavor.

2.6.1.3.4 Factors Influencing the Success of NIR Spectrophotometry

Factors influencing the success of NIR spectrophotometry are very similar for its various applications. Many failures are due to experimental design problems (Prieto et al. (2009a) and to obtain a proper data base, reference samples with large variations in the specific quality characteristics are necessary. Likewise the prediction ability of the NIR spectrophotometry for specific traits may not always be robust enough to distinguish between small differences in measured samples in practice. Ripoll et al. (2008) noted that the *N* content determined by the Kjeldahl method did not correspond with the absorbance of the amide bonds involving *N* by NIR spectroscopy. Sensory traits are even more problematic in this regard as they are very complex, variable, and subjective and the scores used are noncontinuous. Prevolnik et al. (2004) suggested that different statistical approaches such as neural networks may solve this problem and improve predictability by NIR. Mincing or homogenization of the sample for chemical analyses is recommended (Barlocco et al. 2006; Prevolnik et al. 2004) since intact muscle fibers may act as optical fibers that conduct light along their lengths and reduce reflectance. Homogenization will destroy and randomize fiber arrangement and average the effect of scattering apart from presenting the sample more homogenous (Tøgersen et al. 2003) and this limits any online application.

Andersen et al. (1999) and Prieto et al. (2008) found poor predictability for pH using NIR and ascribed this to the loss of information of scattering when samples were ground. Related to this, Andrés et al. (2008) found better predictability when reference data had a wide range of pH values and samples were intact. Hildrum et al. (2004) also demonstrated the use of diode array detectors under industrial conditions that have the ability to scan all wavelengths in a few milliseconds and allow large meat areas to be measured. A larger scan area for intact meat in predictions of shear force was also suggested by Geesink et al. (2003), since subsampling for shear force, that is, using numerous cores, caused variability in the results that did not coincide with a small area of spectroscopy. Differences in the nature of the reference and validation samples due to various reasons may also affect prediction values. In chemical composition predictions using absorption values of O–H bonds can be influenced by water (sample) temperature differences. Likewise the variation in humidity during sampling could affect dry matter/water content analyses and therefore samples for verification and for reference data must be taken simultaneously (Cozzolino et al. 2002). Industrial conditions are also exposed to fluctuations in humidity and temperature and Hoving-Bolink et al. (2005) and Leroy et al. (2003) overcame these difficulties by using fiber optic probes that are less sensitive to these variable conditions on intact carcasses. Certain traits are also scanned (NIR) and measured at different stages postmortem leading to low prediction factors because of changes within the sample over the specific period. Prieto et al. (2009a) suggested that poor predictions of color could be the result of changes taking place in the color pigment between time differences of NIR measurement and actual color measurements. Likewise, online use of NIR spectroscopy to predict of sensory qualities of aged meat gave poorer results ($R^2 = 0.28$ for tenderness) in the study of Prieto et al. (2009b) than that of Ripoll et al. (2008) ($R^2 = 0.98$) where the NIR and sensory analyses were carried out the same day.

For a complex trait such as WHC that is influenced by numerous physicochemical properties of the muscle, NIR measurement of the chemical fractions water, fat, and protein that relate to WHC properties (cooking loss, pressed out fluid, and drip) have shown poor relationships in some studies. Forrest et al. (2000) found that the reference method for samples with low drip loss (<5%) was very precise, but duplicate measurements deviated a lot with samples of high drip loss. Variation in temperature between samples was also suspected for poor results. To improve spectroscopic accuracy, there should be emphasis on improving the repeatability of WHC measurement and understanding the nonspecific (indirect) reference methods used for measuring WHC.

2.6.1.4 Color Measurement

Some aspects of color have been discussed in previous sections. The color of meat can be described purely by instrument or visually with the assistance of color references such as Japanese Pork Color Standards (JPCS) or the National Pork Producers Council (NPPC) Pork Standards (Brewer et al. 2001). Although useful, visual evaluation of meat color can vary with evaluator and may be quite expensive.

Various color evaluation systems or descriptions of color space exist for instrumental measurement of color, but all are based on the three primary colors, red, blue, and green (RBG) or extensions of these (principal hues), while lightness (black to white) is combined with these three components to obtain tristimulus values for color. The operation of these systems relates to the human eye's ability to detect blue, red, and green spectra through the three related types of cones that have peak responses to these three colors (AMSA 2012). Lightness and darkness are detected by rods that are not sensitive to color, but merely sense dark vs. light as black through gray to white. In instrumental measurement of color, additional attributes such as chroma and Hue (angle) are often calculated from the three primal colors to describe the sensation of color in a comprehensive way. Chroma describes the intensity of the typical color of a substance, for example, the brightness of red in meat, while hue angle refers to the deviation from the typical color. Five color systems mentioned by Hunt et al. (1991) are the Munsell color solid, CIE color solid (Commission Internationale de l'Eclairage [CIE] or International Commission on Light), reflectance spectroscopy, tristimulus colorimetry, and Hunter color solid. The most commonly used is the Hunter Lab and CIE $L^*a^*b^*$ using L, a, and b as color coordinates where L, lightness is measured between 0 and 100, where 100 is absolute white and 0 is absolute black. Positive a-values are red, and negative a-values are green, while positive b-values are yellow, and negative b-values are blue. The original two-dimensional Hunter Lab scale was revised by the CIE in 1976 (CIE 1976) to convert it to a three-dimensional space (CIE $L^*a^*b^*$). The two systems also use different calculations for the color coordinates where Hunter uses cube roots of X–Y–Z and CIE $L^*a^*b^*$ the square roots. Both the eye and mechanical color devices analyze color by capturing the reflection of light from objects in various wavelengths and then converting them to an interpretation of total color. Humans are limited to detect light only in the visible spectrum which ranges from 390 to 750 nm. When light strikes an object it will be, either reflected, absorbed, or scattered and the variation of these three components as well as the combination and amount of the wavelengths of light reflected will determine human or mechanical light detection. Light reflected from meat and measured as color will depend on its surface structure and the level and nature of pigmentation. Texture will influence refractive index and light-scattering properties. The ratio of K (absorption) to S (scatter) is used in the calculation or measurement of color (AMSA 2012).

The most important aspect of color measurement is to standardize methods and instruments and describe the specifications in results. Trichromatic colorimeters or spectrophotometers are mostly used to record instrumental color measurements and instruments offer a variety of options to choose from various color systems (Hunter or CIE tristimulus), illuminants (A, C, D65, and Ultralume), observers (2° and 10°), and apertures (0.64–3.2 cm). Spectrophotometers are more complex than colorimeters and supply spectral images in intervals of 1–10 nm and offer more options. While color coordinates (L, a, b) are simple to use, spectral data can provide color coordinates, and also indicate the proportion of Mb derivatives that can be used to estimate or describe discoloration more accurately.

The selection of illuminant, aperture size, and observer angle is guided by the type of light preferred and the properties of the product to be measured, although Tapp et al. (2011) reported that the majority of research studies do not report on these specifications and could be unaware of the effects of variation in these factors on results. Various light sources (sunlight, fluorescent (FL) light, tungsten light) contain different spectral light compositions and are therefore crucial in mechanical measurement and human observation of color, that is, when selecting meat from the shelves. The true color of meat depends on the amount of red light in the light source. Barbut (2001) reported that consumers regarded beef as red, dark brown, and dark red when displayed under incandescent (INC), FL, and metal halide (MH) light, respectively. The differences were mainly due to a lack of redness in the latter two light sources and were more pronounced in darker meat such as beef than in pork. MH and FL light sources are often used in display cabinets due to their low heat emission and efficiency. These differences have relevance when measuring color under experimental conditions vs. conditions where consumers select meat. Illuminants in color meters will therefore also give different results. Illumination A (average INC, 2857 K) is normally used for measuring meat color as it places more emphasis on proportion of red wavelengths than, for example, illuminant C (north sky daylight, 6774 K) and D65 (noon daylight, 6500 K). Values for redness (a^*) will be higher using illuminant A than with the other illuminants and small differences between a^* will also be detected better. Tapp et al. (2011) noted that most scientific studies used illuminants C and D65 that are closer to day light. In contrast, Hunt et al. (1991) stated that illuminant A gave better correlations between objective and subjective measurements due to its higher proportion of red wavelengths and is commonly used for beef. Illuminants C and D65 that are associated with daylight are commonly used for pork (Hunt et al. 1991). According to *AMSA Meat Colour Guidelines* (AMSA 2012) bloomed beef can measure a^* values between 30 and 40 with illuminant A, whereas illuminants C and D65 will record values between 20 and 30. These values will however also depend on the aperture size. As aperture size decreases the percentage of reflected light decreases particularly in the red wavelength range and L^*, a^*, b^*, and saturation index will be lower for various illuminants used (A, C, and D65) (Yancey and Kropf 2008). Hue angle had an inverse relationship with aperture size for all three illuminants mentioned. Honikel (1998) suggested that the largest aperture possible for an instrument should be used to allow the largest possible screening area. Observation angle accounts for the visual field of the average human eye when detecting color. Ten degrees is mostly used, although 0° and 2° are also popular but little research has been done on differences in color measurement due to different view angles (Tapp et al. 2011).

Bloom time is also an important aspect of color measurement, but much controversy exists in literature about optimum bloom time for color measurement. Honikel (1998) suggest that proper blooming of meat takes at least 1 h, while Rentfrow et al. (2004) indicated that color stability on beef *longissimus* will be achieved after 12 min of exposure to air. Wulf and Wise (1999) found that 78 min were necessary for stable a^* and b^* values, but only 30 min for L^* values in beef *longissimus*. Lee et al. (2008a,b) found consensus that 90% stability in all color coordinates and OxyMb is reached after 1 h for beef *longissimus* and *gluteus medius*. Preparation of the sample for color evaluation is critical with regards to fiber orientation. Mohan et al. (2010) report higher OxyMb levels and chroma values in *longissimus* cut perpendicular (PD) to the fibers compared to parallel (PL) cutting in samples stored in high oxygen atmosphere. These differences were also maintained for longer during shelf life in PD samples than in PL samples. Fiber orientation did not affect the amount of OxyMb in vacuum packaged cuts. Farouk et al. (2005) suggested that PD cutting may expose more color pigment bringing about more color than PL cutting. The larger effect of fiber orientation with high oxygen would be a function of oxygen exposure to the pigments forming OxyMb. Biophysical properties, such as water transfusion and surface morphology could also have been affected by fiber orientation (Gou et al. 2002).

An alternative way to determine redness is by using reflectance ratios of 630 nm over 580 nm. This requires with a spectrophotometer, usually using 10 nm increments which gives the ability to analyze color over a broad spectrum of wavelengths with concentration on the red wavelengths (Yancey and Kropf 2008).

Consumers vary in their ability to detect the various colors in the color spectra and this will affect results from human color panels (and even consumers) (AMSA 2012). In addition, consumers will respond more positively to larger cuts than smaller ones since more light is reflected and hence the color is brighter and more vivid. Also, consumers will regard meat color as brighter when packed against a dark packaging background. The angle at which people view the product as well as the position and angle of the light source will influence the color perception. Greene et al. (1971) showed that consumers will make a no-purchase decision when surface MetMb levels reach 30%–40%. However, Hood and Riordin (1973) reported that the discrimination between bright red beef and discolored beef on the shelf will already be evident at surface MetMb levels of 20% which corresponds to chroma values of 18 and less (MacDougall 1982). Chroma values of 14 and <12 were regarded as distinctly brown and gray-greenish, respectively. For lamb loin, Hopkins (1996) and Khliji et al. (2010) found that redness recorded by means of a chromometer as a^* explained most variation in consumer acceptability scores and was threfore regarded as most appropriate to use as a threshold. Khliji et al. (2010) found that for fresh meat when a^* and L^* values were equal or above 9.5 and 34, respectively, on average consumers will consider the meat acceptable. However, to have 95% confidence that a randomly selected consumer will find the sample acceptable, a^* and L^* values of at least 14.5 and 44 will, respectively, be needed. Interestingly, when meat was aged an a^* value equal or greater than 14.8 was regard acceptable, while a value of at least 21.7 was needed to be 95% confident that a randomly selected consumer will consider the sample acceptable.

2.6.1.5 Instruments for Determining Moisture Properties

WHC can be defined as the ability of meat to hold all or part of its own water (Honikel, 1988). Kauffman et al. (1986) described various methods to estimate WHC and their ability to distinguish between DFD, PSE, and normal meat. Two related terms used in the description of WHC are free exudate (or thaw or drip) which refers to amount of intrinsic fluid that exudes from a defined sample under specified conditions, while expressible moisture or pressed out moisture describes the amount of intrinsic fluid expressed from a sample by application of a specified force under specific conditions. Kauffman et al. (1986) categorized methods to estimate WHC in three groups, namely those that measure weight loss (drip), standard laboratory techniques, and filter paper press methods.

The drip method of Honikel (1998) measures the weight loss of a standard weight of sample suspended in net or placed on supporting mesh in a sealed environment, such as an inflated plastic bag that does not come into contact with the sample. After storage of 24 h at a standard temperature, the sample is weighed again and moisture loss is calculated. An alternative to this method is the utilization of samples with standard dimensions, since it is believed that variation in that regard may cause inaccuracies (Honikel 1998). A third alternative involves placing a sample on an absorption pad in a styrofoam tray and overwrapping the sample with PVC (polyvinyl chloride) film. An addition to the drip loss methods could also include weight loss during cooking in order to determine the total loss from processing to consumption. Two centrifuge methods are discussed by Kauffman et al. (1986). In one method minced meat is used to which water is added to measure the amount of water absorption when the sample is centrifuged at low speed. The second method involves a high centrifuge technique to determine how much water is retained in a meat sample under high speed centrifugation. Kauffman et al. (1986) also described a percentage transmission method measuring protein solubility as affected by protein denaturation and a permittivity test in which electrical capacitance and dielectric constant ratios are calculated at two energy frequencies. Capacitance and a combination of conductivity and dielectric loss are measured by a pork quality meter (PQM) and related to WHC. The filter paper press method initially developed by Grau and Hamm (1953), as cited by Kauffman et al. (1986), involves force applied for a specific duration to small sample of meat placed in the middle of a filter paper disk. The ratio of the two areas on the paper, namely the pressed meat and pressed out water is then related to WHC or the meat area divided by the initial sample weight (Kauffman et al. 1986) is used to describe WHC. The areas can be measured by a planimeter or, more recently, by video imaging (Irie et al. 1996). Another filter paper technique involves a filter paper disc or pH paper applied to the cut surface of meat at a specific time after cutting and for a set duration. The amount (weight) of absorbed moisture is then related to the specific cut area (Kauffman et al. 1986). Surface to weight ratio is important in this method as it will affect the relative amount of drip.

Kauffman et al. (1986) report that most of the above methods could distinguish DFD and PSE pork muscle from normal muscle, although the measurement of protein solubility (transmission), imbibition test (absorption through pH paper), and pressed fluid methods could not always distinguish DFD from normal samples. The most reliable methods appeared to be drip loss measurement using a standardized size instead of standardized weight, the swelling of homogenized samples by added water and absorption of surface water by filter paper using the weight of absorbed fluid. The latter method is also inexpensive. With regard to conductivity, PSE meat records higher values than normal meat due to a poor WHC and a high drip loss. Lee et al. (2000) showed that conductivity on its own or in combination with pH_u are accurate predictors of WHC and can be used to classify pork carcasses at 24 h postmortem. Christensen (2003) compared an adapted drip loss method, the EZ-DripLoss method (http://www.dmri.com) with the conventional drip method of Honikel (1987). The EZ-DripLoss method uses smaller standardized sample sizes and has the ability to test transverse variation within a sample. Although samples are removed by a standard coring device, the study did indicate variation in sample size probably due to structure differences and slight variation in transverse sample thickness (Christensen 2003). Therefore, weighing of the initial sample is still needed. No explanation was given for the consistent lower drip values (1.2%) of the EZ-DripLoss method compared with the standard method. The advantage of the EZ method would be to detect PSE spots in individual steaks and the study demonstrated that with a single small sample taken at a specific region of the loin and specific position of the cut, a fairly accurate prediction of consumer drip loss could be calculated (error of prediction = 1.3%).

2.6.1.6 pH Measurement

In simple terms pH measures the acidity of muscle, meat or processed meat. pH is not a quality characteristic by itself, but influences meat color, WHC, flavor, tenderness, and shelf-life (Borgaard 2014). pH measurements are performed on muscle just after slaughter and during the rigor process, on meat postrigor and during product manufacturing. Muscle pH measurements during the rigor process combined with temperature give information about forthcoming or expected qualities of the final product in particular the protein functionality that will influence color and WHC. Ultimate pH is indicative of the final state of the meat and will provide information on WHC (e.g., DFD, acid meat, or normal meat) for the utilization of fresh, and also processed meat.

pH is recommended to be measured at temperatures between 20°C and 25°C (Honikel 2014), but meat temperature varies between 0°C and 43°C. Since calibration of pH meters is normally performed at around 20°C it is recommended to accept pH readings of no more than one decimal place if the calibration temperature and meat temperature differs more than 10°C, even when temperature adjustments are made by means of temperature probes (Honikel 2014). Alternatively, calibrations should be done at the actual temperature of the meat, although temperature drift must be taken into account. For example, a phosphate buffer will deviate from pH 7 by 0.16 units between 0°C and 35°C. However, this deviation is a mere 0.02 between 20°C and 35°C (Honikel 2014). Calibration of a pH probe and proper maintenance are keys to accurate pH recording. Two buffers, usually pH 4 and 7, are used for calibration that should be frequently repeated to make sure the standard of recording is good.

2.6.2 Sensory Evaluation of Meat

Sensory evaluation has been defined as "a scientific method used to evoke, measure, analyze, and interpret those responses to products as perceived through the senses of sight, touch, smell, taste and hearing" (Stone and Sidel, 1993) . Test methods for sensory analysis can be divided into discriminative tests (triangle, duo-trio, paired comparison, ranking), descriptive tests (the quantitative descriptive analyses, the texture profile method, the spectrum method, flash profile), and affective tests. In meat quality evaluation, all these methods are widely used to describe and compare visual quality or product palatability.

Standardization of research procedures is becoming important, while cultural and regional uniqueness of consumers are considered in sensory studies (Ngapo et al. 2003; Sañudo et al. 2007; Hocquette et al. 2011). The Committee on Sensory Evaluation of the American Society for Testing and Material (ASTM) is working on a manual that will cover these issues extensively for different cultures across Europe, North and South America, and Asia.

2.6.2.1 Visual Assessment of Meat

Visual assessment of meat can be done on raw or cooked meat samples. When samples are raw important factors for standardization are the type and intensity of light, the color and type of the packaging or tray on which samples are displayed, the standardization of time between processing and evaluation and the display temperature. The latter two are important as color stability of raw meat is influenced by these. Attributes to be considered are meat and fat color, visual texture, and the amount of surface moisture and/or purge. Digital images can be used if these reflect the characteristics of the original product accurately (Ngapo et al. 2007).

2.6.2.2 Assessment of Eating Quality

Sensory assessment of the eating quality of meat can be done by using a trained panel or untrained consumer panel. By definition, a trained panel will be able to distinguish small differences in a specific attribute independently of other sensory dimensions (Watson et al. 2008a,b). Trained panel results also yield less variation due to a smaller spread of scores as this is the purpose of training these panels. Ultimately less variation in scores means more reliable results, but the result may not be valid as a result of shifting of the mean. In contrast, consumer scores may be more variable, but if obtained correctly may result in more valid answers.

2.6.2.2.1 Trained Sensory Panels

In selection and training of a panel member their ability to distinguish between various levels of basic taste components (sweat, sour, bitter, salt) as well as advanced training to evaluate specific product characteristics should be considered. As a selected and trained panel is used as an instrumental tool in descriptive analysis it is necessary to evaluate the skill of each assessor on a regular basis (Lawless and Heymann 2009). Furthermore, panelists should be familiar with the sensory quality attributes of the product (meat) and consistency and accuracy of each panelist should be monitored statistically (Meilgaard et al. 1999). Panelists should receive specific theoretical and practical training for the methods applied in an experiment, especially for quantitative descriptive analysis (QDA) (ISO 8586-2:1994). Sensory descriptors should be defined according to a specific standard (ISO 13299:2003) (Giboreau et al. 2007).

2.6.2.2.2 Consumer Panels

While trained panel assessment of eating quality is still very valid and utilized globally in research, the MSA's decision to utilize consumer panels on a large scale to engender confidence with both the industry and consumer sectors in the design of a grading scheme has merit. In this approach robust protocols were developed that would accommodate as many factors as possible that would influence the final product as consumed and experienced by the consumer. In the end such results can be used to design a grading system based on specific cuts, instead of the whole carcass related to evaluation of a single muscle (Russell et al. 2005; Thompson et al. 2005a,b; Polkinghorne et al. 2008). To this end the effects of wide range of cattle and sheep management practices, processing systems, cuts, aging periods, and cooking methods have been researched and modeled to predict the contribution of each of the factors (in combination) on eating quality as perceived by the consumer. The final model was then applied commercially to predict consumer satisfaction (see Chapters 11 and 15). In designing these protocols, the objective was to minimize variation for all elements other than the individual consumer assessment and the product being tested. This required highly controlled test procedures. Apart from carefully considering factors that may affect eating quality up to the point of purchase, various preparation methods (slicing and cooking) had to be considered to reflect normal consumer behavior, and also to remove potential variation of preparation, cooking, and serving which is normally very well standardized in trained panel evaluations. In designing the sampling and presentation procedures, carry-over effects of samples previously tested and position of sampling in the muscle is important with trained panel designs, but more so in large consumer studies (Durier et al. 1997; Kunert 1998). Likewise the effect of testing fresh or frozen samples is also of significance in trained and consumer panels and while freezing has an impact on quality, practical considerations normally require that frozen samples are used (Watson et al. 2008a,b). The number of replications is also always critical and although trained panels can often afford less replicates due to less variation expected, consumer panels have to consider the limitations of the physical size of the cut being tested and the cost restrictions. Cost increases linearly with number of repetitions (consumers per sample) although standard error only decreases by the square root of the number of samples (Watson et al. 2008a). To further accommodate the trade-off between cost/physical restrictions and precision, trimmed means of 10 consumers were used in the MSA design for beef evaluations, meaning that the largest and smallest two values of each of the 10 scores per sample were trimmed and a mean of eight values was used. By considering the challenges discussed, MSA designed a reliable prediction model from the results of over 86,000 consumers scoring more than 603,000 beef samples of different cuts sourced from a variety of production, processing, value adding, and cooking treatments (Watson et al. 2008b). Using these results a grading system was designed for each muscle (cut) on eating quality by combining the scores of tenderness, juiciness, flavor, and overall liking into a single score (MQ4) to predict four grades, namely <3, 3, 4, or 5 star indicating unsatisfactory, good every day, better than every day, and premium quality. The scores obtained relate only to the individual consumer and the beef sample, and are not affected by random influences such as irregular thickness or cooking variation. The protocols also detail issues of sample preparation, order and method of serving. For example, every consumer is served seven samples, which include a high- and low-quality product.

Whereas trained panels focus on every attribute independently, consumer panels, such as the MSA system normally seek to find a single indicator to link quality to expected satisfaction. Three of the four variables in the MSA system (tenderness, juiciness, and flavor) describe three quality attributes related to eating satisfaction while "overall acceptability" probably relates most to tenderness at least for beef. Weights were added to each attribute based on its contribution to explaining variation among samples and the lowest weight was awarded to juiciness. The four attributes were developed from a list of 13 initial attributes and reduced to four by grouping attributes that described similar characteristics. Cut-off values for the different star categories could be determined by utilizing the feedback data from consumer panels with regard to preferences. The MSA grading system has been tested in France (Hocquette et al. 2011) and other countries (see Chapter 15) and is an example of globalization testing procedures that can effectively be applied to predict consumer responses.

2.6.2.2.3 Sensory Methods: Discriminative Tests

Discriminative methods investigate sensory differences between samples. The most common methods are the triangle test, duo-trio test, paired comparison test and ranking test. Triangle and duo-trio tests are used to determine whether a sensory difference exists between two samples. A paired comparison test is used to determine in which way a sensory characteristic differs (e.g., choose more or less tender; juicy). In ranking tests a set of samples (e.g., 6) are ranked according to one specific attributes (e.g., tenderness). The ranking test is recommended where expected differences are small (Table 2.3).

2.6.2.2.4 Sensory Methods: Descriptive Tests

Descriptive sensory tests are among the most sophisticated tools available to the sensory professional and involve the detection and description of both qualitative and quantitative sensory components of a studied product by a trained panel. Tests most often used are flavor profiles, free choice profiles, QDA, and the TPA (texture profile analysis). With QDA a detailed description of the aroma, texture, and flavor of the tested product is required. The set of specific features are discussed, defined, and selected by panelists, during a preliminary session (Table 2.3). Examples of flavor descriptors and definitions used in lamb or beef meat have been described by Nute (2008).

2.6.2.2.5 Sensory Methods: Affective Tests

Affective tests are used in consumer studies and the primary objective is to assess the personal response (preference or acceptance) of a potential user of a product (Lawless and Heymann 2009). There are several tests to choose from. Hedonic rating scales are used and the nine-point hedonic scale is probably the most common (Lawless and Heymann 2009) (Table 2.4). For specific description of tenderness and juiciness, an eight-point category scale (Table 2.5) is often used (Miller et al. 2001; Monsón et al. 2005).

TABLE 2.3
Example of Sensory Descriptors and Their Definitions Used in Analysis of Pork Quality

Attribute	Definition	Anchors
	Odor Attributes	
Heated meat	Aroma associated with heated meat	None to very strong
Sour	Basic aroma sensation stimulated by acids	None to very strong
Sharp	Degree of irritation of smell senses	None to very strong
Fatty	Aroma sensation derived from fat	None to very strong
Other	Enter the intensity and name or association	None to very strong
	Texture Attributes	
Tenderness	Degree to which sample holds together on first bite	Not tender to very tender
Juiciness	Impression of moisture release during chewing	Not juicy to very juicy
Chewiness	Amount of chewing before swallowing	Not chewy to very chewy
Heated meat	Flavor associated with heated meat	None to very strong
Sour	The basic taste sensation stimulated by acids	None to very strong
Fatty	Sensation containing derived from fat	None to very strong
Sweet	The basic taste sensation stimulated by sucrose	None to very strong
Salty	The basic taste sensation stimulated by salts	None to very strong
Other	Enter the intensity and name or association	None to very strong
Overall quality	Impression based on all tested attributes	Very low to very high

Note: Descriptor categories between extremes may vary between 6 and 10.

TABLE 2.4
The 9-Point Hedonic Scale and 8-Point Category Scale for Juiciness and Tenderness Evaluation in Affective Test

9-Point Hedonic Scale		8-Point Category Scale		
Point	**Category**	**Point**	**Tenderness**	**Juiciness**
9	Like extremely	8	Extremely tender	Extremely juicy
8	Like very much	7	Very tender	Very juicy
7	Like moderately	6	Moderately tender	Moderately juicy
6	Like slightly	5	Slightly tender	Slightly juicy
5	Neither like nor dislike	4	Slightly tough	Slightly dry
4	Dislike slightly	3	Moderately tough	Moderately dry
3	Dislike moderately	2	Very tough	Very dry
2	Dislike very much	1	Extremely tough	Extremely dry
1	Dislike extremely			

TABLE 2.5
Methods Used to Measure Early Postmortem and Postrigor Pork Quality

Method	Code	Company	Carcass Location
I. Early PM methods			
A. Reflection caused by light scattering and absorption			
1. Fat-o-meter	FOM	SFK, Denmark	4th LV[a]
2. Hennessy grading probe	HGP	Hennessy & Chong, New Zealand	12th TV[b]
3. Sensoptic invasive probe	SIP	Sensoptic, the Netherlands	3rd LV
4. Colormet probe	CMP	Colormet, Canada	1st LV
5. Fiber optic probe	FOP	United Kingdom	5th LV
B. Electrical properties affected by impedance, resistance, and conductivity			
1. Sensoptic conductivity probe	SCP	Sensoptic, the Netherlands	1st LV
2. Pork quality meter	PQM	TecPro, Germany	1st LV
C. pH			
1. Ingold glass electrode	IGE	Ingold, Sweden	15th TV
2. ISFET–REFET electrode	IRE	Sensoptic, the Netherlands	15th TV
D. Temperature			
1. Omega digital thermometer	ODT	Omega, USA	1st LV
E. Muscle stiffness and rigidity			
1. Muscle rigorometer	IVO	IVO-DLO, the Netherlands	Pelvic limb
2. Thoracic limb rigidity	TLR	–	Thoracic limb
II. Postrigor methods			
A. Water-holding capacity (WHC)			
1. Filter paper wetness	FPW		4th LV
2. Percent drip loss	PDL	–	3rd LV
B. Reflection caused by light scattering and absorption			
1. Japanese color standards	JCC	Japan	2nd LV
2. Pork color standards	PCS	NPPC, USA	2nd LV
3. Minolta portable chromameter	MPC	Minolta, Japan	2nd LV
4. Hunter labscan	HLC		2nd LV

Source: Adapted from Kauffman, R.G. et al. 1993. *Meat Sci.* 34:283–300.

[a] Lumbar vertebra.

[b] Thoracic vertebra.

In descriptive tests, a line scale instead of a number scale may be used to indicate the intensity of an attribute (Meilgaard et al. 1999). The most common is an unstructured, linear graphical scale consisting of a 100 mm line that is converted to numerical values (0–10 conventional units c.u.). The length of the line drawn by the panelist is a proportion of the total scale that indicates the intensity of the attribute. MSA utilizes an anchored line scale which represents a blank line with description of the two extremes of a specific attribute and the opposite ends, for example, "not tender" and "very tender" (Watson et al. 2008a).

2.6.2.2.6 Evaluation Conditions

Evaluation of difference tests or descriptive methods should be performed at a sensory analysis laboratory which meets certain standard requirements (e.g., ISO 8589:1998) as described by Meilgaard et al. (1999). Consumer preferences, acceptance, and degree of liking tests can be conducted under laboratory conditions, public places or at a consumer's home. It is considered that results in a laboratory setting are controlled and not typical of the consumer's environment. Public places normally have large numbers of available consumers, but the conditions may make it difficult to focus on the evaluation task. However, if preparation protocols are well designed and adhered to, consumer panels could provide very accurate results. To this end MSA has designed exact testing protocols that describe the preparation and packaging of samples, timing of sample preparation before and during the panel sessions, procedures to manage and limit the maintenance of cooked temperature before serving, the precise cooking methods, serving of samples with maximum variance between samples, the utilization of correct "warm-up" or first link samples, and the pairing of consumers and allocation of muscle positions where applicable (Gee et al. 2005).

2.7 METHODS OF IDENTIFYING FAULTY MEAT AND QUALITY STANDARDS IN DIFFERENT SPECIES

Quality defects in meat can be identified by either visual or instrumental means. Obviously, defects such as bruises are merely observed by visual inspection, but conditions like PSE and DFD will require a description of a variety of quality characteristics against set standards.

Various scoring systems to identify the age of bruises and to quantify occurrence and severity are used. The most recent NBQA data reported by McKeith et al. (2012) used the following scoring system for bruises on beef carcasses: Number of bruises per carcass, location (round, loin, rib, chuck, flank/plate/brisket), and severity described as minimal (1–3), major (4–6), critical (7–9), extreme (10). Other systems physically measure the diameter and depth and by combining these values with the location and the value of the cut, bruises can be categorized according to economic value. The age of a bruise is important when managing the prevention of bruising. Bruises caused by trauma which occurs within the last few hours will display a bright red coloration while bruises resulting from trauma occurring more than 24 h preslaughter will display an orange–red, yellow, or brown pigmentation due to breakdown of various blood pigments (Zachary and McGavin 2012). A similar system used for identifying and describing bruises on cattle carcasses is used to score skin blemishes on pig carcasses. In addition, distinctions are made between "mounting" and "fighting" bruises on pig skins based on the position on the carcass as well as the size and shape of bruises. Photographic standards of the Institut Technique du Porc (ITP 1996) are used in these visual assessments (Faucitano 2001).

For defects like DFD and PSE (and related defects) visual, but preferably instrumental color is used together with pH and drip loss, mostly measured at 24 h postmortem. DFD meat is judged on final pH (pH_u, at rigor mortis; 24 h) and the expected relation that this has with meat color and other properties like WHC.

A high pH_u is associated with tightly packed muscle fibers, due to water retention that causes reduced light scattering and a thin layer of red OxyMb on the surface of the meat that causes the darker color of DFD (Renerre 1990). Normal pH of the loin muscle of unstressed animals is regarded as ranging between 5.4 and 5.7. Early work by Tarrant (1981) recorded various countries' pH standards for categorising carcasses as DFD based on pH of the *longissimus*. Only Australia and two states of the United States indicated a final pH of 5.8 and higher as indicative of DFD in beef. Forty one per cent of the countries used 6.0 as the criteria, while, respectively, 18%, 17%, and 7% used 6.1, 6.2, and 6.3 as the critical value to distinguish DFD carcasses. In general, products with values between 5.8 and 5.9 were rejected on the basis of lower keeping quality (vacuum packaging) and some color defects, but most would not consider these as DFD. These respondents (82%) regarded values above 6 as the critical value for rejection of meat in the market. At present Meat Standards Australia (MLA 2014) uses 5.71 as the critical pH_u, while the AUS-MEAT chiller assessment for color utilizing AUS-MEAT color chips (AUSMEAT 2014) classes carcasses as DFD when color scores are >3. However, Hughes et al. (2014) indicated firstly that the low critical pH value of 5.7 is probably too strict and that color and pH does not always correspond. For carcasses with pH_u values of 5.8, 6.0, and 6.2, 28%, 74%, and 96% of the carcasses had color scores of >3, respectively. More than 70% of the carcasses recording a pH of 5.8 recorded a color score of 3 (47%) or less (30%). This study together with Murray (1989) also indicated that the incidence of DFD carcasses as determined by color scoring declined almost three-fold if chiller assessment is delayed from around 14 h to 23–31 h postmortem. No mechanism could be confirmed for this phenomenon, but time-dependent changes in structure and the Mb status (Young et al. 1999) as well as myofibrillar shrinkage and drip formation is suggested (Offer and Cousins 1992). Page et al. (2001) reported a specific $pH_u = 5.87$ as critical value for distinguishing DFD carcasses when using the USDA chiller assessment color standards (USDA 1997) based on a 92% accuracy to classify carcasses as DFD. In this study 0.6% of carcasses with $pH_u < 5.87$ were identified as DFD. Furthermore intermediate correlations between pH_u and color coordinates L^*, a^*, and b^* of −0.34, −0.53, and −0.48, respectively, were recorded in the study.

pH alone is therefore not absolute and borderline carcasses can be approved as normal while in fact they may possess DFD characteristics (Sornay et al. 1981). In some cases more robust values of ultimate pH < 6.0 are used as cut-off points for DFD, while others distinguish between moderate DFD (pH_u 6.0–6.2) and serious DFD (pHu > 6.2) (Guárdia et al. 2005).

Another problem with identifying DFD is that a single muscle, *M. longissimus* is normally used as reference muscle of the whole carcass, while different muscles do not respond the same to stress conditions (e.g., with regards to pH) and also do not display changes in color as a result of such stress. When Wulf et al. (2002) used a critical value of $pH_u = 6$ for the *longissimus* to differentiate between DFD and normal carcasses, mean pH_u values of seven other muscles (*psoas major, gluteus medius, tensor fasciae latae, rectus femoris, semimembranosus, biceps femoris*, and *semitendinosus*) were lower than 6. The *semimembranosis, semitendinosus*, and *gluteus medius* did show significant pH and color (L^*, a^*, and b^*) differences between the "normal" and DFD groups, but the remaining four muscles did not exhibit color differences and only the *biceps femoris* recorded a higher pH_u for DFD carcasses.

PSE is defined as muscle with pH below 6.0 at 45 min after slaughter. In some cases, less stringent standards are used and pH < 5.9 is regarded as PSE, while in countries where strict management of PSE is needed, pH < 6.2 may be used as criteria (Adzitey and Nurul 2011). The identification of PSE and DFD using color can be achieved by subjective color references such as the Japanese color reference scale (O'Neill et al. 2003) used for pork or the AUS-MEAT color reference standards for beef (Anonymous 2014) or by objective color measurements recorded with colorimeters. PSE is characterized by high L^* (pale meat), low a^* (less red meat), and high b^* (more yellow), but Brewer et al. (2001) found the L^* value to be the best indicator of PSE and DFD in pork. An $L^* < 54$ is characteristic of DFD in pork (Warriss and Brown 1993), while $L^* > 50$ (Gaudré and Vautier 2006) or >57 (Van der Wal et al. (1988) are used to define PSE meat. Warner et al. (1997) used $L^* < 42$ and $L^* > 50$ to distinguish between PSE and DFD pork, respectively, in combination with specific moisture and pH values.

Moisture properties, in particular WHC are usually combined with color and/or pH to identify DFD, but particularly PSE defects in meat, especially pork. Standards for purge and conductivity are often used singularly or in combination with color L^* and pH to categorize meat into the three types. Warner et al. (1997) defined PSE as samples having more than 5% drip in combination with $L^* > 50$ and $pH_u < 6.0$, while DFD samples had drip <5%, $L^* < 42$, and $pH_u > 6.0$ in pork. Van Laack et al. (1994) advised against the use of a single parameter such as light reflectance (L^*) to categorize pork carcasses since pork samples appearing RSE, may record similar reflectance values (L^*) as normal pork (RFN) between 52 and 58 but may have excessive drip (>5%). It should be noted that the L^* value range for normal pork applied by Van Laack et al. (1994) differed from that of Warner et al. (1997) who suggested L^* values between 42 and 50. Van de Perre et al. (2011) regarded conductivity values of pork muscle higher than 3.9 as indicative of PSE.

Kauffman et al. (1993) investigated the utilization of 12 different techniques involving measurement of reflection caused by light scattering, electrical properties, pH, temperature, and muscle stiffness applied early postmortem (30 min) to predict the final quality of pork with regards to moisture properties, color, and firmness (Table 2.5). They concluded that no single or combined technique could accurately predict pork quality before rigor mortis has been finalized. Even pH at 45 min is only effective when extreme values are used for robust categorization of carcasses.

More recent technology to utilize color, specifically reflectance, more accurately to distinguish between PSE and normal pork was described by Chmiel et al. (2011). Chmiel et al. (2011) used color models RGB (red, green, blue), HSV (hue, saturation, value/brightness), and HSL (hue, saturation, and lightness) by means of computer image analyses. Increased accuracy in sorting of samples into normal and defective meat categories was claimed based on the fact that these methods consider the whole surface of the meat, not like chromometers that only measure small areas and assumptions are made that the whole surface area is uniform in color. Based on more enhanced imaging principles, Qiao et al. (2007) used a hyperspectral imaging system whereby spectral information of each pixel in the image is used to form a three-dimensional "hypecube" that provides minor and subtle physical and chemical information of color, shape, texture, etc. Various wavebands were used to form predictions of drip loss, pH, and L^*. Accuracies of between 0.3 and close to 0.8 (R^2) were achieved for the various parameters.

Xing et al. (2007) found that spectroscopy in the visible range could be used with 85% accuracy to distinguish red from pale classes of pork, but only gave 61% accuracy to distinguish between firm, nonexudative and soft, exudative groups. Likewise, Barbin et al. (2012) used spectra in the near-infrared range to classify pork into quality grades without any additional physical information. Initially the high dimensionality of spectral data could be a barrier for the utilizfation of this method, but can be overcome by the ideal selection of most important wavelengths. Bauer et al. (2013) demonstrated that Raman spectroscopy applied at 24 h postmortem can also be used as a noninvasive method to predict drip loss and pH_{24}. The review of Damez and Clerjon (2013) summarizes a wide range of techniques utilising the principal of physical interaction between electromagnetic waves and the muscle sample (high and low frequency impedance, NMR, infrared and ultaviolet light, x-ray). Some of these methods are still experimental and others are already used in practice.

2.8 MEAT PROCESSING METHODS OF FAULTY MEAT

Both DFD and PSE meat are not well suited for fresh meat utilization and preparation. Excessive moisture loss in PSE meat will result in a dry cooked product, while DFD meat is tasteless due to a lack of acidity and can in some cases be tough. However, neither of the two conditions make them unfit for human consumption and PSE and DFD meat can be used in processed meats when blended with normal meat and PSE meat can be added to meat products where water losses are desirable, such as dry-fermented sausages, while DFD meat can be used for raw-cooked products such as frankfurters where high water binding is required as well as in ready-to-eat dishes and canned beef.

2.9 CONCLUSIONS

Meat defects have always been present in the commercial production of red meat as a result of weak links in the meat value-chain. These weak links may be of extrinsic (management, processing) or intrinsic (genetics) origin, but the nature and impact on meat quality may change as production strategies change. Over the past decade, increased focus on efficiency and higher yields to provide for an increasing demand of red meat has brought new challenges in the management of meat quality. Most of the changes in production strategies for yield and efficiency impact on the harvesting process where increased variation in metabolism during the conversion of muscle to meat is experienced. A classic example of this phenomenon is the occurrence of high rigor temperature in beef experienced as a result of high carcass weights.

Noninvasive online measurement of quality is experiencing new interest as advances technological and analytical abilities are taking place. In this regard video imaging, spectroscopy, and techniques involving electromagnetic waves are experiencing much attention.

REFERENCES

Aalhus, J.L., and Dugan, M.E.R. 2014. Factors affecting spoilage: Oxidative and enzymatic. In Dikeman, M. and Devine, C. (eds.), *Encyclopaedia of Meat Sciences*, 2nd edition. Elsevier Science Ltd., Oxford, England, pp. 394–400.

Adzitey, F., and Nurul, H. 2011. Pale soft exudative (PSE) and dark firm dry (DFD) meats: Causes and measures to reduce these incidences—A mini review. *Int. Food Res. J.* 18:11–20.

Alsmeyer, R.H., Thornton, J.W., Hiner, R.L. et al. 1966. Beef and pork tenderness measured by the press, Warner–Bratzler and STE methods. *Food Technol.* 20:115.

AMSA. 2012. Meat colour measurement guidelines. American Meat Science Association, Champaign, Illinois, USA. http://54.173.5.58/publications-resources/printed-publications/amsa-meat-color-measurement-guidelines (Accessed May 26, 2015).

Andersen, J.R., Borggaard, C., Rasmussen, A.J. et al. 1999. Optical measurements of pH in meat. *Meat Sci.* 53:135–141.

Andrés, S., Silva, A., Soares-Pereira, A.L. et al. 2008. The use of visible and near infrared reflectance spectroscopy to predict beef *M. Longissimus thoracis et lumborum* quality attributes. *Meat Sci.* 78:217–224.

Anonymous. 2014. Australian Beef Chiller Assessment. Meat and Livestock Australia. http://www.australian-meat.com/Foodservice/Proteins/Beef/Australian_Beef_Chiller_Assessment/ (Accessed May 25, 2015).

Arnold, R.N., Scheller, K.K., Arp, S.C. et al. 1993. Dietary a-tocopheryl acetate enhances beef quality in Holstein and beef breed steers. *J. Food Sci.* 58:28–33.

AUSMEAT. 2014. Australian Beef Carcase Evaluation: Chiller Assessment Language. Accessed October 8, 2014, http://www.ausmeat.com.au/media/1711/Chiller%2010%20Low.pdf.

Banovic, M., Grunert, K.G., Barreira, M.M. et al. 2009. Beef quality perception at the point of purchase: A study from Portugal. *Food Qual. Prefer.* 20:335–342.

Barbin, D., Elmasry, G., Sun, D.-W. et al. 2012. Near-infrared hyperspectral imaging for grading and classification of pork. *Meat Sci.* 90:259–268.

Barbut, S. 2001. Effect of illumination source on the appearance of fresh meat cuts. *Meat Sci.* 59:187–191.

Barlocco, N., Vadell, A., Ballesteros, F. et al. 2006. Predicting intramuscular fat, moisture and Warner–Bratzler shear force in pork muscle using near infrared reflectance spectroscopy. *Anim. Sci.* 82:111–116.

Barnes, K., and Barnes, N. 2000. *Ecchymosis (Blood Splash) in Deer Carcasses. Influence of Pre-Slaughter Conditions.* Rural industries research and development corporation, Publication number 00/69. Canprint, Canberra, Australia.

Bauer, A., Petzer, A., Schwägele, F. et al. 2013. Towards an online assessment of meat quality in pork. In *The Proc. 59th Int. Congr. Meat Sci. Tech.*, S7B-20, Izmir, Turkey.

Bee, G., Anderson, A.L., Lonergan, S.M., and Huff-Lonergan, E. 2007. Rate and extent of pH decline affect proteolysis of cytoskeletal proteins and water-holding capacity in pork. *Meat Sci.* 76:359–365.

Borch, E., and Arinder, P. 2002. Bacteriological safety issues in red meat and ready-to-eat meat products, as well as control measures. *Meat Sci.* 62:381–390.

Borch, E., Kant-Muermans, M.-L., and Blixt, Y. 1996. Bacterial spoilage of meat and cured meat products. *Int. J. Food Microbiol.* 33:103–120.

Borgaard, C. 2014. On-line measurement of meat quality. In Dikeman, M. and Devine, C. (eds.), *Encyclopaedia of Meat Sciences*, 2nd edition. Elsevier Science Ltd., Oxford, England, pp. 489–497.

Bourne, M.C. 2002. *Food Texture and Viscosity: Concept and Measurement*, 2nd edition. Academic Press, San Diego, USA.

Bouton, P.E., and Harris, P.V. 1972. The effects of cooking temperature and time on some mechanical properties of meat. *J. Food Sci.* 37:140–144.

Bouton, P.E., Harris, P.V., and Shorthose, W.R. 1976. Thermal contraction of meat during cooking and its possible influence on tenderness. *J. Text. Stud.* 7:193–203.

Bratzler, L.J. 1932. Measuring the tenderness of meat by means of a mechanical shear. MSc thesis, Kansas State College, Manhattan, KS.

Brewer, M.S., Zhu, L.G., Bidner, B. et al. 2001. Measuring pork color: Effects of bloom time, muscle, pH and relationship to instrumental parameters. *Meat Sci.* 57:169–176.

Bryden, W.L. 2012. Mycotoxin contamination of the feed supply chain: Implications for animal productivity and feed security. *Anim. Food Sci. Tech.* 173:134–158.

Calkins, C.R., and Hodgen, J.M. 2007. A fresh look at meat flavour. *Meat Sci.* 77:63–80.

Campion, D.R., Crouse, J.D., and Dikeman, M.E. 1975. Predictive value of USDA beef quality grade factors for cooked meat palatability. *J. Food Sci.* 40:1225.

Campo, M.M., Nute, G.R., Hughes, S.I. et al. 2006. Flavour perception of oxidation in beef. *Meat Sci.* 72:303–311.

Carpenter, C.E., Cornforth, D.P., and Whittier, D. 2001. Beef quality grade standards—Need for modifications? *Proc. Reciprocal Meat Conf.* 27:122.

Carpenter, Z.L. 1974. Consumer preferences for beef color and packaging did not affect eating satisfaction. *Meat Sci.* 57:359–363.

Carpenter, Z.L., Kauffman, R.G., Bray, R.W. et al. 1965. Objective and subjective measures of pork quality. *Food Technol.* 19:118.

Channon, H.A., Payne, A.M., and Warner, R.D. 2000. Halothane genotype, pre-slaughter handling and stunning methods all influence pork quality. *Meat Sci.* 56:291–299.

Chmiel, M., Słowiński, M., and Krzysztof, D. 2011. Lightness of the colour measured by computer image analysis as a factor for assessing the quality of pork meat. *Meat Sci.* 88:566–570.

Cho, S., Park, B., Ngapo, T. et al. 2007. Effect of meat appearance on South Korean consumers' choice of pork chops determined by image methodology. *J. Sens. Stud.* 22:99–114.

Christensen, L.B. 2003. Drip loss sampling in porcine *M. longissimus dorsi. Meat Sci.* 63:469–477.

CIE (Commission Internationale de l'Eclairage). 1976. Recommendations on uniform color spaces-color difference equations, Psychometric Color Terms. Supplement No. 2 to CIE Publication No. 15 (E-1.3.1.) 1978, 1971/(TC-1-3). Commission Internationale de l'Eclairage, Paris, France.

Ciobanu, D.C., Bastiaansen, J., Lonergan, S.M. et al. 2004. New alleles in calpastatin gene are associated with meat quality traits in pigs. *J. Anim. Sci.* 82:2829–2839.

Clausen, I., Jakobsen, M., Ertbjerg, P. et al. 2009. Modified atmosphere packaging affects lipid oxidation, myofibrillar fragmentation index and eating quality of beef. *Packag. Technol. Sci.* 22:85–96.

Cockram, M.S., and Lee, R.A. 1991. Some preslaughter factors affecting the occurrence of bruising in sheep. *Brit. Vet. J.* 147:120–125.

Correa, J.A., Gonyou, H.W., Torrey, S. et al. 2013. Welfare and carcass and meat quality of pigs being transported for two hours using two vehicle types during two seasons of the year. *Can. J. Anim. Sci.* 93:43–55.

Cozzolino, D., Barlocco, N., Vadell, A. et al. 2003. The use of visible and near-infrared reflectance spectroscopy to predict colour on both intact and homogenised pork muscle. *Lebensm.-Wiss. Technol.* 36:195–202.

Cozzolino, D., De Mattos, D., and Martins, V. 2002. Visible/near infrared reflectance spectroscopy for predicting composition and tracing system of production of beef muscle. *Anim. Sci.* 74:477–484.

Cozzolino, D., and Murray, I. 2002. Effect of sample presentation and animal muscle species on the analysis of meat by near infrared reflectance spectroscopy. *J. Near Infrared Spectrosc.* 10:37–44.

Dalla Costa, O.A., Faucitano, L., Coldebella, A. et al. 2007. Effects of the season of the year, truck type and location on truck on skin bruises and meat quality in pigs. *Live Sci.* 107:29–36.

Damez, J.-L., and Clerjon, S. 2013. Quantifying and predicting meat and meat products quality attributes using electromagnetic waves: An overview. *Meta Sci.* 95:879–896.

Darby, M.R., and Karni, E. 1973. Free competition and the optimal amount of fraud. *J. Law Econ.* 16:67–88.

Davis, K.J., Sebranek, J.G., Huff-Lonergan, E. et al. 2004. The effects of ageing on moisture-enhanced pork loins. *Meat Sci.* 66:519–524.

Destefanis, G., Brugiapaglia, A., Barge, M.T. et al. 2008. Relationship between beef consumer tenderness perception and Warner–Bratzler shear force. *Meat Sci.* 78:153–156.

DiGiacomo, K., Leury, B.J., and Dunshea, F.R. 2014. Potential nutritional strategies for the amelioration or prevention of high rigor temperature in cattle—A review. *Anim. Prod. Sci.* 54:430–443.

Dransfield, E., Ngapo, T.M., Nielsen, N.A. et al. 2005. Consumer choice and suggested price for pork as influenced by its appearance, taste and information concerning country of origin and organic pig production. *Meat Sci.* 69:61–70.

Duarte, S.C., Lino, C.M., and Pena, A. 2012. Food safety implications of ochratoxin A in animal-derived food products. *Vet. J.* 192:286–292.

Dunne, P.G., Keane, M.G., O'Mara, F.P. et al. 2004. Colour of subcutaneous adipose tissue and *M. longissimus dorsi* of high index dairy and beef × dairy cattle slaughtered at two live weights as bulls and steers. *Meat Sci.* 68:97–106.

Durier, C., Monod, H., and Bruetschy, A. 1997. Design and analysis of factorial sensory experiments with carry-over effects. *Food Qual. Prefer.* 8:141–149.

Elmore, J.S., and Mottram, D.S. 2009. Flavour development in meat. In Kerry, J.P. and Ledward, D. (eds.), *Improving the Sensory and Nutritional Quality of Fresh Meat.* Woodhead Publishing Limited, Cambridge, pp. 111–146.

Enser, M., and Wood, J.D. 1993. Effect of time of year on fatty acid composition and melting point of UK lamb. In *The Proceedings of the 39th International Congress of Meat Science and Technology*, Calgary, Canada, pp. 74.

Farouk, M.M., Zhang, S.X., and Cummings, T. 2005. Effects of muscle-fibre/fibre bundle alignment on physical and sensory properties of restructured beef steaks. *J. Muscle Foods* 16:256–273.

Faucitano, L. 2001. Causes of skin damage to pig carcasses. *Can. J. Anim. Sci.* 81:39–45.

Ferguson, D.M., Bruce, H.L., Thompson, J.M. et al. 2001. Factors affecting beef palatability—Farm gate to chilled carcass. *Aust. J. Exp. Agric.* 41:879–891.

Ferguson, D.M., and Gerrard, D.E. 2014. Regulation of post-mortem glycolysis in ruminant muscle. *Anim. Prod. Sci.* 54:464–481.

Fraqueza, M.J., Roseiro, L.C., Almeida, J. et al. 1998. Effects of lairage temperature and holding time on pig behavior and on carcass and meat quality. *Appl. Anim. Behav. Sci.* 60:317–330.

Forrest, J.C., Morgan, M.T., Borggaard, C. et al. 2000. Development of technology for the early post mortem prediction of water holding capacity and drip loss in fresh pork. *Meat Sci.* 55:115–122.

Forrest, R.J. 1981. Effect of high concentrate feeding on carcass quality and fat coloration of grass-reared steers. *Can. J. Anim. Sci.* 61:575–580.

Gajadhar, A.A., Pozio, E., Gamble, H.R. et al. 2009. *Trichinella* diagnostics and control: Mandatory and best practices for ensuring food safety. *Vet. Parasitol.* 159:197–205.

Gallo, C., Caro, M., and Villarroel, C. 1999. Characteristics of cattle slaughtered within the Xth Region (Chile) according to the terms stated by the official Chilean standards for classification and carcass grading. *Arch. Med. Vet.* 31:81–88.

Gaudré, D., and Vautier, A. 2006. Incidence zootechnique d'un taux de complementation vitaminique élevé en engraissement [The zootechnical impact of high rates of vitamin-rich nutrient supplementation during finishing]. *Techniporc* 29:19–26.

Gee, A., Porter, M., and Polkinghorne, R. 2005. Protocols for the thawing, preparation and serving of beef for MSA trials for 5 different cooking methods. Meat and Livestock Australia: North Sydney. http://www.mla.com.au/msa (accessed October 2014).

Geesink, H., Schreutelkamp, F.H., Frankhuizen, R. et al. 2003. Prediction of pork quality attributes from near infrared reflectance spectra. *Meat Sci.* 65:661–668.

George, M.H., Tatum, J.D., Dolezal, H.G. et al. 1997. Comparison of USDA quality grade with tendertec for the assessment of beef palatability. *J. Anim. Sci.* 75:1538–1546.

Giboreau, A., Dacremont, C., and Egoroff, C. 2007. Defining sensory descriptors: Towards writing guidelines based on terminology. *Food Qual. Pref.* 18:265–274.

Gilbert, K.V., and Devine, D.E. 1982. Effect of electrical stunning methods on petechial haemorrhages and on the blood pressure of lambs. *Meat Sci.* 7:197–207.

Gill, C.O., and Harrison, J.C.L. 1982. Microbiological and organanoleptic qualities of bruised meat. *J. Food Prot.* 45:646–649.

Gill, C.O., and Newton, K.G. 1978. The ecology of bacterial spoilage of fresh meat at chill temperatures. *Meat Sci.* 2:207–217.

Gou, P., Comaposada, J., and Arnau, J. 2002. Meat pH and meat fibre direction effects on moisture diffusivity in salted ham muscles dried at 5°C. *Meat Sci.* 61:25–31.

Graafhuis, A.E., Honikel, K.O., Devine, C.E. et al. 1991. Meat tenderness of different muscles cooked to different temperatures and assessed by different methods. In *The Proceedings of the 37th International Congress of Meat Science and Technology*, Kulmbach, Germany, pp. 365–368.

Grau, R., and Hamm, R. 1953. Eine einfache methode zur bestimmung der wasserbindung im muskel [A simple method to determine water binding in muscle]. *Naturwissenschaften* 40:29–30.

Greene, B.E., Hsin, I., and Zipser, M.W. 1971. Retardation of oxidative color changes in raw ground beef. *J. Food Sci.* 36:940–942.

Gregory, N.G. 1985. Stunning and slaughter of pigs. *Pig News Inf.* 6:407–413.

Grunert, K.G., Bredahl, L., and Brunsø, K. 2004. Consumer perception of meat quality and implications for product development in the meat sector—A review. *Meat Sci.* 66:259–272.

Grunert, K.G., Larsen, H.H., Madsen, T.K. et al. 1996. *Market Orientation in Food and Agriculture.* Kluwer Academic Press, Boston.

Guárdia, M.D., Estany, J., Balasch, S. et al. 2005. Risk assessment of DFD meat due to pre-slaughter conditions in pigs. *Meat Sci.* 70:709–716.

Hamdy, M.K., and Carpenter, J.A. 1974. Bacterial persistence in animal tissues. *Poult. Sci.* 63:577–585.

Harris, S.E., Huff-Lonergan, E., Lonergan, S.M. et al. 2001. Antioxidant status affects colour stability and tenderness of calcium chloride-injected beef. *J. Anim. Sci.* 79:666–677.

Hildrum, K.I., Nilsen, B.N., Westad, F. et al. 2004. In-line analysis of ground beef using a diode array near infrared instrument on a conveyor belt. *J. Near Infrared Spectrosc.* 12:367–376.

Hill, F. 1968. Quality in meat and quality control of meat products. *J. Inst. Meat* 59:6–16.

Hocquette, J.F., Legrand, I., Jurie, C. et al. 2011. Perception in France of the Australian system for the prediction of beef quality (Meat Standards Australia) with perspectives for the European beef sector. *Anim. Prod. Sci.* 51:30–36.

Hoffman, L.C., and Luhl, J. 2012. Causes of cattle bruising during handling and transport in Namibia. *Meat Sci.* 92:115–124.

Hoffman, D.E., Spire, M.F., Schwenke, J.R. et al. 1998. Effect of source of cattle and distance transported to a commercial slaughter facility on carcass bruises in mature beef cows. *J. Am. Med. Vet. Assoc.* 212:668–672.

Honikel, K.O. 1987. The water binding of meat. *Fleischwirtschaft* 67:1098–1102.

Honikel, K.O. 1988. The water binding of meat. *Fleischwirtsch. Int.* 1:14–22.

Honikel, K.O. 1998. Reference methods for the assessment of physical characteristics of meat. *Meat Sci.* 49:447–457.

Honikel, K.O. 2014. Chemical and physical characteristics of meat: pH measurement. In Dikeman, M. and Devine, C. (eds.), *Encyclopaedia of Meat Sciences*, 2nd edition. Elsevier Science Ltd., Oxford, England, pp. 262–266.

Honikel, K.O., Kim, C.J., Hamm, R. et al. 1986. Sarcomere shortening of prerigor muscles and its influence on drip loss. *Meat Sci.* 16:267–282.

Hood, D.E., and Riordan, E.B. 1973. Discolouration in pre-packaged beef: Measurement by reflectance spectrophotometry and shopper discrimination. *J. Food Technol.* 8:333–343.

Hope-Jones, M., Strydom, P.E., Frylinck, L. et al. 2012. Effect of dietary beta-agonist treatment, vitamin D_3 supplementation and electrical stimulation of carcasses on colour and drip loss of steaks from feedlot steers. *Meat Sci.* 90:607–612.

Hopkins, D.L. 1996. Assessment of lamb meat colour. *Meat Focus Int.* 5:400–401.

Hopkins, D.L., Hegarty, R.S., Walker, P.J. et al. 2006. Relationship between animal age, intramuscular fat, cooking loss, pH, shear force and eating quality of aged meat from young sheep. *Aus. J. Exp. Agric.* 46:878–884.

Hopkins, D.L., Lamb, T.A., Kerr, M.J. et al. 2013. The interrelationship between sensory tenderness and shear force measured by the G2 tenderometer and a Lloyd texture analyser fitted with a Warner–Bratzler head. *Meat Sci.* 93:838–842.

Hopkins, D.L., Toohey, E.S., Kerr, M.J. et al. 2011. Comparison of the G2 tenderometer and the Lloyd texture analyser for measuring shear force in sheep and beef meats. *Anim. Prod. Sci.* 51:71–76.

Hopkins, D.L., Toohey, E.S., Warner, R.D. et al. 2010. Measuring the shear force of lamb meat cooked from frozen samples: Comparison of two laboratories. *Anim. Prod. Sci.* 50:382–385.

Hoving-Bolink, A.H., Vedder, H.W., Merks, J.W.M. et al. 2005. Perspective of NIRS measurements early post mortem for prediction of pork quality. *Meat Sci.* 69:417–423.

Huff-Lonergan, E., and Lonergan, S.M. 2005. Mechanisms of water-holding capacity of meat: The role of postmortem biochemical and structural changes. *Meat Sci.* 71:194–204.

Huff-Lonergan, E., Mitsuhashi, T., Beekman, D.D. et al. 1996. Proteolysis of specific muscle structural proteins by mu-calpain at low pH and temperature is similar to degradation in post mortem bovine muscle. *J. Anim. Sci.* 74:993–1008.

Huffman, K.L., Miller, M.F., Hoover, L.C. et al. 1996. Effect of beef tenderness on consumer satisfaction with steaks consumed in the home and restaurant. *J. Anim. Sci.* 74:91–97.

Hughes, J., Oiseth, S., Purslow, P. et al. 2014. A structural approach to understanding the interactions between colour, water-holding capacity and tenderness. *Meat Sci.* 98:520–532.

Hunt, M.C., Acton, J.C., Benedict, R.C. et al. 1991. American Meat Science Association committee on guidelines for meat color evaluation. In *The Proceedings of the 44th Reciprocal Meat Conference,* Kansas State University, Manhattan, KS.

Immonen, K., Kauffman, R.G., Schaefer, D.M. et al. 2000. Glycogen concentrations in bovine *longissimus dorsi* muscle. *Meat Sci.* 54:163–167.

Irie, M., Izumo, A., and Mohri, S. 1996. Rapid method for determining water-holding capacity in meat using video image analysis and simple formulae. *Meat Sci.* 42:95–102.

Issanchou, S. 1996. Consumer expectations and perceptions of meat and meat product quality. *Meat Sci.* 43: S5–S19.

ITP. 1996. Notation des hématomes sur couenne: porcs vivant ou carcasse. Istitute Technique du Porc, Rennes, France.

Jacob, R.H., and Hopkins, D.L. 2014. Techniques to reduce the temperature of beef muscle early in the post mortem period—A review. *Anim. Prod. Sci.* 54:482–493.

Jago, O., Hargreaves, A.L. Harcourtb, R.G. et al. 1996. Risk factors associated with bruising in red deer at a commercial slaughter plant. *Meat Sci.* 44:181–191.

Jay, J.M. 2000. *Modern Food Microbiology*. Aspen Publishers, Inc., Gaithersburg, MD.

Joo, S.T., Kaufman, R.G., Kim, B.C. et al. 1999. The relationship of sarcoplasmic and myofibrillar protein solubility to colour and water-holding capacity in porcine *longissimus* muscle. *Meat Sci.* 52:291–297.

Kanda, H., and Kancchika, T. 1992. Some properties of abnormal porcine muscles (PFE, PFD) differed from PSE and DFD. In *The Proceedings 38th International Congress of Meat Science and Technology*, Clermont-Ferrand, France, pp. 919–922.

Kauffman, R.G., Eikelenboom, G., Van der Wal, P.G. et al. 1986. A comparison of methods to estimate water-holding capacity in post-rigor porcine muscle. *Meat Sci.* 18:307–322.

Kauffman, R.G., Sybesma, W., Smulders, F.J.M. et al. 1993. The effectiveness of examining early post-mortem musculature to predict ultimate pork quality. *Meat Sci.* 34:283–300.

Khliji, S., van de Ven, R., Lamb, T.A. et al. 2010. Relationship between consumer ranking of lamb colour and objective measures of colour. *Meat Sci.* 85:224–229.

Killinger, M., Calkins, C.R., Umberger, W.J. et al. 2004. Consumer sensory acceptance and value for beef steaks of similar tenderness, but differing in marbling level. *J. Anim. Sci.* 82:3294–3301.

Kim, Y.H., Huff-Lonergan, E., Sebranek, J.G. et al. 2010. High-oxygen modified atmosphere packaging system induces lipid and myoglobin oxidation and protein polymerization. *Meat Sci.* 85:759–767.

Kim, Y.H.B., Warner, R.D., and Rosenvold, K. 2014. Influence of high pre-rigor temperature and fast pH fall on muscle proteins and meat quality: A review. *Anim. Prod. Sci.* 54:375–395.

Kirton, A.H., Bishop, W.H., Mullord, M.M. et al. 1978. Relationships between time of stunning and time of throat cutting and their effect on blood pressure and blood splash in lambs. *Meat Sci.* 2:199–206.

Knowles, T.G. 1999. A review of the road transport of cattle. *Vet. Rec.* 144:197–201.

Knowles, T.G., Maunder, D.H.L., and Warriss, P.D. 1994. Factors affecting the incidence of bruising in lambs arriving at one slaughterhouse. *Vet. Rec.* 134:44–45.

Koohmaraie, M. 1992. Effect of pH, temperature, and inhibitors on autolysis and catalytic activity of bovine skeletal muscle mu-calpain. *J. Anim. Sci.* 70:3071–3080.

Koohmaraie, M. 1996. Biochemical factors regulating the toughening and tenderisation processes of meat. *Meat Sci.* 43:193–201.

Koohmaraie, M., Kent, M.P., Shackelford, S.D. et al. 2002. Meat tenderness and muscle growth: Is there any relationship? *Meat Sci.* 62:345–352.

Koohmaraie, M., Seideman, S.C., Schollmeyer, J.E. et al. 1987. Effect of post-mortem storage on Ca^{++}-dependent proteases, their inhibitor and myofibril fragmentation. *Meat Sci.* 19:187–196.

Kristensen, L., and Purslow, P.P. 2001. The effect of ageing on the water-holding capacity of pork: Role of cytoskeletal proteins. *Meat Sci.* 58:17–23.

Kunert, J. 1998. Sensory experiments as crossover studies. *Food Qual. Prefer.* 9:243–253.

Lambert, M.G., Knight, T.W., Cosgrove, G.P. et al. 1998. Exercise effects on muscle glycogen concentration in beef cattle. *N. Z. Soc. Anim. Prod.* 60:243–244.

Lawless, H.T., and Heymann, H. 2009. *Sensory Evaluation of Food: Principles and Practices.* Springer, New York.

Lawrie, R.A., and Ledward, D.A. 2006. *Lawrie's Meat Science*, 7th edition. Woodhead Publishing Limited, Cambridge.

Lee, M.S., Apple, J.K., Yancey, J.W.S. et al. 2008a. Influence of wet-aging on bloom development in the *longissimus thoracis. Meat Sci.* 80:703–707.

Lee, M.S., Apple, J.K., Yancey, J.W.S. et al. 2008b. Influence of vacuum-aging period on bloom development of the beef *gluteus medius* from top sirloin butts. *Meat Sci.* 80:592–598.

Lee, S., Norman, J.M., Gunasekaran, S., van Laack, R.L.J.M. et al. 2000. Use of electrical conductivity to predict water-holding capacity in post-rigor pork. *Meat Sci.* 55:385–389.

Leet, N.G., Devine, C.E., and Gavey, A.B. 1977. The histology of blood splash in lambs. *Meat Sci.* 1:229–234.

Leroy, B., Lambotte, S., Dotreppe, O., Lecocq, H., Istasse, L., and Clinquart, A. 2003. Prediction of technological and organoleptic properties of beef longissimus thoracis from near-infrared reflectance and transmission spectra. *Meat Sci.* 66:45–54.

Lorenzen, C.L., Hale, D.S., Griffin, D.B. et al. 1993. National beef quality audit: Survey of producer-related defects and carcass quality and quantity attributes. *J. Anim. Sci.* 71:1495–1502.

Lund, M.N., Lametsch, R., Hviid, M.S. et al. 2007. High oxygen packaging atmosphere influences protein oxidation and tenderness of porcine *longissimus dorsi* during chill storage. *Meat Sci.* 77:295–303.

MacDougall, D.B. 1982. Changes in the colour and opacity of meat. *Food Chem.* 9:74–88.

Maddock, K.R., Huff-Lonergan, E., Rowe, L.J. et al. 2005. Effect of pH and ionic strength on l- and m-calpain inhibition by calpastatin. *J Anim. Sci.* 83:1370–1376.

Mancini, R.A., and Hunt, M.C. 2005. Current research in meat color. *Meat Sci.* 71:100–121.

Mann, N.J., Ashton, Y., O'Connell, S. et al. 2006. Food group categories used in dietary analysis can misrepresent the amount and type of fat present in foods. *Nutr. Diet.* 63:69–78.

Martinaud, A., Mercier, Y., Marinova, P. et al. 1997. Comparison of oxidative processes on myofibrillar proteins from beef during maturation and by different model oxidation systems. *J. Agric. Food Chem.* 45:2481–2487.

McCausland, I.P., and Dougherty, R. 1978. Histological ageing of bruises in lambs and calves. *Aust. Vet. J.* 54:525–527.

McKeith, R.O., Gray, G.D., Hale, D.S. et al. 2012. National beef quality audit-2011: Harvest-floor assessments of targeted characteristics that affect quality and value of cattle, carcasses, and byproducts. *J. Anim. Sci.* 90:5135–5142.

McKenna, D.R., Roebert, D.L., Bates, P.K. et al. 2002. National beef quality audit-2000: Survey of targeted cattle and carcass characteristics related to quality, quantity, and value of fed steers and heifers. *J. Anim. Sci.* 80:1212–1222.

McVeigh, J.M., and Tarrant, P.V. 1983. Effect of propranolol on muscle glycogen metabolism during social regrouping of young bulls. *J. Anim. Sci.* 56:71–80.

Meilgaard, M., Civille, G.V., and Carr, B.T. 1999. *Sensory Evaluation Techniques*, 3rd edition. CRC Press, Boca Raton, FL.

Melody, J.L., Lonergan, S.M., Rowe, L.J. et al. 2004. Early post mortem biochemical factors influence tenderness and water-holding capacity of three porcine muscles. *J. Anim. Sci.* 82:1195–1205.

Miller, M.F., Carr, M.A., Ramsey, C.B. et al. 2001. Consumer thresholds for establishing the value of beef tenderness. *J. Anim. Sci.* 79:3062–3068.

Miller, R.K. 2002. Factors affecting the quality of raw meat. In Kerry, J.P., Kerry, J.F. and Ledward, D. (eds.), *Meat Processing—Improving Quality.* Woodhead Publishing Co, Cambridge, England, pp. 27–63.

Minka, N.S., and Ayo, J.O. 2007. Effects of loading behaviour and road transport stress on traumatic injuries in cattle transported by road during the hot-dry season. *Livest. Sci.* 107:91–95.

MLA (Meat and Livestock Australia). 2014. The effect of pH on beef eating quality. Meat Standards Australia, Tips and Tools. http://www.mla.com.au/Marketing-beef-and-lamb/Meat-Standards-Australia/MSA-beef/Grading (Accessed May 26, 2015).

Mohan, A., Hunt, M.C., Barstow, T.J. et al. 2010. Effects of fibre orientation, myoglobin redox form, and postmortem storage on NIR tissue oximeter measurements of beef *longissimus* muscle. *Meat Sci.* 84:79–85.

Monin, G. 1981. Muscle metabolic type and the DFD condition. In Hood, D.E. and Tarrant, P.V. (eds.), *Current Topics in Veterinary Medicine and Animal Science*, Vol. 10*: The Problem of Dark Cutting in Beef.* Martinus Nijhoff Publishers, The Hague, pp. 63–81.

Monsón, F., Sañudo, C., and Sierra, I. 2005. Influence of breed and ageing time on the sensory meat quality and consumer acceptability in intensively reared beef. *Meat Sci.* 71:471–479.

Mulley, R.C., and Falepau, D.F. 1999. Ecchymosis—What causes it? In *Rural Industries Research and Development Corporation*, Publication number 99/48. CanPrint, Canberra, Australia.

Mulley, R.C., Falepau, D.F., Flesch, J.S. et al. 2010. Rate of blood loss and timing of exsanguination on prevalence of ecchymosis in fallow deer (*Dama dama*). *Meat Sci.* 85:21–25.

Murray, A.C. 1989. Factors affecting colour at the time of grading. *Can. J. Anim. Sci.* 69:247–355.

Newton, K.G., and Gill, C.O. 1981. The microbiology of DFD fresh meats: A review. *Meat Sci.* 5:223–232.

Ngapo, T.M., and Dransfield, E. 2006. British consumers preferred fatness levels in beef: Surveys from 1955, 1982 and 2002. *Food Qual. Prefer.* 17:412–417.

Ngapo, T.M., Dransfield, E., Martin, J.-F. et al. 2003. Consumer perceptions: Pork and pig production. Insights from France, England, Sweden and Denmark. *Meat Sci.* 66:125–134.

Ngapo, T.M., Martin, J.-F., and Dransfield, E. 2004. Consumer choices of pork chops: Results from three panels in France. *Food Qual. Pref.* 15:349–359.

Ngapo, T.M., Martin, J.-F., and Dransfield, E. 2007. International preferences for pork appearance: I. Consumer choices. *Food Qual. Prefer.* 18:26–36.

Nute, G.R. 2008. Sensory Descriptors. In Nollet, L.M.L. and Toldra, F. (eds.), *Handbook of Muscle Foods Analysis*. CRC Press, Boca Raton, FL, pp. 513–524.

Nute, G.R., Richardson, R.I., Wood, J.D. et al. 2007. Effect of dietary oil source on the flavour and the colour and lipid stability of lamb meat. *Meat Sci.* 77:547–555.

Offer, G. 1991. Modelling of the formation of pale, soft and exudative meat: Effects of chilling regime and rate and extent of glycolysis. *Meat Sci.* 30:157–184.

Offer, G., and Cousins, T. 1992. The mechanism of drip production—Formation of 2 compartments of extracellular-space in muscle post mortem. *J. Sci. Food Agric.* 58:107–116.

Offer, G., and Knight, P. 1988. The structural basis of water-holding capacity in meat. Part 2: Drip losses. In Lawrie, R. (ed.), *Developments in Meat Science*. Elsevier Science Publications, London, Vol. 4, pp. 173–243.

Offer, G., and Trinick, J. 1983. On the mechanism of water holding in meat: The swelling and shrinking of myofibrils. *Meat Sci.* 8:245–281.

O'Neill, D.J., Lynch, P.B., Troy, D.J. et al. 2003. Influence of time of year on the incidence of PSE and DFD in Irish pig meat. *Meat Sci.* 64:105–111.

Page, J.K., Wulf, D.M., and Schwotzer, T.R. 2001. A survey of beef muscle color and pH. *J. Anim. Sci.* 79:678–687.

Pearce, K.L., Rosenvold, K., Andersen, H.J. et al. 2011.Water distribution and mobility in meat during the conversion of muscle to meat and ageing and the impacts on fresh meat quality attributes—A review. *Meat Sci.* 89:111–124.

Pearson, A.J., Kilgour, R., Langen, D.E. et al. 1977. Hormonal responses of lambs to trucking, handling and electrical stunning. *Proc. N. Z. Soc. Anim. Prod.* 3:243–248.

Perry, D., Thompson, J.M., Hwang, I.H. et al. 2001. Relationship between objective measurements and taste panel assessment of beef quality. *Aust. J. Exp. Agric.* 41:981–989.

Petersen, G.V., Carr, D.H., Davies, A.S. et al. 1986. The effect of different methods of electrical stunning of lambs on blood pressure and muscular activity. *Meat Sci.* 16:1–15.

Pighin, D.G., Brown, W., Ferguson, D.M. et al. 2014. Relationship between changes in core body temperature in lambs and post-slaughter muscle glycogen content and dark-cutting. *Anim. Prod. Sci.* 54:459–463.

Platter, W.J., Tatum, J.D., Belk, K.E. et al. 2003. Relationships of consumer sensory ratings, marbling score, and shear force value to consumer acceptance of beef strip loin steaks. *J. Anim. Sci.* 81:2741–2750.

Polkinghorne, R., Thompson, J.M., Watson, R. et al. 2008. Evolution of the Meat Standards Australia (MSA) beef grading system. *Aust. J. Exp. Agric.* 48:1351 – 1359.

Prevolnik, M., Candek-Potokar, M., and Škorjanc, D. 2004. Ability of NIR spectroscopy to predict meat chemical composition and quality: A review. *Czech. J. Anim. Sci.* 49:500–510.

Prieto, N., Andrés, S., Giráldez, F.J. et al. 2008. Ability of near infrared reflectance spectroscopy (NIRS) to estimate physical parameters of adult steers (oxen) and young cattle meat samples. *Meat Sci.* 79:692–699.

Prieto, N., Roehe, R., Lavín, P. et al. 2009a. Application of near infrared reflectance spectroscopy to predict meat and meat products quality: A review. *Meat Sci.* 83:175–186.

Prieto, N., Ross, D.W., Navajas, E.A. et al. 2009b. On-line application of visible and near infrared reflectance spectroscopy to predict chemical–physical and sensory characteristics of beef quality. *Meat Sci.* 83:96–103.

Priolo, A., Prache, S., Micol, D. et al. 2002. Reflectance spectrum of adipose tissue to trace grass feeding in sheep: Influence of measurement site and shrinkage time after slaughter. *J Anim. Sci.* 80:886–891.

Purchas, R.W. 2014. Tenderness measurement. In Dikeman, M. and Devine, C. (eds.), *Encyclopaedia of Meat Sciences*, 2nd edition. Elsevier Science Ltd., Oxford, England, pp. 452–459.

Qiao, J., Wang, N., Ngadi, M.O. et al. 2007. Prediction of drip-loss, pH, and color for pork using a hyperspectral imaging technique. *Meat Sci.* 76:1–8.

Reis, M.M., and Rosenvold, K. 2014. Early on-line classification of beef carcasses based on ultimate pH by near infrared spectroscopy. *Meat Sci.* 96:862–869.

Renerre, M. 1990. Factors involved in the discoloration of beef meat. *Int. J. Food Sci. Technol.* 25:613–630.

Rentfrow, M.L., Linville, M.L., Stahl, C.A. et al. 2004. The effects of the antioxidant lipoic acid on beef longissiumus bloom time. *J. Anim. Sci.* 82:3034–3037.

Rezac, D.J. 2013. Gross pathology monitoring of cattle at slaughter. PhD thesis. Department of Diagnostic Medicine/Pathobiology, College of Veterinary Medicine, Kansa State University, Manhattan, KS.

Rhee, M.S., Wheeler, T.L., Shackelford, S.D. et al. 2004. Variation in palatability and biochemical traits within and among eleven beef muscles. *J. Anim. Sci.* 82:534–550.

Ripoll, G., Albertí, P., Panea, B. et al. 2008. Near-infrared reflectance spectroscopy for predicting chemical, instrumental and sensory quality of beef. *Meat Sci.* 80:697–702.

Rogers, S.A., Hollywood, N.W., and Mitchell, G.E. 1992. The microbiological and technological properties of bruised beef. *Meat Sci.* 32:437–447.

Romero, M.H., Uribe-Velásquez, L.F., Sánchez, J.A. et al. 2013. Risk factors influencing bruising and high muscle pH in Colombian cattle carcasses due to transport and preslaughter operations. *Meat Sci.* 95:256–263.

Rosenvold, K., and Andersen, H.J. 2003. Factors of significance for pork quality—A review. *Meat Sci.* 64:219–237.

Roserriro, L.C., Santos, C., and Melo, R.S. 1993. Muscle pH 60, colour (*L;a;b*) and water holding capacity and the influence of post-mortem meat temperatures. In *The Proceedings of the 39th International Congress of Meat Science and Technology* S4P21, WP, Calgary, Alberta, Canada.

Russell, B.C., McAlister, G., Ross, I.S. et al. 2005. Lamb and sheepmeat quality—Industry and scientific issues and the need for integrated research. *Aust. J. Exp. Agric.* 45:465–467.

Sañudo, C., Alfonso, M., San Julián, R. et al. 2007. Regional variation in the hedonic evaluation of lamb meat from diverse production systems by consumers in six European countries. *Meat Sci.* 75:610–621.

Sayre, R.N., and Briskey, E.J. 1963. Protein solubility as in sequenced by physiological conditions in muscle. *J. Food Sci.* 28:675–679.

Schäfer, A.L., Knight, P.J., Wess, T.J. et al. 2000. Influence of sarcomere length on the reduction of myofilament lattice spacing post-mortem and its implications on drip loss. In *The Proceedings of the International Congress of Meat Science and Technology*, Buenos Aires, Argentina, pp. 434–435.

Shackelford, S.D., Wheeler, T.L., and Koohmaraie, M. 1999. Evaluation of slice shear force as an objective method of assessing beef longissimus tenderness. *J. Anim. Sci.* 77:2693–2699.

Sheard, P.R., Enser, M., Wood, J.D. et al. 2000. Shelf life and quality of pork and pork products with raised n-3 PUFA. *Meat Sci.* 55:213–221.

Smith, G.C., Savell, J.W., Cross, H.R. et al. 1983. The relationship of USDA quality grade to beef flavour. *Food Technol.* 37:233–238.

Smith, G.C., Tatum, J.D., and Belk, K.E. 2008. International perspective: Characterisation of United States Department of Agriculture and Meat Standards Australia systems for assessing beef quality. *Aust. J. Exp. Agric.* 48:1465–1480.

Smulders, F.J.M., Eikelenboom, G., Lambooy, E. et al. 1989. Electrical stimulation during exsanguination: Effects on the prevalence of blood splash and on sensory characteristics in veal. *Meat Sci.* 26:89–99.

Snyman, J.D., Frylinck, L., and Strydom, P.E. 2008. Comparison of two methods to determine myofibril fragmentation post mortem. *54th Int. Congre. Meat Sci. Technol.*, Cape Town, South Africa, pp. 8.19.

Sofos, J.N. 2008. Challenges to meat safety in 21st century. *Meat Sci.* 78:3–13.

Sornay, J., Dumont, B.L., and Founaud, J. 1981. Practical aspects of dark, firm, dry (DFD) meat. In Hood, D.E. and Tarrant, P.V. (eds.), Current Topics in Veterinary Medicine and Animal Science, *Vol. 10: The Problem of Dark Cutting in Beef.* Martinus Nijhoff Publishers, The Hague, pp. 362–373.

Stone, H., and Sidel, J.L. 1993. *Sensory Evaluation Practice.* Academic Press, Inc., San Diego, pp. 50–59.

Strappini, A.C., Frankena, K., Metz, J.H.M. et al. 2010. Characteristics of bruises in carcasses of cows sourced from farms or from livestock markets. *Animal* 6:502–509.

Strydom, P.E., Frylinck, L., Montgomery, J.L. et al. 2009. The comparison of three β-agonists for growth performance, carcass characteristics and meat quality of feedlot cattle. *Meat Sci.* 81:557–564.

Strydom, P.E., and Rosenvold, K. 2014. Muscle metabolism in sheep and cattle in relation to high rigor temperature—Overview and perspective. *Anim. Prod. Sci.* 54:510–518.

Suman, S.P., and Joseph, P. 2014. Chemical and physical characteristics of meat: Color and pigment, In Dikeman, M. and Devine, C. (eds.), *Encyclopaedia of Meat Sciences*, 2nd edition. Elsevier Science Ltd., Oxford, England, pp. 244–251.

Tapp, W.H., Yancey, Y.W.S., and Apple, J.K. 2011. How is the instrumental color of meat measured? *Meat Sci.* 89:1–5.

Tarladgis, B.G., Watts, B.M., and Yonathan, M. 1960. Distillation method for the determination of malonaldehyde in rancid foods. *J. Am. Oil Chem. Soc.* 37:44–48.

Tarrant, P.V. 1981. The occurrence, causes and economic consequences of dark-cutting in beef—A survey of current information. In Hood, D.E. and Tarrant, P.V. (eds.), *Current Topics in Veterinary Medicine and Animal Science*, Vol. 10*: The Problem of Dark Cutting in Beef.* Martinus Nijhoff Publishers, The Hague, pp. 3–36.

Tatum, J.D., Smith, G.C., and Carpenter, Z.L. 1982. Interrelationships between marbling, subcutaneous fat thickness and cooked beef palatability. *J. Anim. Sci.* 54:777–784.

Thompson, J. 2002. Managing meat tenderness. *Meat Sci.* 62:295–308.

Thompson, J.M. 2004. The effects of marbling on flavour and juiciness of cooked beef, after adjusting to a constant tenderness. *Aust. J Exp. Agric.* 44:645–652.

Thompson, J.M., Gee, A., Hopkins, D.L. et al. 2005a. Development of sensory protocol for testing palatability of sheep meats. *Aust. J. Exp. Agric.* 45:469–476.

Thompson, J.M., Pleasants, A.B., and Pethick, D.W. 2005b. The effect of demographic factors on consumer sensory scores. *Aust. J. Exp. Agric.* 45:477–482.

Tøgersen, G., Arnesen, J.F., Nielsen, B.N. et al. 2003. On-line prediction of chemical composition of semi-frozen ground beef by non-invasive NIR spectroscopy. *Meat Sci.* 63:515–523.

Tornberg, E., Wahlgren, M., Brondum, J. et al. 2000. Pre-rigor conditions in beef under varying temperature and pH falls studied with rigormeter, NMR and NIR. *Food Chem.* 69:407–418.

Tweed, W., Clark, G.A., and Edington, J.W. 1931. Splashing of meat in the slaughter of animals. *Vet. Rec.* 11:23–27.

USDA. 1997. United States Standards for Grades of Carcass Beef. AMS, USDA, Washington, DC. Available from http://www.ams.usda.gov/AMSv1.0/getfile?dDocName=STELDEV3002979. Accessed October 8, 2014.

Van de Perre, V., Driessen, B., Van Thielen, J. et al. 2011. Comparison of pig behaviour when given a sequence of enrichment objects or a chain continuously. *Anim. Welfare* 20:641–649.

Van der Wal, P.G., Bolink, A.H., and Merkus, G.S.M. 1988. Differences in quality characteristics of normal, PSE and DFD pork. *Meat Sci.* 24:79–84.

Van Laack, R.L.J.M., Kauffman, R.G., Sybesma, W. et al. 1994. Is colour brightness (L-value) a reliable indicator of water-holding capacity in porcine muscle? *Meat Sci.* 38:193–201.

Velarde, A., Gispert, M., Faucitano, L. et al. 2000. The effect of stunning method on the incidence of PSE meat and haemorrhages in pork carcasses. *Meat Sci.* 55:309–314.

Verbeke, W., Van Wezemael, L., and de Barcellos, M.D. 2010. European beef consumers' interest in a beef eating-quality guarantee. Insights from a qualitative study in four EU countries. *Appetite* 54:289–296.

Viljoen, H.F., de Kock, H.L., and Webb, E.C. 2002. Consumer acceptability of dark, firm and dry (DFD) and normal pH beef steaks. *Meat Sci.* 61:181–185.

Walker, P.J., Warner, R.D., and Winfield, C.G. 1990. Sources of variation in subcutaneous fat colour of beef carcasses. *Proc. Aust. Soc. Anim. Prod.* 18:416–419.

Warner, R.D., Dunshea, F.R., Gutzke, D. et al. 2014a. Factors influencing the incidence of high rigor temperature in beef carcasses in Australia. *Anim. Prod. Sci.* 54:363–374.

Warner, R.D., Kauffman, R.G., and Greaser, M.L. 1997. Muscle protein changes post mortem in relation to pork quality traits. *Meat Sci.* 45:339–352.

Warner, R.D., Kauffman, R.G., and Russell, R.L. 1993. Quality attributes of major porcine muscles: A comparison with the *longissimus lumborum*. *Meat Sci.* 33:359–372.

Warner, R.D., Thompson, J.M., Polkinghorne, R. et al. 2014b. A consumer sensory study of the influence of rigor temperature on eating quality and ageing potential of beef strip loin and rump. *Anim. Prod. Sci.* 54:396–406.

Warren, H.E., Scollan, N.D., Nute, G. et al. 2008. Effects of breed and a concentrate or grass silage diet on beef quality in cattle of 3 ages. II: Meat stability and flavour. *Meat Sci.* 78:270–278.

Warriss, P.D. 2003. Optimal lairage times and conditions for slaughter pigs: A review. *Vet. Rec.* 153:170–176.

Warriss, P.D., Bevis, E.A., and Ekins, P.J. 1989. The relationships between glycogen stores and muscle ultimate pH in commercially slaughtered pigs. *Br. Vet. J.* 145:378–383.

Warriss, P.D., and Brown, S.N. 1993. Relationships between the subjective assessments of pork quality and objective measures of colour. In Wood, J.D. and Lawrence, T.L.J. (eds.), *Safety and Quality of Food from Animals. Occasional Publication of the British Society of Animal Production no.* 17. Edinburgh, UK, pp. 98–101.

Watson, R., Gee, A., Polkinghorne, R., and Porter, M. 2008a. Consumer assessment of eating quality—Development of protocols for Meat Standards Australia (MSA) testing. *Aust. J. Exp. Agric.* 48:1360–1367.

Watson, R., Polkinghorne, R., and Thompson, J.M. 2008b. Development of the Meat Standards Australia (MSA) prediction model for beef palatability. *Aust. J. Exp. Agric.* 48:1368–1379.

Weeks, C.A., McNally, P.W., and Warriss, P.D. 2002. Influence of the design of facilities at auction markets and animal handling procedures on bruising in cattle. *Vet. Rec.* 150:743–748.

Wheeler, T.L., Shackelford, S.D., Johnson, L.P. et al. 1997. A comparison of Warner–Bratzler shear force assessment within and among institutions. *J. Anim. Sci.* 75:2423–2432.

Wheeler, T.L., Shackelford, S.D., and Koohmaraie, M. 2005. Shear force procedures for meat tenderness measurement. http://ars.usda.gov/SP2UserFiles/Place/54380530/protocols/ShearForceProcedures.pdf. (accessed October 2, 2014).

Wheeler, T.L., Vote, D., Leheska, J.M. et al. 2002. The efficacy of three objective systems for identifying beef cuts that can be guaranteed tender. *J. Anim. Sci.* 80:3315–3327.

Wood, J.D., Enser, M., Fisher, A.V. et al. 2008. Fat deposition, fatty acid composition and meat quality: A review. *Meat Sci.* 78:343–358.

Wood, J.D., Richardson, R.I., Nute, G.R. et al. 2003. Effects of fatty acids on meat quality: A review. *Meat Sci.* 66:21–23.

Wulf, D.M., Emnett, R.S., Leheska, J.M. et al. 2002. Relationships among glycolytic potential, dark cutting (dark, firm, and dry) beef, and cooked beef palatability. *J. Anim. Sci.* 80:1895–1903.

Wulf, D.M., and Wise, J.W. 1999. Measuring muscle color on beef carcasses using the *L*a*b** color space. *J. Anim. Sci.* 77:2418–2427.

Wythes, J.R., Kaus, R.K., and Newman, G.A. 1985. Bruising in cattle slaughtered at an abattoir in Southern Queensland. *Aust. J. Exp. Agric.* 25:727–733.

Wythes, J.R., and Shorthose, W.R. 1991. Chronological age and dentition effects on carcass and meat quality of cattle in Northern Australia. *Aust. J. Exp. Agric.* 31:145–152.

Xing, J., Ngadi, M., Gunenc, A. et al. 2007. Use of visible spectroscopy for quality classification of pork meat. *J. Food Eng.* 82:135–141.

Xiong, Y.L. 2014. Protein functionality. In Dikeman, M. and Devine, C. (eds.), *Encyclopaedia of Meat Sciences*, 2nd edition. Elsevier Science Ltd., Oxford, England, pp. 267–273.

Xiong, Y.L., and Decker, E.A. 1995. Alterations of muscle protein functionality by oxidative and antioxidative processes. *J. Muscle Foods* 6:139–160.

Yancey, J.W.S., and Kropf, D.H. 2008. Instrumental reflectance values of fresh pork are dependent on aperture size. *Meat Sci.* 79:734–739.

Yang, A., Larsen, T.W., and Tumne, R.K. 1992. Carotenoid and retinol concentrations in serum, adipose tissue and liver and carotenoid transport in sheep, goats and cattle. *Aust. J. Agric. Res.* 43:1809–1817.

Yeh, E., Anderson, B., Jones, P.N., and Shaw, F.D. 1978. Bruising in cattle transported over long distances. *Vet. Rec.* 103:117–119.

Young, O.A., Berdagué, J.-L., Viallon, C. et al. 1997. Fat-borne volatiles and sheepmeat odour. *Meat Sci.* 45:183–200.

Young, O.A., Priolo, A., Simmons, N.J. et al. 1999. Effects of rigor attainment temperature on meat blooming and colour on display. *Meat Sci.* 52:47–56.

Zachary, J.F., and McGavin, M.D. 2012. *Pathologic Basis of Veterinary Disease.* Elsevier, Mosby, Missouri.

3 Conversion of Muscle to Meat

Qingwu W. Shen and Min Du

CONTENTS

3.1 INTRODUCTION

The conversion of muscle to meat is a complex process. Muscle does not die immediately after animal harvesting. Instead, there are complicated physiological changes which occur during and shortly after harvesting, followed by additional physical and chemical changes. Exsanguination of animals marks the beginning of a series of postmortem changes. These changes include the loss of homeostasis, cell death, proteolysis, oxidation, and others, among which the loss of homeostatic control is central to the conversion of muscle to meat. These changes have a profound impact on the palatability of meat. Postmortem glycolysis and proteolysis are the two most important postmortem changes affecting meat quality attributes. In this chapter, biochemical changes which occur during the conversion of muscle to meat, factors affecting this process, and the effect of these factors on meat quality are discussed, focusing on postmortem glycolysis and proteolysis.

3.2 CHEMICAL AND PHYSICAL CHANGES IN POSTMORTEM MUSCLE

After an animal is slaughtered, blood circulation stops and oxygen supply ceases. The cessation of blood circulation initiates a series of postmortem changes in muscle. These changes finally lead to the death of muscle cells, disruption of cell integrity, and the conversion of muscle to meat.

3.2.1 Muscle pH Decline

After exsanguination, muscle is no longer able to use oxygen to generate adenosine triphosphate (ATP). To maintain homeostasis, energy metabolism is shifted to anaerobic glycolysis to generate ATP. In living animals, lactic acid produced by glycolysis is transported to the liver, where it is resynthesized into glucose and glycogen, or to heart, where it is metabolized to carbon dioxide and water (Spahr et al. 1985). However, in the absence of blood circulation in postmortem muscle, lactic acid produced through glycolysis accumulates inside muscle, which leads to a decrease of muscle pH.

The declining rates of glycolysis and pH in postmortem muscle, especially during the early postmortem stage, coupled with the high temperature, have a primary effect on water holding capacity (WHC) and other properties, such as tenderness, color, and cooking yield, as well as muscle protein functionality that is important in meat processing (Jacob and Hopkins 2014; Kim et al. 2014). Abnormal glycolysis in postmortem muscle can lead to inferior meat quality, such as pale, soft, and exudative (PSE) meat, acid meat (Hampshire effect), and dark, firm, and dry (DFD) meat. The rate and extent of glycolysis and pH decline in postmortem muscle are highly variable.

The normal pH decline in pork starts from approximately pH 7.4 in living muscle to a pH of about 5.6–5.7 within 6–8 h postmortem, and then to an ultimate pH (reached at about 24 h postmortem) of about 5.3–5.7 (Briskey and Wismer-Pedersen 1961). Improper handling or genetic factors can cause accelerated lactic acid accumulation and pH decline in postmortem muscle. In some carcasses, muscle pH drops rapidly to around 5.4–5.5 within the first hour after exsanguination (Pardi et al. 1993). In some extreme instances, the ultimate pH can be reached in 15 min postmortem (Solomon et al. 1998). This rapid pH decline early postmortem, while muscle temperature is still high (more than 36°C), plus heat generation due to glycolysis, leads to PSE meat (Briskey and Wismer-Pedersen 1961; Schwagele et al. 1994; Solomon et al. 1998; Hopkins et al. 2014). A similar, but different case is the red, soft, and exudative (RSE) meat. Pigs carrying the Rendement Napole gene (*RN*$^-$) have 70% more glycogen storage in skeletal muscle (Estrade et al. 1993; Milan et al. 2000). Muscles from these pigs have normal glycolytic and pH rates of decline early postmortem (Copenhafer et al. 2006), but glycolysis continues to a lower than normal ultimate pH, which could be attributed to an active AMP-activated protein kinase (AMPK) (see discussion later in this chapter). The *RN⁻* gene is primarily observed in purebred or crossbred Hampshire populations, so it is also commonly called the "Hampshire effect." While, traditionally, PSE is a problem mainly associated with pork, recent studies from Australia show that a similar condition, defined by the pH less than 6 and temperature greater than 35°C, negatively affects beef quality (Jacob and Hopkins 2014). On the other hand, if glycogen storage in muscle is exhausted before slaughter, the muscle pH will be abnormally high (pH 6.0–7.0), generating DFD meat (Warriss et al. 1989; Viljoen et al. 2002). The high incidence of DFD meat in beef cattle is closely associated with the high ratio of oxidative muscle fibers (Zerouala and Stickland 1991). Without sufficient glycogen to supply ATP, muscle goes into rigor rapidly and this meat is more susceptible to bacterial spoilage than normal or PSE meat.

3.2.2 Heat Dissipation and Carcass Temperature

After slaughter, the circulatory system of animals stops, which hinders rapid dissipation of heat from deep parts of the carcass. Therefore, a rise in muscle temperature may occur soon after exsanguination because of the heat generated by glycolysis (Briskey and Wismer-Pedersen 1961). Rapid glycolysis early postmortem will inevitably heighten muscle temperature, together with lactic acid accumulation, causing muscle protein denaturation and the PSE syndrome (Scheffler and Gerrard 2007), and recently a similar condition in beef has been studied (Warner et al. 2014).

External factors associated with the slaughter process and carcass chilling influence heat production and dissipation postmortem. Slaughter room temperature, duration of the slaughter and dressing operation, and temperature of the initial chiller all influence the rate of carcass temperature decline (Jacob and Hopkins 2014). Furthermore, preslaughter factors and several gene mutations influence postmortem glycolysis, affecting heat production and carcass temperature postmortem. For example, muscle from stress positive pigs, which carry a mutation in the ryanodine receptor, exhibits accelerated glycolysis resulting in greater heat production, leading to PSE meat (Fujii et al. 1991; Wendt et al. 2000). In beef carcasses, hyperthermia ante mortem is an important contributing factor to elevated carcass temperature (DiGiacomo et al. 2014).

3.2.3 Rigor Mortis

Right after exsanguination, the muscle of animals is soft and extensible, but within hours, it starts to become rigid, a phenomenon called *rigor mortis* (Latin for "stiffness of death"). Stiffness of muscle observed in rigor mortis is due to the formation of permanent strong cross-bridges between actin and myosin as ATP is depleted in postmortem muscle.

Skeletal muscle has an ATP reserve at several micromoles per gram of muscle, which can supply enough energy for only a few twitches. During the early postmortem stage, creatine phosphate (CP) is used for the re-phosphorylation of ADP to ATP by creatine kinase. Additionally, myokinase catalyzes the conversion of two ADP molecules to AMP and ATP. Together these reactions allow muscle to maintain cellular ATP homeostasis temporally. Consistently, the consumption of CP is a sensitive parameter to judge the metabolic process in postmortem muscle (Scheffler et al. 2013, 2014). As the CP store depletes, muscle resorts to anaerobic glycolysis to generate ATP, however, the amount of ATP that can be generated from stored glycogen metabolism is limited. As ATP levels decline, an increasing level of cross-bridge formation occurs and the cells lose the ability to sequestrate calcium from the cytosol. Calcium ions leak from the sarcoplasmic reticulum and mitochondria into the cytosol, to induce the formation of cross-bridges (Nogueira et al. 2013). As actomyosin bridges form, the muscle becomes less extensible gradually, marking the onset of rigor mortis and the muscle contracts and shortens and muscle tension increases. These events reach a maximum when ATP is exhausted and all binding sites form actomyosin bridges. Because the development of rigor mortis is due to the depletion of ATP, factors affecting ATP consumption, especially myosin ATPase activity, profoundly affects rigor development and subsequent meat quality (Bowker et al. 2005). Electric stimulation and high carcass temperature dramatically accelerates ATP consumption and the development of rigor (Kim et al. 2012; Hopkins et al. 2014).

3.2.4 Disruption of Muscle Structure

During development and completion of rigor mortis, the microscopic structure of muscle has much the same appearance as living muscle except shorter sarcomere lengths. Noticeable changes of muscle properties and ultrastructure are evident during the "resolution" of rigor mortis. The "resolution" is not the result of the dissociation of myosin from actin, but rather due to proteolytic degradation of myofibrillar proteins, especially intermediate filaments, which weakens the structure of Z disks and impairs muscle ultrastructural integrity (Koohmaraie et al. 1991; Weaver et al. 2009).

During aging (extended storage at chilled temperatures), the first observable change in the ultrastructure of postmortem muscle is the disruption at Z disks and the fragmentation of myofibrils. As intermediate filaments surrounding the Z disk degrade, the tension developed as the result of rigor mortis causes myofibrils to break at the Z disk-I band junction, producing fragments of myofibrils (Geesink and Koohmaraie 1999). At this point, muscle regains softness and extensibility, and meat tenderness increases. The progressive loss of Z disk structure that occurs during aging is due to proteolytic degradation of specific myofibrillar proteins within this structure, primarily desmin and titin, not due to the hydrolysis of α-actinin, which is the major component of Z disk and quite stable (Taylor et al. 1995). In addition, several other myofibrillar proteins are also degraded in the aging process of meat, which include nebulin and TnT (Hopkins and Thompson 2002).

There are three protease systems proposed to be responsible for postmortem proteolysis, of which the caplain system is the major player (Figure 3.1). The involvement of cathepsins and caspases in postmortem proteolysis remains controversial.

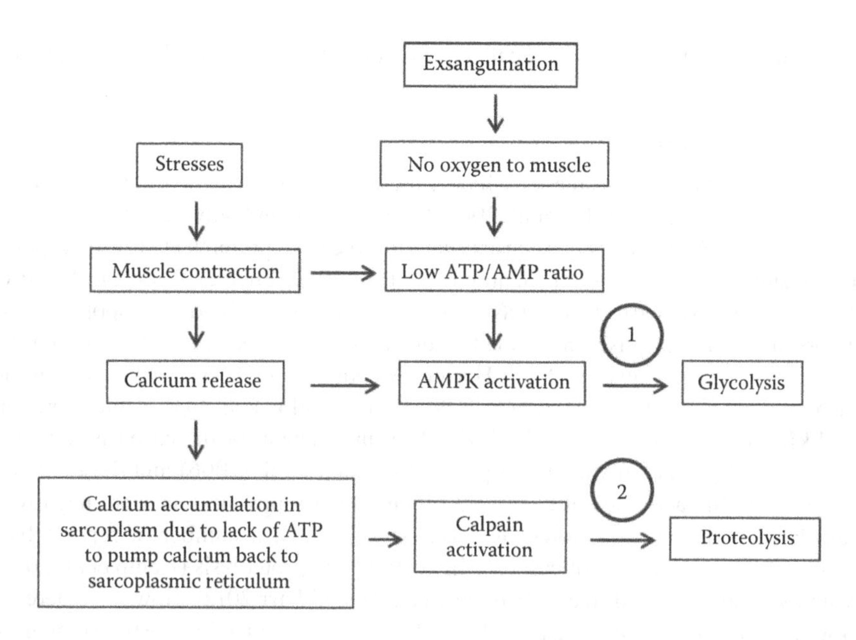

FIGURE 3.1 Postmortem glycolysis and proteolysis are major changes in postmortem muscle which impact meat quality. (1) Glycolysis: preslaughter stress activates AMPK to accelerate glycolysis, which may lead to PSE pork; excessive electric stimulation in beef carcass may overly activate AMPK and generate PSE-like meat; on the other hand, long-term stress which depletes glycogen leads to DFD meat. (2) μ-Calpain is mainly responsible for postmortem proteolysis and meat tenderization; besides calpains, cathepsin and caspases may also involve in postmortem proteolysis but need to be further studied.

3.2.4.1 Caplain System

The calpain/calpastatin system is composed of calpain I, calpain II, and calpastatin, which has been credited as the proteolytic system responsible for postmortem degradation of myofibrillar proteins (Koohmaraie et al. 1991; Uytterhaegen et al. 1994; Goll et al. 2003), though calpain and calpastatin only explain 30% of the variability in tenderness (Morgan et al. 1993; Koohmaraie 1995). The remaining variation could be due to the background toughness contributed by collagen and connective tissues (Du et al. 2013). The calpain system has been extensively reviewed (Koohmaraie 1992, 1994; Goll et al. 2003; Goll et al. 2008; Hopkins and Geesink 2009) and thus is not discussed in detail here.

3.2.4.2 Cathepsins

Cathepsins are located inside lysosomes, which have an acidic environment. As a result, cathepsins are not active in the cytoplasm of cells, and thus, cathepsins are unlikely to contribute significantly to postmortem proteolysis (Goll et al. 2008). However, in meat products with long aging periods such as dry-cured country style hams, these cathepsins are released from lysosomes and the fermentation reduces meat pH, both of which makes cathepsins, including cathepsin B and L, major contributors to proteolysis of meat proteins (Toldra and Flores 1998; Sentandreu et al. 2007).

3.2.4.3 Caspases

Caspases are cysteine proteases which function in apoptosis or programmed cell death (Suzuki et al. 2001; Du et al. 2004; Fuentes-Prior and Salvesen 2004; Herrera-Mendez et al. 2006). These proteases function to cleave proteins at specific aspartic acid residues upon activation (Sentandreu et al. 2002; Fuentes-Prior and Salvesen 2004; Herrera-Mendez et al. 2006). Caspases can be classified into apoptotic initiators including caspases 8, 9, or 10, and apoptotic effectors such as caspase 3 (Fuentes-Prior and Salvesen 2004; Herrera-Mendez et al. 2006). In skeletal muscle, caspase 3 exists in its pro-form which is inactive, and is activated by cleavage into a 14 kD caspase 3 (Turpin et al. 2006). The involvement of the caspase system in proteolytic degradation was first reported by Kemp et al. (2006) and these authors showed that the activities of caspase 3, 7, and 9 early postmortem were negatively correlated with the shear force in pork. Subsequently, a number of reports have observed the involvement of caspases in postmortem proteolysis (Kemp et al. 2009), which was summarized in a later review (Kemp and Parr 2012). However, there are other reports showing that caspase 3 is unlikely to be involved in postmortem aging (Underwood et al. 2008; Mohrhauser et al. 2011, 2013). A key question for the possible involvement of caspases in postmortem proteolysis is that the initiation of apoptosis requires a surge in the intracellular ATP level (Tsujimoto 1997). In postmortem muscle, this surge is unlikely to occur and thus is expected to follow the necrosis pathway (Tsujimoto 1997). Nevertheless, albeit the lack of active apoptosis process, the activation of individual apoptosis components remains possible, which likely contributes to postmortem proteolysis and tenderization (Kemp and Parr 2012).

3.2.5 Changes in Physical Properties of Muscle

WHC and tenderness are the two most important quality attributes of meat. Water accounts for 75% on average of total muscle mass. In living muscle fibers, much of the water is tightly bound to various proteins. Changes in water binding during the conversion of muscle to meat are largely dependent on the rate and extent of pH decline and the amount of protein denaturation (Offer and Trinick 1983). When the pH of postmortem muscle is high (pH > 6.5), the water-binding properties of meat are similar to those of living muscle. Studies carried out in the past decade propose that protein degradation postmortem influences the WHC of meat. Proteolysis of key muscle proteins (including desmin, vinculin, and talin) minimizes the loss of water-holding capacity caused by lateral shrinkage of myofibrils in postmortem muscle (Melody et al. 2004; Huff-Lonergan and Lonergan 2005). It is known that μ-calpains degrade intermediate filament proteins and costameric proteins (including desmin, vinculin, and talin) in postmortem muscle (Taylor et al. 1995; Goll et al. 2008). Thus it is suggested that factors regulating calpain activity impact the water-holding capacity of meat. These factors include calpastatin content, pH, and protein oxidation. A low pH early postmortem and oxidation of μ-calpain inhibit μ-calpain activity and its hydrolysis of muscle proteins (Rowe et al. 2004; Carlin et al. 2006).

Overall the enzymatic systems currently recognized to be involved in meat tenderization are primarily the calpains. Limited studies indicate that caspases, key mediators of cell apoptosis, contribute to the initiation of the conversion of muscle into meat (Ouali et al. 2006), but further studies are needed (Ouali et al. 2013).

Pigments are the most important contributors to meat color, mainly referring to hemoglobin and myoglobin. In well-bled muscle, myoglobin accounts for 80%–90% of the total pigment. Meat color is the first quality characteristic perceived by consumers when purchasing the meat, which is defined by the extent of myoglobin oxygenation and the oxidative status of the heme iron (Ponnampalam et al. 2012; Suman and Joseph 2013). In living animals, muscle with sufficient oxygen appears red. In postmortem muscle, as oxygen is used in metabolism during the early postmortem period, muscle becomes a dark purplish red due to deoxygenation of myoglobin. When fresh meat is cut, the exposed surface has a dark red appearance immediately after cutting, but after exposure to the atmosphere for a few minutes, the myoglobin becomes oxygenated and the surface changes to a brighter red color. Oxidation of the heme iron of myoglobin from the ferrous state to the ferric state makes meat look brown during storage as a result of metmyoglobin formation (Jeong et al. 2009).

Conversion of muscle into meat is complicated process and many aspects of this process are far from being well defined. Up to date, the conversion of muscle into meat is assumed to occur through three steps: the prerigor step, the rigor step, and the tenderization step. Many events, especially those at the first step, are still unclear, but have major impacts on the eating quality of meat because this step sets the basis for the conversion of muscle to meat.

3.3 POSTMORTEM GLYCOLYSIS AND MEAT QUALITY

Abnormal glycolysis in postmortem muscle can cause meat defects, including PSE, RSE, and DFD meat. Overall, excessive postmortem glycolysis causes PSE or acid meat, while abbreviated glycolysis results in DFD meat (Figure 3.2).

3.3.1 PSE Meat

Since the first description of the PSE condition in 1953 (Ludvigsen 1953), the PSE problem in meat has been recognized for decades, but remains unsolved. PSE is characterized by a low pH, a pale and exudative appearance, and a soft texture. Due to its inferior quality, PSE meat is not preferred by consumers. Intensive selection for lean growth and against fat deposition in pigs has contributed to the increased incidence of PSE meat (Solomon et al. 1998; Lee and Choi 1999). A PSE-like condition is also observed in beef and sheep carcasses, primarily associated with rapid pH decline coupled with high carcass temperature (Hopkins et al. 2014; Jacob and Hopkins 2014; Strydom and Rosenvold 2014).

RSE pork has a red color that consumers desire, but is soft and exudative, which is undesirable. RSE meat has a lower than normal ultimate pH, and thus, is commonly

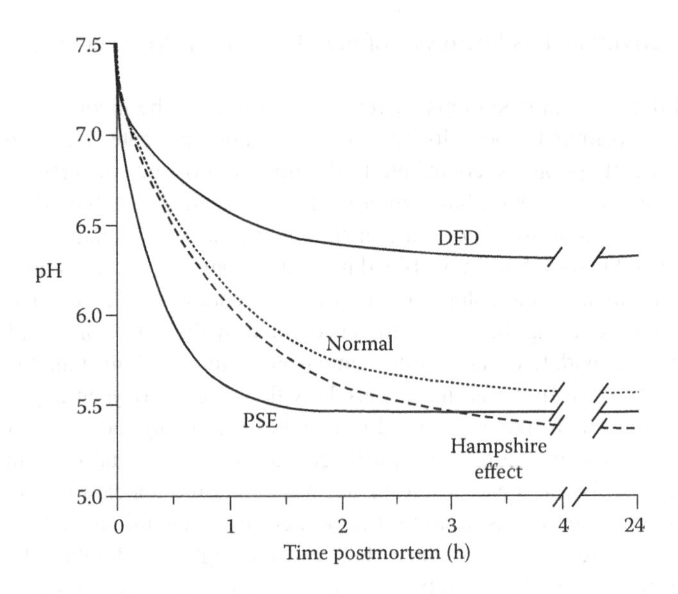

FIGURE 3.2 Postmortem pH decline curves. (Modified from Murray, 1995. *Quality and Grading of Carcasses of Meat Animals*, CRC Press. New York, pp. 234.)

referred as "acid meat." Acid meat has increased drip loss and therefore a lower water-holding capacity. Rendement Napole gene (*RN*$^-$) is believed to be the cause of RSE meat. When compared to noncarriers, drip loss and cooking loss increased by 21% and 12%, respectively, in meat of pigs carrying Napole gene (Lundstrom et al. 1996).

3.3.1.1 Mechanisms Leading to PSE Meat

Currently, it is well accepted that fast and/or excessive glycolysis and a buildup of lactic acid (a rapid pH drop) in early-stage postmortem muscle is the cause of the PSE syndrome in the meat of pigs, turkeys, and chicken, which is also related to the incidence of "acid meat" in Hampshire pigs (Enfalt et al. 1997; Hamilton et al. 2000; Moeller et al. 2003). Rapid glycolysis also generates significant amounts of heat with increasing carcass temperature. In combination with low pH early postmortem this temperature rise denatures muscle proteins, resulting in an undesirable pale color and low water-holding capacity of meat (Owens et al. 2000a,b; Sams and Alvarado 2004; Freise et al. 2005; Hopkins et al. 2014). Along with the conditions listed above, myosin denatures and shrinks, causing a reduction of the filament spacing. Thus, water is expelled from the cells and lost in the form of purge or drip (Offer et al. 1989). In porcine muscle that becomes PSE meat, the pH declines much faster than that in normal muscle, with a speed of 1.04 versus 0.65 units/hour at 37°C (Bendall et al. 1963). Similarly, a rapid pH decline when carcass temperature remains high is also observed in beef (Jacob and Hopkins 2014; Strydom and Rosenvold 2014). Therefore, slowing down glycolysis in early postmortem muscle is the key to solving the PSE problem in pork and similarly in beef (Jacob and Hopkins 2014).

3.3.1.2 PSE Meat Induce a Huge Loss to the Meat Industry

PSE meat has a high drip loss, a low cooking yield, and a tough texture after cooking (Golding-Myers et al. 2010). Due to the inferior quality, the incidence of PSE and RSE meat causes a huge economic loss to animal slaughter and the meat processing industry (Kauffman et al. 1992; Woelfel et al. 2002; Kim et al. 2014). One of the major reasons for the increased incidence of PSE pork is intensive selection for rapid lean growth, which results in an increase in the number of fast-twitch glycolytic and a concomitant reduction in slow-twitch oxidative muscle fibers. Glycolytic muscle fibers are more efficient in growth and thus, the high ratio of glycolytic fibers enhance lean growth, but the glycolytic nature of energy metabolism in these fibers increases PSE incidence (Vanhoof 1979; Golding-Myers et al. 2010).

Recently, PSE-like conditions were reported in beef and sheep carcasses, primarily due to elevated carcass temperature, accompanied by fast pH decline (Strydom and Rosenvold 2014). Therefore, electric stimulation, which can further increase carcass temperature up to 4°C (Davey and Pham 1997), is expected to worsen this problem. However, the rapid formation of actomyosin due to electrical stimulation may protect myosin from denaturation, which might actually protect beef quality (Rosenvold et al. 2008). On the other hand, high temperature denatures μ-calpain, which is important for postmortem proteolysis and impairs postmortem tenderization (Kim et al. 2012). There is clearly a need to further elucidate how stimulation might interact with a propensity for a fast pH decline.

3.3.2 DFD Meat

DFD meat occurs mainly in beef, which is more commonly referred to as dark cutting beef. Although it was recognized more than 50 years ago (Munns and Burrell 1965), DFD meat is still a significant cause of financial loss to the meat industry. DFD is associated with a high pH in postmortem muscle, which occurs when the pH postmortem measured after 12–48 h remains ≥6 (Adzitey and Nurul 2011). A pH value of 5.87 in the *longissimus* muscle is the approximate cutoff between normal and DFD (Page et al. 2001), though DFD may occur at even lower pH. At high pH, the respiration of mitochondria is maintained, which consumes oxygen and causes the formation of deoxy-myoglobin, resulting in dark red color (Ashmore et al. 1971, 1973). The high pH in postmortem muscle of DFD is due to glycogen deficiency in the muscle (Mcveigh et al. 1982; Warriss et al. 1984). It appears that a glycolytic potential at 100 μmol/g muscle is the threshold for the occurrence of DFD (Wulf et al. 2002). Factors influencing glycogen content in muscle at slaughter directly affect the incidence of DFD (Kenny and Tarrant 1988; Bartos et al. 1993; Kreikemeier et al. 1998; Scanga et al. 1998). Therefore, to solve DFD, it is necessary to increase glycogen content in muscle by reducing stresses before slaughter (see Chapter 2). In addition, nutritional supplements can also increase muscle glycogen content (Lowe et al. 2002; Dellar et al. 2006; Knee et al. 2007). Besides this, the muscle fiber composition of beef cattle affects DFD incidence (Zerouala and Stickland 1991) and, thus, genetic selection for low oxidative muscle fiber ratio may be another approach to reduce DFD incidence.

3.3.3 AMPK Regulates Glycolysis and Glycogen Accumulation in Muscle

3.3.3.1 AMPK, Postmortem Glycolysis, and PSE Meat

AMPK is a serine/threonine kinase composed of a catalytic subunit (α) and two regulatory subunits (β and γ). The key function of AMPK is to regulate the energy balance within cells, and is referred to as the intracellular energy sensor and the master of energy metabolic switch. AMPK is activated by an increase in AMP/ATP ratio (Corton et al. 1994). Once activated, AMPK then phosphorylates downstream substrates and the biological effects of AMPK are pleiotropic: switching off ATP-consuming processes such as biosynthetic pathways, and switching on catabolic processes to generate ATP; including glycolysis and β-oxidation of fatty acids (Hardie et al. 2003, 2006; Carling 2004). AMPK is activated in ischemic heart and hypoxic skeletal muscle. Activated AMPK increases glycolysis *in vivo* by two signaling pathways: (1) AMPK phosphorylates and activates phosphorylase kinase, which then phosphorylates and activates glycogen phosphorylase (GP), an enzyme controlling glycogenelysis and catalyzing the production of substrate for glycolysis (Young et al. 1997; Fraser et al. 1999; Russell et al. 1999); (2) AMPK phosphorylates and activates phosphofructokinase-2 (PFK-2) (Marsin et al. 2000). PFK-2 catalyzes the production of fructose-2, 6-biphosphate, a potent allosteric activator of phosphofructokinase-1 (PFK-1). PFK-1 is the most important rate-controlling enzyme in glycolysis. *In vitro* phosphorylation of PFK-2 by purified AMPK further confirmed a direct involvement of AMPK in increasing glycolysis (Marsin et al. 2000). Indeed, AMPK was activated earlier and AMPK activity was much higher early postmortem, especially within half an hour after exsanguination, in the muscle that became PSE meat (Shen et al. 2006b). Preslaughter stress as well as AMPK activity in pork loin early postmortem (Shen et al. 2006a) increased the risk of PSE meat. Activation of AMPK led to a faster pH decline and lower ultimate pH values in postmortem muscle (Shen and Du 2005). Conversely, inhibition of AMPK by compound C resulted in lower AMPK activity, a higher pH value, and less lactic acid accumulation in postmortem muscle when compared to controls (Shen et al. 2008). Consistently, in the absence of AMPK activity, the glycolytic rate was dramatically reduced (Shen and Du 2005; Liang et al. 2013). All these data clearly demonstrate that AMPK regulates postmortem glycolysis in muscle and suggest that PSE meat might be prevented by inhibiting AMPK activity (Figure 3.1).

3.3.3.2 AMPK Regulation of Glycogen Accumulation in Muscle and Meat Quality

As an energy sensor that regulates cellular metabolism, AMPK controls whole-body glucose homeostasis by regulating metabolism in multiple peripheral tissues, such as skeletal muscle, liver, adipose tissues, and pancreatic β cells. AMPK activity is in parallel with increased glucose uptake (Hutber et al. 1997). AMPK phosphorylates Akt substrate 160 (AS160/TBC1D1), promoting its association with the 14-3-3 scaffolding protein which promotes glucose transporter 4 (GLUT4) translocation to plasma membranes and glucose uptake in skeletal muscle (Geraghty et al. 2007; Taylor et al. 2008).

Glycogen is the major source of stored glucose within cells. The dynamics of stored glycogen is determined by the action of glycogen synthase (GS) and GP which catalyzes glycogen synthesis and degradation, respectively. It has long been recognized that AMPK phosphorylates muscle isoforms of GS at Ser7, leading to its inactivation (Carling and Hardie 1989). This phosphorylation was confirmed later *in vivo* by others (Jorgensen et al. 2004). Further, a body of literature has reported that GS activity decreases in response to acute 5-aminoimidazole-4-carboxamide-1-β-4-ribofuranoside (AICAR) treatment in cultured muscle-like cells (Halse et al. 2003), isolated and perfused skeletal muscle (Aschenbach et al. 2002; Wojtaszewski et al. 2002; Miyamoto et al. 2007), and fast twitch, but not slow twitch, muscle *in vivo* (Aschenbach et al. 2002), indicating that AMPK activation may, under some conditions, decrease glycogen synthesis. While AMPK activation acutely inhibits glycogen synthesis, promotes glycogenolysis and glycolysis as stated above, it concomitantly promotes glucose uptake (Aschenbach et al. 2002). Glucose transported into cells is rapidly converted to glucose-6-phosphate (G6P) by hexokinase. A gross accumulation of G6P can allosterically activate GS to such an extent to override the phosphate's inhibition (Aschenbach et al. 2002). Therefore, an acute activation of AMPK depletes glycogen through its action on GS and GP, but a chronic activation can lead to accelerated glycogen storage by the increased levels of G6P within cells.

AMPK regulation of glycogen storage within muscle is clearly evidenced by AMPK mutant animals. One of the best examples is RN^- pigs. Pigs carrying the RN^- gene have 70% more glycogen storage in skeletal muscle and have a high incidence of acid meat. The "Hampshire effect" in RN^- pigs results from a point mutation, R200Q mutation, in AMPK γ3 subunit (Milan et al. 2000). This mutation increases AMPK activity and enhances muscle glucose uptake, leading to enhanced glycogen accumulation (Milan et al. 2000; Andersson 2003; Yu et al. 2006; Sanders et al. 2007). In humans, a homologous mutation has been found in genetic studies of the lean and obese human population. The R225W mutation on the human γ3 subunit results in increased glucose uptake in exercised muscle, increased glycogen storage and decreased triglyceride levels in muscle, and a resistance to muscle fatigue (Costford et al. 2007; Crawford et al. 2010). Much like the mutation found in RN^- pigs, the R225W mutation increases basal activity of AMPK in human skeletal muscle (Costford et al. 2007). Furthermore, a point mutation (R302Q) in γ2 subunit, the highly expressed isoform in cardiac muscle, also results in amplified glycogen accumulation in the heart (Gollob et al. 2001; Gollob 2003). AMPK has two α catalytic subunits, α1 and α2. It was recently showed that AMPKα2, not α1, regulates postmortem glycolysis and meat quality (Liang et al. 2013).

3.3.4 Factors Related to Abnormal Postmortem Glycolysis

Improper handling leads to PSE or DFD meats. The most common factor leading to both PSE and DFD in meat is stress antemortem. Generally speaking, acute stress just before slaughtering leads to PSE whereas chronic or long-term stress before slaughter increases DFD occurrence (see Chapter 2). There is no single method that exists for eliminating PSE pork from stress-susceptible pigs, however minimization of stress before or during slaughter reduces PSE occurrence. Resting after transportation reduces the incidence of PSE loins (Fortin 2002). Proper postslaughter handling, such as rapid chilling, reduces the frequency of PSE meat, because low temperature decreases glycolytic enzyme activity and glycolytic rate postmortem, as well as slows protein denaturation (Sayre et al. 1963; Briskey 1964; Bendall and Swatland 1988; Solomon et al. 1998; Lee and Choi 1999). Similarly, in beef and lamb, reducing carcass temperature is critical for preventing the PSE-like condition (Kim et al. 2014; Strydom and Rosenvold 2014). However, unlike pig carcasses, rapid chilling may induce cold shortening in beef and lamb carcasses, which increase toughness (Jacob and Hopkins 2014).

3.4 CONCLUSIONS

The conversion of muscle to meat is a complex process. Major postmortem changes include glycolysis and proteolysis, both of which dramatically affect the quality of meat (Figure 3.1). Proteolysis in postmortem muscle is highly correlated with beef tenderness. Calpains, especially μ-calpains, are mainly responsible for postmortem proteolysis and meat tenderization. Cathepsins, proteases located inside lysosomes, are unlikely a major player in postmortem proteolysis, but have critical roles in the proteolysis of muscle proteins in dry cured hams and other meat products requiring long aging durations. Limited data suggest caspases may be involved in postmortem proteolysis, but there are also contradictory reports and more studies are needed to analyze their role in postmortem proteolysis.

For pigs, the incidence of PSE meat is a major quality problem; PSE occurs due to the rapid glycolysis shortly after harvesting. Recent studies show a PSE-like condition in beef and sheep, which is primarily associated with high carcass temperature and accelerated pH decline. AMPK regulates postmortem glycolysis and pH decline, thus having a critical role in the occurrence of PSE meat. AMPK is activated due to preslaughter stresses, which initiates glycolysis before or right after slaughter, eliciting rapid glycolysis and heat generation in postmortem muscle. Therefore, proper preslaughter handling which reduces stress to animals, and rapid postmortem chilling which reduces glycolytic rate, can effectively reduce PSE incidence. Though the PSE-like condition can occur in beef carcasses due to high carcass temperature and excessive electric stimulation, the incidence of DFD meat is the major concern in beef.

REFERENCES

Adzitey, F., and Nurul, H. 2011. Pale soft exudative (PSE) and dark firm dry (DFD) meats: Causes and measures to reduce these incidences—A mini review. *Int. Food Res. J.* 18:11–20.

Andersson, L. 2003. Identification and characterization of AMPK gamma 3 mutations in the pig. *Biochem. Soc. Trans.* 31:232–235.

Aschenbach, W.G., Hirshman, M.F., Fujii, N., Sakamoto, K., Howlett, K.F., and Goodyear, L.J. 2002. Effect of AICAR treatment on glycogen metabolism in skeletal muscle. *Diabetes* 51:567–573.

Ashmore, C.R., Carroll, F., Doerr, L., Tompkins, G., Stokes, H., and Parker, W. 1973. Experimental prevention of dark-cutting meat. *J. Anim. Sci.* 36:33–36.

Ashmore, C.R., Doerr, L., Foster, G., and Carroll, F. 1971. Respiration of mitochondria isolated from dark-cutting beef. *J. Anim. Sci.* 33:574–577.

Bartos, L., Franc, C., Rehak, D., and Stipkova, A. 1993. A practical method to prevent dark-cutting (Dfd) in beef. *Meat Sci.* 34:275–282.

Bendall, J.R., Hallund, O., and Wismer-Penersen, J. 1963. Postmortem changes in the muscles of Landrace pigs. *J. Food Sci.* 28:156–162.

Bendall, J.R., and Swatland, H.J. 1988. A review of the relationship of pH with physical aspects of pork quality. *Meat Sci.* 24:85–96.

Bowker, B.C., Swartz, D.R., Grant, A.L., and Gerrard, D.E. 2005. Myosin heavy chain isoform composition influences the susceptibility of actin-activated S1 ATPase and myofibrillar ATPase to pH inactivation. *Meat Sci.* 71:342–350.

Briskey, E.J. 1964. Etiological status and associated studies of pale, soft, exudative porcine musculature. *Adv. Food Res.* 13:89–178.

Briskey, E.J., and Wismer-Pedersen, J. 1961. Biochemistry of pork muscle structure. I. Rate of anaerobic glycolysis and temperature change versus the apparent structure of muscle tissue. *J. Food Sci.* 26:297–305.

Carlin, K.R., Huff-Lonergan, E., Rowe, L.J., and Lonergan, S.M. 2006. Effect of oxidation, pH, and ionic strength on calpastatin inhibition of mu- and m-calpain. *J Anim. Sci.* 84: 925–937.

Carling, D. 2004. The AMP-activated protein kinase cascade—A unifying system for energy control. *Trends Biochem. Sci.* 29:18–24.

Carling, D., and Hardie, D.G. 1989. The substrate and sequence specificity of the AMP-activated protein kinase. Phosphorylation of glycogen synthase and phosphorylase kinase. *Biochim. Biophys. Acta* 1012:81–86.

Copenhafer, T.L., Richert, B.T., Schinckel, A.P., Grant, A.L., and Gerrard, D.E. 2006. Augmented postmortem glycolysis does not occur early postmortem in AMPKgamma3-mutated porcine muscle of halothane positive pigs. *Meat Sci.* 73:590–599.

Corton, J.M., Gillespie, J.G., and Hardie, D.G. 1994. Role of the AMP-activated protein kinase in the cellular stress response. *Curr. Biol.* 4:315–324.

Costford, S.R., Kavaslar, N., Ahituv, N., Chaudhry, S.N., Schackwitz, W.S., Dent, R., Pennacchio, L.A., McPherson, R., and Harper, M.E. 2007. Gain-of-function R225W mutation in human AMPK gamma(3) causing increased glycogen and decreased triglyceride in skeletal muscle. *PLoS One* 2:e903.

Crawford, S.A., Costford, S.R., Aguer, C. et al. 2010. Naturally occurring R225W mutation of the gene encoding AMP-activated protein kinase (AMPK) gamma(3) results in increased oxidative capacity and glucose uptake in human primary myotubes. *Diabetologia* 53:1986–1997.

Davey, L.M., and Pham, Q.T. 1997. Predicting the dynamic product heat load and weight loss during beef chilling using a multi-region finite difference approach. *Int. J. Refrig.* 20:470–482.

Dellar, S.J., Accioly, J.M., McIntyre, B.L., Williams, I., and Pethick, D.W. 2006. Effect of lupin supplementation and phenotypic characteristics on the performance of pastoral cattle grazing tagasaste. *Aust. J. Exp. Agric.* 46:947–950.

DiGiacomo, K., Leury, B.J., and Dunshea, F.R. 2014. Potential nutritional strategies for the amelioration or prevention of high rigor temperature in cattle—A review. *Anim. Prod. Sci.* 54:430–443.

Du, J., Wang, X.N., Miereles, C., Bailey, J.L., Debigare, R., Zheng, B., Price, S.R., and Mitch, W.E. 2004. Activation of caspase-3 is an initial step triggering accelerated muscle proteolysis in catabolic conditions. *J. Clin. Invest.* 113:115–123.

Du, M., Huang, Y., Das, A.K., Yang, Q., Duarte, M.S., Dodson, M.V., and Zhu, M.J. 2013. Meat science and muscle biology symposium: Manipulating mesenchymal progenitor cell differentiation to optimize performance and carcass value of beef cattle. *J. Anim. Sci.* 91:1419–1427.

Enfalt, A.C., Lundstrom, K., Hansson, I., Johansen, S., and Nystrom, P.E. 1997. Comparison of non-carriers and heterozygous carriers of the *RN*⁻ allele for carcass composition, muscle distribution and technological meat quality in Hampshire-sired pigs. *Livest. Prod. Sci.* 47:221–229.

Estrade, M., Vignon, X., Rock, E., and Monin, G. 1993. Glycogen hyperaccumulation in white muscle fibres of *RN*⁻ carrier pigs. A biochemical and ultrastructural study. *Comp. Biochem. Physiol. B* 104:321–326.

Fortin, A. 2002. The effect of transport time from the assembly yard to the abattoir and resting time at the abattoir on pork quality. *Can. J. Anim. Sci.* 82:141–150.

Fraser, H., Lopaschuk, G.D., and Clanachan, A.S. 1999. Alteration of glycogen and glucose metabolism in ischaemic and post-ischaemic working rat hearts by adenosine A(1) receptor stimulation. *Br. J. Pharmcol.* 128:197–205.

Freise, K., Brewer, S., and Novakofski, J. 2005. Duplication of the pale, soft, and exudative condition starting with normal postmortem pork. *J. Anim. Sci.* 83:2843–2852.

Fuentes-Prior, P., and Salvesen, G.S. 2004. The protein structures that shape caspase activity, specificity, activation and inhibition. *Biochem. J.* 384:201–232.

Fujii, J., Otsu, K., Zorzato, F., de Leon, S., Khanna, V.K., Weiler, J.E., O'Brien, P.J., and MacLennan, D.H. 1991. Identification of a mutation in porcine ryanodine receptor associated with malignant hyperthermia. *Science* 253:448–451.

Geesink, G.H., and Koohmaraie, M. 1999. Postmortem proteolysis and calpain/calpastatin activity in callipyge and normal lamb biceps femoris during extended postmortem storage. *J. Anim. Sci.* 77:1490–1501.

Geraghty, K.M., Chen, S., Harthill, J.E., Ibrahim, A.F., Toth, R., Morrice, N.A., Vandermoere, F., Moorhead, G.B., Hardie, D.G., and MacKintosh, C. 2007. Regulation of multisite phosphorylation and 14-3-3 binding of AS160 in response to IGF-1, EGF, PMA and AICAR. *Biochem. J.* 407:231–241.

Golding-Myers, J.D., Showers, C.D., Shand, P.J., and Rosser, B.W.C. 2010. Muscle fiber type and the occurrence of pale, soft, exudative pork. *J. Muscle Foods* 21:484–498.

Goll, D.E., Neti, G., Mares, S.W., and Thompson, V.F. 2008. Myofibrillar protein turnover: The proteasome and the calpains. *J. Anim. Sci.* 86:E19–E35.

Goll, D.E., Thompson, V.F., Li, H., Wei, W., and Cong, J. 2003. The calpain system. *Physiol. Rev.* 83:731–801.

Gollob, M.H. 2003. Glycogen storage disease as a unifying mechanism of disease in the PRKAG2 cardiac syndrome. *Biochem. Soc. Trans.* 31:228–231.

Gollob, M.H., Seger, J.J., Gollob, T.N., Tapscott, T., Gonzales, O., Bachinski, L., and Roberts, R. 2001. Novel PRKAG2 mutation responsible for the genetic syndrome of ventricular preexcitation and conduction system disease with childhood onset and absence of cardiac hypertrophy. *Circulation* 104:3030–3033.

Halse, R., Fryer, L.G., McCormack, J.G., Carling, D., and Yeaman, S.J. 2003. Regulation of glycogen synthase by glucose and glycogen: A possible role for AMP-activated protein kinase. *Diabetes* 52:9–15.

Hamilton, D.N., Ellis, M., Miller, K.D., McKeith, F.K., and Parrett, D.F. 2000. The effect of the Halothane and Rendement Napole genes on carcass and meat quality characteristics of pigs. *J. Anim. Sci.* 78:2862–2867.

Hardie, D.G., Hawley, S.A., and Scott, J. 2006. AMP-activated protein kinase—Development of the energy sensor concept. *J. Physiol.* 574:7–15.

Hardie, D.G., Scott, J.W., Pan, D.A., and Hudson, E.R. 2003. Management of cellular energy by the AMP-activated protein kinase system. *FEBS Lett.* 546:113–120.

Herrera-Mendez, C.H., Becila, S., Boudjellal, A., and Ouali, A. 2006. Meat ageing: Reconsideration of the current concept. *Trends Food Sci. Technol.* 17:394–405.

Hopkins, D.L., and Geesink, G.H. 2009. Protein degradation post mortem and tenderisation. In: M. Du and J.R. McCormick (eds.), *Applied Muscle Biology and Meat Science.* CRC Press, Taylor & Francis Group, USA, pp. 149–173.

Hopkins, D.L., Ponnampalam, E.N., van de Ven, R.J., and Warner, R.D. 2014. The effect of pH decline rate on the meat and eating quality of beef carcasses. *Anim. Prod. Sci.* 54:407–413.

Hopkins, D.L., and Thompson, J.M. 2002. The degradation of myofibrillar proteins in beef and lamb using denaturing electrophoresis—An overview. *J. Muscle Foods* 13:81–102.

Huff-Lonergan, E., and Lonergan, S.M. 2005. Mechanisms of water-holding capacity of meat: The role of postmortem biochemical and structural changes. *Meat Sci.* 71:194–204.

Hutber, C.A., Hardie, D.G., and Winder, W.W. 1997. Electrical stimulation inactivates muscle acetyl-CoA carboxylase and increases AMP-activated protein kinase. *Am. J. Physiol.* 272:E262–E266.

Jacob, R.H., and Hopkins, D.L. 2014. Techniques to reduce the temperature of beef muscle early in the post mortem period—A review. *Anim. Prod. Sci.* 54:482–493.

Jeong, J.Y., Hur, S.J., Yang, H.S., Moon, S.H., Hwang, Y.H., Park, G.B., and Joo, S.T. 2009. Discoloration characteristics of 3 major muscles from cattle during cold storage. *J. Food Sci.* 74:C1–C5.

Jorgensen, S.B., Nielsen, J.N., Birk, J.B. et al. 2004. The alpha2-5'AMP-activated protein kinase is a site 2 glycogen synthase kinase in skeletal muscle and is responsive to glucose loading. *Diabetes* 53:3074–3081.

Kauffman, R.G., Cassens, R.G., Scherer, A., and Meeker, D.L. 1992. *Variations in Pork Quality.* National Pork Producers Council Bulletin, Des Moines, IA.

Kemp, C.M., Bardsley, R.G., and Parr, T. 2006. Changes in caspase activity during the postmortem conditioning period and its relationship to shear force in porcine longissimus muscle. *J. Anim. Sci.* 84:2841–2846.

Kemp, C.M., King, D.A., Shackelford, S.D., Wheeler, T.L., and Koohmaraie, M. 2009. The caspase proteolytic system in callipyge and normal lambs in longissimus, semimembranosus, and infraspinatus muscles during postmortem storage. *J. Anim. Sci.* 87: 2943–2951.

Kemp, C.M., and Parr, T. 2012. Advances in apoptotic mediated proteolysis in meat tenderisation. *Meat Sci.* 92:252–259.

Kenny, F.J., and Tarrant, P.V. 1988. The effect of estrus behavior on muscle glycogen concentration and dark-cutting in beef heifers. *Meat Sci.* 22:21–31.

Kim, Y.H.B., Stuart, A., Nygaard, G., and Rosenvold, K. 2012. High pre rigor temperature limits the ageing potential of beef that is not completely overcome by electrical stimulation and muscle restraining. *Meat Sci.* 91:62–68.

Kim, Y.H.B., Warner, R.D., and Rosenvold, K. 2014. Influence of high pre-rigor temperature and fast pH fall on muscle proteins and meat quality: A review. *Anim. Prod. Sci.* 54:375–395.

Knee, B.W., Cummins, L.J., Walker, P.J., Kearney, G.A., and Warner, R.D. 2007. Reducing dark-cutting in pasture-fed beef steers by high-energy supplementation. *Aust. J. Exp. Agric.* 47:1277–1283.

Koohmaraie, M. 1992. The role of Ca(2+)-dependent proteases (calpains) in post mortem proteolysis and meat tenderness. *Biochimie* 74:239–245.

Koohmaraie, M. 1994. Muscle proteinases and meat aging. *Meat Sci.* 36:93–104.

Koohmaraie, M. 1995. The biological basis of meat tenderness and potential genetic approaches for its control and prediction. In: *Reciprocal Meat Conference,* San Antonio, TX, pp. 69–76.

Koohmaraie, M., Whipple, G., Kretchmar, D.H., Crouse, J.D., and Mersmann, H.J. 1991. Postmortem proteolysis in longissimus muscle from beef, lamb and pork carcasses. *J. Anim. Sci.* 69:617–624.

Kreikemeier, K.K., Unruh, J.A., and Eck, T.P. 1998. Factors affecting the occurrence of dark-cutting beef and selected carcass traits in finished beef cattle. *J. Anim. Sci.* 76:388–395.

Lee, Y.B., and Choi, Y.I. 1999. PSE (pale, soft, exudative) pork: The causes and solutions—Review. *Asian-Australas. J. Anim. Sci.* 12:244–252.

Liang, J., Yang, Q., Zhu, M.J., Jin, Y., and Du, M. 2013. AMP-activated protein kinase (AMPK) alpha2 subunit mediates glycolysis in postmortem skeletal muscle. *Meat Sci.* 95:536–541.

Lowe, T.E., Peachey, B.M., and Devine, C.E. 2002. The effect of nutritional supplements on growth rate, stress responsiveness, muscle glycogen and meat tenderness in pastoral lambs. *Meat Sci.* 62:391–397.

Ludvigsen, J. 1953. Muscular degeneration in pigs. *Int. Vet. Congre. 15th Congr.* Stockholm, Sweden. 1:602.

Lundstrom, K., Andersson, A., and Hansson, I. 1996. Effect of the RN gene on technological and sensory meat quality in crossbred pigs with Hampshire as terminal sire. *Meat Sci.* 42:145–153.

Marsin, A.S., Bertrand, L., Rider, M.H., Deprez, J., Beauloye, C., Vincent, M.F., Van den Berghe, G., Carling, D., and Hue, L. 2000. Phosphorylation and activation of heart PFK-2 by AMPK has a role in the stimulation of glycolysis during ischaemia. *Curr. Biol.* 10:1247–1255.

Mcveigh, J.M., Tarrant, P.V., and Harrington, M.G. 1982. Behavioral stress and skeletal-muscle glycogen-metabolism in young bulls. *J. Anim. Sci.* 54:790–795.

Melody, J.L., Lonergan, S.M., Rowe, L.J., Huiatt, T.W., Mayes, M.S., and Huff-Lonergan, E. 2004. Early postmortem biochemical factors influence tenderness and water-holding capacity of three porcine muscles. *J. Anim. Sci.* 82:1195–1205.

Milan, D., Jeon, J.T., Looft, C. et al. 2000. A mutation in PRKAG3 associated with excess glycogen content in pig skeletal muscle. *Science* 288:1248–1251.

Miyamoto, L., Toyoda, T., Hayashi, T. et al. 2007. Effect of acute activation of 5′-AMP-activated protein kinase on glycogen regulation in isolated rat skeletal muscle. *J. Appl. Physiol.* 102:1007–1013.

Moeller, S.J., Baas, T.J., Leeds, T.D., Emnett, R.S., and Irvin, K.M. 2003. Rendement Napole gene effects and a comparison of glycolytic potential and DNA genotyping for classification of rendement Napole status in Hampshire-sired pigs. *J. Anim. Sci.* 81:402–410.

Mohrhauser, D.A., Kern, S.A., Underwood, K.R., and Weaver, A.D. 2013. Caspase-3 does not enhance *in vitro* bovine myofibril degradation by mu-calpain. *J. Anim. Sci.* 91:5518–5524.

Mohrhauser, D.A., Underwood, K.R., and Weaver, A.D. 2011. *In vitro* degradation of bovine myofibrils is caused by μ-calpain, not caspase-3. *J. Anim. Sci.* 89:798–808.

Morgan, J.B., Wheeler, T.L., Koohmaraie, M., Savell, J.W., and Crouse, J.D. 1993. Meat tenderness and the calpain proteolytic system in longissimus muscle of young bulls and steers. *J. Anim. Sci.* 71:1471–1476.

Munns, W.O., and Burrell, D.E. 1965. Use of rib-eye Ph for detecting dark-cutting beef. *Food Technol.* 19:1432–1434.

Murray, 1995. The evaluation of muscle quality. In M. Jones (ed.), *Quality and Grading of Carcasses of Meat Animals*, CRC Press. New York, pp. 234.

Nogueira, L., Shiah, A.A., Gandra, P.G., and Hogan, M.C. 2013. Ca2+-pumping impairment during repetitive fatiguing contractions in single myofibers: Role of cross-bridge cycling. *Am. J. Physiol.-Reg. Int. Comp. Physiol.* 305:R118–R125.

Offer, G., Knight, P., Jeacocke, R., Almond, R., Cousins, T., Elsey, J., Parsons, N., Sharp, A., Starr, R., and Purslow, P. 1989. The structural basis of the water-holding, appearance and toughness of meat and meat-products. *Food Microstruct.* 8:151–170.

Offer, G., and Trinick, J. 1983. On the mechanism of water holding in meat: The swelling and shrinking of myofibrils. *Meat Sci.* 8:245–281.

Ouali, A., Gagaoua, M., Boudida, Y., Becila, S., Boudjellal, A., Herrera-Mendez, C.H., and Sentandreu, M.A. 2013. Biomarkers of meat tenderness: Present knowledge and perspectives in regards to our current understanding of the mechanisms involved. *Meat Sci.* 95:854–870.

Ouali, A., Herrera-Mendez, C.H., Coulis, G., Becila, S., Boudjellal, A., Aubry, L., and Sentandreu, M.A. 2006. Revisiting the conversion of muscle into meat and the underlying mechanisms. *Meat Sci.* 74:44–58.

Owens, C.M., Hirschler, E.M., McKee, S.R., Martinez-Dawson, R., and Sams, A.R. 2000a. The characterization and incidence of pale, soft, exudative turkey meat in a commercial plant. *Poult. Sci.* 79:553–558.

Owens, C.M., McKee, S.R., Matthews, N.S., and Sams, A.R. 2000b. The development of pale, exudative meat in two genetic lines of turkeys subjected to heat stress and its prediction by halothane screening. *Poult. Sci.* 79:430–435.

Page, J.K., Wulf, D.M., and Schwotzer, T.R. 2001. A survey of beef muscle color and pH. *J.Anim. Sci.* 79:678–687.

Pardi, C.P., Santos, I.F., Souza, E.R., and Pardi, H.S. 1993. Fundamentals of science of meat. In: *Science and Technology of Meat Hygiene*. EDUFF, Goiânia, Chapter 2, pp. 35–127.

Ponnampalam, E.N., Butler, K.L., McDonagh, M.B., Jacobs, J.L., and Hopkins, D.L. 2012. Relationship between muscle antioxidant status, forms of iron, polyunsaturated fatty acids and functionality (retail colour) of meat in lambs. *Meat Sci.* 90:297–303.

Rosenvold, K., North, M., Devine, C., Micklander, E., Hansen, P., Dobbie, P., and Wells, R. 2008. The protective effect of electrical stimulation and wrapping on beef tenderness at high pre rigor temperatures. *Meat Sci.* 79:299–306.

Rowe, L.J., Maddock, K.R., Lonergan, S.M., and Huff-Lonergan, E. 2004. Oxidative environments decrease tenderization of beef steaks through inactivation of mu-calpain. *J. Anim. Sci.* 82:3254–3266.

Russell, R.R., Bergeron, R., Shulman, G.I., and Young, L.H. 1999. Translocation of myocardial GLUT-4 and increased glucose uptake through activation of AMPK by AICAR. *Am. J. Physiol.-Heart Circ. Physiol.* 277:H643–H649.

Sams, A.R., and Alvarado, C.Z. 2004. Turkey carcass chilling and protein denaturation in the development of pale, soft, and exudative meat. *Poult. Sci.* 83:1039–1046.

Sanders, M.J., Grondin, P.O., Hegarty, B.D., Snowden, M.A., and Carling, D. 2007. Investigating the mechanism for AMP activation of the AMP-activated protein kinase cascade. *Biochem. J.* 403:139–148.

Sayre, R.N., Briskey, E.J., and Hoekstra, W.G. 1963. Effect of excitement, fasting and sucrose feeding on porcine muscle phosphorylase and postmortem glycolysis. *J. Food Sci.* 28:472–477.

Scanga, J.A., Belk, K.E., Tatum, J.D., Grandin, T., and Smith, G.C. 1998. Factors contributing to the incidence of dark cutting beef. *J. Anim. Sci.* 76:2040–2047.

Scheffler, T.L., and Gerrard, D.E. 2007. Mechanisms controlling pork quality development: The biochemistry controlling postmortem energy metabolism. *Meat Sci.* 77:7–16.

Scheffler, T.L., Kasten, S.C., England, E.M., Scheffler, J.M., and Gerrard, D.E. 2014. Contribution of the phosphagen system to postmortem muscle metabolism in AMP-activated protein kinase gamma 3 R200Q pig Longissimus muscle. *Meat Sci.* 96:876–883.

Scheffler, T.L., Rosser, A.L., Kasten, S.C., Scheffler, J.M., and Gerrard, D.E. 2013. Use of dietary supplementation with beta-guanidinopropionic acid to alter the muscle phosphagen system, postmortem metabolism, and pork quality. *Meat Sci.* 95:264–271.

Schwagele, F., Lopez, P., Haschke, C., and Honikel, K.O. 1994. Rapid pH drop in PSE-muscles—Enzymological investigations into the causes. *Fleischwirtschaft* 74:95.

Sentandreu, M.A., Armenteros, M., Calvete, J.J., Ouali, A., Aristoy, M.C., and Toldra, F. 2007. Proteomic identification of actin-derived oligopeptides in dry-cured ham. *J. Agric. Food Chem.* 55:3613–3619.

Sentandreu, M.A., Coulis, G., and Ouali, A. 2002. Role of muscle endopeptidases and their inhibitors in meat tenderness. *Trends Food Sci. Technol.* 13:400–421.

Shen, Q.W., and Du, M. 2005. Role of AMP-activated protein kinase in the glycolysis of postmortem muscle. *J. Sci. Food Agric.* 85:2401–2406.

Shen, Q.W., Means, W.J., Thompson, S.A., Underwood, K.R., Zhu, M.J., McCormick, R.J., Ford, S.P., and Du, M. 2006a. Pre-slaughter transport, AMP-activated protein kinase, glycolysis, and quality of pork loin. *Meat Sci.* 74:388–395.

Shen, Q.W., Means, W.J., Underwood, K.R., Thompson, S.A., Zhu, M.J., McCormick, R.J., Ford, S.P., Ellis, M., and Du, M. 2006b. Early post-mortem AMP-activated protein kinase (AMPK) activation leads to phosphofructokinase-2 and -1 (PFK-2 and PFK-1) phosphorylation and the development of pale, soft, and exudative (PSE) conditions in porcine longissimus muscle. *J. Agric. Food Chem.* 54:5583–5589.

Shen, Q.W.W., Gerrard, D.E., and Du, M. 2008. Compound C, an inhibitor of AMP-activated protein kinase, inhibits glycolysis in mouse longissimus dorsi postmortem. *Meat Sci.* 78:323–330.

Solomon, M.B., Van Laack, R.L.J.M., and Eastridge, J.S. 1998. Biophysical basis of pale, soft, exudative (PSE) pork and poultry muscle: A review. *J. Muscle Foods* 9:1–11.

Spahr, R., Probst, I., and Piper, H.M. 1985. Substrate utilization of adult cardiac myocytes. *Basic Res. Cardiol.* 80(Suppl 1):53–56.

Strydom, P.E., and Rosenvold, K. 2014. Muscle metabolism in sheep and cattle in relation to high rigor temperature overview and perspective. *Anim. Prod. Sci.* 54:510–518.

Suman, S.P., and Joseph, P. 2013. Myoglobin chemistry and meat color. *Annu. Rev. Food Sci. Technol.* 4:79–99.

Suzuki, K., Kostin, S., Person, V., Elsasser, A., and Schaper, J. 2001. Time course of the apoptotic cascade and effects of caspase inhibitors in adult rat ventricular cardiomyocytes. *J. Mol. Cell Cardiol.* 33:983–994.

Taylor, E.B., An, D., Kramer, H.F. et al. 2008. Discovery of TBC1D1 as an insulin-, AICAR-, and contraction-stimulated signaling nexus in mouse skeletal muscle. *J. Biol. Chem.* 283:9787–9796.

Taylor, R.G., Geesink, G.H., Thompson, V.F., Koohmaraie, M., and Goll, D.E. 1995. Is Z-disk degradation responsible for postmortem tenderization. *J. Anim. Sci.* 73:1351–1367.

Toldra, F., and Flores, M. 1998. The role of muscle proteases and lipases in flavor development during the processing of dry-cured ham. *Crit. Rev. Food Sci. Nutr.* 38:331–352.

Tsujimoto, Y. 1997. Apoptosis and necrosis: Intracellular ATP level as a determinant for cell death modes. *Cell Death Differ.* 4:429–434.

Turpin, S.M., Lancaster, G.I., Darby, I., Febbraio, M.A., and Watt, M.J. 2006. Apoptosis in skeletal muscle myotubes is induced by ceramides and is positively related to insulin resistance. *Am. J. Physiol. Endocrinol. Metab.* 291:E1341–E1350.

Underwood, K.R., Means, W.J., and Du, M. 2008. Caspase 3 is not likely involved in the postmortem tenderization of beef muscle. *J. Anim. Sci.* 86:960–966.

Uytterhaegen, L., Claeys, E., and Demeyer, D. 1994. Effects of exogenous protease effectors on beef tenderness development and myofibrillar degradation and solubility. *J. Anim. Sci.* 72:1209–1223.

Vanhoof, J. 1979. Influence of ante-mortem and peri-mortem factors on biochemical and physical characteristics of Turkey breast muscle. *Vet. Q.* 1:29–36.

Viljoen, H.F., de Kock, H.L., and Webb, E.C. 2002. Consumer acceptability of dark, firm and dry (DFD) and normal pH beef steaks. *Meat Sci.* 61:181–185.

Warner, R.D., Dunshea, F.R., Gutzke, D., Lau, J., and Kearney, G. 2014. Factors influencing the incidence of high rigor temperature in beef carcasses in Australia. *Anim. Prod. Sci.* 54:363–374.

Warriss, P.D., Bevis, E.A., and Ekins, P.J. 1989. The relationships between glycogen stores and muscle ultimate pH in commercially slaughtered pigs. *Br. Vet. J.* 145:378–383.

Warriss, P.D., Kestin, S.C., Brown, S.N., and Wilkins, L.J. 1984. The time required for recovery from mixing stress in young bulls and the prevention of dark cutting beef. *Meat Sci.* 10:53–68.

Weaver, A.D., Bowker, B.C., and Gerrard, D.E. 2009. Sarcomere length influences mu-calpain-mediated proteolysis of bovine myofibrils. *J. Anim. Sci.* 87:2096–2103.

Wendt, M., Bickhardt, K., Herzog, A., Fischer, A., Martens, H., and Richter, T. 2000. Porcine stress syndrome and PSE meat: clinical symptoms, pathogenesis, aetiology and aspects of animal welfare. *Berl. Muench. Tieraerztl. Wochenschr.* 113:173–190.

Woelfel, R.L., Owens, C.M., Hirschler, E.M., Martinez-Dawson, R., and Sams, A.R. 2002. The characterization and incidence of pale, soft, and exudative broiler meat in a commercial processing plant. *Poult. Sci.* 81:579–584.

Wojtaszewski, J.F., Jorgensen, S.B., Hellsten, Y., Hardie, D.G., and Richter, E.A. 2002. Glycogen-dependent effects of 5-aminoimidazole-4-carboxamide (AICA)-riboside on AMP-activated protein kinase and glycogen synthase activities in rat skeletal muscle. *Diabetes* 51:284–292.

Wulf, D.M., Emnett, R.S., Leheska, J.M., and Moeller, S.J. 2002. Relationships among glycolytic potential, dark cutting (dark, firm, and dry) beef, and cooked beef palatability. *J. Anim. Sci.* 80:1895–1903.

Young, L.H., Renfu, Y., Russell, R., Hu, X.Y., Caplan, M., Ren, J.M., Shulman, G.I., and Sinusas, A.J. 1997. Low-flow ischemia leads to translocation of canine heart GLUT-4 and GLUT-1 glucose transporters to the sarcolemma *in vivo*. *Circulation* 95:415–422.

Yu, H., Hirshman, M.F., Fujii, N., Pomerleau, J.M., Peter, L.E., and Goodyear, L.J. 2006. Muscle-specific overexpression of wild type and R225Q mutant AMP-activated protein kinase gamma3-subunit differentially regulates glycogen accumulation. *Am. J. Physiol. Endocrinol. Metab.* 291:E557–E565.

Zerouala, A.C., and Stickland, N.C. 1991. Cattle at risk for dark-cutting beef have a higher proportion of oxidative muscle fibres. *Meat Sci.* 29:263–270.

4 Impact of Animal Nutrition on Muscle Composition and Meat Quality

Eric N. Ponnampalam, Benjamin W.B. Holman, and Joseph P. Kerry

CONTENTS

4.1 INTRODUCTION

With growing populations and the increase in the socioeconomic status of Asian populations (e.g., China and India), it is expected that there will be an increasing demand for red meat as a source of animal protein. It is predicted that the consumption of red meat from cattle (bovine) and sheep (ovine) will increase approximately 200% by 2050 while pork (porcine) will increase by 158% (Alexandratos and Bruinsma 2012). In this context, global animal production will face significant challenges over the coming decades in order to meet the increasing demand for protein. Sustainable production, efficient use of natural resources, and improvement of animal welfare all need to share the focus when attempting to meet this heightened demand. Hence, animal nutrition can be employed to play a major role in achieving these requirements. Consequential benefits related to increased efficiency of animal production and improved body composition associated with animal nutrition includes improvements in carcass traits, muscle composition, and meat quality. The maintenance of dietary nutrition, feed, or ration during the critical times and stages of animal life provides several advantages to the producer, processor, and consumer (Demeyer and Doreau 1999). For example, the management of dietary nutrition can alter lean and fat deposition in the animal carcass and avoid complications typical of an unbalanced diet wherein animals are sent to slaughter either underweight or overweight with an excess of fat, thus incurring a price penalty. Furthermore optimizing the dietary nutrition of animals can impact meat quality by optimizing the intrinsic and extrinsic characteristics of muscle.

When pasture is available for sheep and cattle, extensive grazing systems are often the cheapest strategy. However, seasonal change and climate variation affect the nutrient value of pasture under such systems (Nardone et al. 2010). Under these circumstances, when the diet is lacking in protein, energy, minerals, or vitamins, supplementary feeding is often required. Feeding systems around the world are different from country to country where animals are raised partially or completely intensively or extensively. Yet, all producers will often routinely use supplements either as energy, protein, or micronutrients during periods when traditional diets are in short supply or in low quality and the duration of supplementation will vary based on the requirement of animals and the nutritive characteristics of the base feed (Beever and Thorp 1997; Dixon and Stockdale 1999).

With advances in technology and research, producers and retailers are well informed about the importance of nutrition and food quality for consumer growth, development, and overall health (Bermingham et al. 2008). There is an increasing trend for consumers and food manufacturers to choose clean and green meat products grown naturally due to concerns about food safety, impacts on the environment, the welfare of animals, and health status (Baghurst 2004; Daley et al. 2010; McAfee et al. 2010). Carcass yield, appearance (mainly color), organoleptic characters (aroma and flavor), texture (tenderness), and nutritive value are properties that influence consumer perception, decision making, and consequently the sales of meat (Tatum et al. 1999). These properties are affected by the structural, functional, and biochemical components of muscle or meat (Gatellier et al. 2005; Dunne et al. 2006; Luciano et al. 2009a; Ponnampalam et al. 2012, 2014b). The nutritional characteristics of diets (e.g., protein, energy, vitamins, and minerals) can all impact on value-adding properties of meat and therefore meat quality.

4.2 PROTEIN

4.2.1 Dietary Protein and Its Role in Animal Production

The amount of daily protein consumption is important as is the source and the quality of the protein in the diet (Mitchell 2007). Proteins in feed can be classified as complete or incomplete depending on their amino acid availability, with complete proteins being those that contain all nine essential amino acids in sufficient concentrations to meet the requirements of animals or humans. The proteins and their components such as amino acids serve as building blocks for the synthesis of proteins into skeletal muscle, which in turn are used to repair, rebuild, and grow muscle. Monogastric animals, such as pigs and poultry, rely on mammalian enzymatic digestion for absorption of nutrients (proteins) from the diets. The nutrients contained in forages are largely unavailable to monogastric species as opposed to ruminants (e.g., cattle, sheep, and goats) whose large microbial populations in their gastrointestinal track (rumen) permits the extensive digestion of fibrous components of forages (Beever and Thorp 1997). This releases the nutrients from diets including proteins and soluble sugars for subsequent digestion and absorption by the host animal.

Ruminants obtain a significant amount of proteins from forages and when the nutritive characteristics of forages is inadequate or the production potential of animal is high, producers tend to feed protein-rich concentrated supplements and alternate forage sources to match their nutrient requirement (Dixon and Stockdale 1999). According to microbe susceptibilities to rumen proteolytic activity, dietary protein will be degraded to peptides and amino acids. Protein available in ruminant diets can be categorized as rumen degradable and digestible undegraded protein (see Figure 4.1). The ammonia produced from degradable protein is utilized by microbes to form microbial protein while digestible undegraded protein moves to the small intestine and undergoes enzymatic digestion prior to utilization by the host animal. Urea can also be added to diets of low quality as a source of nonprotein nitrogen for ruminants. Rumen microbes convert urea to microbial protein and ammonia, where the

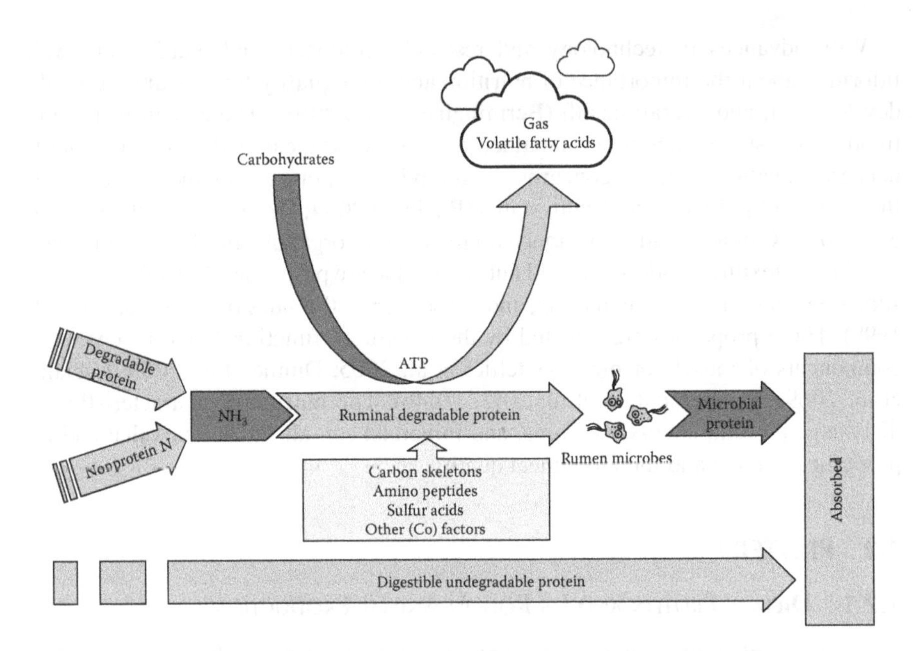

FIGURE 4.1 Microbial protein synthesis within a rumen—ruminal degradable and digestible undegradable protein pathways.

microbial protein which passes the rumen will be digested and utilized by the host animal for the synthesis of nucleic acid (Beever and Thorp 1997; Dixon et al. 2003). When diets are low in nitrogen content, it is common to add supplements such as silage, legume grains, and oilseeds rich in protein to improve the growth performance and the muscle productivity of animals.

4.2.2 Effect of Dietary Protein on Protein Synthesis and Muscle Deposition

Optimization of dietary protein in animal production is always essential for maximizing muscle productivity (Montossi et al. 2013). Any ration providing above or below the requirement can influence the profitability of the enterprise and the quality of meat. For example, when dietary protein is offered in surplus this will result in excess ammonia production within the gastrointestinal tract of ruminants. This ammonia will be absorbed by the rumen, reticulum, and omasum, carried in the portal vein to the liver and subsequently converted to urea and ultimately excreted in the urine (Siddons et al. 1985). This can be detrimental to animal performance and cause nutrient wastage in a production system, which in turn can contribute to environmental pollution. Paradoxically, suboptimal dietary protein provision can cause an imbalance in the protein:energy ratio of the diet and the amount of protein (amino acids) absorbed in the intestine, which in turn may affect the rate and efficiency of protein synthesis and muscle deposition.

4.2.3 Muscle Composition

In order to optimize protein synthesis and muscle production, animals should be offered recommended levels of a daily allowance for maintenance and muscle deposition. In terms of muscle production and meat quality, it is sensible to apply a cost:benefit assessment as achieving less than maximum performance may be the best practice if that level of performance does not justify the cost. The production system should be economically sound in terms of cost of production and product quality (National Research Council 1981). Two points should be considered when converting plant and other feed protein (resources) in to animal product, such as lean and fat content; (1) there should be benefit both for the producer and processor and (2) meat product development should meet consumer needs in terms of human health (low fat, high protein) and eating quality (flavor/taste and tenderness), as this has been of increased focus over recent years (Scollan et al. 2006; Daley et al. 2010).

In growing animals, improved muscle gain is accompanied by an increase in both whole body (Lobley 1993) and muscle protein synthesis (Dawson et al. 1991). Protein gain for both the whole animal and individual tissues (muscle) is the outcome of protein synthesis and protein degradation. Improved muscle deposition can be achieved by altering the synthesis and degradation process in the animal (MacRae and Lobley 1991), where the balance of dietary protein and energy is important. When the levels of protein and energy do not match the physiological status of the animal, the excess protein can be deposited in the body as fat or excreted with urine. The increase in muscle deposition in response to increased dietary protein in domestic animals is well documented (see Lobley 1998; Sillence 2004). In pigs, improved muscle deposition and energy metabolism in response to increased dietary protein have been shown previously (Campbell et al. 1991; Dunshea et al. 1993).

The value of the carcass is largely determined by the yield of lean meat, where the meat is distributed in the carcass and the quality of the meat (Johnson et al. 2005). Differences between genotype, sex, stage of maturity, and age on carcass composition or muscle composition have been reported. However, the influence of dietary protein on lean meat content or muscle deposition is always interrelated with the energy content of the diet. Either restriction or excessive dietary protein can affect muscle composition through increased deposition of fat in the carcass and skeletal muscle diverted via the excess energy available for fat synthesis (Beauchemin et al. 1995; Chiba et al. 1999; Fabian et al. 2002). The latter can influence lean content and meat quality of the carcass. Beauchemin et al. (1995) reported that there was no response to increasing crude protein content of high concentrate diets beyond 15% crude protein (CP) as CP level and degradability had little effect on growth, lean deposition, and carcass traits of lambs. This study also indicated that when lambs were compared on a constant weight basis; high-energy diets with high protein promote kidney fat deposition and to a small extent increase subcutaneous fat deposition. Therefore, any changes in muscle leanness or carcass due to dietary protein quality and quantity may influence meat quality traits.

4.2.4 Influence of Dietary Protein on Meat Quality

By manipulating the dietary protein content of animals (or applying other media) it may be desirable to reduce protein degradation and increase synthesis for better growth and muscle deposition in live animals. However, in terms of development of tender meat during the conversion of muscle to meat after death, the process of protein degradation by calpain activated systems are vital (Rowe et al. 2004). Research has aimed to identify the causes of variations in both the rate and extent of tenderization in meat during postmortem aging. Most studies have related tenderization to the amount and strength of connective tissue, sarcomere length, and the amount of proteolysis that occurs during postmortem (Hopkins et al. 2011). Connective tissue and sarcomere length provide background toughness while proteolysis of key myofibrillar proteins is the most cited event that attributes to an increase in meat tenderness (Huff-Lonergan et al. 1996; Melody et al. 2004). Protein content of a diet or dietary regimen can influence the tenderness of meat by altering either the components of connective tissues or the enzymatic processes responsible for proteolysis. The connective tissues in all muscles are found in the epimysial, perimysial, and endomysial and provide strength to the muscle fibers from their collagen content (see Chapter 1).

Free hydroxyproline has been found to be an indicator of collagen turnover and may influence meat tenderness through alterations to intramuscular collagen stability (Bailey and Light 1989). Bruce et al. (1991) reported that steers fed a protein supplemented alfalfa/grass silage diet had lower concentrations of hydroxyproline in their plasma than steers fed corn silage, although growth rates were not different. This study however, found no differences in the level of soluble collagen as a percent of total collagen due to diet. Although differences in collagen turn over and cross-linkage formation may have occurred as suggested, by changes in blood hydroxyproline, the cross-linkage changes were not sufficient to change the heat labile status of collagen. Shear force was lower for steers fed the corn-silage diet than those fed alfalfa/grass diets. This suggests that shear force is influenced by dietary energy, being positively associated with red (beta) myofibril diameter, in turn influencing shear force (Moody et al. 1980), but not by protein supplementation. Furthermore, it was found that sarcoplasmic protein solubility was significantly correlated to shear force and hence the increased sarcoplasmic protein solubility was indicative of heightened shear force, which is a proxy of decreased tenderness of meat (Bruce et al. 1991). Other research has reported that following 7 days of aging, undernourished, and older crossbred steers that were used for draught service produced extraordinarily high shear force values—indicative of very low tenderness of meat and this was also reflected by poor sensory grading for tenderness, both of which were associated with extremely low collagen heat stabilities (Jaturasitha et al. 2004). Muscle tenderness can be affected by the increase in cross-linking associated with an expansion of dense connective tissues and the insolubility of collagen that increases with age. Therefore, when considering the collagen content, types of collagen, and cross-linking, younger animals will produce more tender meats than will older animals (Gerrard and Grant 2003).

Muscle proteins are categorized as sarcoplasmic, myofibrillar, and stromal based on their solubility and stromal protein content that includes proteins of connective tissues which are very fibrous and insoluble (Aberle et al. 2001). Insoluble proteins are 50% collagen, 3% elastin, and the remaining is a mixture of various proteins. Nutrition leading to higher growth rates produces lean meat which has lower shear force values and is therefore tender (Harper 1999; Vestergaard et al. 2000). There is evidence that faster growth rates during the later stages of production, prior to slaughter can affect tenderness and cooking loss of meat (Lawrence and Fowler 2002). Compensatory growth through altered nutrition also improves meat tenderness in bulls (Hornick et al. 1998) and in pigs (Kristensen et al. 2004). This could be due to a faster growth rate enhancing the deposition of younger muscles with less structured connective tissues. The influence of growth rate on meat tenderness depends mainly on changes in muscle protein turnover. A general increase in total nutrient supply stimulates both muscle protein synthesis and degradation at intakes above maintenance (Lobley 1998). Thomson et al. (1999) reported that the increased growth rate was accompanied by higher μ-calpain activities. This calpain is a key protease that is responsible for meat tenderization postmortem (Koohmaraie 1996; Dransfield and Sosnicki 1999). So when a higher rate of muscle protein degradation is observed this is likely to result in a higher postslaughter rate of muscle protein degradation and better meat tenderness.

Muscle fiber characteristics (fiber thickness and numbers) are affected by nutrition and can influence meat quality. In neonatal lambs (14 days), the proportion of fast fibers were reduced and slow fibers were increased when nutrient restriction was imposed immediately prior to myogenesis (days 30–70) (Fahey et al. 2005). In postnatal animals, muscle hypotrophy is more associated with fast fibers (Type II) than slow myofibers (type I) (Sartorius et al. 1998). Meat tenderness is also influenced by the ratio of slow twitch oxidative (type I) and fast twitch glycolytic (type II) muscle fibers. Beef tenderness is positively associated to type I muscle, with differences linked to a higher ratio of protein turnover in tender muscles and a greater level of calpain activity (Lawrence and Fowler 2002). Dietary supplementation of protein sources of medium quality (Ponnampalam et al. 2003) or cereal grains of low quality (Ponnampalam et al. 2004) with a roughage-based diet significantly improved protein gain (lean content) in the carcass. However, lamb meat quality traits such as tenderness, sarcomere length or meat color were unaffected. The data presented above prompt the inclusion of collagen content, sarcomere length, and the protease enzyme into any model aimed at explaining the dietary effect of protein on meat tenderness.

4.3 ENERGY

4.3.1 INGREDIENTS OF FAT (ENERGY) USED IN ANIMAL FEEDING SYSTEM AND THE PURPOSE

There are two types of feeding commonly practiced with ruminants, based on; (1) grain feeding or (2) forage feeding. These practices are not exclusive and can be combined depending on the availability of feed resources, the purpose of the production system, and the nutritive characteristics of feeds. Forage-based diets provide more proteins either as microbial degradable or undegraded digestible protein while grain-based diets provide more energy, as short chain volatile fatty acids such as acetate, propionate, and butyrate. Propionate is the major precursor for glucose production in ruminants being utilized for glycogen synthesis in the liver and muscles of growing animals or lactose synthesis in lactating animals (Cerrilla and Martínez 2003). Glucose from the circulatory system can also be utilized by the cells for fat synthesis, through the provision of glycerol phosphate and NADPH (nicotinamide adenine dinucleotide phosphate) (Beever and Thorp 1997). Fatty acid biosynthesis occurs largely with the transformation of acetate within ruminant adipose tissue or from glucose within swine adipose tissue. Acetate and glucose carbons enter fatty acid biosynthesis via the malonyl-CoA and acetyl-CoA carboxylase reaction cycles, and then palmitic acid is produced by fatty acid synthase (Rule et al. 1995). On the other hand, lipid supplements are used to increase the energy concentration of diets (Chilliard 1993) or improve the nutritive value of meat (Ponnampalam 1999; Wood et al. 2003; Scollan et al. 2006). Lipids in animal diets can be included as natural/protected fats (Scott et al. 1995; Scollan et al. 2006), oils (Ponnampalam et al. 2001; Nute et al. 2007) or oilseeds, and meals (Demeyer and Doreau 1999; Bolte et al. 2002), and their use is dependent upon feed availability, price of supplements, and the purpose of production.

It is widely accepted that fats are the storage form of fatty acids (FA) in living organisms. Fats in a human and animal body provide structure, support for organs, energy for work, act as a media for transporting nutrients (vitamins and carotenoids), act on the transcriptions of genes, and maintain reproduction and health. Individual fatty acids serve different purposes in the body: some are oxidized for energy, some are structural features of cell membranes, others are converted to different fatty acids or substances (e.g., sterols), while others perform special functions in muscle systems and nerve cells. Ruminant diets generally contain 2%–5% lipids of which 50% consist of fatty acids. Lipids provide more energy value than carbohydrates and protein, but the dietary level of fat should be less than 6%–7% of dry matter to optimize animal performance (Doreau et al. 1997). The PUFA:SFA (polyunsaturated fatty acid:saturated fatty acid) ratios should be at least 0.4, which is a challenge for ruminants, even with forage-based diets. Some studies have reported that more desirable PUFA:SFA ratios through increasing dietary PUFA can consequently depress meat SFA content (Mir et al. 2002; Aurousseau et al. 2004; Noci et al. 2007). In contrast, others reported minimal changes to the PUFA:SFA ratio when dietary PUFA were elevated because meat SFA content remained unchanged (Santos-Silva et al. 2002; Realini et al. 2004).

4.3.2 Influence of Dietary Energy (Fat) on Muscle Composition: Fatness and Leanness

Feedlots or concentrate diets containing proportionally higher amounts of grain are typical high-energy diets that can provide greater levels of dietary SFA (palmitic and stearic) when compared with a forage diet. The literature indicates that cattle and sheep fed high-energy diets will produce heavier body weights comprised of fatter carcasses with higher levels of intramuscular fat content. Yang et al. (2002) found that with cattle slaughtered at the same time from the beginning of the experiment, those fed grain (mainly sorghum fed feedlot ration) had 80 kg heavier carcass weights compared to those grass fed (mainly Rhodes grass). Carcass fatness (subcutaneous fat depth at the 8th rib site) was also affected by the feeding system, where pasture-fed cattle produced carcasses with 3 mm fat depth compared with 18 mm for grain-fed cattle (Yang et al. 2002). Long-term grain feeding produces carcasses with greater fat thickness and harder fat due to high levels of saturated fats and removal of hard fat at boning often reduces the premium price, and there is also proportionally a reduction in the lean content in the carcass. Meanwhile, recent study conducted in goat kids supplemented with low levels of palm oil, soybean oil, and fish oil showed no differences in primal cuts or muscle composition (moisture, protein, ether extract, and ash contents in the *longissimus lumborum* [LL]), but fatty acid composition in the intramuscular fat was significantly altered—with fish oil supplementation resulting in highest omega-3 content (Najafi et al. 2012).

In many countries, concentrate feeding is practiced due to seasonal limitations in pasture quantity and quality available for animal production. The benefits of feeding high-energy diets appear to be the ability to achieve higher growth rates and greater carcass weights over a shorter period. In this context, grain feeding may be advantageous for some meat quality aspects associated with animal age or fat thickness—for example, tenderness. High-energy diets are desirable in countries where larger meat cuts with high levels of marbling fat (intramuscular fat) are preferred (North America) unlike some European countries which prefer lighter weight animals wherein low-energy diets are applied during finishing. Muir et al. (1998) found that when cattle are fed pasture to achieve similar growth rates and similar carcass weights at finishing, it is likely that the beef eating quality from animals finished on pasture is comparable to those finished on grain. When animals of similar genetics that were maintained on similar diets were compared, it was reported that faster growing lambs had lower lean and greater fat deposits in their carcasses compared with their slower growing counterparts (Ponnampalam et al. 2008). It appears that faster growing animals possibly harvest more diet, therefore consume more energy and other nutrients compared with slower growing animals (Hopkins et al. 2007a,b). As such, the excess energy available above that used for maintenance and lean deposition in the body is utilized toward fat deposition in the carcass. A recent study also demonstrated that animals with heavier body weight at the start of the experiment had greater muscle fat content and higher levels of antioxidant in the muscle at the end of the feeding study when compared with smaller body weight animals at the commencement having lower muscle fat and less muscle antioxidant in muscle at finishing (Ponnampalam et al. 2014b).

The use of green forages or high-quality conserved forages may provide nutritionally balanced ruminant diets, rich in antioxidants, minerals, vitamins, and essential fatty acids that contribute to high value meat products or heightened meat quality. Conversely, feedlotting or long-term grain supplementation, a common practice during dry seasons, produces meat with higher fat and lower nutritive characteristics (Ponnampalam et al. 2006; Descalzo and Sancho 2008; Daley et al. 2010). The influence of the dietary energy source on fatty acid profiles of animals can be more significant than genotype or breed effects (Bas and Morand-Fehr 2000; Garcia et al. 2008; Ponnampalam et al. 2014a). Pasture finishing systems produce higher quantities of n-3 PUFA in the muscle than concentrate feeding, which can improve both

PUFA:SFA and n-6:n-3 ratios (Wood and Enser 1997; Ponnampalam 1999; Gatellier et al. 2005; Rochfort et al. 2008). For example, lambs (Ponnampalam et al. 2002a) and beef cattle (Ponnampalam et al. 2006) fed concentrate diets produced meat with significantly higher n-6:n-3 ratios in comparison to the favorable levels achieved by forage-fed animals.

4.3.3 The Impact of Dietary Energy (Fat) on Meat Eating Quality

4.3.3.1 Flavor, Aroma, and Taste

The mechanisms that underlie the effects of diet on the sensory attributes of lamb meat are complex and not fully understood (Resconi et al. 2009). The differences in sensory characteristics of ruminant meat are mediated by animal growth rate determining carcass leanness, animal age, and muscle fat content and composition which are in turn influenced by the diet (Calkins and Hodgen 2007). Long-chain PUFAs are preferentially deposited in the phospholipids of muscle membranes, which can develop desirable flavor during processing and cooking through oxidation products, but at certain concentrations also contribute to undesirable off-flavors (Mottram 1998; Campo et al. 2006). Nute et al. (2007) evaluated the flavor of meat (loin) in lambs fed concentrate-based diets intensively with flax oil, fish oil, protected lipid supplement (PLS), fish oil with marine algae and PLS with marine algae. Lambs fed the flax oil had the highest concentration of α-linolenic acid (ALA) (18:3n-3) in the phospholipid fraction of FA and gave the highest rating for lamb flavor, overall liking and a lower rating for abnormal lamb flavor, which was similar to the findings of Choi et al. (2000) and Vatansever et al. (2000). Feeding a protected lipid supplement dramatically increased muscle LA (18:2n-6) which in turn reduced the lamb flavor and increased the abnormal lamb flavor compared to the flax oil diet, an outcome reported previously (Ford et al. 1975; Fisher et al. 2000). Fish oil and PLS with marine algae supplemented diets offered meat with higher concentrations of long-chain C20–C22 carbon PUFA, had increased levels of fishy and abnormal flavors, respectively while reducing the rating of lamb flavor. However, studies conducted by Ponnampalam et al. (2002b) in lambs and Najafi et al. (2012) in goats showed that the significant increase in long-chain omega-3 fatty acids (C20:5n-3 and C22:6n-3) in meat with 2% fish oil feeding in a forage-based diet or concentrate diet, respectively, compared with similar basal diets containing oilseed meal or oil supplements did not affect the sensory properties or other meat quality aspects such as the color of the meat.

Using correlation coefficients between FA proportions and sensory ratings, Nute et al. (2007) suggested that dietary effects on FA composition were largely responsible for the changes in flavor, but a more rigorous statistical approach is required to confirm causation. The study of Nute et al. (2007) showed that ALA (18:3n-3) had the strongest correlation with lamb flavor and the strongest negative correlation with abnormal lamb flavor, which supports the finding of Sanudo et al. (2000) that flavor intensity and flavor liking in lamb meat is strongly associated with ALA concentrations. Sanudo et al. (2000) found positive correlations between meat aroma, flavor, and n-3 fatty acid (ALA) contents in meat from lambs fed forage-based diets and negative correlations of flavor and aroma with n-6 fatty acid content in the meat of lambs fed concentrate-based diets. Fisher et al. (2000) also found that the concentration of

ALA in lamb meat increased by grass feeding compared with a concentrate and the score for lamb flavor intensity increased while the abnormal flavor is reduced. Taken together, these findings indicate that fatty acid type in the diet is more influential in the flavor and aroma of cooked meat than the level of fat in the diet and that n-3 fatty acid as ALA provides better flavor to meat.

Lipid oxidation is an undesirable process for meat as it occurs continuously during storage and leads to the development of rancid odors and off flavors. The rate and extent of lipid peroxidation is more significant with increasing unsaturated chain length (Scislowski et al. 2005). It was reported that the oxidation of linolenic acid was 20–30 times faster than linoleic acid, which was 10 times faster than that for oleic acid (Li and Liu 2012). The oxidation of muscle lipids after slaughter can adversely affect the flavor and nutritive value of fresh, frozen, and cooked meat and meat products (Morrissey et al. 1998). O'Grady et al. (2001) reported that oxymyoglobin (OMb) oxidation is enhanced by the presence of highly oxidizable lipids and a significant increase in lipid oxidation could occur before significant OMb oxidation is detected. The susceptibility of muscle to lipid oxidation varies depending on animal species and muscle type (Rhee et al. 1996). Park et al. (2004) has shown that beef lipids have a greater susceptibility to damage caused by oxygen free radicals than those in pork.

Hayes et al. (2007) found higher levels of lipid oxidation in beef compared to pork muscle, and indicated that beef muscle was 4-fold more susceptible to oxidative changes than pork. It is reasonable to summarize that this may not be the same for lamb and goat meat because both species are mostly reared under extensive grazing that offers opportunities for the consumption of green feed that is high in antioxidants, carotenoids, and flavonoids. These compounds have the capacity to protect or delay meat from oxidation processes and quality deterioration (Ponnampalam et al. 2012a). Research shows that the influence of animal genetics on lamb flavor is minor when compared with nutritional background before finishing (Daley et al. 2010; Watkins et al. 2013), which has a greater impact on carcass fat, muscle fat content (intramuscular fat), and muscle fat composition—among these muscle fat composition affects meat flavors most.

4.3.3.2 Tenderness, Texture, and Juiciness

The consumer perception of taste, as reflected by overall like/dislike ratings, is associated with differences in juiciness, flavor, and tenderness (Tatum et al. 1999). Juiciness of meat is determined by the amounts of water and fat remaining in the meat after the product has been cooked (Smith 1997) and is strongly influenced by the degree of doneness (Lorenzen et al. 1999). Therefore, any changes in water content (increased lean content) or fat content in the muscle through dietary energy (fat) variations may ultimately influence meat juiciness and/or flavor. Luchak et al. (1998) reported cooked beef cuts having high intramuscular fat and/or water binding capacity are juicier than those having low levels of intramuscular fat and/or low water binding capacity. The flavor of meat varies with the amount and composition of intramuscular fat that can be affected by the feeding system (grain versus grass) and by the number of days animals are fed concentrate diets (Tatum et al. 1999). The review by Harper (1999) discussed all the factors influencing beef tenderness in detail and describes how the features of an animals' growth path (e.g., compensatory growth or feeding a high-energy diet) during its development can significantly affect meat toughness through interactions with connective tissue component properties.

4.3.4 Nutritional Value

There is considerable evidence among human nutritionists that imbalances of dietary fatty acids and cholesterol are the primary cause for heart disease, obesity, vascular disease, inflammation, and several cancers (Broughton and Wade 2002; Calder 2004). A high dietary SFA intake is associated with elevated low-density lipoproteins (LDL) and total cholesterol production. Replacing 1% of SFAs with polyunsaturated FA is thought to lower LDL cholesterol by 2%–3% (Astrup et al. 2011) and has led nutritionists to recommend a PUFA:SFA ratio of 0.4 or higher (Wood and Enser 1997). Red meat is recognized as one of the main sources of SFA and typically has a PUFA:SFA ratio of 0.1 (Geay et al. 2001; Luciano et al. 2009a). However, not all SFAs have an equal effect and research has demonstrated stearic acid (18:0) has no influence on cholesterol (Hu et al. 2001; Hunter et al. 2010). Meat derived from grass-fed ruminants tends to contain more stearic acid and less myristic (C14:0) and palmitic acids (C16:0) than grain-fed animals (Daley et al. 2010), however under the current recommendations this is not distinguished. These conclusions were reinforced by a large-scale study conducted in Australia covering approximately 6000 lambs over three consecutive years across eight production sites that indicated that C14:0 and C16:0 were lower in lambs mainly finished on grass/pasture-based diets than those finished mostly on grain or concentrate diets (Ponnampalam et al. 2014c). Due to the recommendations of health professionals around the world, a net reduction in the overall consumption of saturated FA, trans fatty acids and cholesterol is evident, with an emphasis on increasing the dietary intake of omega-3 PUFAs (Simopoulos 1991; Kris-Etherton and Innis 2007). This has promoted anti-saturated fat campaigns that have reduced consumption of dietary fats—particularly animal proteins, such as meat, processed meat products, dairy products, and eggs, over the last three decades (Putnam et al. 2002; Baghurst 2004).

Red meat produced from any feeding system is nutrient dense and regarded as an important source of essential amino acids, vitamin A, B, D, E, and minerals including iron, zinc, and selenium (Biesalski 2005; Williamson et al. 2005). In addition, meat provides considerable amounts of fats that provide energy and media facilitating the transportation of fat-soluble vitamins. U.S. research reports that animal fat contributes approximately 60% of the SFAs in the American diet (Smith 1995; Daley et al. 2010), most of which are palmitic (16:0) and stearic (18:0) acid, the latter having a neutral effect on cholesterol levels. Monounsaturated FA represent 30% of the total dietary FA, with the main one being oleic acid (18:1) which alleviates blood cholesterol, elevated blood pressure, and stroke risk (Kris-Etherton et al. 1999). A healthy and balanced diet should contain roughly 1:4 ratio of omega-3 to omega-6 PUFA. However, a typical western diet contains 11–30 times more omega-6 fats than the omega-3 fats. This has been hypothesized as a significant factor contributing to the rising rate of inflammatory disorders and obesity (Calder 2004; Palmquist 2009).

The nutritional value of ruminant meats to human health has received considerable emphasis in the last two decades (Scollan et al. 2006). In this context, increased levels of PUFA to SFA and omega-3 to omega-6 (n-3/n-6) in animal muscle foods are reported as beneficial for overall health and well-being by lowering the incidence of degenerative diseases, such as heart disease and arthritis, and improving visual and mental health (Connor 2000; Anderson and Ma 2009). Among land- and sea-based animal foods (seafood, poultry, and meat), current Food Standards favor fish over red meat due to higher levels of EPA (eicosapentaenoic acid) and DHA (docosahexaenoic acid). However, red meat contains relatively high levels of DPA (docosapentaenoic acid) compared to fish and plays a major part in many diets, contributing micronutrients and essential omega-3 PUFAs (Scollan et al. 2006; Williams 2007). On average, Australians consume more red meat than fish and DPA has been shown to account for 29% of the total long-chain omega-3 PUFA intake in Australian adults (Howe et al. 2007). Yet, at present the health benefits or effects on consumers from DPA-rich diets remains relatively unexplored—with outcomes of considerable importance to lamb and other red meat management strategies.

Epidemiologic studies have shown that consumption of fish oil rich in omega-3 PUFA reduces the risk of cardiovascular disease, arthritis, and obesity in human. This effect results in part from an inhibition of lipogenesis and stimulation of fatty acid oxidation in the liver (Brown and Hu 2001) as well as in muscle tissues (Calder 2004). It is worth noting that like other PUFA-rich sources, fish oil is readily peroxidized to form hydroxyperoxides that increase oxidative stress. However, research has shown that in contrast to omega-6 PUFAs, omega-3 PUFAs in the muscle membrane systems can inhibit free radical generation. For example, Takahashi et al. (2002) showed that feeding mice diets rich in fish oil increases the expression of antioxidant genes in the liver and up-regulates the expression of lipid catabolism genes. This indicates that omega-3 PUFA in the body (blood and tissues) assists in cardiovascular and inflammation disease prevention by inhibiting free radical production, which is believed to be through a lowering of the oxidative stress in the body. In support, a recent study investigating lambs fed different diets indicated a significant increase in n-3 fatty acid, which in turn improves gene expression and enzymatic activity, all of which indicate a lowering of oxidative stress in tissue systems (Vahedi et al. 2014).

4.3.5 Color Associated with Lipid (Fat) Oxidation

The demand and consumer perception of meat is influenced by nutritional value, color, texture, and flavor, much of which deteriorates by the effects of lipid oxidation. There have been several reports on the level of PUFA in meat and its association with the flavor and color of meat. Lipid oxidation in muscle foods leads to off-flavor development (Asghar et al. 1990), and contributes to increased myoglobin (Mb) oxidation, that leads to meat discoloration (Greene 1969; Faustman et al. 2010). Studies have also shown that substrates from pigment oxidation can promote lipid oxidation and vice versa although the strength of the relationship between these two on the shelf life of meat is low (Renerre 2000). The stability of meat with respect to oxidation is a function of the balance between pro-oxidants and antioxidants, and the composition of muscle substrates, such as PUFA, proteins, pigments, and cholesterol, which can be affected by the diet quality and the immune status of the animal (Bekhit et al. 2013). Studies have reported that elevated meat PUFA content can reduce shelf life and color stability. A recent study indicated that more so than the level of PUFA or form of iron in meat, it is vitamin E concentration in muscle that contributes to the maintenance of meat color under retail display (Ponnampalam et al. 2012a). Nevertheless, including 3% canola oil into goat diets, as an energy source, was found to have beneficial effects on meat quality by improving health benefits through significantly increasing the muscle omega-3 PUFA concentration while reducing the lipid oxidative substances in meat (Karami et al. 2013).

Consumers prefer meat having a bright red color. This is produced from the initial oxidation of Mb to OMb when meat is sliced, bloomed in the presence of oxygen, and presented at retail display. However, further oxidation produces metmyoglobin (MMb) when the loss of an electron results in the conversion of Mb's central heme iron from its ferrous state to a ferric state, which in turn instills an undesirable brown color (Geay et al. 2001). Since the conversion of OMb to MMb is reliant on electron transfer, the oxidation potential due to the level of unsaturated fatty acids largely influences color stability (Min and Ahn 2005). Earlier studies have proposed that the color of meat was determined by the total amount of omega-3 PUFA, as adverse effects on lipid and Mb oxidation are only seen once α-linolenic acid approaches 3% of lipids (Wood et al. 2003). However, a recent study suggested that the level of omega-3 PUFA has an indirect effect, but more important is the relationship with heme iron and α-tocopherol—with these compounds mediating muscle redness (Ponnampalam et al. 2012a). Both dietary iron and PUFA sourced from meat or other food-types are necessary for better maintenance of human health. But, evidence shows that if vitamin E concentration of meat is not adequate, these may negatively influence meat color and color stability, and therefore saleability (Bekhit et al. 2013). Hence, careful thought should be applied when diets are formulated for meat producing animals so as to offer optimum eating quality and nutritional value while considering consumer attraction in appearance.

4.3.6 Diet, Muscle Glycogen, and Meat Quality

In the muscles of living animals or humans, the complete oxidation of carbohydrate (i.e., glycogen) to CO_2 and water, under aerobic conditions, releases a substantial amount of energy that is quickly applied to transform ADP (adenosine triphosphate) to ATP (adenosine diphosphate)—the fuel source necessary for muscular contraction. After a series of steps in the glycolytic pathway in the cytoplasm, molecules with six carbon atoms derived from the glucose units of glycogen are split to produce two molecules of pyruvate (see Figure 4.2). Pyruvate formed in the cytosol of the muscle fiber enters into a mitochondrion and is converted to acetyl CoA, which becomes fused to oxaloacetate to form citrate. The citrate then oxidized in the Kreb's cycle with the regeneration of oxaloacetate. The continuous activity of Kreb's cycle is fuelled by a range of carbohydrates, fatty acids, and amino acids and is the primary system for the ATP. Glycogen is the primary storage carbohydrate in muscle fibers and located in sarcoplasm between myofibrils and under the cell membrane (Chapter 3). Muscle contraction is a primary user of ATP in the living animal, but

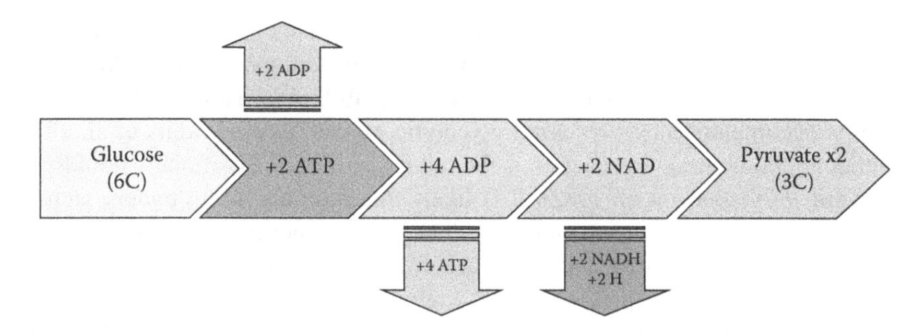

FIGURE 4.2 A summation of the glycolytic pathway.

substantial amounts of ATP are used by the membranes around and within the fiber for maintaining ionic concentration gradients. Typical skeletal muscle fibers contain glycogen reserves around 1.5% of the total muscle weight. If glycogen reserves are low, the muscle fiber can also break down other substrates such as lipids and amino acids as an alternative fuel.

Skeletal muscle structure and its biochemical components influence its transformation to meat affecting quality attributes, including tenderness, juiciness, color, and flavor. In ruminants, nutrition can influence the metabolic activity, structure and composition of muscles, and thereby affect meat quality (Geay et al. 2001). Muscle glycogen concentration preslaughter is one of the most important factors affecting meat quality via the potential downgrading of carcasses due to quality defects, such as pale, soft, and exudative (PSE) or dark, firm, and dry (DFD) meat (see Chapter 2; Schaefer et al. 1997; Immonen et al. 2000; Bowker et al. 2005).

Variation in muscle glycogen levels in live animals caused by antemortem activities or nutritional regulation can contribute to changes in postmortem muscle pH (ultimate) that can lead to a degradation of meat quality (Schaefer et al. 2001). The depletion and repletion of glycogen levels in skeletal muscle is mainly influenced by the amount and the metabolic rate of type IIB fibers (fast-twitch fibers) due to their higher ATPase capacity and glycolytic nature (Schaefer et al. 1997; Bowker et al. 2005). This can influence the lactic acid production in preslaughter and postslaughter muscles and contribute to variation in pH and protein denaturation in postmortem muscles (Immonen et al. 2000; Bowker et al. 2005; Foury et al. 2005). Any activities that change the number and size of the muscle fiber types in live animals may lead to changes in metabolic rate and energy production of skeletal muscle which in turn may affect postmortem glycolysis and associated meat quality characteristics.

A rise in glycolytic activity with age has been observed in several species including piglets (Lefaucheur and Vigneron 1986), rabbits (Briand et al. 1993), and Limousin calves (Jurie et al. 1995). In young cattle, aged between 2 and 6 months, muscle growth is characterized by an increase in the proportion of type I fibers, and the conversion of type IIA into type IIB, a rise in glycolytic activity, and an increase in fiber area (Picard et al. 1995; Greenwood et al. 2007). This conversion of IIA fibers into IIB fibers has been observed in a number of species, including cattle (Spindler et al. 1980). Seideman and Crouse (1986) and Beverly et al. (1991) both reported that postnatal energy restriction could result in a decrease in the proportion of type IIB fibers and an increase in type IIA fibers. This is supported by several studies which have shown that low-energy intake due to undernutrition slows down the maturing process of muscle fibers (Lanz et al. 1992; Ward and Stickland 1993), which is accompanied by decreased glycolytic metabolism (Howells et al. 1978). Another effect of energy restriction is the reduction in the size of fibers (Goldspink and Ward 1979; Bedi et al. 1982). It is likely that animals living under extensive conditions or facing undernutrition may change their energy metabolism pathway from glycolytic to oxidative and therefore utilize more fat and less carbohydrate resources for fuel (ATP) production, through a shift in aerobic metabolism. This strategy is possibly used for their survival due to lower glycogen and glucose availability in the body (muscle and blood) that is associated with the availability of diet and extended travelling.

4.4 VITAMINS

While much is understood about the important roles that vitamins play in the healthy biochemical and nutritional functioning of living vertebrates, little is still understood with respect to the impact of vitamins on muscle derived from some members of this group for meat consumption purposes, with the noted exception being vitamin E. More often than not, the impact of vitamins on meat quality has focused primarily on antioxidant properties that influence positively meat color, lipid oxidation, and on the latterly linked and very specific sensory attribute of warmed-over flavor (WOF). Consequently, if a vitamin does not function nutritionally or biochemically as an antioxidant, then there has been a tendency to dismiss such vitamins from dietary study. This would account for the relatively few studies carried out to date for some vitamins. While it may indeed be the case that many of these vitamins play no functional role whatsoever with respect to any meat quality attributes, whether offered in the diet or processed directly into muscle-based products, studies are required to definitively demonstrate potential capacity.

Additionally, most studies examining responses of growing animals to dietary supplementation with vitamins (and other nutrients) have concentrated on the economically important traits of growth rate, feed intake, feed conversion efficiency, fertility, health, and gross carcass composition (fat, lean, bone content), while measurement of tissue levels of vitamin additives have tended to be confined to blood and sometimes organs (Lynch and Kerry 2000). In addition, many studies report only total dietary levels of the nutrient under study and give no information on bioavailability or report on interactions with other nutrients or dietary components which may also complicate any response (Lynch and Kerry 2000).

There is growing consumer resistance to the incorporation of additives into foods, especially where the additives are of synthetic origin, even when they have a nutritional or health advantage. Dietary supplementation of the growing animal provides a unique method of manipulating the content of some micronutrients, like vitamins, and other non-nutrient bioactive compounds in meat, with a view to improving the nutrient intake and health status of the living animal in the first instance and that of the consumer in the second instance following consumption of meat and meat-based products from the supplemented animal.

4.4.1 Fat Soluble Vitamins

4.4.1.1 Vitamin E

Of all the vitamins available, vitamin E or α-tocopherol has been the most widely studied vitamin within a meat or muscle-based food perspective. Consequently, it is not surprising that the effect of vitamin E on meat or muscle-food quality has been comprehensively described and for a wide and diverse range of muscle-food sources including meat, poultry, and fish.

Vitamin E is a natural antioxidant and all tocopherols are plant-derived phenolic compounds that are deposited in skeletal muscle via the diet (Parker 1989). Tocopherols, which are lipid soluble, function as antioxidants by preventing free radical-induced oxidative processes and are described as free radical scavengers (Decker et al. 2000).

Tissue accumulation of vitamin E during dietary supplementation appears to generally occur in a dose- and duration-dependent manner by most muscle-food producing animals, however, the rate of vitamin E uptake by different species varies considerably as does that for various muscles and other tissues contained within the carcass. For example, Arnold et al. (1993) noted that liver and plasma reached equilibrium faster than muscle (m. *longissimus dorsi*) derived from cattle; equilibrium times being <42 days and 84–126 days, respectively. For purely muscle-derived samples, O'Sullivan et al. (1997) showed that vitamin E accumulation in porcine muscle was highest in the thoracic limb, neck, and thorax followed by the pelvic limb, with lowest levels found in the back.

Irrespective of dose and duration of feeding, once vitamin E levels accumulate to 3.0 mg α-tocopherol/kg muscle, a noted delay in lipid oxidation and meat discoloration occurred in retailed displayed beef (Faustman et al. 1989). Arnold et al. (1993) concluded that a level of 3.5 μg α-tocopherol/g muscle was required to provide optimum protection against beef discoloration, but that this was dependent on the muscle in question. Liu et al. (1996) suggested that vitamin E levels in muscle tissues between 3.0 and 3.5 μg α-tocopherol/g muscle may be minimal requirements and these authors suggested an optimal target level of 5.27 μg α-tocopherol/g muscle be achieved. However, these authors cautioned that in order to reach these levels in meat, the economic implications and the obtained benefits from supplementing animal diets to achieve these levels in meat would have to be carefully considered.

Supplementing animals with vitamin E in order to increase tissue levels described above make sense when animals are intensively reared in feedlot-type production systems and have access only to defined, nutritionally balanced diets based on concentrate or cereal-based offerings. For example, Kerry et al. (2000) reported the impact that feeding beef cattle different grades of forage had on vitamin E tissue levels and these authors reported the high levels of vitamin E that could be achieved through solely offering high-quality green forage without supplementing with vitamin E at all. This observation was reflected in research with lambs. After back-grounding on annual pasture for 2 weeks, lambs (7 months old) reared under extensive conditions and offered hay and barley grain diets for 5 weeks produced muscles containing 1.69 mg vitamin E/kg meat, while similar lambs that had been grazing annual ryegrass pasture had a muscle vitamin E concentration of 3.42 mg/kg meat (Ponnampalam et al. 2012a). Similarly, lambs (7 months old) grazed on low nutritive annual pasture and supplemented with cereal grains or flaxseed/meal supplements for 7 weeks, with 1 week of introduction to supplements during the autumn season, produced muscles which contained between 3.2 and 3.6 mg of vitamin E/kg of meat. However, lambs grazed on perennial pastures (alfalfa) of low nutritive quality produced muscles which contained 5.9 mg vitamin E/kg of meat (Ponnampalam et al. 2012b). In a recent study, Ponnampalam et al. (2013) reported that when 10-week old weaned lambs were fed either a (1) feedlot diet, (2) feedlot plus ryegrass pasture diet, (3) ryegrass pasture diet, and (4) a lucerne (alfalfa) pasture diet for 6 weeks, with 2 weeks of adaptation to diets, the muscle vitamin E concentrations were 0.73, 2.16, 2.91, and 2.10 mg/kg meat, respectively. These findings indicate that when animals are grown under extensive conditions, the replenishment of vitamin E concentration in muscles is dependent upon the basal forage diets offered, the type of supplements used, feeding history, and the location from where the animals originated and the age of animals.

Faustman et al. (2010) cautioned the overreliance on targeting tissue vitamin E accumulation levels alone in terms of guaranteeing oxidative stability in meat, particularly if other factors such as; PUFA accumulation in muscle, for example, is ignored. There was clear indication of complex biochemical reactions of PUFA and iron on the maintenance of redness (see Figure 4.3; Ponnampalam et al. 2012a) and lipid oxidation (see Figure 4.4; Ponnampalam et al. 2014b) of meat at simulated retail display, but vitamin E plays a dominant role above the PUFA or iron (see Table 4.1). The most comprehensive and still the most relevant report on the impact of vitamin E on meat quality generally is the book entitled *Antioxidants in Muscle Foods* (Decker et al. 2000). This resource, in combination with recent observations, allows dietary supplementation of vitamin E to muscle-food producing animals' benefits to be summarized as

- Oxidative stability of meat by reducing lipid oxidation
- Meat flavor by removing WOF specifically
- Muscle-food color, especially in red pigmented meat systems
- Reduced drip losses from muscle foods over time in some instances
- Overall quality of muscle-based foods as packaging, restructuring, cooking, and other processing factors were applied to vitamin E supplemented muscle but did not improve
- Microbiological stability, quality, or safety of muscle-based food products

The type and level of research carried out on vitamin E in meat-producing animal experiments serves as a template for studies which need to be carried out for other under-investigated vitamin forms.

4.4.1.2 Vitamin A and the Carotenoids

Carotenoids constitute a very diverse grouping of plant-derived yellow-to-red pigmented polyenes, with greater than 600 compounds having been identified as members of this chemical grouping. A select number of these have nutritional,

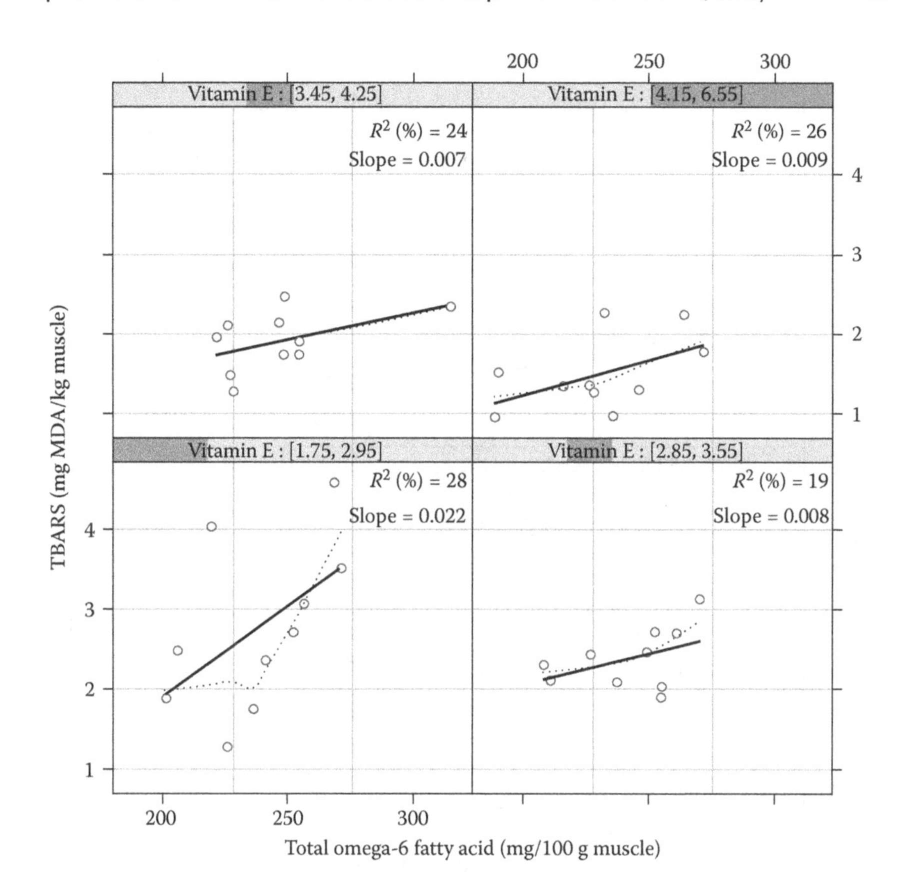

FIGURE 4.3 Lipid oxidation (TBARS) versus total omega-6 fatty acid condition on vitamin E. The vitamin E was conditioned by creasing four shingles or classes of equal count, with 10% overlap between adjacent intervals. The shingle intervals were 1.75–2.95 (shingle 1), 2.85–3.55 (shingle 2), 3.45–4.25 (shingle 3), and 4.15–6.55 (shingle 4), respectively. The R^2 (%) values in each of the panels are to be read as percent. The slopes are dependent on the scales between the *x* and *y* variables. The emphasis is on the magnitude of slope of the straight-line in each panel. The R^2 and slope values are subjected to some dependency between the overlaps (10% overlap) from one panel to the next. (Adapted from Ponnampalam, E.N. et al. 2014b. *Lipids* 49:757–766.)

as well as aesthetic properties and are biochemically converted to vitamin A following ingestion, the most notable carotenoid being β-carotene. Like vitamin E, carotenoids and, in particular, β-carotene and lycopene have been suggested to be important lipid-soluble dietary antioxidants which are thought to play a role in controlling oxidatively induced diseases (Palozza and Krinsky 1992). However, the antioxidant role played by nutritionally relevant carotenoids is complex and controversial as antioxidant activity appears to be affected by the biochemical environment (e.g., presence and type of chemical catalysts, oxygen concentration, presence of other antioxidants, etc.) within which the carotenoid in question finds itself. Jørgensen and Skibsted (1993) demonstrated the potential that carotenoids

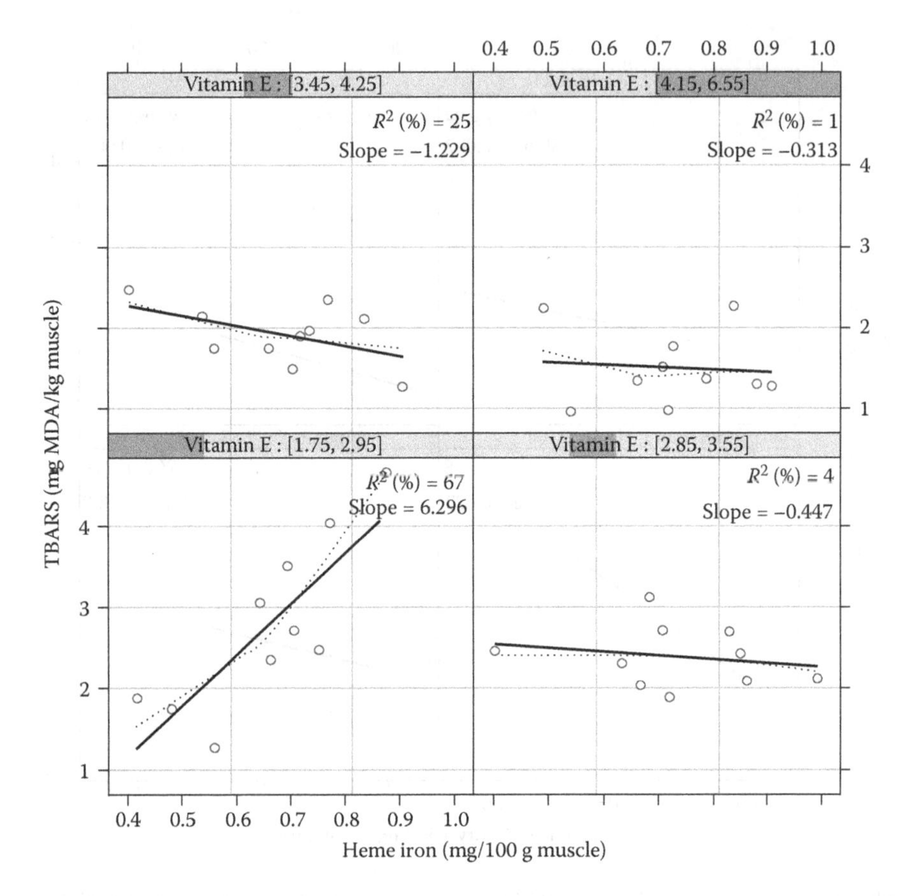

FIGURE 4.4 Lipid oxidation (TBARS) versus total heme iron conditioned on vitamin E. The vitamin E was conditioned by creasing four shingles or classes of equal count, with 10% overlap between adjacent intervals. The shingle intervals were 1.75–2.95 (shingle 1), 2.85–3.55 (shingle 2), 3.45–4.25 (shingle 3), and 4.15–6.55 (shingle 4), respectively. The R^2 (%) values in each of the panels are to be read as percent. The slopes are dependent on the scales between the x and y variables. The emphasis is on the magnitude of slope of the straight-line in each panel. The R^2 and slope values are subjected to some dependency between the overlaps (10% overlap) from one panel to the next. (Adapted from Ponnampalam, E.N. et al. 2014b. *Lipids* 49:757–766.)

have to be either antioxidants or pro-oxidants, with antioxidant activity being more pronounced at low oxygen partial pressure.

The supplementation of animal diets with carotenoids has been investigated in an attempt to explore their antioxidant potential, as described above, but also to utilize carotenoids for their potential to pigment muscle of certain animal species. For example, fish such as Atlantic salmon and rainbow trout have been supplemented with carotenoids in an attempt to produce the stable pink coloration in the muscle of aquaculturally produced fish that consumers expect to see in wild fish equivalents. Pasture-fed cattle had significantly higher levels of β-carotene in plasma,

TABLE 4.1
Residual Standard Deviation and Percentage Variance Accounted for Some Competing Models to Explain Redness of Meat at Days 3 and 4 of Retail Display

Terms in Model	Residual Standard Deviation	Percentage Variance Accounted For
None	1.55	–
Diet	1.07	39.3
Total n-3	1.26	16.2
Total n-6	1.04	42.7
Total PUFA	1.04	42.4
Vitamin E	0.99	48.2
Heme iron	1.21	21.9
Heme iron and diet	1.01	45.7
Heme iron and total n-3	1.13	32.3
Heme iron and total n-6	0.86	61.3
Heme iron and total PUFA	0.89	58.2
Vitamin E and total n-3	1.01	57.5
Vitamin E and total n-6	0.99	59.1
Vitamin E and total PUFA	0.99	59.4
Vitamin E and diet	0.94	53.7
Vitamin E and heme iron	0.84	70.2

Source: Adapted from Ponnampalam, E.N. et al. 2012a. *Meat Sci.* 90:297–303.

liver, adipose tissue, and muscle tissues than grain-fed cattle and the levels were reduced in all tissues by supplementation of vitamin E in the diet (Yang et al. 2002). Generally, the feeding of carotenoids to fish and poultry has been successful in terms of generating desirable visual hues in the final muscle, as well as egg quality in the case of chickens. However, what is less clear is the antioxidant activity associated with carotenoids as described previously, and further investigation is required to understand how nutritionally important carotenoids need to be manipulated in order to enhance antioxidant activity, especially with respect to balancing animal diets with other antioxidants, like vitamin E, to deliver greater oxidative stability in muscle systems.

In relation to the impact of supplementing animal diets with carotenoids, little, if anything has been done to investigate the effects of carotenoids on other muscle-food quality attributes, therefore, it is difficult to say whether carotenoid supplementation has any effect on sensory attributes other than color, drip loss, and microbiology. Further studies are required to better understand the biochemical functioning of dietary carotenoid following supplementation to a wider range of muscle-food producing animals and the full extent of how carotenoids impact on product quality.

4.4.1.3 Vitamin D

Vitamin D is essential in the maintenance of calcium and phosphate levels in circulating blood which facilitates bone mineralization, muscle contraction, nerve signaling, and cellular function. Muscle-food producing do not maintain body stores of vitamin D as this vitamin is typically synthesized in the skin on exposure to sunlight. In the absence of or suboptimal availability of sunlight, vitamin D can be supplemented in the diet. Vitamin D is not an active vitamin form and must be biochemically converted to the active form 1.25 dihydroxyvitamin D or calcitriol. It is this active form that is responsible for all necessary physiological functions.

The supplementation of vitamin D to animals through the diet has been investigated, especially for poultry (Mattila et al. 2011), however most of the studies have focused on vitamin D accumulation in tissues and the ramifications of supplementation on animal health, physiology, and safety of use. For example, a recent report by the European Food Safety Authority Panel on Additives and Products or Substances used in Animal Feed—FEEDAP (2013) assessed the safety and efficacy of vitamin D_3 (cholecalciferol) as a feed additive for pigs, piglets, bovines, ovines, calves, equines, broiler chickens, turkeys, other poultry, and fish. This work reported that elevated levels below safety threshold levels brought about increased vitamin D_3 levels in these meat producing animals without promoting any side effects. However, like this study indicates, little has been reported on the effects of vitamin D and its derivatives on meat quality with one exception. Some research has focused on determining if vitamin D supplementation has any positive impact on beef tenderness values.

It is now well established that if calcium levels are increased in muscle postmortem (typically through calcium injections into the muscle) the process of meat tenderization is enhanced (Wheeler et al. 1994). However, this practice is not widely accepted industrially. An alternative approach to increasing calcium levels in muscle might be achieved through the dietary supplementation of beef with vitamin D. In an early study by Hibbs et al. (1951), the authors demonstrated that oral vitamin D supplementation to lactating cows for 2 weeks prepartum increased serum calcium. Additionally, injections of vitamin D metabolites increased serum calcium levels in cows (Bar et al. 1985; Sachs et al. 1987). From these studies, obvious questions were raised with respect to how such approaches and effects might improve beef tenderness. In his review of the area, Morgan (2007) concluded that published and unpublished work, in general, indicated that the elevation of muscle calcium concentration as a result of vitamin D supplementation promotes meat tenderness, but that the impact is inconsistent and the factors involved in obtaining a positive effect are poorly understood.

4.4.1.4 Vitamin K

Vitamin K is required for the manufacture of blood plasma clotting factors. There are two major natural sources of vitamin K and these are the phylloquinones (vitamin K found in plants) and the menaquinones (vitamin K produced by bacterial rumen microflora) (Rasby et al. 2011). These authors also highlight the fact that because vitamin K is synthesized in large quantities by ruminal bacteria and is abundant in both pasture and green roughages, dietary supplementation is for the most part, not required. From extensive review of the scientific literature, no studies were found with respect to the dietary supplementation of animals diets on resulting muscle-food quality.

4.4.2 Water Soluble Vitamins

4.4.2.1 B Vitamins

The B vitamins are comprised of eight separate vitamin forms, namely; B_1 (thiamine), B_2 (riboflavin), B_3 (niacin or niacinamide), B_5 (pantothenic acid), B_6 (comprised of pyridoxine, pyridoxal, pyridoxamine, pyridoxine hydrochloride), B_7 (biotin), B_9 (folic acid), and B_{12} (comprised of various cobalamins). All eight B vitamins together are described as the vitamin B complex and all eight B vitamins have their own specific nutritional and biochemical functions in the body. B vitamins are derived from feed sources such as cereals and milk and B vitamins in ruminants are synthesized by rumen microorganisms. As all B vitamins are water soluble, unlike fat-soluble vitamins, animal stores of B vitamins are minimal and excess levels of B vitamins are rapidly excreted by the animal, with the exception of B_{12} which is stored in the liver (Lynch and Kerry 2000). Therefore, supplementing animals with B vitamins in an attempt to improve meat quality consequently might prove unsuccessful and this may be a reason for the lack of experiments carried out to investigate such an impact. However, one of the nutritionally linked benefits to consuming meat is that meat provides one of the best sources of B vitamins generally, especially for vitamins B_1, B_6, and B_{12}.

It has been clearly shown that of all meat types, pork was the best provider of thiamine (USDA 1998). Thiamine is found in the body only in a form attached to enzymes and consequently, excess is excreted through the urine (Windham et al. 1990). Kirchgessner et al. (1995) noted that thiamine content of lean beef declined with increasing weight and was higher when animals were fed a restricted level of feed, which slowed the growth rate. The effect of feeding was not consistent for the sexes (bulls, steers, and heifers) and differences due to gender were greater than differences due to feeding level.

Vitamin B_6 is also supplied by meat and Sauberlich (1990) reported that meat in the United States supplied about 40% of the dietary intake of vitamin B_6. In contrast with B_1, Kirchgessner et al. (1995) reported that the vitamin B_6 content of lean beef increased with increasing weight. At a common slaughter weight tissue B_6 was higher when animals were fed a restricted level of feed which slowed growth rate. Just as in the case for vitamin B_1, differences due to animal gender (e.g., hormonal variation) were greater than differences due to feeding level.

Vitamin B_{12} is almost exclusively supplied to the human diet by meat and liver consumption, providing almost 75% of the dietary intake in the United States (Sauberlich 1990). Ruminant meats tend to be higher in vitamin B_{12} than chicken and pork, probably reflecting the abundant dietary supply from rumen bacteria (Lynch and Kerry 2000).

Other than the mere presence of B vitamins in muscle-based foods, little, if anything, has been reported on their impact on meat quality attributes.

4.4.2.2 Vitamin C

Vitamin C or ascorbic acid (the salt of which is ascorbate) is a reversible biologic reductant, and as such, provides reducing equivalents for a variety of biochemical reactions and is considered the most important antioxidant in extracellular fluid (Sies et al. 1992). Ascorbate is thermodynamically close to the bottom of the list of one electron reducing potentials of oxidizing free radicals ($E^{o\prime}$ = +282 mV). For this reason, ascorbate is the first line of defense against oxidizing free radical species in plasma (Briviba and Sies 1994). Ascorbate, which like B vitamins is water soluble, is reactive enough to effectively interrupt oxidants in the aqueous phase before they can attack and cause detectable oxidative damage to DNA and lipids (Morrissey and Kerry 2004).

One of the functions of ascorbic acid in living animals is to regenerate α-tocopherol from the α-tocopheroxyl radical (Packer and Kagan 1993). This suggests that supplementation of animal diets with ascorbic acid might enhance the protection afforded by α-tocopherol against lipid oxidation in meats after slaughter. Lahučký et al. (2005) reported that vitamin C supplementation to pigs improved meat pH, reduced drip loss from meat, reduced the rate of oxidation (malondialdehyde [MDA] production) by stimulation with Fe^{2+}/ascorbate (incubation of *longissimus* muscle for 0 and 30 min) in supplemented pigs compared to control groups, but also reported no improvement in the oxidative stability of meat when thiobarbituric acid reactive substances (TBARS) were measured between both dietary groups. In beef cattle, jugular infusion of ascorbate (1.7 mol) 10 min prior to slaughter delayed OMb oxidation and extended the color display life of the *psoas*, *gluteus*, and *longissimus* muscles (Schaefer et al. 1995), but the authors concluded that dietary supplementation would probably not be as effective given the rapid rate at which ascorbate disappears from plasma. Data on the ability of ascorbate to regenerate α-tocopherol *in vivo* are conflicting and Wenk et al. (2000) concluded that the "sparing" action of ascorbate on α-tocopherol may be of negligible importance *in vivo* in animals that are not oxidatively stressed.

4.5 TRACE ELEMENT/MINERALS

4.5.1 Calcium

The role of calcium (Ca) in meat quality has been extensively discussed in a previous chapter (Chapter 3). Therefore, the following is a brief overview to provide context.

Ca ions play a central role in determining meat tenderness. This occurs via its function within the calpain system which acts to degrade and autolyze muscle proteins through synthesizing protease enzymes (Kemp et al. 2010). There are three categorizes of calpains; (1) μ-calpains; (2) m-calpains; and (3) calpains-3. Calpastatins and the availability of Ca are important for regulating calpain function (Kemp et al. 2010). Consequently, increased muscle and plasma Ca concentrations equate to parallel increases in calpain activity and improved meat quality, with greater myofibrillar fragmentation and tenderness following postmortem protein degradation (Zinn and Shen 1996; Tizioto et al. 2014). Muscle Ca content is known to vary, and management strategies can be applied to improve Ca levels. Genetic management of calpain genes, such as the Ca-activated neutral protease 1 gene (CAPN1) has been shown to increase calpain activity (Tizioto et al. 2014). Yet, nutritional tools are often proven to be more effective and practical than genetic management approaches.

Ca absorption has been found to be relatively minor in rumens, albeit with Ca supplementation absorption can increase provided lipid content of the diet is monitored. Post-rumen and total tract Ca absorption is comparatively efficient and unaffected by lipid supplementation, which can have a limiting effect (Zinn and Shen 1996). However, dietary Ca absorption cannot be discussed without mentioning vitamin D, as both compounds are intrinsically linked. Several studies have reported vitamin D supplementation improves meat tenderness through promoting Ca content in animal plasma and muscles (Montgomery et al. 2000; Karges et al. 2001; Foote et al. 2004). As vitamin D is freely provided with exposure to sunlight, Lobo-Jr et al. (2012) tested the merits of supplementing vitamin D to cattle under normal and shaded sunlight exposure. It was found that vitamin D supplementation only influenced plasma and muscle Ca content when animals were shaded from direct sunlight. From this, producers can determine the advantages of vitamin D supplementation on increasing Ca content as per individual production environments.

4.5.2 Magnesium

Magnesium (Mg) is an essential cofactor in several enzymatic reactions necessary in energy and protein metabolism (Duffine and Volpe 2013). It also can antagonize Ca within metabolic pathways and hence, reduce neurotransmission, neuromuscular stimulation, and overall muscle activity (D'Souza et al. 1999). A combination of these functions results in dietary Mg being implicated in several areas of meat quality.

Meat water holding capacity (WHC) has severe economic considerations, based on its ultimate relation to yield and eating quality. WHC develops from the release of ions from the sarcolemma and this can detrimentally affect the ability of muscle to retain fluid via its contribution to the myofibril lattice swelling and contraction (Duffine and Volpe 2013). The lost fluid can be defined as drip loss. Increased Mg content within myofibrils postmortem can improve WHC through the disruption of ion flux mechanisms and the improvement of osmolarity potentials (Bond and Warner 2007). Previous research supports this, with animals supplemented with increased Mg levels over basal diet levels producing meat with higher Mg content, lower drip loss, and better WHC comparative to unsupplemented animals (D'Souza et al. 1999; Lahučký et al. 2004; Bond and Warner 2007). The cost-effectiveness of Mg supplementation has also been found improved by appropriate Mg source selection (D'Souza et al. 1999).

Muscle color provides the initial cue for consumer discrimination toward a quality product, and Mg supplementation contributes to color stability. D'Souza et al. (1999) describes Mg supplementation to swine as resulting in more desirable and darker meat coloring. This effect was not shared by ruminant species, with Apple et al. (2000) reporting Mg inclusion into a lamb diet did not influence meat color regardless of muscle cut, Mg supplementation dosage or Mg source. This difference could stem from digestive system differences, with Mg reported to have an antioxidant effect preventing lipid peroxidation in swine (Lahučký et al. 2004) which may have been disabled while transversing the ruminal system. This effect could also explain the observed delay in pork loin rancidity while stored in refrigerated vacuum packaging with Mg supplementation (Apple et al. 2001). Another explanation involves the reduction of meat intramuscular fat percentage observed with Mg supplementation (Apple et al. 2000). This, in turn, could contribute to the increases in meat toughness associated to Mg supplementation via meat shear force (Ramirez et al. 1998), as a proxy to tenderness.

Postmortem carcass pH values also influence meat tenderness (Hopkins et al. 2011), and supplementation of swine with Mg for the 5 days preceding slaughter was found to increase pH measurements taken at 45 min postslaughter. This is not unique to swine (Bass et al. 2010). Gardner et al. (2001) linked Mg supplementation with lesser glycogen loss and more replenishment within muscle postmortem. This could contribute to the incidence of dark cutting. However, this effect has also been shown to stabilize the effects of stress-induced catecholamine secretion within swine, which can lead to lactic acid accumulation, lower pH, pale muscle color, and increased drip loss (D'Souza et al. 2000).

4.5.3 Selenium

Dietary selenium (Se) has been associated with improved cardiovascular health, amelioration from UV damage or inflammation, and protection against heavy metal toxicity and cancer (Navarro-Alarcon and Cabrera-Vique 2008). These benefits result from Se contributions to the synthesis of several antioxidants (Vignola et al. 2009), including glutathione peroxidases (GPx). In this regard, Se complements the cellular protective function of vitamin E, eliminating any residual metabolic peroxides as a virtual "second-defense line."

It is recommended to consume at least 40 μg of Se per day to maintain GPx expression and avoid deficiency (Combs 2001); hence prevent Keshan or Kaschin–Beck disorders. Levels up to 300 μg of Se per day are necessary to ensure that these aforementioned health benefits are expressed (Navarro-Alarcon and Cabrera-Vique 2008). Fortunately, red meat is a viable source of Se, with over 20% of daily recommended intake provided by a 100 g serve (Williams 2007) (equating to approximately 10 μg/100 g). Se sourced from meat occurs within amino acid compounds; Se-Methionine (Se-Met) and Se-Cysteine, as Se is interchangeable with sulfur. Se-Met is readily absorbed, demonstrating similar properties to plant-sourced Se (Navarro-Alarcon and Cabrera-Vique 2008). However, substantial variations exist in Se availability as per geographic location and animal diet composition (Combs 2001)—prompting research focused on improving meat Se content.

Supplementation of animals with Se has been found to result in improved muscle content. For instance, Vignola et al. (2009) found supplementing lambs with Se-enriched yeasts improved Se accumulation within muscle tissue. Juniper et al. (2008) found that deposition of Se in the muscle of Limousin x Holstein-Friesian cattle was source and dose dependent and using organic Se feed sources resulted in greater accumulation in the muscle compared to inorganic Se, such as sodium selenite. In swine, Se supplementation improved muscle Se content regardless of Se supplement type (Goehring et al. 1984; Kim and Mahan 2001; Zhan et al. 2007). Improving meat Se content does not solely contribute to nutritional enrichment, but can directly influence meat quality. Cozzi et al. (2011) reported that Charolais bulls supplemented Se-enriched yeast developed meat with greater tenderness and lighter color (as per CIE colorimetric values) when compared with unsupplemented animals. Ripoll et al. (2011) found Se supplementation increased the lightness of lamb meat following vacuum packaging; however other studies show Se supplementation did not affect meat color in veal (Skřivanová et al. 2007) and lamb (Vignola et al. 2009). The basis of this discrepancy was thought to be differences in dosage levels, being insufficient to promote muscle GPx activity and limit oxidation and color decay (Skřivanová et al. 2007). Se contributions to meat pH and WHC have been found insignificant (Wolter et al. 1999).

4.5.4 Iron

4.5.4.1 Implications on Health

Iron is essential in any balanced human diet. It contributes to numerous cellular processes, and possibly most importantly facilitates the movement of oxygen throughout the cardiovascular system. Consequently, insufficient dietary iron can result in anemia and other health complications (Gibson and Ashwell 2003), with women and persons of strict vegan and vegetarian diets most at risk (Cosgrove et al. 2005).

Red meat is an excellent source of dietary iron. For instance, a 100 g serve of beef or lamb can provide an adult human with over one-quarter of their daily recommended intake—albeit requirements can differ depending on age, gender and status (Williams 2007). Red meat provides dietary iron as heme iron, its predominant form especially in muscle Mb. Humans can better absorb heme iron at between 20% and 30% of that available, compared with non-heme iron, (Williamson et al. 2005) because it is less influenced by other dietary constituents which may otherwise enhance or inhibit absorption (Gibson and Ashwell 2003). This proves advantageous over other iron sources, such as plant based, which are generally absorbed at only 7% (Williamson et al. 2005). Iron absorption can be promoted by the presence of glutathione, being a component of glutathione peroxidase enzymes. This further promotes meat as a source of iron, as red meats have been estimated to contain between 12 and 20 mg of glutathione per 100 g of meat (Williams 2007). Hence, Gibson and Ashwell (2003) have reported individuals consuming regular moderate serves of red meat as having a lower risk of developing disorders associated with iron deficiencies. This outcome is complemented by other studies (Cosgrove et al. 2005; Wyness et al. 2011; Sharma et al. 2013).

There is a flip side to sourcing dietary iron from heme iron. Research has linked consumption of red meats and specifically heme iron with increased risk of congenital heart disease (Sempos et al. 1996) and colorectal cancer development, the latter specifically with processed red meats (Huang 2003; Corpet 2011; Oostindjer et al. 2014). These associations are founded on heme iron consumption being shown to promote cellular proliferation within the gastric mucosa through lipid oxidation (peroxidation of unsaturated fatty acids) to release carcinogenic reactive oxygen species—potentially damaging DNA (Sesink et al. 1999; Ijssennagger et al. 2013).

However, provided that red meat is consumed as a part of a balanced diet (Wyness et al. 2011); the benefits from dietary iron and anticarcinogenic compounds sourced within red meat promote its continued consumption.

4.5.4.2 Iron: Implications on Color and Quality

Initial appraisal of meat quality is largely based upon its appearance and color. Consumers prefer a bright red colored meat surface—that having heightened CIE L^* and a^* values—as being a quality product (Khliji et al. 2010). Therefore, it is within retailers' best interests to market meats which can maintain this desirable color over prolonged periods of retail display.

Many factors can influence meat color, most through their interplay with muscle Mb. Mb is the predominant coloring pigment compound within muscle tissue and functions in oxygen transport and distribution (Baron and Andersen 2002). Unsurprisingly then, Mb is highly reactive to oxygen. This is due to the central role iron has within each Mb complex, also referred to as heme iron (Baron and Andersen 2002). Studies have associated heme iron with meat color, with increased levels found to correspond with increased pigmentation content and muscle redness or a^* values (Allen and Cornforth 2006; Ponnampalam et al. 2012). However, it is the chemical structure of heme iron which most varies Mb contributions to muscle color.

Muscle Mb forms deoxymyoglobin (DMb) within anaerobic conditions, and following exposure to oxygen as per exposing a fresh meat surface, causes DMb to bind with oxygen to form OMb (AMSA 2012). OMb is responsible for the desirable bright red meat surface color and its central iron atom shifts to a ferrous isotope (Fe^{2+}) (AMSA 2012). The transformation of DMb to OMb is reversible with the removal of the oxygen atom, but OMb is a volatile compound (low redox stability) and can further oxidize to form MMb (Faustman et al. 2010). This reaction involves the reaction of OMb ferrous iron atom with superoxide (O^{2-}) to become ferric iron (Fe^{3+}), the central iron isotope within MMb. Research has found this reaction can even occur in very low concentrations of oxygen, down to 1%–2% availability levels (AMSA 2012). MMb content is acknowledged to contribute to meat discoloration—increasing brownness instead of the desirable redness (Mancini and Hunt 2005). Consequently, MMb formation is detrimental to meat quality and retail value.

The ferrous iron (Fe^{2+}) in OMb can drive other reactions within meat. The Fenton reaction is the most prolific of these and involves Fe^{2+} reacting with hydrogen peroxide (H_2O_2, often sourced from O^{2-} dismutation) to produce hydroxyl radicals (OH^-) (Allen and Cornforth 2006; Chaijan 2008). OMb can also autoxidize to release reactive species (Baron and Andersen 2002). These reactive species can permeate into muscle lipid regions inaccessible to many other reactive compounds, and prompt lipid oxidation. Through this reaction, OMb oxidation to MMb has been closely akin with lipid oxidation in several scientific studies (e.g., Allen and Cornforth 2006).

Lipid oxidation, or peroxidation, involves the subtraction of a lipid electron by a free radical species, such as OH^- as produced by the Fenton reaction. This results in unsaturated fatty acids (UFA) being especially prone to oxidation because of their possession of multiple and highly reactive double bonds between carbon atoms. Lipid oxidation occurs in three stages; (1) initiation; (2) propagation; and (3) termination reaction (Chaijan 2008). These each release chemical species, for example, alkyl, alkoxy, and peroxy groups, with the latter possibly the most common secondary compound which can be broken down further into aldehydes, ketones, epoxides, and malondialdehydes (Chaijan 2008; Faustman et al. 2010). Malondialdehydes can be measured in TBARS analysis which is widely used to quantify lipid oxidation level and rancidity, in combination with the aforementioned secondary derivatives. These same compounds are thought to contribute to OMb oxidation to MMb (Gorelik and Kanner 2001; Faustman et al. 2010), contributing to the cyclic relationship between Mb and lipid oxidation centered by a shared role of iron.

Supplementation of dietary iron has been investigated as a tool for managing muscle iron content and follow-on effects on meat color and rancidity characteristics. For instance, O'Sullivan et al. (2003) found swine muscle iron content increased with supplementation, although varying dependent on anatomical position. This was thought to contribute to the observed WOF variations between muscle types and supplementation level. Macdougall et al. (1973) reported veal calf Mb content increased with iron supplementation, and the response was dose dependent as reflected in color changes. Ponnampalam et al. (2012) found lamb meat redness (a^*) increased with heme-iron content, which was diet dependent. Likewise, Apple et al. (2007) demonstrated swine supplemented with 100 ppm iron produced chops with greater redness (a^*) and vividness over 7 days on display than those supplemented at 50 ppm, and compared to unsupplemented controls.

4.6 HERBS AND BIOACTIVE COMPOUNDS

Many plants produce secondary derivatives that are not directly involved in plant survival, but instead contribute to their adaption and defense against biological and environmental threats (Rochfort et al. 2008; Durmic and Blache 2012). For countless years these plants have been employed as health promoting agents, classified as herbs and cultivated to promote expression of these secondary derivatives. The effect of these same compounds on meat quality is only beginning to be explored (see Table 4.2), in association with improvements to animal health and productivity (Vasta and Luciano 2011).

Tannins are large polyphenolic compounds, phenolytic compounds characterized by their hydroxylated aromatic ring chemical structures and natural occurrence in botanical species. The use of tannins within animal diets has been the topic of several investigations, many of which note direct effects on bypassing dietary proteins in the rumen leading to increased muscle deposition and improvements in meat quality traits. For instance, Luciano et al. (2009a) reported improved color (redness) in *semimembranosus* (SM) from lambs supplemented with tannin-rich quebracho (*schinopsis lorentizii*). This same study found increased MMb formation and heme pigment concentration in SM refrigerated for 2 weeks from lamb supplemented with tannins compared to unsupplemented controls. Further research discovered lamb color stability was extended over several days on display with tannin-rich quebracho supplementation; as MMb percentage decreased and total muscle phenols increased (Luciano et al. 2011), the latter possibly due to an improvement in antioxidant activity (Rochfort et al. 2008).

TABLE 4.2
A Summary Table of the "Influence of Trace Element/Minerals" and "Herbs and Botanical Secondary-Derivatives" Sections

Compound	Meat Quality	Reference
Calcium	Regulates tenderness	Zinn and Shen (1996), Kemp et al. (2010)
Magnesium	Improves water holding capacity	Bond and Warner (2007)
	Contributes to color stability	D'Souza et al. (1999)
	Limits peroxidation/rancidity	Lahučký et al. (2004)
	pH decline rate	Gardner et al. (2001), Bass et al. (2010)
Selenium	Improves tenderness	Cozzi et al. (2011)
	Promotes lighter color	Ripoll et al. (2011)
Iron	Influences color	Allen and Cornforth (2006), Ponnampalam et al. (2012)
	Contributes to peroxidation/ rancidity	Gorelik and Kanner (2001), Faustman et al. (2010)
Tannins	Increase redness (color)	Luciano et al. (2009a), Luciano et al. (2011)
	Reduce peroxidation/rancidity	Larraín et al. (2008)
Saponins	Fatty acid composition	Brogna et al. (2011)
Betaine	Improves water holding capacity	Rochfort et al. (2008)

As phenols are associated with antioxidant properties, tannin effects on lipid oxidation have been considered. Luciano et al. (2011) observed MMb percentages on both meat surfaces and the level was lower following nitrite-induced oxidation when lambs were supplemented with tannins. Similarly, Larraín et al. (2008) found *gluteus medius* (GM), TBARS, a by-product of lipid oxidation, concentration was decreased when sourced from steers supplemented with tannins. Interestingly, this finding was exclusive to the GM, with the *longissimus* muscle removed from tannin supplemented steers found to exhibit increased discoloration and TBARS accumulation upon the meat surface. These findings suggest tannin supplementation to ruminants will affect lipid oxidation, possibly via reducing aerobic storage oxidation potential while accelerating oxidation during display. Studies of the actual influence of dietary tannins on meat fatty acid profiles have shown minimal changes, albeit lambs fed condensed tannins from carob pulp were observed to have lower conjugated linoleic acid and trans-vaccenic acid content in the *longissimus* muscle (Vasta et al. 2007).

The type and method of including specific herbs and plant secondary derivatives into animal diets can affect meat quality. Dietary feeding trials where poultry, pig, and beef cattle diets were supplemented with green tea catechins have been reported in the scientific literature. Dietary supplementation of pig diets (Mason et al. 2005) and beef diets (O'Grady et al. 2006) with tea catechins showed low or no effect with respect to the oxidative stability of pork and beef, respectively. Vasta et al. (2013) demonstrated that the *longissimus* muscle fatty acid profile is affected by white wormwood (*Arthemisia herba alba*) supplementation in lambs, resulting in increased vaccenic, rumenic, and linolenic acids and PUFA content. This effect was not repeated with rosemary essential oil supplementation, nor was any effect on volatile organic compound profiles observed. Lambs fed cinnamon (cinnamaldehyde), garlic, or juniper berry extract at 2 g per kg of basal diet were reported to demonstrate no change in meat aroma or flavor traits (Chaves et al. 2008). As consumers are generally sensitive to shifts in food sulfur compounds, of which garlic is a rich source, the absence of change in aroma with garlic supplementation suggests possible partitioning of these sulfur compounds away from tissue sinks.

Meat color has been identified as affected by herb supplementation. Simitzis et al. (2008) fed lambs alfalfa hay sprayed with oregano extract (at 1 mL/kg) over a 2-month period and observed *longissimus* muscle pH and *a** (redness) and *b** (yellowness) values varied in comparison to unsupplemented lambs. Fresh vetch (*Vicia sativa*) also altered color, with SM in supplemented lambs observed to have increased redness and lesser yellowness and hue angle values when compared to the control group (Luciano et al. 2009b). These shifts are thought to be based upon supplemented herbs containing phenolic compounds (secondary derivatives) which delay or retard oxidation following ingestion. Furthermore, there is a possibility this effect can resonate across generations. Nieto et al. (2011) fed ewes with distilled rosemary leaf extract throughout a pregnancy (approximately 240 days), the progeny were then fattened using commercial pellets (no rosemary leaf extract) and found to produce *longissimus* muscle with delayed rancidity as per TBARS analysis and shelf-life tests. Lamb meat color and juiciness were also reported as affected by dam supplementation, as determined by a trained sensory panel, when compared to control counterparts (Nieto et al. 2011).

Meat quality effects from herb supplementation rely upon their content of various phytonutrient compounds, some of which are more officious than others. Saponins are such a secondary derivative and typified by their sugar moiety of either triterpenoid or steroid nature (Brogna et al. 2011). Supplementing lambs with saponins—extracted from the soapbark tree (*Quillaja soponaria L.*)—could improve arachidonic acid (C20:4n-6) content in lamb meat provided dosage did not exceed 60 ppm (Brogna et al. 2011). Yet, other effects on meat fatty acid profiles or meat quality aspects proved minimal. Betaine is an amino acid derivative and another phytonutrient of interest, with supplementation linked with improved WHC and drip loss in swine (Rochfort et al. 2008), although no effect was observed in ovine (Pearce et al. 2008). However, more research is required in the fields of individual plant secondary-derivative supplementation, especially in improving the cost-effectiveness when compared to supplementation of complete herb or basic extract.

4.7 CONCLUSIONS

Nowadays, consumers are more selective about their food and are well aware of the effect of meat on their health and wellness. The development of high value flavorsome nutritious lamb and beef with value added properties, such as good appearance and tenderness, are the ways to improve consumer appeal and market competitiveness. However, proper nutritional strategies need to be applied, particularly where animals are grown extensively with limited control on feed consumption and, where the cost of dietary supplementation becomes a restrictive factor. This chapter aimed at clarifying these strategies in regard to immediate effects upon meat production and quality—with energy (fat) and protein supplementation relating primarily to the former; vitamin, mineral, and herbs and secondary-botanical compounds to the latter; and lipid supplementation contributing to both focuses.

From the literature it can be concluded that the amount and type of FA in meat can be managed utilizing tailored feeding systems which, in turn affect the functional properties of meat. Muscle pigment (Fe) and lipid oxidation are related, and contribute to meat discoloration and shifts in consumer perceived meat qualities. Furthermore, feeding systems influence the animal muscle tissue content of iron, FA, and antioxidant and vitamin status. This highlights a necessity to quantify animal dietary effects on meat quality and health benefits while simultaneously investigating new and novel feed types, including herbs and secondary plant compounds.

REFERENCES

Aberle, E.D., Forrest, J.C., Gerrard, D.E., and Mills, E.W. 2001. *Principles of Meat Science*. Hunt Publishing, Kendal.

Alexandratos, N., and Bruinsma, J. 2012. World agriculture towards 2030/2050: The 2012 revision. ESA Working Paper No. 12-03. FAO, Rome, Italy.

Allen, K.E., and Cornforth, D.P. 2006. Myoglobin oxidation in a model system as affected by non-heme iron and iron chelating agents. *J. Agric. Food Chem.* 54:10134–10140.

AMSA. 2012. Meat color measurement guidelines. Hunt, M. and King, D. (eds.), December 2012 edition. American Meat Science Association (ASMA), Champaign, IL.

Anderson, B.M., and Ma, D.W.L. 2009. Are all n-3 polyunsaturated fatty acids created equal? *Lipids Health Dis.* 8:33.

Apple, J.K., Davis, J.R., Rakes, L.K., Maxwell, C.V., Stivarius, M.R., and Pohlman, F.W. 2001. Effects of dietary magnesium and duration of refrigerated storage on the quality of vacuum-packaged, boneless pork loins. *Meat Sci.* 57:43–53.

Apple, J.K., Wallis-Phelps, W.A., Maxwell, C.V., Rakes, L.K., Sawyer, J.T., Hutchison, S., and Fakler, T.M. 2007. Effect of supplemental iron on finishing swine performance, carcass characteristics, and pork quality during retail display. *J. Anim. Sci.* 85:737–745.

Apple, J.K., Watson, H.B., Coffey, K.P., Kegley, E.B., and Rakes, L.K. 2000. Comparison of different magnesium sources on lamb muscle quality. *Meat Sci.* 55:443–449.

Arnold, R.N., Scheller, K.K., Arp, S.C., Williams, S.N., and Schaefer, D.M. 1993. Tissue equilibrium and subcellular distribution of vitamin E relative to myoglobin and lipid oxidation of displayed beef. *J. Anim. Sci.* 71:105–118.

Asghar, A., Lin, C.F., Gray, J.I., Buckley, D.J., Booren, A.M., and Flegal, C.J. 1990. Effects of dietary oils and α-tocopherol supplementation on membranal lipid oxidation in broiler meat. *J. Food Sci.* 55:46–50.

Astrup, A., Dyerberg, J., Elwood, P. et al. 2011. The role of reducing intakes of saturated fat in the prevention of cardiovascular disease: Where does the evidence stand in 2010? *Am. J. Clin. Nutr.* 93:684–688.

Aurousseau, B., Bauchart, D., Calichon, E., Micol, D., and Priolo, A. 2004. Effect of grass or concentrate feeding systems and rate of growth on triglyceride and phospholipid and their fatty acids in the m. *longissimus thoracis* of lambs. *Meat Sci.* 66:531–541.

Baghurst, K. 2004. Dietary fats, marbling and human health. *Aust. J. Anim. Sci.* 44:635–644.

Bailey, A.J., and Light, N.D. 1989. *Connective Tissue in Meat and Meat Products*. Elsevier Applied Sciences, London, pp. 195–224.

Bar, A., Perlman, R., and Sachs, M. 1985. Observation on the use of 1ahydroxyvitamin D_3 in the prevention of bovine parturient paresis: The effect of a single injection on plasma 1ahydroxyvitamin D_3, calcium, and hydroxyproline. *J. Dairy Sci.* 68:1952–1958.

Baron, C.P., and Andersen, H.J. 2002. Myoglobin-induced lipid oxidation. A review. *J. Agric. Food Chem.* 50:3887–3897.

Bas, P., and Morand-Fehr, P. 2000. Effect of nutritional factors on fatty acid composition of lamb fat deposits. *Livest. Prod. Sci.* 64:61–79.

Bass, P.D., Engle, T.E., Belk, K.E., Chapman, P.L., Archibeque, S.L., Smith, G.C., and Tatum, J.D. 2010. Effects of sex and short-term magnesium supplementation on stress responses and *longissimus* muscle quality characteristics of crossbred cattle. *J. Anim. Sci.* 88:349–360.

Beauchemin, K.A., McClelland, L.A., Jones, S.D.M., and Kozub, G.C. 1995. Effects of crude protein content, protein degradability and energy concentration of the diet on growth and carcass characteristics of market lambs fed high concentrate diets. *Can. J. Anim. Sci.* 75:387–395.

Bedi, K.S., Birzgalis, A.R., Mahon, M., Smart, J.L., and Wareham, A.C. 1982. Early life under nutrition in rats. 1. Quantitative histology of skeletal muscles from underfed young and refed adult animals. *Br. J. Nutr.* 47:417–431.

Beever, D.E., and Thorp, C.L. 1997. Supplementation of forage diets. In Welch, R.A.S., Burns, D.J.W., Davis, S.R., Popay, A.I., Prosser, C.G. (eds.), *Milk Composition, Production and Biotechnology*. CAB International, Oxon, UK, p. 419.

Bekhit, A., Hopkins, D.L., Fahri, F., and Ponnampalam, E.N. 2013. Oxidative processes in muscle systems and fresh meat: Sources, markers and remedies. *Compr. Rev. Food Sci. Food Saf.* 12:565–597.

Bekhit, A.E.D., Geesink, G.H., Morton, J.D., and Bickerstaffe, R. 2001. Metmyoglobin reducing activity and colour stability of ovine *longissimus* muscle. *Meat Sci.* 57:427–435.

Bermingham, E.N., Roy, N.C., Anderson, R.C., Barnett, M.P.G., Knowles, S.O., and McNabb, W.C. 2008. Smart foods from the pastoral sector—Implications for meat and milk producers. *Aust. J. Exp. Agric.* 48:726–734.

Beverly, S.B., Daood, M.J., Laframboise, W.A., Watchko, J.F., Foley, T.P., Butler-Browne, G.S., Whalen, R.G., Guthrie, R.D., and Ontell, M. 1991. Effects on perinatal undernutrition on elimination of immature myosin isoforms in the rat diaphragm. *Lung Cell. Mol. Physiol.* 5:49–54.

Biesalski, H.K. 2005. Meat as a component of a healthy diet—Are there any risks or benefits if meat is avoided in the diet? *Meat Sci.* 70:509–524.

Bolte, M.R., Hess, B.W., Means, W.J., Moss, G.E., and Rule, D.C. 2002. Feeding lambs high-oleate or high-linoleate safflower seeds differentially influences carcass fatty acid composition. *J. Anim. Sci.* 80:609–616.

Bond, J.J., and Warner, R.D. 2007. Ion distribution and protein proteolysis affect water holding capacity of *longissimus thoracis et lumborum* in meat of lamb subjected to antemortem exercise. *Meat Sci.* 75:406–414.

Bowker, B.C., Swartz, D.R., Grant, A.L., and Gerrard, D.E. 2005. Myosin heavy chain isoform composition influences the susceptibility of actin-activated S1 ATPase and myofibrillar ATPase to pH inactivation. *Meat Sci.* 71:342–350.

Briand, M., Boisonnet, G., Laplace-Marieze, V., and Briand, Y. 1993. Metabolic and contractile differentiation of rabbit muscles during growth. *Int. J. Biochem.* 25:1881–1887.

Briviba, K., and Sies, H. 1994. Nonenzymatic antioxidant defence systems. In: Frei, B. (ed.), *Natural Antioxidants in Human Health and Disease*. Academic Press, London, pp. 107–128.

Brogna, D.M.R., Nasri, S., Salem, H.B., Mele, M., Serra, A., Bella, M., Priolo, A., Makkar, H.P.S., and Vasta, V. 2011. Effect of dietary saponins from *Quillaja saponaria* L. on fatty acid composition and cholesterol content in muscle *longissimus dorsi* of lambs. *Animal* 5:1124–1130.

Broughton, K.S., and Wade, J.W. 2002. Total fat and (n-3/n-6) fat ratios influence eicosanoid production in mice. *J. Nutr.* 132:88–94.

Brown, A.A., and Hu, F.B. 2001. Dietary modulation of endothelial function: Implications for cardiovascular disease. *Am. J. Clin. Nutr.* 73:673–686.

Bruce, H.L., Mowat, D.N., and Ball, R.O. 1991. Effects of compensatory growth on protein metabolism and meat tenderness of beef steers. *Can. J. Anim. Sci.* 71:659–668.

Calder, P.C. 2004. n-3 Fatty acids, inflammation, and immunity—Relevance to postsurgical and critically ill patients. *Lipids* 39:1147–1161.

Calkins, C.R., and Hodgen, J.M. 2007. A fresh look at meat flavour. *Meat Sci.* 77:63–80.

Campbell, R.G., Johnson, R.J., Taverner, M.R., and King, R.H. 1991. Interrelationships between exogenous porcine somatotropin (PST) administration and dietary protein and energy intake on protein deposition capacity and energy metabolism of pigs. *J. Anim. Sci.* 69:1522–1531.

Campo, M.M., Nute, G.R., Hughes, S.I., Enser, M., Wood, J.D., and Richardson, R.I. 2006. Flavour perception of oxidation in beef. *Meat Sci.* 72:303–311.

Cerrilla, M.E.O., and Martínez, G.M. 2003. Starch digestion and glucose metabolism in the ruminant: A review. *Interciencia* 28:380–386.

Chaijan, M. 2008. Review: Lipid and myoglobin oxidations in muscle foods. *Songklanakarin J. Sci. Tech.* 30:47–53.

Chaves, A.V., Stanford, K., Dugan, M.E.R., Gibson, L.L., McAllister, T.A., Van Herk, F., and Benchaar, C. 2008. Effects of cinnamaldehyde, garlic and juniper berry essential oils on rumen fermentation, blood metabolites, growth performance, and carcass characteristics of growing lambs. *Livest. Sci.* 117:215–224.

Chiba, L.I., Ivey, H.W., Cummins, K.A., and Gamble, B.E. 1999. Growth performance and carcass traits of pigs subjected to marginal dietary restrictions during the grower phase. *J. Anim. Sci.* 77:1769–1776.

Chilliard, Y. 1993. Dietary fat and adipose tissue metabolism in ruminants, pigs, and rodents: A review. *J. Dairy Sci.* 76:3897–3931.

Choi, N.J., Enser, M., Wood, J.D., and Scollan, N.D. 2000. Effect of breed on the deposition in beef and adipose tissue of dietary n-3 polyunsaturated fatty acids. *Anim. Sci.* 71:509–519.

Combs, G. F. 2001. Selenium in global food systems. *Br. J. Nutr.* 85:517–547.

Connor, W.E. 2000. Importance of n-3 fatty acids in health and disease. *Am. J. Clin. Nutr.* 71:171–175.

Corpet, D.E. 2011. Red meat and colon cancer: Should we become vegetarians, or can we make meat safer? *Meat Sci.* 89:310–316.

Cosgrove, M., Flynn, A., and Kiely, M. 2005. Consumption of red meat, white meat and processed meat in Irish adults in relation to dietary quality. *Br. J. Nutr.* 93:933–942.

Cozzi, G., Prevedello, P., Stefani, A., Piron, A., Contiero, B., Lante, A., Gottardo, F., and Chevaux, E. 2011. Effect of dietary supplementation with different sources of selenium on growth response, selenium blood levels and meat quality of intensively finished Charolais young bulls. *Animal* 5:1531–1538.

Daley, C.A., Abbott, A., Doyle, P.S., Nader, G.A., and Larson, S. 2010. A review of fatty acid profiles and antioxidant content in grass-fed and grain-fed beef. *Nutr. J.* 10:53.

Dawson, J.M., Buttery, P.J., Lammiman, M.J., Soar, J.B., Essex, C.P., Gill, M., and Beever, D.E. 1991. Nutritional and endocrinological manipulation of lean deposition in forage-fed steers. *Br. J. Nutr.* 66:171–185.

Decker, E.A., Livisay, S.A., and Zhou, S. 2000. Mechanisms of endogenous skeletal muscle antioxidants: Chemical and physical aspects. In: Decker, E.A., Faustman, C., Lopez-Bote, C.J. (eds.), *Antioxidants in Muscle Foods*. John Wiley and Sons, Inc., New York, pp. 25–60.

Demeyer, D., and Doreau, M. 1999. Targets and procedures for altering ruminant meat and milk lipids. *Proc. Nutr. Soc.* 58:593–607.

Descalzo, A.M., and Sancho, A.M. 2008. A review of natural antioxidants and their effects on oxidative, odor and quality of fresh beef produced in Argentina. *Meat Sci.* 78:423–436.

Dixon, R.M., Hosking, B.J., and Egan, A.R. 2003. Effects of oilseed meal and grain-urea supplements fed infrequently on digestion in sheep. 1. Low quality grass hay diets. *Anim. Feed Sci. Technol.* 110:75–94.

Dixon, R.M., and Stockdale, C.R. 1999. Associative effects between forages and grains: Consequences for feed utilisation. *Aust. J. Agric. Res.* 50:757–774.

Doreau, M., Demeyer, D.I., and Van Nevel, C. 1997. Transformations and effects of unsaturated fatty acids in the rumen: Consequences on milk fat secretion. In: Welch, R.A.S., Burns, D.J.W., Davis, S.R., Popay, A.I., Prosser, G.G. (eds.), *Milk Composition, Production and Biotechnology.* CAB International, Wallingford, CT, pp. 73–92.

Dransfield, E., and Sosnicki, A. 1999. Relationship between muscle growth and poultry meat quality. *Poult. Sci.* 78:743–746.

D'Souza, D.N., Warner, R.D., Dunshea, F.R., and Leury, B.J. 1999. Comparison of different dietary magnesium supplements on pork quality. *Meat Sci.* 51:221–225.

D'Souza, D.N., Warner, R.D., Leury, B.J., and Dunshea, F.R. 2000. The influence of dietary magnesium supplement type, and supplementation dose and duration, on pork quality and the incidence of PSE pork. *Aust. J. Agric. Res.* 51:185–190.

Duffine, A.E., and Volpe, S.L. 2013. Magnesium and metabolic disorders. In: Watson, R., Preedy, Y.V.R., Zibadi, S. (eds.), *Magnesium in Human Health and Disease.* Springer Science, London, UK.

Dunne, P.G., O'Mara, F.P., Monahan, F.J., and Moloney, A.P. 2006. Changes in colour characteristics and pigmentation of subcutaneous adipose tissue and m. *longissimus dorsi* of heifers fed grass, grass silage or concentrate-based diets. *Meat Sci.* 74:231–241.

Dunshea, F.R., King, R.H., and Campbell, R.G. 1993. Interrelationships between dietary protein and ractopamine on protein and lipid deposition in finishing gilts. *J. Anim. Sci.* 71:2931–2941.

Durmic, Z., and Blache, D. 2012. Bioactive plants and plant products: Effects on animal function, health and welfare. *Anim. Feed Sci. Technol.* 176:150–162.

European Food Safety Authority Panel on Additives and Products or Substances Used in Animal Feed (FEEDAP) 2013. Scientific opinion on the safety and efficacy of vitamin D3 (cholecalciferol) as a feed additive for pigs, piglets, bovines, ovines, calves, equines, chickens for fattening, turkeys, other poultry, fish and other animal species or categories, based on a dossier submitted by Fermenta Biotech Ltd. *Eur. Food Saf. Auth. J.* 11:1–26.

Fabian, J., Chiba, L.I., Kuhlers, D.L., Frobish, L.T., Nadarajah, K., Kerth, C.R., McElhenney, W.H., and Lewis, A.J. 2002. Degree of amino acid restrictions during the grower phase and compensatory growth in pigs selected for lean growth efficiency. *J. Anim. Sci.* 80:2610–2618.

Fahey, A., Brameld, J., Parr, T., and Buttery, P. 2005. The effect of maternal undernutrition before muscle differentiation on the muscle fiber development of the newborn lamb. *J. Anim. Sci.* 83:2564–2571.

Faustman, C., Cassens, R.G., Schaefer, D.M., Buege, D.R., Williams, S.N., and Scheller, K.K. 1989. Improvement of pigment and lipid stability in Holstein steer beef by dietary supplementation with vitamin E. *J. Food Sci.* 54:485–486.

Faustman, C., Sun, Q., Mancini, R., and Suman, S.P. 2010. Myoglobin and lipid oxidation interactions: Mechanistic bases and control. *Meat Sci.* 86:86–94.

Fisher, A.V., Enser, M., Richardson, R.I., Wood, J.D., Nute, G.R., Kurt, E., Sinclair, L.A., and Wilkinson, R.G. 2000. Fatty acid composition and eating quality of lamb types derived from four diverse breed x production systems. *Meat Sci.* 55:141–147.

Foote, M.R., Horst, R.L., Huff-Lonergan, E.J., Trenkle, A.H., Parrish, F.C., and Beitz, D.C. 2004. The use of vitamin D3 and its metabolites to improve beef tenderness. *J. Anim. Sci.* 82:242–249.

Ford, A.L., Park, R.J., and McBride, R.L. 1975. Effect of a protected lipid supplement on flavour properties of sheep meat. *J. Food Sci.* 40:236–239.

Foury, A., Devillers, N., Sanchez, M.P., Griffon, H., Le Roy, P., and Mormede, P. 2005. Stress hormones, carcass composition and meat quality in large white x Duroc pigs. *Meat Sci.* 69:703–707.

Garcia, P., Pensel, N., Sancho, A., Latimori, N., Kloster, A., Amigone, M., and Casal, J. 2008. Beef lipids in relation to animal breed and nutrition in Argentina. *Meat Sci.* 79:500–508.

Gardner, G.E., Jacob, R.H., and Pethick, D.W. 2001. The effect of magnesium oxide supplementation on muscle glycogen metabolism before and after exercise and at slaughter in sheep. *Aust. J. Agric. Res.* 52:723–729.

Gatellier, P., Mercier, Y., Juin, H., and Reneere, M. 2005. Effect of finishing mode (pasture- or mixed-diet) on lipid composition, color stability and lipid oxidation in meat from Charolais cattle. *Meat Sci.* 69:175–186.

Geay, Y., Bauchart, D., Hocquette, J., and Culioli, J. 2001. Effect of nutritional factors on biochemical, structural and metabolic characteristics of muscles in ruminants: Consequences on dietetic value and sensorial qualities of meat. *Reprod. Nutr. Dev.* 41:1–26.

Gerrard, D.E., and Grant, A.L. 2003. *Principles of Animal Growth and Development.* Kendall/Hunt Publishing, USA.

Gibson, S., and Ashwell, M. 2003. The association between red and processed meat consumption and iron intakes and status among British adults. *Public Health Nutr.* 6:341–350.

Goehring, T.B., Palmer, I.S., Olson, O.E., Libal, G.W., and Wahlstrom, R.C. 1984. Effects of seleniferous grains and inorganic selenium on tissue and blood composition and growth performance of rats and swine. *J. Anim. Sci.* 59:725–732.

Goldspink, G., and Ward, P.S. 1979. Changes in rodent muscle fibre types during post-natal growth, undernutrition and exercise. *J. Physiol.* 296:453–469.

Gorelik, S., and Kanner, J. 2001. Oxymyoglobin oxidation and membranal lipid peroxidation initiated by iron redox cycle. *J. Agric. Food Chem.* 49:5939–5944.

Greene, B.E. 1969. Lipid oxidation and pigment changes in raw beef. *J. Food Sci.* 34:110–113.

Greenwood, P.L., Harden, S., and Hopkins, D.L. 2007. Myofibre characteristics of ovine *longissimus* and *semitendinosus* muscles are influenced by sire-breed, gender, rearing type, age, and carcass weight. *Aust. J. Exp. Agric.* 47:1137–1146.

Harper, G.S. 1999. Trends in skeletal muscle biology and understanding of toughness in beef. *Aust. J. Agric. Res.* 50:1105–1129.

Hayes, J.E., Stepanyan, V., Allen, P., O'Grady, M.N., O'Brien, N.M., and Kerry, J.P. 2007. The effect of lutein, sesamol, ellagic acid and olive leaf extract on lipid oxidation and oxymyoglobin oxidation in bovine and porcine muscle model systems. *Meat Sci.* 83:201–208.

Hibbs, J.W., Pounden, W.D., and Krauss, W.E. 1951. Studies on milk fever in dairy cows. III. Further studies on the effect of vitamin D on some of the blood changes at parturition and the composition of colostrum in normal and milk-fever cows. *J. Dairy Sci.* 34:855–864.

Hopkins, D.L., Stanley, D.F., Martin, L.C., Ponnampalam, E.N., and van de Ven, R. 2007a. Sire and growth path effects on sheep meat production. 1. Growth and carcass characteristics. *Anim. Prod. Sci.* 47:1208–1218.

Hopkins, D.L., Stanley, D.F., Toohey, E.S., Gardner, G.E., Pethick, D.W., and Van De Ven, R. 2007b. Sire and growth path effects on sheep meat production. 2. Meat and eating quality. *Anim. Prod. Sci.* 47:1219–1228.

Hopkins, D.L., Toohey, E.S., Lamb, T.A., Kerr, M.J., van de Ven, R.J., and Refshauge, G. 2011. Explaining the variation in the shear force of lamb meat using sarcomere length, the rate of rigor onset and pH. *Meat Sci.* 88:794–796.

Hornick, J.L., Van Eenaeme, C., Clinquart, A., Diez, M., and Istasse, L. 1998. Different periods of feed restriction before compensatory growth in Belgian Blue bulls: I. Animal performance, nitrogen balance, meat characteristics, and fat composition. *J. Anim. Sci.* 76:249–259.

Howe, P., Buckley, J., and Meyer, B.J. 2007. Long-chain omega-3 fatty acids in red meat. *Nutr. Diet.* 64:135–139.

Howells, K.F., Mathews, D.R., and Jordan, T.C. 1978. Effects of pre- and perinatal malnutrition on muscle fibres from fast and slow rat muscles. *Res. Exp. Med.* 173:35–40.

Hu, F., Manson, J., and Willett, W. 2001. Types of dietary fat and risk of coronary heart disease: A critical review. *J. Am. Coll. Nutr.* 20:5–19.

Huang, X. 2003. Iron overload and its association with cancer risk in humans: Evidence for iron as a carcinogenic metal. *Mutat. Res./Fundam. Mol. Mech. Mutagen.* 533:153–171.

Huff-Lonergan, E., Mitsuhashi, T., Beekman, D.D., Parrish, F.C., Olson, D.G., and Robson, R.M. 1996. Proteolysis of specific muscle structural proteins by mucalpain at low pH and temperature is similar to degradation in post mortem bovine muscle. *J. Anim. Sci.* 74:993–1008.

Hunter, J., Zhang, J., and Kris-Etherton, P. 2010. Cardiovascular disease risk of dietary stearic acid compared with trans-, other saturated, and unsaturated fatty acids: A systematic review. *Am. J. Clin. Nutr.* 91:46–63.

Ijssennagger, N., Rijnierse, A., de Wit, N.J.W., Boekschoten, M.V., Dekker, J., Schonewille, A., Müller, M., and van der Meer, R. 2013. Dietary heme induces acute oxidative stress, but delayed cytotoxicity and compensatory hyperproliferation in mouse colon. *Carcinogen* 34:1628–1635.

Immonen, K., Ruusunen, M., and Puolanne, E. 2000. Some effects of residual glycogen concentration on the physical and sensory quality of normal pH beef. *Meat Sci.* 55:33–38.

Jaturasitha, S., Thirawong, P., Leangwunta, V., and Kreuzer, M. 2004. Reducing toughness of beef from Bos indicus draught steers by injection of calcium chloride: Effect of concentration and time post mortem. *Meat Sci.* 68:61–69.

Johnson, P.L., Puchas, R.W., and Blair, H.T. 2005. Effect of slaughter group and sire on carcass composition and meat quality characteristics of Texel-sired lambs. *Proc. N. Z. Soc. Anim. Prod.* 65:241–246.

Jørgensen, K., and Skibsted, L.H. 1993. Carotenoid scavenging of radicals. Effect of carotenoid structure and oxygen partial pressure on antioxidative activity. *Z. Lebensm. Unters. Forsch.* 196:423–429.

Juniper, D.T., Phipps, R.H., Ramos-Morales, E., and Bertin, G. 2008. Effect of dietary supplementation with selenium-enriched yeast or sodium selenite on selenium tissue distribution and meat quality in beef cattle. *J. Anim. Sci.* 86:3100–3109.

Jurie, C., Robelin, J., Picard, B., and Geay, Y. 1995. Post-natal changes in biological characteristics of semitendinosus muscle of Limousin male cattle. *Meat Sci.* 41:125–135.

Karami, M., Ponnampalam, E.N., and Hopkins, D.L. 2013. The effect of palm oil and canola oil (saturated- versus polyunsaturated-fatty acids) on feedlot performance, plasma and tissue fatty acid profiles and meat quality in goats. *Meat Sci.* 94:165–169.

Karges, K., Brooks, J.C., Gill, D.R., Breazile, J.E., Owens, F.N., and Morgan, J.B. 2001. Effects of supplemental vitamin D3 on feed intake, carcass characteristics, tenderness, and muscle properties of beef steers. *J. Anim. Sci.* 79:2844–2850.

Kemp, C.M., Sensky, P.L., Bardsley, R.G., Buttery, P.J., and Parr, T. 2010. Tenderness—An enzymatic view. *Meat Sci.* 84:248–256.

Kerry, J.P., Buckley, D.J., and Morrissey, P.A. 2000. Improvement of oxidative stability of beef and lamb with vitamin E. In: Decker, E.A., Faustman, C., Lopez-Bote, C.J. (eds.), *Antioxidants in Muscle Foods*. John Wiley and Sons, Inc., New York, pp. 229–261.

Khliji, S., van de Ven, R.J., Lamb, T.A., Lanza, M., and Hopkins, D.L. 2010. Relationship between consumer ranking of lamb colour and objective measures of colour. *Meat Sci.* 85:224–229.

Kim, Y., and Mahan, D. 2001. Effects of high dietary levels of selenium-enriched yeast and sodium selenite on macro and micro mineral metabolism in grower-finisher swine. *Asian-Australas. J. Anim. Sci.* 14:243–249.

Kirchgessner, M., Roth-Maier, D.A., Heindl, U., and Schwarz, F.J. 1995. B-vitamine (thiamine, vitamin B_6 and pantothensaure) im muskelgewebe wachsender rinder der rasse Deutsches Fleckviek in abhangigkeit von mastendemasse und futterungsintensität [B-vitamins (thiamin, vitamin B6, pantothenic acid) in lean tissue of growing cattle of German Simmental breed under different feeding intensities]. *Z. Lebensm. Unters. Forsch.* 201:20–24.

Koohmaraie, M. 1996. Biochemical factors regulating the toughening and tenderization processes of meat. *Meat Sci.* 43:193–201.

Kris-Etherton, P.M., and Innis, S. 2007. Position of the American Dietetic Association and Dietitians of Canada: Dietary fatty acids. *J. Am. Diet. Assoc.* 107:1599–1611.

Kris-Etherton, P.M., Poth, S.Y., Sabate, J., Ratcliffe, H.E., Zhao, G., and Etherton, T.D. 1999. Nuts and their bioactive constituents: Effects on serum lipids and other factors that affect disease risk. *Am. J. Clin. Nutr.* 70:504–511.

Kristensen, L., Therkildsen, M., Aaslyng, M.D., Oksbjerg, N., and Ertbjerg, P. 2004. Compensatory growth improves meat tenderness in gilts but not in barrows. *J. Anim. Sci.* 82:3617–3624.

Lahučký, R., Bahelka, L., Novotnā, K., and Vašīčovā, K. 2005. Effect of dietary vitamin E and vitamin C supplementation on the level of α-tocopherol and L-ascorbic acid in muscle and on the antioxidative status and meat quality of pigs. *Czech J. Anim. Sci.* 50:175–184.

Lahučký, R., Nurnberg, K., Kuchenmeister, U., Bahelka, I., Mojto, J., Nurnberg, G., and Ender, K. 2004. The effect of dietary magnesium oxide supplementation on fatty acid composition, antioxidative capacity and meat quality of heterozygous and normal malignant hyperthermia (MH) pigs. *Arch. Anim. Breed.* 47:183–191.

Lantz, V., Ambrosio, L., and Schedl, P. 1992. The *Drosophila* orb gene is predicted to encode sex-specific germline RNA-binding proteins and has localized transcripts in ovaries and early embryos. *Development* 115:75–88.

Lanz, J.K., Donahoe, M., Rogers, R.M., and Ontell, M. 1992. Effects of growth hormone on diaphragmatic recovery from malnutrition. *J. Appl. Physiol.* 73:801–805.

Larraín, R.E., Schaefer, D.M., Richards, M.P., and Reed, J.D. 2008. Finishing steers with diets based on corn, high-tannin sorghum or a mix of both: Color and lipid oxidation in beef. *Meat Sci.* 79:656–665.

Lawrence, T.L.J., and Fowler, V.R. 2002. *Growth of Farm Animals*. CABI Publishing, Wallingford, Oxon, UK.

Lefaucheur, L., and Vigneron, P. 1986. Post-natal changes in some histochemical and enzymatic characteristics of three pig muscles. *Meat Sci.* 16:199–216.

Li, Y., and Liu, S. 2012. Review: Reducing lipid peroxidation for improving colour stability of beef and lamb: On-farm considerations. *J. Sci. Food Agric.* 92:719–726.

Liu, Q., Scheller, K.K., Arp, S.C., Schaefer, D.M., and Williams, S.N. 1996. Titration of fresh colour stability and malondialdehyde development with Holstein steers fed vitamin E-supplemented diets. *J. Anim. Sci.* 74:106–116.

Lobley, G.E. 1993. Species comparison of tissue protein metabolism: Effects of age and hormonal action. *J. Nutr.* 123:337–343.

Lobley, G.E. 1998. Nutritional and hormonal control of muscle and peripheral tissue metabolism in farm species. *Livest. Prod. Sci.* 56:91–114.

Lobo-Jr, A.R., Delgado, E.F., Mourão, G.B., Pedreira, A.C.M.S., Berndt, A., and Demarchi, J.J.A.A. 2012. Interaction of dietary vitamin D_3 and sunlight exposure on *B. indicus* cattle: Animal performance, carcass traits, and meat quality. *Livest. Sci.* 145:196–204.

Lorenzen, C.L., Neely, T.R., Miller, R.K., Tatum, J.D., Wise, J.W., Taylor, J.F., Buyck, M.J., Reagan, J.O., and Savell, J.W. 1999. Beef customer satisfaction: Cooking method and degree of doneness effects on the top loin steak. *J. Anim. Sci.* 77:637–644.

Luchak, G.L., Miller, R.K., Belk, K.E., Hale, D.S., Michaelson, S.A., Johnson, D.D., West, R.L., Leak, F.W., Cross, H.R., and Savell, J.W. 1998. Determination of sensory, chemical and cooking characteristics of retail beef cuts differing in intramuscular and external fat. *Meat Sci.* 50:55–72.

Luciano, G., Monahan, F.J., Vasta, V., Biondi, L., Lanza, M., and Priolo, A. 2009a. Dietary tannins improve lamb meat colour stability. *Meat Sci.* 81:120–125.

Luciano, G., Monahan, F.J., Vasta, V., Pennisi, P., Bella, M., and Priolo, A. 2009b. Lipid and colour stability of meat from lambs fed fresh herbage or concentrate. *Meat Sci.* 82:193–199.

Luciano, G., Vasta, V., Monahan, F.J., López-Andrés, P., Biondi, L., Lanza, M., and Priolo, A. 2011. Antioxidant status, colour stability and myoglobin resistance to oxidation of *longissimus dorsi* muscle from lambs fed a tannin-containing diet. *Food Chem.* 124:1036–1042.

Lynch, B., and Kerry, J.P. 2000. Utilising diet to incorporate bioactive compounds and improve the nutritional quality of muscle foods. In: Decker, E.A., Faustman, C., Lopez-Bote, C.J. (eds.), *Antioxidants in Muscle Foods*. John Wiley and Sons, Inc., New York, pp. 455–480.

Macdougall, D.B., Bremner, I., and Dalgarno, A.C. 1973. Effect of dietary iron on the colour and pigment concentration of veal. *J. Sci. Food Agric.* 24:1255–1263.

MacRae, J.C., and Lobley, G.E. 1991. Physiological and metabolic implications of conventional and novel methods for the manipulation of growth and production. *Livest. Prod. Sci.* 27:43–59.

Mancini, R.A., and Hunt, M.C. 2005. Current research in meat color. *Meat Sci.* 71:100–121.

Mason, L.M., Hogan, S.A., Lynch, A., O'Sullivan, K., Lawlor, P.G., and Kerry, J.P. 2005. Effects of restricted feeding and antioxidant supplementation on pig performance and quality characteristics of *longissimus dorsi* muscle from Landrace and Duroc pigs. *Meat Sci.* 70:307–317.

Mattila, P.H., Valkonen, E., and Valaja, J. 2011. Effect of different vitamin D supplementation in poultry feed on vitamin D content of eggs and chicken meat. *J. Agric. Food Chem.* 59:8298–8303.

McAfee, A.J., McSorley, E.M., Cuskelly, G.J., Moss, B.W., Wallace, J.M.W., Bonham, M.P., and Fearon, A.M. 2010. Red meat consumption: An overview of the risks and benefits. *Meat Sci.* 84:1–13.

Melody, J.L., Lonergan, S.M., Rowe, L.J., Huiatt, T.W., Mayes, M.S., and Huffer-Lonergan, E. 2004. Early post mortem biochemical factors influencing tenderness and water-holding capacity of three porcine muscles. *J. Anim. Sci.* 82:1195–1205.

Min, B., and Ahn, D. 2005. Mechanism of lipid peroxidation in meat and meat products—A review. *Food Sci. Biotechnol.* 14:152–163.

Mir, P., Mir, Z., Kuber, P. et al. 2002. Growth, carcass characteristics, muscle conjugated linoleic acid (CLA) content, and response to intravenous glucose challenge in high percentage Wagyu, Wagyu x Limousin, and Limousin steers fed sunflower oil-containing diets. *J. Anim. Sci.* 80:996–1081.

Mitchell, A.D. 2007. Impact of research with cattle, pigs, and sheep on nutritional concepts: Body composition and growth. *J. Nutr.* 137:711–714.

Montgomery, J.L., Parrish, F.C., Beitz, D.C., Horst, R.L., Huff-Lonergan, E.J., and Trenkle, A.H. 2000. The use of vitamin D3 to improve beef tenderness. *J. Anim. Sci.* 78:2615–2621.

Montossi, F., Font-i-Furnols, M., Campo, M., San Julian, R., Brito, G., and Sanudo, C. 2013. Sustainable sheep production and consumer preference trends: Compatibilities, contradictions, and unresolved dilemmas. *Meat Sci.* 95:772–789.

Moody, W.G., Kemp, J.D., Mahyuddin, M., Johnston, D.M., and Ely, D.G. 1980. Effect of feeding systems, slaughter weight and sex on histological properties of lamb carcasses. *J. Anim. Sci.* 50:249–256.

Morgan, J.B. 2007. Does vitamin D_3 improve beef tenderness? Beef report produced for the National Cattlemen's Beef Association on behalf of The Beef Checkoff, USA.

Morrissey, P.A., and Kerry, J.P. 2004. Lipid oxidation and shelf-life of muscle foods In: Steele, R. (ed.), *Understanding and Measuring the Shelf-Life of Food.* Woodhead Publishing Ltd., Cambridge, England, pp. 357–395.

Morrissey, P.A., Sheehy, P.J.A., Galvin, K., Kerry, J.P., and Buckley, D.J. 1998. Lipid stability in meat and meat products. *Meat Sci.* 49:573–586.

Mottram, D.S. 1998. Flavour formation in meat and meat products: A review. *Food Chem.* 62:415–424.

Muir, P.D., Smith, N.B., Wallace, G.J., Cruickshank, G.J., and Smith, D.R. 1998. The effect of short-term grain feeding on liveweight gain and beef quality. *N. Z. J. Agric. Res.* 41:517–526.

Najafi, M.H., Zeinoaldini, S., Ganjkhanlou, M., Mohammadi, H., Hopkins, D.L., and Ponnampalam, E.N. 2012. Performance, carcass traits, muscle fatty acid composition and meat sensory properties of male Mahabadi goat kids fed palm oil, soybean oil or fish oil. *Meat Sci.* 92:848–854.

Nardone, A., Ronchi, B., Lacetera, N., Ranieri, M.S., and Bernabucci, U. 2010. Effects of climate changes on animal production and sustainability of livestock systems. *Livest. Sci.* 130:57–69.

National Research Council. 1981. *Effect of Environment on Nutrient Requirements of Domestic Animals.* National Academy Press, Washington, USA.

Navarro-Alarcon, M., and Cabrera-Vique, C. 2008. Selenium in food and the human body: A review. *Sci. Total Environ.* 400:115–141.

Nieto, G., Estrada, M., Jordán, M.J., Garrido, M.D., and Bañón, S. 2011. Effects in ewe diet of rosemary by-product on lipid oxidation and the eating quality of cooked lamb under retail display conditions. *Food Chem.* 124:1423–1429.

Noci, F., French, P., Monahan, F., and Moloney, A. 2007. The fatty acid composition of muscle fat and subcutaneous adipose tissue of grazing heifers supplemented with plant oil-enriched concentrates. *J. Anim. Sci.* 85:1062–1073.

Nute, G.R., Richardson, R.I., Wood, J.D., Hughes, S.I., Wilkinson, R.G., Cooper, S.L., and Sinclair, L.A. 2007. Effect of dietary oil source on the flavour and the colour and lipid stability of lamb meat. *Meat Sci.* 77:547–555.

O'Grady, M.N., Maher, M., Troy, D.J., Moloney, A.P., and Kerry, J.P. 2006. An assessment of dietary supplementation with tea catechins and rosemary extract on the quality of fresh beef. *Meat Sci.* 73:132–143.

O'Grady, M.N., Monahan, F.J., and Mooney, M.T. 2001. Oxymyoglobin in bovine muscle systems as affected by oxidizing lipids, vitamin E and metmyoglobin reductase activity. *J. Muscle Foods* 12:19–35.

Oostindjer, M., Alexander, J., Amdam, G.V. et al. 2014. The role of red and processed meat in colorectal cancer development: A perspective. *Meat Sci.* 97:583–596.

O'Sullivan, M.G., Byrne, D.V., Nielsen, J.H., Andersen, H.J., and Martens, M. 2003. Sensory and chemical assessment of pork supplemented with iron and vitamin E. *Meat Sci.* 64:175–189.

O'Sullivan, M.G., Kerry, J.P., Buckley, D.J., Lynch, P.B., and Morrissey, P.A. 1997. The distribution of dietary vitamin E in the muscles of the porcine carcass. *Meat Sci.* 45:297–305.

Packer, L., and Kagan, N.E. 1993. The antioxidant harvesting centre of membranes and lipoproteins. In: Packer, L., Fuchs, J. (eds.), *Vitamin E in Health and Disease.* Marcel Dekker, Inc., New York, pp. 172–192.

Palmquist, D.L. 2009. Omega-3 fatty acids in metabolism, health and nutrition and for modified animal product foods. *Prof. Anim. Sci.* 25:207–249.

Palozza, P., and Krinsky, N.I. 1992. Antioxidant effects of carotenoids *in vivo* and *in vitro*: An overview. *Methods Enzymol.* 213:403–420.

Park, H.R., Ahn, H.J., Kim, J.H., Yook, H.S., Kim, S., Lee, C.H., and Byun, M.W. 2004. Effects of irradiated phytic acid on antioxidation and color stability in meat models. *J. Agric. Food Chem.* 52:2572–2576.

Parker, R.S. 1989. Dietary and biochemical aspects of vitamin E. *Adv. Food Nutr. Res.* 33:157–232.

Pearce, K.L., Masters, D.G., Jacob, R.H., Hopkins, D.L., and Pethick, D.W. 2008. Effects of sodium chloride and betaine on hydration status of lambs at slaughter. *Anim. Prod. Sci.* 48:1194–1200.

Picard, B., Gagniere, H., Geay, Y., Hocquette, J.F., and Robelin, J. 1995. Study of the influence of age and weaning on the contractile and metabolic characteristics of bovine muscle. *Reprod. Nutr. Dev.* 35:71–84.

Ponnampalam, E.N. 1999. *Nutritional Modification of Muscle Long Chain Omega-3 Fatty Acids in Lamb.* PhD thesis. The University of Melbourne, Melbourne, Australia.

Ponnampalam E.N., Bekhit, A.A., Fahri F.T., Linden, N., and Hopkins, D.L. 2013. Muscle antioxidant potential and oxidative process of meat in lambs fed traditional diets. Nutrition Society of Australia. Brisbane, Australia, December 4–6, vol. 37. pp. 25.

Ponnampalam E.N., Burnett, V.F., Norng, S., Warner, R.D., and Jacobs J.L. 2012b. Vitamin E and fatty acid content of lamb meat from perennial or annual pasture systems with supplementation. *Anim. Prod. Sci.* 52:255–262.

Ponnampalam, E.N., Butler, K.L., Hopkins, D.L., Kerr, M.G., Dunshea, F.R., and Warner, R.D. 2008. Genotype and age effects on sheep meat production. 5. Lean meat and fat content in the carcasses of Australian sheep genotypes at 20, 30 and 40 kg carcass weights. *Aust. J. Exp. Agric.* 48:893–897.

Ponnampalam, E.N., Butler, K.L., McDonagh, M.B., Jacobs, J.L., and Hopkins, D.L. 2012a. Relationship between muscle antioxidant status, forms of iron, polyunsaturated fatty acids and functionality (retail colour) of meat in lambs. *Meat Sci.* 90:297–303.

Ponnampalam, E.N., Butler, K.L., Pearce, K.M., Mortimer, S.I., Pethick, D.W., Ball, A.J., and Hopkins, D.L. 2014a. Sources of variation of health claimable long chain omega-3 fatty acids in meat from Australian lamb slaughtered at similar weights. *Meat Sci.* 96:1095–1103.

Ponnampalam, E.N., Dixon, R.M., Hosking, B.J., and Egan, A.R. 2004. Intake, growth and carcass characteristics of lambs consuming low digestible hay and cereal grain. *Anim. Feed Sci. Technol.* 114:31–41.

Ponnampalam, E.N., Giri, K., Pethick, D.W., and Hopkins, D.L. 2014c. Nutritional background, sire type and dam type affect saturated and monounsaturated (oleic) fatty acid concentration of lambs reared for meat production in Australia. *Anim. Prod. Sci.* 54:1358–1362.

Ponnampalam, E.N., Hosking, B.J., and Egan, A.R. 2003. Rate of carcass components gain, carcass characteristics, and muscle *longissimus* tenderness in lambs fed dietary protein sources with a low quality roughage diet. *Meat Sci.* 63:143–149.

Ponnampalam, E.N., Mann, N.J., and Sinclair, A.J. 2006. Effect of feeding systems on omega-3 fatty acids, conjugated linoleic acid and trans fatty acids in Australian beef cuts, potential impacts on human health. *Asian Pac. J. Clin. Nutr.* 15:21–29.

Ponnampalam, E.N., Norng, S., Burnett, V.F., Dunshea, F.R., Jacobs, J.L., and Hopkins, D.L. 2014b. The synergism of biochemical components controlling lipid oxidation in lamb muscle. *Lipids* 49:757–766.

Ponnampalam, E.N., Sinclair, A.J., Egan, A.R., Ferrier, G., and Leury, B.J. 2002b. Dietary manipulation of muscle long-chain omega-3 and omega-6 fatty acids and sensory properties of lamb meat. *Meat Sci.* 60:125–132.

Ponnampalam, E.N., Sinclair, A.J., Hosking, B.J., and Egan, A.R. 2002a. Effects of dietary lipid type on muscle fatty acid composition, carcass leanness, and meat toughness in lambs. *J. Anim. Sci.* 80:628–636.

Ponnampalam, E.N., Trout, G.R., Sinclair, A.J., Egan, A.R., and Leury, B.J. 2001. Comparison of the color stability of fresh and vacuum packaged lamb muscle containing elevated omega-3 and omega-6 fatty acid levels from dietary manipulation. *Meat Sci.* 58:151–161.

Putnam, J.J., Allshouse, J.E., and Kantor, L.S. 2002. US per capita food supply trends: More calories, refined carbohydrates, and fats. *Food Rev.* 25:2–15.

Ramirez, J.E., Alvarez, E.G., Montano, M., Shen, Y., and Zinn, R.A. 1998. Influence of dietary magnesium level on growth-performance and metabolic responses of Holstein steers to laidlomycin propionate. *J. Anim. Sci.* 76:1753–1759.

Rasby, R.J., Berger, A.L., Bauer, D.E., and Brink, D.R. 2011. Minerals and vitamins for beef cows. Report published by the Board of Regents of the University of Nebraska on behalf of the University of Nebraska-Lincoln Extension, USA.

Realini, C., Duckett, S., Brito, G., Dalla Rizza, M., and De Mattos, D. 2004. Effect of pasture vs. concentrate feeding with or without antioxidants on carcass characteristics, fatty acid composition, and quality of Uruguayan beef. *Meat Sci.* 66:567–577.

Renerre, M. 2000. Oxidative processes and myoglobin. In: Deker, E., Faustman, C., Lopez-Bote, C.J. (eds.), *Antioxidants in Muscle Foods*. John Wiley and Sons, Inc., New York, NY, pp. 113–133.

Resconi, V.C., Campo, M.M., Furnols, M., Montossi, F., and Sanudo, C. 2009. Sensory evaluation of castrated lambs finished on different proportions of pasture and concentrate feeding systems. *Meat Sci.* 83:31–37.

Rhee, K.S., Anderson, L.M., and Sams, A.R. 1996. Lipid oxidation potential of beef, chicken and pork. *J. Food Sci.* 61:8–12.

Ripoll, G., Joy, M., and Muñoz, F. 2011. Use of dietary vitamin E and selenium (Se) to increase the shelf life of modified atmosphere packaged light lamb meat. *Meat Sci.* 87:88–93.

Rochfort, S., Parker, A.J., and Dunshea, R. 2008. Plant bioactives for ruminant health and productivity. *Phytochemistry* 69:299–322.

Rowe, L.J., Maddock, K.R., Lonergan, S.M., and Huff-Lonergan, E. 2004. Oxidative environment decrease tenderization of beef steaks through inactivation of μ-calpain. *J. Anim. Sci.* 82:3254–3266.

Rule, D.C., Smith, S.B., and Romans, J.R. 1995. Fatty acid composition of muscle and adipose tissue of meat animals. In: Smith, S.B., Smith, D.R. (eds.), *The Biology of Fat in Meat Animals: Current Advances*. American Society of Animal Science, Champaign, IL, p. 144.

Sachs, M., Perlman, R., and Bar, A. 1987. Use of 1ahydroxyvitamin D_3 in the prevention of bovine parturient paresis. IX. Early and late effects of a single injection. *J. Dairy Sci.* 70:1671–1675.

Santos-Silva, J., Bessa, R., and Santos-Silva, F. 2002. Effect of genotype, feeding system and slaughter weight on the quality of light lambs. II. Fatty acid composition of meat. *Livest. Prod. Sci.* 77:187–194.

Sanudo, C., Enser, M., Campo, M.M., Nute, G.R., Maria, G., Sierra, I., and Wood, J.D. 2000. Fatty acid composition and fatty acid characteristics of lamb carcasses from Britain and Spain. *Meat Sci.* 54:339–346.

Sartorius, C.A., Lu, B.D., Acakpo-Satchivi, L., Jacobsen, R.P., Byrnes, W.C., and Leinwand, L.A. 1998. Myosin heavy chains IIa and IId are functionally distinct in the mouse. *J. Cell. Biol.* 141:943–953.

Sauberlich, H.E. 1990. Vitamin B_6, vitamin B_{12} and folate. In: Perarson, A.M., Dutson, T.R. (eds.), *Meat and Health: Advances in Meat Research.* Elsevier Applied Science, London, pp. 461–495.

Schaefer, A.L., Dubeski, P.L., Aalhus, J.L., and Tong, A.K.W. 2001. Role of nutrition in reducing antemortem stress and meat quality aberrations. *J. Anim. Sci.* 79:91–101.

Schaefer, A.L., Jones, S.D., and Stanley, R.W. 1997. The use of electrolyte solutions for reducing transport stress. *J. Anim. Sci.* 75:258.

Schaefer, D.M., Liu, Q., Faustman, C., and Yin, M.C. 1995. Supranutritional administration of vitamin E and C improves oxidative stability of beef. *J. Nutr.* 125:1792–1798.

Scislowski, V., Bauchart, D., Gruffat, D., Laplaud, P., and Durand, D. 2005. Effects of dietary n-6 or n-3 polyunsaturated fatty acids protected or not against ruminal hydrogenation on plasma lipids and their susceptibility to peroxidation in fattening steers. *J. Anim. Sci.* 83:2162–2174.

Scollan, N., Hocquette, J.F., Nuernberg, K., Dannenberger, D., Richardson, I., and Moloney A. 2006. Innovations in beef production systems that enhance the nutritional and health value of beef lipids and their relationship with meat quality. *Meat Sci.* 74:17–33.

Scott, T.W., Ashes, J.R., Rich, J.C., and Gulati, S.K. 1995. Effects of protected lipids on meat quality parameters and characteristics. In: *Proceedings of CSIRO Meat Industry Research Conference*, September 10–12, 1995, Gold Coast, Australia, pp. 7–9.

Seideman, S.C., and Crouse, J.D. 1986. The effects of sex condition, genotype and diet on bovine muscle fiber characteristics. *Meat Sci.* 17:55–72.

Sempos, C.T., Looker, A.C., and Gillum, R.F. 1996. Iron and heart disease: The epidemiologic data. *Nutr. Rev.* 54:73–84.

Sesink, A.L.A., Termont, D.S.M.L., Kleibeuker, J.H., and Van der Meer, R. 1999. Red meat and colon cancer the cytotoxic and hyperproliferative effects of dietary heme. *Cancer Res.* 59:5704–5709.

Sharma, S., Sheehy, T., and Kolonel, L.N. 2013. Contribution of meat to vitamin B12, iron and zinc intakes in five ethnic groups in the USA: Implications for developing food-based dietary guidelines. *J. Hum. Nutr. Diet.* 26:156–168.

Siddons, R.C., Nolan, J.V., Beever, D.E., and MacRae, J.C. 1985. Nitrogen digestion and metabolism in sheep consuming diets containing different forms and levels of N. *Br. J. Nutr.* 54:175–187.

Sies, H., Stahl, W.M., and Sandquist, A.R. 1992. Antioxidant functions of vitamins. Vitamins E and C, β-carotene and other carotenoids. *Ann. N. Y. Acad. Sci.* 669:7–20.

Sillence, M.N. 2004. Technologies for the control of fat and lean deposition in livestock. *Vet. J.* 167:242–257.

Simitzis, P.E., Deligeorgis, S.G., Bizelis, J.A., Dardamani, A., Theodosiou, I., and Fegeros, K. 2008. Effect of dietary oregano oil supplementation on lamb meat characteristics. *Meat Sci.* 79:217–223.

Simopoulos, A.P. 1991. Omega-3 fatty acids in health and disease and in growth and development. *Am. J. Clin. Nutr.* 54:438–463.

Skřivanová, E., Marounek, M., De Smet, S., and Raes, K. 2007. Influence of dietary selenium and vitamin E on quality of veal. *Meat Sci.* 76:495–500.

Smith, G.C. 1997. Beef quality and palatability: How veterinarians can help producers improve the quality of their cattle and carcasses. In *Proc. 59th Ann. Conf. for Veterinarians, Coll. of Vet. Med.* Kansas State Univ., Manhattan, NY, pp. 295–302.

Smith, S.B. 1995. Substrate utilization in ruminant adipose tissues. In: Smith, S.B., Smith, D.R. (eds.), *The Biology of Fat in Meat Animals: Current Advances.* American Society of Animal Science, Champaign, IL, p. 166.

Spindler, A.A., Mathias, M.M., and Cramer, D.A. 1980. Growth changes in bovine muscle fiber types as influenced by breed and sex. *J. Food Sci.* 45:25–31.

Takahashi, M., Tsuboyama-Kasaoka, N., Nakatani, T., Ishii, M., Tsutsumi, S., Aburatani, H., and Ezaki, O. 2002. Fish oil feeding alters liver gene expressions to defend against PPARα activation and ROS production. *Am. J. Physiol. Gastrointest. Liver Physiol.* 28:338–348.

Tatum, J.D., Smith, G.C., and Belk, K.E. 1999. New approaches for improving tenderness, quality, and consistency of beef. *J. Anim. Sci.* 77:1–10.

Thomson, B.C., Muir, P.D., and Dobbie, P.M. 1999. Effect of growth path and breed on the calpain system in steers finished in a feedlot. *J. Agric. Sci.* 133:209–215.

Tizioto, P.C., Gromboni, C.F., de Araujo Nogueira, A.R., de Souza, M.M., de Alvarenga Mudadu, M., Tholon, P., Rosa, A.N., Tullio, R.R., Medeiros, S.R., and Nassu, R.T. 2014. Calcium and potassium content in beef: Influences on tenderness and associations with molecular markers in Nellore cattle. *Meat Sci.* 96:436–440.

USDA. 1998. USDA Nutrient Database for Standard Reference, Release 12. U.S. Department of Agriculture, Agricultural Research Service. Nutrient Data Laboratory Home Page, http://www.nal.usda.gov/fnic/foodcomp

Vahedi, V., Lewandowski, P., Dunshea, F.R., Hopkins, D.L., and Ponnampalam, E.N. 2014. Effect of diets on antioxidant enzyme activity and gene expression of longissimus muscle in lambs. *NSA 2014 Annual Scientific Meeting.* Nutrition Society of Australia, Hobart, Tasmania, Australia, Vol. 38, November 26–28, 2014, pp. 58.

Vasta, V., Aouadi, D., Brogna, D.M.R., Scerra, M., Luciano, G., Priolo, A., and Ben Salem, H. 2013. Effect of the dietary supplementation of essential oils from rosemary and artemisia on muscle fatty acids and volatile compound profiles in Barbarine lambs. *Meat Sci.* 95:235–241.

Vasta, V., and Luciano, G. 2011. The effects of dietary consumption of plants secondary compounds on small ruminants' products quality. *Small Ruminant Res.* 101:150–159.

Vasta, V., Pennisi, P., Lanza, M., Barbagallo, D., Bella, M., and Priolo, A. 2007. Intramuscular fatty acid composition of lambs given a tanniniferous diet with or without polyethylene glycol supplementation. *Meat Sci.* 76:739–745.

Vatansever, L., Kurt, E., Enser, M., Nute, G.R., Scollan, N.D., Wood, J.D., and Richardson, R.I. 2000. Shelf life and eating quality of beef from cattle of different breeds given diets differing in n-3 polyunsaturated fatty acid composition. *Anim. Sci.* 71:471–482.

Vestergaard, M., Oksbjerg, N., and Henckel, P. 2000. Influence of feeding intensity, grazing and finishing feeding on muscle fibre characteristics and meat colour of semitendinosus, *longissimus dorsi* and supraspinatus muscles of young bulls. *Meat Sci.* 54:177–185.

Vignola, G., Lambertini, L., Mazzone, G., Giammarco, M., Tassinari, M., Martelli, G., and Bertin, G. 2009. Effects of selenium source and level of supplementation on the performance and meat quality of lambs. *Meat Sci.* 81:678–685.

Ward, S.S., and Stickland, N.C. 1993. The effect of undernutrition in the early post-natal period on skeletal muscle tissue. *Br. J. Nutr.* 69:141–150.

Watkins, P.J., Frank, D., Singh, T.K., Young, O.A., and Warner, R.D. 2013. Sheepmeat flavor and the effect of different feeding systems: A review. *J. Agric. Food Chem.* 61:3561–3579.

Wenk, C., Leonhardt, M., and Scheeder, M.L. 2000. Monogastric nutrition and potential for improving muscle quality. In: Decker, E.A., Faustman, C., Lopez-Bote, C.J. (eds.), *Antioxidants in Muscle Foods.* John Wiley and Sons, Inc., New York, pp. 199–227.

Wheeler, T.L., Koohmaraie, M., and Shackelford, S.D. 1994. Reducing inconsistent beef tenderness with calcium-activated tenderization. In: *Proceedings of the Meat Industry Research Conference*, Chicago, IL, USA, pp. 109–130.

Williams, P. 2007. Nutritional composition of red meat. *Nutr. Diet.* 64:113–119.

Williamson, C.S., Foster, R.K., Stanner, S.A., and Buttriss, J.L. 2005. Red meat in the diet. *Nutr. Bull.* 30:323–355.

Windham, C.T., Wyse, B.W., and Hansen, R.G. 1990. Thiamin, riboflavin, niacin and pantothenic acid. In: Perarson, A.M., Dutson, T.R. (eds.), *Meat and Health: Advances in Meat Research.* Elsevier Applied Science, London, pp. 401–459.

Wolter, B., Ellis, M., McKeith, F.K., Miller, K.D., and Mahan, D.C. 1999. Influence of dietary selenium source on growth performance, and carcass and meat quality characteristics in pigs. *Can. J. Anim. Sci.* 79:119–121.

Wood, J.D., and Enser, M. 1997. Factors influencing fatty acids in meat and the role of antioxidants in improving meat quality. *Br. J. Nutr.* 78:49–60.

Wood, J.D., Richardson, R., Nute, G., Fisher, A., Campo, M., Kasapidou, E., Sheard, P., and Enser, M. 2003. Effects of fatty acids on meat quality: A review. *Meat Sci.* 66:21–32.

Wyness, L., Weichselbaum, E., O'Connor, A., Williams, E.B., Benelam, B., Riley, H., and Stanner, S. 2011. Red meat in the diet: An update. *Nutr. Bull.* 36:34–77.

Yang, A., Lanari, M.C., Brewster, M., and Tume, R.K. 2002. Lipid stability and meat colour of beef from pasture-and grain-fed cattle with or without vitamin E supplement. *Meat Sci.* 60:41–50.

Zhan, X., Wang, M., Zhao, R., Li, W., and Xu, Z. 2007. Effects of different selenium source on selenium distribution, loin quality and antioxidant status in finishing pigs. *Anim. Feed Sci. Technol.* 132:202–211.

Zinn, R.A., and Shen, Y. 1996. Interaction of dietary calcium and supplemental fat on digestive function and growth performance in feedlot steers. *J. Anim. Sci.* 74:2303–2309.

5 Terms of Farming and Animal Welfare and Meat Quality

Roman Kołacz and Robert Kupczyński

CONTENTS

5.1 INTRODUCTION

Animal welfare-related problems have been discussed in international forums for nearly 40 years. A range of definitions and criteria of farm animal welfare protection have been elaborated over that time. Animal welfare is the first and foremost important thing for the animal. However, better livestock management and care can improve the productivity level and food quality. Consideration of environmental factors and behavioral patterns is an essential issue in the development of animal welfare (Kołacz and Bodak 1999; EFSA 2012). This provides the basis for the creation of animal-friendly production systems. In addition to the optimization of production conditions taking into account space allocation and physical factors (e.g., temperature, light, and noise), the protection of animals against excessive stress is also an important issue (Kołacz and Dobrzański 2006; Kilchsperger et al. 2010; Velarde and Dalmau 2012). Moreover, disease prevention activities and farm biosafety are significant in terms of reducing the risk of production losses, which affects the quality and safety of animal-based products.

The relationships among animal welfare, animal health, and food safety are strongly emphasized in the legal regulations of numerous countries. Housing conditions and systems have a specified effect on meat quality and its safety for the consumer. Improving an animal's welfare can positively affect numerous aspects of product quality (e.g., reducing the occurrence of tough or watery meat as well as the incidence of bruising, bone breakage, and blood spots) and disease resistance, decreasing the immunosuppressive effect of chronic stress and the need for antibiotics (Blokhuis et al. 2008). This may affect diminishing potential for drug residues. Increasing attention is being paid to consumers' opinions and preferences, as they wish to have knowledge about the farms and the conditions under which animals are raised, as well as their diet composition (PPG 2007). There is also a growing interest in meat derived from organic farms, characterized by features concerning its nutritional value (low fat content, high content of polyunsaturated fatty acids and bioactive compounds) as well as the organoleptic value (Rembiałkowska and Badowski 2012).

Modern systems of animal-origin food safety control penetrate each element of the production chain, starting from herd health status monitoring, through welfare protection strategy, biosafety, intervention prevention-treatment programs, to final product quality control. This chapter discusses aspects of animal welfare related to the conditions of raising animals, breeding systems, biosafety, and microclimate factors.

5.2 DEFINITIONS AND WELFARE ASSESSMENT

Animal welfare, especially for farm animals, is quite a broad and difficult issue. Animal welfare may be defined as the fulfillment of physiological and behavioral needs, and the assurance of living comfort and high care levels. Physical health affects welfare to a large degree, and the animal concurrently feels positive emotions (pleasure and satisfaction) and negative ones such as fear or frustration (Kołacz and Bodak 2002). Physical health means not only an absence of clinical disease symptoms but also the lack of disorders without clear general or local symptoms. The position of welfare in animal health protection is shown in Figure 5.1. According to Kołacz and Bodak (1999), the welfare of an animal is its "state" in terms of its ability to cope with its environment. These authors describe welfare as an organism's ability for homeostasis maintenance in circumstances of endogenous and exogenous factors under constant fluctuation. Welfare disturbance is observed when the intensity of stimuli affecting an organism exceeds its ability to maintain balance.

This concept was adopted by the World Organization of Animal Health (OIE) that defines animal welfare as: (i) how well an animal copes with the conditions in which it lives; (ii) an animal experiences good welfare if, as indicated by scientific evidence, it is healthy, comfortable, well nourished, safe, able to express key aspects of behavior, and if it is not suffering from unpleasant states, such as pain, fear, and distress; and (iii) good animal welfare requires disease prevention and veterinary treatment for illness and injuries, appropriate shelter, management, nutrition,

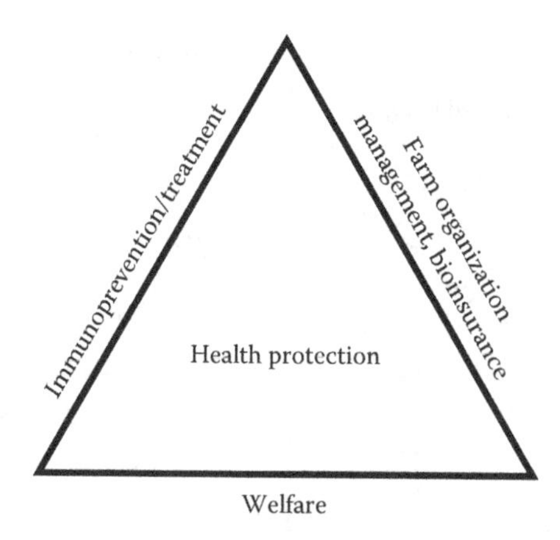

FIGURE 5.1 Animal health protection—welfare significance.

humane handling, and humane slaughter/killing (OIE 2011). While the term "animal welfare" refers to the state of an individual animal, in practical circumstances, these individual measurements are used to assess the mean welfare in a group or herd (Truszczyński and Kołacz 2009). The European Food Safety Authority's (EFSA) view of welfare is based on a multidimensional concept that includes both the physical health and the emotional state of the animal, determined mainly based on behavioral observations (Figure 5.2).

The main assumptions of farm animal welfare were elaborated by the Farm Animals Welfare Council (FAWC) and are included in the so-called Farm Animals Welfare Code (FAWC 1992). These assumptions may be expressed using the "five freedoms":

1. Freedom from hunger and thirst—by ready access to fresh water and a diet to maintain full health and fitness
2. Freedom from discomfort—by providing an appropriate environment, including shelter and a comfortable resting area
3. Freedom from pain, injury, or disease—by prevention or rapid diagnosis and treatment
4. Freedom to express normal behavior—by providing sufficient space, proper facilities, and the company of the animal's own kind
5. Freedom from fear and distress—by ensuring conditions and treatment that avoid mental suffering

Welfare may be divided into various levels—from good (*good welfare*) to poor (*poor welfare*). The methods of its evaluation may be based on objective criteria (clinical, laboratory diagnostics, ethological examinations, statistical meta-analyses) and on subjective criteria (observations of animal behavior, individual perception of environmental state) (Kołacz and Bodak 1999, 2002). In broader terms, these criteria involve:

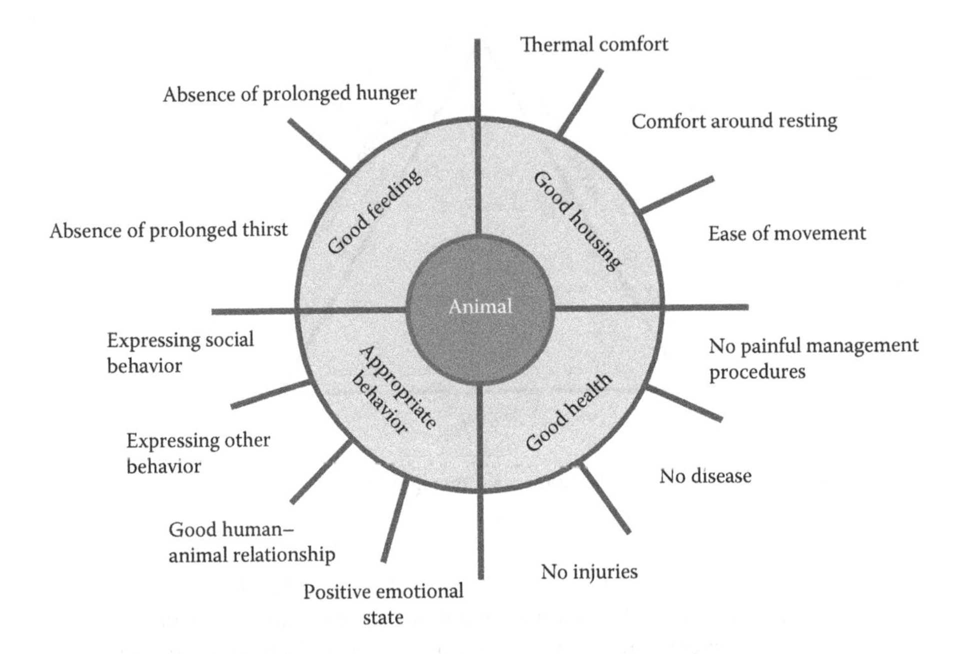

FIGURE 5.2 The four principles and 12 animal-based criteria used as guidelines for good welfare according to the Welfare Quality® project. (Modified from EFSA Panel on Animal Health and Welfare (AHAW). 2012. *EFSA J.* 10(6):2767. doi:10.2903/j.efsa.2012.2767.)

- On-farm monitoring of health and welfare by producers, including targets for key welfare indicators
- Restrictions on transport duration (and limitations of live export)
- Specifications and monitoring of the stunning and slaughter process to ensure effective stunning and unconsciousness until death

The aim of numerous legislative, research, and practical activities undertaken within welfare is the development of alternative production systems, in which profit maximization, that is, production outcomes, are the result of fulfilling animals' biological needs. Therefore, extreme technological solutions are more often replaced by sustainable production programs, often organic ones, which maintain and obey animal welfare rules. Methods of animal production that take into account the welfare rules differ from methods used in intensive systems, since they are closely focused on animals' state of being in their environment, which reflects the tolerance to applied technological solutions. However, intensive production systems are also obliged to assure the welfare of animals, and behavior and mood must be an effect of tolerance to applied technological solutions (Kołacz and Bodak 2002).

The conditions and systems of housing have a specified effect on animal-origin products (including meat) and their safety. According to Becker (2000), meat quality is defined by the traits that the consumer perceives as desirable, including both visual and sensory traits as well as credence traits of safety, health, and more intangible traits such as "clean" or the welfare status of the production system. The conditions under which the animals are maintained on farms affect their health state and presence of diseases. Practices related to animal transport to slaughterhouses and pre-slaughter procedures cause stress in animals and violation of internal organism homeostasis (see Chapter 6). However, they may also be modulated by several intrinsic animal factors (e.g., genetics, sex, age, and physiological state) and by past experiences and acquired learning. At this stage, the animals are exposed to a range of stimuli that may eliminate the effects of breeding practices improving performance results (Ferguson and Warner 2008).

5.3 LEGAL PROTECTION OF ANIMALS

In Europe, animal protection laws are issued and formulated by national governments. However, specific initiatives are produced by supranational institutions, such as the Council of Europe and the European Union (EU), which stipulate minimum requirements that need to be adopted by all member states (European Commission 2012). The EU Commission has the power to create legislative texts. Animal protection is under the responsibility of the General Directorate for Health and Consumer Protection (DG-SANCO). When a decision is taken about setting up a new piece of legislation to protect animals, DG-SANCO consults a scientific committee of the EFSA.

In turn, the scientific committee appoints an ad hoc working group consisting of scientific experts recognized for their experience in the specific topic covered. This working group produces a report on existing systems and procedures for farming (or transport or slaughter), reviews the scientific evidence on the effects of any aspect that may affect animal welfare, and provides recommendations on how to protect animals. The Economic and Social Committee, made up of representatives of major societal stakeholders, can be consulted to give its opinion on a draft directive. A draft directive is submitted to the Council of Ministers of the EU and becomes a Council Directive only after receiving their approval. In parallel with the European legislation, all member states have their own national legislation. This legislation must at least conform to European regulations, but may also define more stringent measures (Veissier et al. 2008).

Since the creation of EFSA, there has been an Animal Health and Welfare (AHAW) panel. EFSA provides independent scientific advice regarding risks associated with food and feed, plant health, environment, animal health, and animal welfare. The development of EFSA activity areas is aimed at the determination of relationships among animal welfare, animal diseases, and food safety. The possible implications for food safety involving other areas of expertise in EFSA, such as biohazards, contaminants, and plant health, have been considered in most of the scientific opinions on animal welfare (Blokhuis et al. 2008). In its opinions and reports, EFSA methodologically takes into account risk assessment (RA) evaluation. The outcomes of the methodology and the identification of welfare indicators may allow the establishment of control and monitoring plans at farm level, ensuring a high-quality "farm to table" food chain.

The relationships among animal welfare, animal health, and food safety are strongly emphasized in the EU legislation. Currently, according to cross-compliance rules, satisfying animal welfare requirements is one of the subsidy conditions for agricultural farms.

In the international aspect, animal welfare in particular countries is affected by the OIE. This is a global range, since the OIE associates 178 member countries. The OIE Animal Welfare Working Group was only established in 2002 (OIE 2013a). Since then, it has adopted 11 animal welfare standards, addressing areas such as transport of animals by land, the use of animals in research and education, and the slaughter of animals for human consumption (OIE 2013b). Currently, this organization has welfare standards for slaughter, transport, and killing of animals for disease control. However, these standards were primarily designed to assure sanitary and safe international trade in animals, and they cannot be treated as a global policy toward animal welfare (Otter et al. 2012).

The Brambell Report, which was published in 1965, included two federal laws in the United States regulating the treatment of farm animals. Given the legislative climate surrounding farm animal welfare issues in Europe, there has been almost no change in U.S. federal legislation related to farm animals in the last few years (Mench 2008). Care of the animals on the farm, during transport, or at slaughter are included in the guidelines of the Animal Welfare Act elaborated by the U.S. Department of Agriculture (Mench 2008). Apart from these standards, recommendations included in the "accepted farm practices" are also taken into account in Canada and the United States (Whiting 2013). In Australia, the strategies of animal welfare were elaborated by the National Consultative Committee on Animal Welfare (NCCAW). The Ministry of Agriculture, Fisheries and Forestry developed Model Codes of Practice for FAW in cooperation with the Commonwealth Scientific and Industrial Research Organization (CSIRO) and with feedback from the other stakeholders. A welfare code is intended as a guide for all people responsible for the welfare of a particular animal species (MAFF 2007).

International actions aimed at animal welfare are also undertaken by the World Society for the Protection of Animals (WSPA). The elaborated Universal Declaration on Animal Welfare is propagated by the animal welfare groups in many countries, including New Zealand, Switzerland, and the EU (WSPA 2007). According to Fraser (2008), improvements to animal welfare may still be achieved (i) through the basic economic incentive to reduce losses caused by injury, stress, and malnutrition, (ii) through disease control programs, and (iii) by international corporations applying their existing animal welfare standards on a more global basis.

5.4 CONSUMER VERSUS ANIMAL WELFARE

A general trend toward the increasing importance of health, safety, environmental issues, and animal welfare may be observed over the last few years (Hocquette et al. 2014). There is a growing concern by consumers with regard to how meat is produced, especially in relation to animal welfare and organic/natural production. Consumers demand that animals are reared, transported, and slaughtered under humane conditions (Troy and Kerry 2010). The European market for animal-friendly products is still largely fragmented and the differences between European countries are considerable (Ingenbleek et al. 2013). Animal health status and absence of diseases are also significant issues (Rushen et al. 2011). Meat and meat products are important transmission of zoonotic pathogens (food-borne illnesses) is considerable in terms of the resultant public health. Pathogens can survive in the environment and in feces for extended periods ranging from several weeks to many months, providing an important transmission route for pathogens (*Salmonella*, *Escherichia coli* O157:H7, *Listeria monocytogenes*, and *Campylobacter*) within herds and farms (Troy and Kerry 2010). Therefore, it is very important to maintain animal hygiene and prophylaxis.

Labeling regimens, which assure consumers that certain animal welfare standards have been complied with, can have an important role in providing consumers with information about how farm animals are raised (Veissier et al. 2008; Ingenbleek et al. 2013). The aspiration to improve animal production conditions is not inconsistent with final product quality. However, an improvement in animal housing conditions is usually related to investment.

Consumers are increasingly concerned about the welfare of food-producing animals. In numerous EU countries, over 50% of consumers state that they would prefer to purchase products of animal origin from sources where they are produced without animal welfare violations (PPG 2007). The Plough to Plate Group acting in the United Kingdom ranked "raising standards of animal welfare" as their top future priority, ahead of the environment, local sourcing, and fairer prices for producers (PPG 2007).

An increasing number of consumers link animal welfare with better product quality or product safety. Sometimes, critical areas of the "scientific" assessment of welfare, such as lameness or biting, resonate less with consumers than more anthropomorphic concerns such as "freedom," being outdoors, and "acting naturally," all of which are considerably more difficult to "quantify" and thereby translate into standards appropriate to formal, comparative labeling. Eurobarometer (2005) reported that 74% of the 1000 participants interviewed in 30 European countries believed that buying welfare-friendly products could have a positive effect on the quality of animals' life, and 60% declared a willingness to pay a higher price for products sourced from more animal welfare-friendly production systems. There is however a limitation in these beliefs, caused by a lack of knowledge of specific issues. The next report of Eurobarometer (2007) survey demonstrated that 85% of respondents said that they knew little about farming practice, and 54% said it was not easy to find information on welfare provenance when shopping. The views expressed in response to questions are not always reflected in actual purchasing behavior. A part of the consumers prefer to transfer the responsibility for the assurance of animal welfare others, for example, suppliers, or treat this problem as purely one of ethical considerations.

Also, high standards of animal welfare are significant in order to improve an effectiveness and profitability of business (e.g., International Financing Corporation), in order to fulfill consumer expectations and cover domestic and international markets (Velarde and Dalmau 2012), accepting that animal welfare is gaining increased recognition as an important element of commercial livestock operations around the world.

5.5 BIOINSURANCE AND QUALITY SYSTEMS IN PRIMARY PRODUCTION

Lowered animal welfare levels are always accompanied by a lowered health status, and vice versa, a sick animal is always characterized by lowered welfare. Thus, when we accept that safe food may only be obtained from healthy animals, then the relationship between animal welfare and food safety is obvious. The basic issues of animal health protection include bioinsurance. In terms of large livestock farms, bioinsurance may be defined as a strategy of activities and undertakings aimed at the prevention of the introduction of pathogenic factors (bacteria, viruses) onto a farm, and in the event of their occurrence, the control of their spread and activities aimed at their complete elimination. Bioinsurance generally involves two key activities. First are the strategies for the reduction of the risk of the appearance of infection factors on a farm (reduction of pathogenic threats). Second is the limitation of the transmission of pathogenic threats to animals and preventing their spreading in a region, including to other farms (Mee et al. 2012). The purpose of bioinsurance is thus the prevention of disease occurrence on a farm by the elimination or maximal limitation of potential disease sources.

In large-scale animal production systems, the occasional occurrence of some diseases is difficult to avoid, but it is possible to effectively prevent their mass occurrence and spread in a herd. Bioinsurance programs should be developed individually for each farm, taking into account critical risk points. Such programs should include: farm location, buildings and climatic conditions, hygiene of technology and personnel, quarantine application, control of visits on a farm, control of undesirable animal presence, control of vaccinations and treatment programs, utilization of dead animals, and supervision in animal turnover. The significant issue is the systematic room disinfection performed after the end of each production cycle, which should eliminate specific pathogenic factors (Kołacz and Dobrzański 2006).

The guiding principle in safe food production in the EU is the fact that the nutritional chain "from farm to table" constitutes an integral and inseparable whole in legal, control, and information aspects. The roles of all participants in the food chain, starting from fodder producers, farmers, food processors, traders, and finally controllers, are specified in the European law regulations. Current systems of animal-origin product safety control analyze each element of the production chain, starting from herd health status, through welfare protection strategy, to bioinsurance, for final product quality control. Accepted level of product quality and their safety on a global scale is established by the FAO. The accepted Code of Practice for Good Animal Feeding concerns the safety of food for human consumption through adherence to good animal feeding practice at the farm level and good manufacturing practices (GMP), but it does not concern animal welfare (FAO and IFIF 2010). It contains the recommendations regarding adequate nutrition at every stage of growth and production.

A complex approach to hygienic standards imposes an obligation on all operators dealing with food production to implement good hygienic practice (GHP) and Hazard Analysis and Critical Control Points (HACCP). The fundamental role of the system of food safety supervision and control performed within the HACCP system is control over those elements in the food chain that concern primary production, that is, on the farm level. This part of control in the HACCP system, often called "preharvester food safety," is defined as a complex of constant activities on the farm level, including fodder materials' field production, their storage, fodder manufacturing and distribution, bioinsurance of farms and animal welfare, environmental protection, and animal transport to slaughterhouses. The purpose of these activities is to prevent or minimize the levels of chemical, physical, and microbiological contamination in humans from food transported via the alimentary tract by animals or animal products, which would constitute a risk for human health.

Farms specializing in animal production are obliged to respect relevant legal regulations concerning the control of threats in primary production and related activities, including

- Means of controlling pollution from air, soil, water, fodder, fertilizers, veterinary medicinal products, plant protection sources, biocides, as well as waste storage, processing, and utilization
- Control of animal health and welfare, which affect human health, including supervision and control of zoonotic disease factors

Critical control points should be determined for the assessment of welfare level threats and a method for their monitoring with a remedial plan should be included in the system of herd health supervision. Those elements of environmental and production technology determining the level of animal welfare, for which critical control points should be created (Table 5.1), include

- Area of pen, stand, cage, yard per animal
- Movement freedom (tethers, individual/group pens, yards, pasture)
- Kind of floor, lair for the animals
- Ergonomics of livestock building equipment
- Microclimatic conditions (lighting, temperature, humidity, air movement, dust level, harmful gaseous admixtures)
- Building ventilation (regulation system, emergency system)
- Practices on animals (castrations, shortening of tails, fangs, etc.)
- Isolation and care of sick animals
- Monitoring of behavioral stereotypes
- Conditions of loading, unloading, and transport of animals
- Supervision of animal production (animal identification, everyday animal control, qualifications of animal handling staff, documentation)

These tasks related to animal welfare supervision and control are entrusted to veterinary inspection units in many European countries.

5.6 SYSTEMS OF ANIMAL HOUSING

Materials used for livestock buildings and pen construction cannot be harmful for animals or food products. They should be easy to clean and disinfect.

TABLE 5.1
Problems in Cattle, Pig, Sheep, and Welfare

Specification	Critical Points and Welfare Problems
Cattle	Immobilization, suitable surface, avoiding tethering and individual housing, space, lighting requirements, litter presence and wire floors, availability of yards and pastures, weaning period, stable groups to avoid aggressive behaviors, precise establishment of nutritional requirements (e.g., roughage), limitation of some practices (dehorning, castration), electric stimulator application
Pigs	Litter availability, pen area, wire floors, provision with complete fodder, avoiding individual housing, age of piglet weaning, enriched environment, hormonal treatments, adequate anesthesia for castration, electric stimulator application
Sheep	Pen area, provision with complete fodder, limitation of some practices (castration), transported long distances, early weaning, heat stress (in some countries), isolation, early diagnosis of diseases
Transport	Time of transport, litter in vehicles for young animals, loading ramps and their slope, separation of unfamiliar groups, interdiction of sedatives/tranquilizers (not allowed in organic husbandry), electric stimulation during loading, animal watering before transport, personnel education
Slaughter	More lairage requirements (start of lairage, space, lighting, floors, etc.), avoidance of group mixing, electric stimulation of live animals, time between stunning and bleeding, specific education of the staff

Source: Adapted from Kilchsperger, R., Schmid, O., and Hecht, J. 2010. Animal welfare initiatives in Europe—Technical report on grouping method for animal welfare standards and initiatives. Deliverable No. 1.1 of EconWelfare Project. Research Institute of Organic Agriculture FiBL, Frick, Switzerland; Averós, X. et al. 2013. *Animals* 3:786–807.

The rooms and their equipment should be designed and constructed so that they do not cause injuries to animals. Technologies applied to animal housing create various levels of welfare; however, no ideal system of production that would take into account all the needs of the animals (physiological, behavioral, and nutritional) and maintain environmental protection and the economic requirements of production has been established so far (Figure 5.3). Farm animals may be housed in an open system or in livestock buildings (e.g., cattle, tether, or free-stall system).

The limited space in intensive systems affects behaviors such as fodder consumption, rest, and so on. In this context, biological costs may be high, and aggression, and thus stress responses, may occur. All hierarchical changes (e.g., introduction of new individuals into a herd) always cause a fight for hierarchy. The individuals placed low in the hierarchy have limited access to resources such as food, space for rest and shade, sexual behavior, and so on. In contrast, more dominant animals generally have priority access to limited resources (Barroso et al. 2000). Additionally, social stress can affect muscle color (Miranda-de la Lama et al. 2013) and also meat tenderness and flavor (Andrighetto et al. 1999).

FIGURE 5.3 Simmental cows feeding in a free-stall barn. (Photo: R. Kupczyński.)

Stress situations related to hierarchy creation and other social interactions of beef cattle housed in indoor systems affect production results (daily gains), and also to a lesser degree meat quality (Miranda-de la Lama et al. 2013). Minimizing aggressive interactions will potentially improve performance and productivity. This is why it is important to maintain social stability throughout the fattening period. It has also been observed that farm animals that experience bad treatment from staff in farm conditions are characterized by poorer meat quality. Additionally, the animal's preslaughter experience, including acute stress and physical activity, has been shown to markedly influence the toughness of beef (Gruber et al. 2010) and lamb (Warner et al. 2005).

The following kinds of floors are used in livestock buildings: litter floors, litterless floors (with shallow litter, deep litter, and self-cleaning floors of a slope of 2–3°) as well as wire floors (slotted). Litter floors are considered to be one of the best kinds of floors with regard to animal hygiene and welfare. Litterless systems are predominant in large-scale pig and poultry (laying hens) production. Litterless and wire floors are a bigger problem. The wires, via proper edge rounding and antiskidding protection, limit injuries in animals, and the materials used (plastic and material coated with plastic) are resistant and heat-insulating, and do not cause organism hypothermia.

Farm animal diseases related to floor defects in livestock buildings are often called floor diseases. In the case of pigs, sharp wire edges or large gaps between the girders may be a cause of mechanical injuries and inflammatory states within the limbs. Claw and limb diseases in cows, apart from milk yield decrease, also cause reproduction disorders and predispose animals to mammary gland diseases (van Gastelen et al. 2011), as well as meat quality deterioration.

Three basic systems of cattle and sheep finishing systems are distinguished: intensive, semi-intensive, and extensive. In an intensive system, the animals are kept indoors, that is, in closed rooms. Pasture forage, agricultural-food industry by-products, and less valuable farming fodders are used in an extensive system. Semi-intensive farming involves young bulls, steers, and heifers of meat, meat-dairy breeds as well as crossbreds with meat breeds. This system requires large amounts of fodder; however, in a period of heifers and steers rearing, complete fodders are replaced with pasture (Węglarz 2010).

Goats reared under an extensive system produces kid meat that is similar in terms of physical characteristics and proximal chemical composition to meat produced by goats produced intensively with natural and artificial rearing systems (Zurita-Herrera et al. 2013). However, fatty acid composition is affected by the management systems of feeding, with the meat from the extensive system being the one with the highest oleic fatty acid content and lowest saturated fatty acids content (see Section 5.7), as well as the lowest atherogenicity index (Zurita-Herrera et al. 2013).

In case of pigs, the environmental effects on product quality are the combined result of both the farming system and the feeding regimen (Kołacz et al. 2004). Intensive finishing systems are predominant in Europe; however, extensive systems may also be observed, especially in southern Europe, including nearly entire utilization of natural environment resources. In pigs fed on pastures, the polyunsaturated fatty acid (PUFA) level in intramuscular fat (including eicosapentaenoic acid [C20:5n-3, EPA] and docosahexaenoic acid [C22:6n-3, DHA]) is higher compared to the conventional system (Pugliese and Sirtori 2012). Also, α-tocopherol is provided by the free-range system and it prevents lipid oxidation (see Chapter 4). However, no significant differences in terms of aroma intensity, fragility, and the taste of the meat derived from pigs kept under an indoor system or with yard access were observed (Ventanas et al. 2007).

5.7 ORGANIC FARMS

Animal production is of fundamental significance in the agricultural production of organic farms, since it provides organic matter and nutrients for cultivated soil, contributing thus to an improvement in soil state and sustainable agriculture development. The standards of animal welfare in organic farms, as in conventional ones, were created for transparency in food production processes. The monitoring systems in both cases are similar and concern technical elements (space, feeding systems), human contact and attitude (Figure 5.4), management factors (climatic and hygienic conditions, routine practices on a farm), and behavioral, health-related, and physiological criteria. In the case of some species (e.g., sheep), studies point to a lack of differences between the welfare of these animals when housed in conventional and organic farms (Braghieri et al. 2007).

Traditional food products of high quality, such as those obtained from animals reared outdoors, are in high demand. The free-range system increases the value of animal products due to the influence of outdoor rearing on the chemical, physical, and organoleptic characteristics of the product (Pugliese and Sirtori 2012).

Depending on the cattle farming system, there are some differences in fat amount, sensory value, and fatty acid content in beef meat (Warren et al. 2008). Meat derived from grass cattle is typically lower in total fat as compared to that of grain-fed cattle (Nuernberg et al. 2005; Alfaia et al. 2009). However, intramuscular fat content, which

FIGURE 5.4 Water buffalo organic breeding in Poland. (Photo: M. Zabłocka.)

determines the amount of marbling, is higher in beef from organic farming systems (Rembiałkowska and Wiśniewska 2010; Rembiałkowska and Badowski 2012). The diet of the cattle on organic farms is mainly based on pasture. Grass-based diets result in significantly higher levels of omega-3 acids within the lipid fraction of the meat (Table 5.2). Grass-fed beef has a more desirable saturated fatty acid (SFA) lipid profile (more C18:0 cholesterol neutral SFA and less C14:0 and C16:0 cholesterol-elevating SFAs) as compared to grain-fed beef. Grass-finished beef is also higher in total conjugated linoleic acids (*cis*-9, *trans*-11 C18:2, CLA) isomers, vaccenic acid (*trans*-11 C18:1, TVA), and n-3 FA. This results in a better n-6:n-3 ratio that is preferred by the nutritional community (Daley et al. 2010). Grass beef consistently produces a higher concentration of n-3 FAs, without affecting the n-6 FA content (Daley et al. 2010). This is caused by a high content of linolenic acids in grass. Some studies have not revealed any differences in the level of proteins, β-carotene, α-tocopherol, retinol, or fatty acids in the meat of organic and conventionally reared steers (Walshe et al. 2006). The fatty acid composition of organic and conventional mutton differs analogically as in the case of beef (Rembiałkowska and Wiśniewska 2010).

The differences in the quality of meat from both systems (organic and conventional) are possible in case of pigs. A large differentiation related to the breed and high variability in fodders used is observed in the organic farming system (Rembiałkowska and Badowski 2012). Generally, diet composition in organic farms to a high degree affects intramuscular fat content. Kouba (2003) suggested that the exclusion of synthetic amino acid supplementation in organic pig production resulted in an increase of intramuscular fat content, which is an important positive aspect of food quality characteristics. The intramuscular fat content is directly influenced by the level of limited amino acids, whereas the total crude protein content is of minor importance (Sundrum et al. 2000). Pig carcasses from organic farming usually have a higher muscle weight in breasts and thighs, a lower total fat content in the carcass, a higher content of intramuscular fat, and a different composition of fatty acids

TABLE 5.2
Comparison of Polyunsaturated Fatty Acid Composition of the Meat between Different Systems of Feeding Cattle

	Fatty Acid Composition								
Treatment (Breed)	Total SFA	Total UFA	Total PUFA	C18:2 n-6 Linoleic	CLA	C18:3 n-3 Linolenic	C20:5 n-3 EPA	C22:6 n-3 DHA	Reference
Grass	43.91	56.09	14.29	6.56	0.87	2.22	0.94	0.17	Nuernberg
Grain	44.49	55.51	9.07	5.22	0.72	0.46	0.08	0.05	et al. (2005)
(Simmental bulls) % of FAs									
Grass	38.76	27.86	28.99	12.55	5.14[a]	5.53	2.13	0.20	Alfaia et al.
Grain	42.08	33.69	19.06	11.95	2.65[a]	0.48	0.47	0.11	(2009)
(Crossbred steers)									
g/100 g of FAs									
Pasture	49.08	50.92	9.96	3.29	0.41	1.34	0.69	0.09	Realinia et al.
Concentrate	47.62	52.38	6.02	2.84	0.23	0.35	0.30	0.09	(2004)
(Uruguayan beef)									
% of FAs									
High concentrate	45.77	50.91	3.71	2.54	0.22	0.19	0.15	0.00	Wistuba et al.
High concentrate +3% fish oil	51.13	42.99	4.01	1.93	0.26	0.32	0.24	0.003	(2007)
(Angus crossbred steers)									
% of FAs									

Note: SFA, saturated fatty acids; UFA, unsaturated fatty acids; PUFA, polyunsaturated fatty acids; CLA, conjugated linoleic acids (*cis*-9, *trans*-11 C18:2).
[a] Total CLA.

(considerably lower n-6/n-3 fatty acid ratio) compared to carcasses from conventional farms (Nilzen et al. 2001; Rembiałkowska and Wiśniewska 2010).

Pigs given access to pasture produce meat characterized by higher levels of polyunsaturated fatty acids, n-3 fatty acids, and vitamin E in the muscle compared to pigs fed indoors (Edwards 2005). The differences in fatty acid composition may cause the lowered technological quality of organic pork due to increased lipid oxidation, which deteriorates storage value. Generally, organic pig production can yield high-quality pork, but information on feed, feed intake, and pig characteristics is important to control the production process.

The organic standards have a substantial "welfare potential." No study has found more overall health problems in organic herds than in conventional herds. However, all parasitological studies show a higher prevalence of parasites in organic herds (Lund 2006). The general tendency in the reviewed papers was that health in organic herds was the same or better than that in conventional herds, with the exception of parasite-related diseases that were more frequent in organic farming (Lund and Algers 2003; Lund 2006). Life outdoors has a range of advantages, such as space, fresh air, and the possibility of movement. This however may constitute a threat to animal health. Keeping calves outdoors all the year round favors the occurrence of enteric Cryptosporidia (Kváč et al. 2006). More infections with *Salmonella* bacteria have been noted in large herds (>100 cows) compared to small ones (Fossler et al. 2005). In turn, Iberian free-range pigs can be a potential reservoir of epidemic antimicrobial-resistant strains of Clostridium difficile, showing a prevalence rate similar to that found for intensively raised animals (Álvarez-Pérez et al. 2013).

It should be taken into account that some pathogenic factors constitute a danger for humans. The infection risk may be minimized, for example, by providing the animals with sufficient space in the outdoor area, clean and dry bedding in indoor facilities or under a roof, dietary supplements if necessary (including natural herbal preparations), minerals, clean water *ad libitum*, and appropriate flock management, for example, closed flocks (Vaarst et al. 2005).

5.8 LIVESTOCK BUILDING MICROCLIMATE

Microclimatic conditions in buildings are important factors affecting farm animal welfare. They include air temperature, humidity, the rate of air movement, and the content of harmful gaseous admixtures in air, especially ammonia, hydrogen sulfide, and dust. The building microclimate depends on their localization, kind of construction, materials used for construction and related heat resistance, system of animal housing, as well as the kind and efficiency of ventilation and sewage devices, and the size and kind of animal density.

5.8.1 Light and Radiation

The effect of sun radiation on animal organisms depends on the kind of rays, as well as on the duration and intensity of the radiation. In farm production systems, not all animals have access to radiation, which negatively affects the state of health and indirectly also production. Among UV fractions, a profitable effect is demonstrated for UVB (ultraviolet B, wavelength 315–280 nm). Transformation of 7-dehydrocholesterol into vitamin D_3 in skin is affected by this fraction (Slominski et al. 2004). This vitamin is significant for a proper calcium–phosphate balance. Ultraviolet radiation causes an increase in oxygen secreted from hemoglobin, and thus cell respiration increases. The levels of hemoglobin, as well as the number of red blood cells, and immunological bodies in blood are subject to increases (Kołacz and Dobrzański 2006).

Lighting at a minimum level of 40 lux for 8 h per day is required in buildings for pigs. Partial or complete sterility has been observed in sows housed with limited light access, while in cows, it resulted in the manifestation of poor estrus symptoms, decreased fertilization index, elongated intercalving intervals, and a milk yield decrease. Lack of light in males always leads to spermatogenesis disorders and lowered sexual activities (Kołacz and Dobrzański 2006). Light is also a stimulator of the immunological processes of an organism, which is clearly observed in young animals. Some animals, such as fatteners, may be housed with limited lighting, but not in complete darkness. This technological practice is even profitable, since it prevents cannibalism, and the limited movement of the animals favors better fodder utilization for production purposes.

5.8.2 Temperature

Farm animals are endothermic (homoiothermic) animals and maintain a stable internal temperature using thermoregulation mechanisms, which adjust the amount of heat produced (shivering thermogenosis, nonshivering thermogenosis, brown tissue) and removed from an organism to balance heat in cold and in hot environments.

There is a tendency in production conditions to maintain the animals at temperatures close to the thermal neutral zone (TNZ). This zone is defined as a range of environmental temperatures in which heat losses from an organism are in equilibrium with minimum heat production, that is, the production that is observed in basic transformation conditions (Kołacz and Dobrzański 2006). The range of temperatures limiting the TNZ is not stable and depends on breed, age, body weight, physiological state, external insulation, acclimation and acclimatization, and animal condition and nutrition level (Table 5.3).

Harmful factors in the case of heat stress include beyond-optimum temperature and humidity of air and resulting high values of the temperature–humidity index (THI), as well as sun radiation. The activity of these factors leads to neurohormonal reactions. Crucial differences are observed with respect to the time of temperature action. Different hormonal responses are observed during short-term exposure (4–5 h), and in the case of long-term exposure (several days). THI classification may be presented in the following ranges:

- Proper value <70–74
- 75–78 heat stress
- >79 strong heat stress (animals are not able to maintain internal body temperature within the standards, thermoregulation mechanisms are incompetent)

TABLE 5.3
Optimum Temperature and Relative Humidity Ranges in Buildings for Pigs

Animal Category	Temperature (°C)			Relative Humidity (%)		
	Minimum	Optimum	Maximum	Minimum	Optimum	Maximum
Breeding young boars and gilts	14	17	23	60	70	80
Herd boars	12	15	20	60	75	85
Sows:						
Loose and low-pregnant	12	15	20	60	70	80
High-pregnant	15	19	25	60	70	80
Lactating	18	20	27	60	70	80
Piglets in heated pen:						
1–3 days	25	32	34	50	60	70
4–14 days	24	28	32	50	60	70
15–21 days	18	23	27	50	60	70
22–28 days	18	22	25	50	60	70
28–56 days	18	21	25	50	60	70
Weaners 57–84 days	17	19	25	50	60	70
Fatteners:						
65 kg	15	18	22	60	70	80
95 kg	15	17	20	60	70	80
115 kg	12	16	20	60	70	80

Source: Adapted from Kołacz, R., and Dobrzański, Z. 2006. *Animal Hygiene and Welfare*. AR Wrocław, Poland [in Polish].

There is a range of formulas for THI calculation; below, we present the rules governing its calculation according to Mader et al. (2002):

$$\mathrm{THI} = 0.8 \times \mathrm{T} + [\mathrm{RH} \times (\mathrm{T} - 14.3)/100] + 46.3$$

where

T—ambient temperature

RH—relative humidity

The result of heat stress in animals is an increased internal temperature, dehydration, increased number of breaths, and lowered resistance. Moreover, high air temperatures cause disorders in hormonal and electrolytic balance, and increased secretion of potassium, sodium, and magnesium from an organism (sweating, panting).

The effect of beyond-optimum temperatures on meat quality is to a high degree affected by animal genotype. Extreme heat provokes an adrenergic stress response, and adrenaline stimulates peripheral vasodilatation and muscle glycogenolysis (Lowe et al. 2002). Additionally, physical activity (anaerobic metabolism) and heat load may mean that meat would be characterized by higher hardness (Gregory 2010). In the case of pigs and turkeys, the meat may also be paler in color with more drip forming when presented as cuts (Gregory 2010). Providing shade in a feedlot can reduce the frequency of dark cutters (Mitlohner et al. 2002). Some markets prefer higher marbling, and high ambient temperatures can favor greater muscle marbling and fat deposition in internal depots, in place of the subcutaneous depot (Nardone et al. 2006). Dehydration accompanying heat stress also negatively affects meat quality. Lamb meat may then have a dark color through shrinkage of the myofibers, and because of its dryness, it has less weight loss during cooking (Jacob et al. 2006). In turn, elongated cooling before slaughter may cause increased pH and dark meat (Gregory 2010).

There is no clear evidence concerning increased infection of pigs with *Salmonella* or *E. coli* strains in heat stress conditions; however, such situations should also be taken into account. Fossler et al. (2005) noted a higher number of *Salmonella* infections in winter periods compared to summer periods.

In practice, more cases of pale, soft, and exudative syndrome (PSE) syndrome are noted in summer periods compared to winter periods. This is confirmed in the study by Santos et al. (1997) who noted as much as 58% of PSE occurrence in heterozygotic pigs having a load of high temperature and relative humidity (35°C and 85% RH). With temperature at a level of 12°C and 85% RH, this frequency was 38%. Cattle of European breeds better tolerate lower environmental temperatures than high temperatures; however, animals kept in lower ambient temperatures lose some heat from their bodies, which may be balanced by increased heat production from their metabolism. A decrease in fodder intake is clearly visible in cattle kept above 25°C (Kołacz and Dobrzański 2006). A negative effect of heat stress on cow fertility is manifested in a lowered fertilization index, elongated postpartum interval, intercalving period as a result of poorly expressed estrus symptoms, disorders in follicle growth and development during estrus, and increased embryonic mortality (Hansen et al. 2001; Paula-Lopes et al. 2001).

In pig fatteners, the highest body weight gains were noted at temperatures of 12–17°C, with a decrease in body weight at temperatures lower than 10°C and higher than 20°C. In lactating sows, high temperatures cause a decrease in γ-globulin fraction concentrations in colostrum, and lowered sow milk yield. Pig overheating is quite often observed in summer periods. An additional factor favoring overheating is an excessive concentration of animals in the building and insufficient ventilation. Overheating leading to the PSE syndrome may also be observed in summer periods during the transport of animals (Van de Perre et al. 2010). High temperatures in breeding sows, above 25°C, cause estrus inhibition or shortening. High temperatures during the first period of pregnancy cause increased embryonic mortality, which is proved by the fact that farrows from sows mated in summer periods are less abundant compared to farrows from sows mated in autumn–winter periods (Knox and Zas 2001). Fertilization effectiveness is considerably lowered as a result of high temperatures (Kołacz and Dobrzański 2006).

5.8.3 Noise

The acoustic environments of animals are formed by audible, perceptible sounds, and also by sounds that are not registered by the senses. Noise may be defined as the subjective feeling of unpleasant and undesirable sounds at a given moment. It has also been accepted that noise at a level from 30 dB may cause specified mental reactions, while in the case of higher values—from 65 dB—apart from mental reactions, vegetative ones may also be observed. It is recommended that constant noise in livestock buildings should not exceed 65 dB, and temporary noise should not exceed 85 dB (Kołacz and Dobrzański 2006).

Noise is included among stress-causing factors, and depending on sound intensity, exposure time, and individual susceptibility, it causes unprofitable mental and vegetative reactions (effect on nerves that innervate internal organs) and negatively affects production outcomes (Kołacz and Dobrzański 2006). The changes in physiological parameters such as increases in heart rate, increased breath rate, decreased breath deepness, increased skin temperature, and decreases in rumen movements have been noted in cattle as a result of noise. Pig fatteners exposed to noise demonstrate typical stress reactions (increases in adrenal gland hormone secretion, increased heart action, even for ischemic heart disease), which cause severe physiological disturbances and lowered productivity, for example, body weight gains (Kołacz and Dobrzański 2006). Prevention of excessive noise in buildings should involve minimization of technological noise and acoustic protection of the buildings (van de Weerd et al. 2003; Kołacz and Dobrzański 2006).

5.8.4 Air Pollution

Harmful gaseous admixtures in the air, especially ammonia and hydrogen sulfide, have unprofitable effects on animals. These mixtures are formed as a result of protein substance decomposition from animal manure, rotting litter, and fodder. Their concentration in the air of the livestock building depends on the temperature, humidity, ventilation system, manner of feeding, as well as on the type of animal housing and their density. The concentration of gases in air also depends on the frequency of manure removal, sewage system and state, and the quality of litter applied, since rotting processes may increase gas levels in air. Another important aspect is ammonia and hydrogen sulfide emission and their effect on the natural environment. The concentration of gases in livestock buildings should not exceed the following values: NH_3—20 ppm, H_2S—5 ppm, and CO_2—3000 ppm (Kołacz and Dobrzański 2006).

Substances such as odors, ammonia, or dust are transmitted through inlet air from livestock buildings to the atmosphere. Their significance in the context of emission is very high. They result in increased requirements concerning environmental protection to such a degree that animal production units are obliged to actively participate in undertaking steps aimed at air purification. Ammonia is considered a primary gas produced in livestock buildings. Negative results of ammonia and hydrogen sulfide activity on animals depend on their concentration in the air, exposure time, and modes of transfer to an organism (Done et al. 2005; Kavolelis 2006). The respiratory tract is exposed to the activity of this gas to the highest degree (Done et al. 2005).

Harmful effects of ammonia on animals start from acute irritation of eye conjunctiva and respiratory tract mucous membranes. The secondary changes caused by increased ammonia concentrations involve blood-related changes (hemoglobin is changed into hematin), which in consequence cause decreased hemoglobin levels, poorer oxygen transfer, and lower levels of γ-globulin protein fractions, which decreases organism immunity. High concentrations of ammonia, exceeding 100 ppm, cause tangible damages in lung structure, lead to a decline in alveoli walls, collagenization of lung capillaries, hemorrhages to the windpipe, as well as paralysis of the trigeminal nerve and central nervous system (Kołacz and Dobrzański 2006).

Also, in the case of hydrogen sulfide, the respiratory tract is endangered by its activity to the highest degree. After transfer to the blood, iron is blocked in respiratory enzymes (cytochrome oxidase), which causes a paralysis of intracellular respiration. The next problem is sulfur compound penetration to the blood, since they join methaemoglobin to form sulfamethaemoglobin, which is an inactive compound in oxygen uptake and transport (Kołacz and Dobrzański 2006). Moreover, hydrogen sulfide damages the cells of the central nervous system, peripheral nerves, and circulation system; irritates the respiratory tract and eyes; and causes coughing attacks, increased heart rate, increased blood pressure, nausea, vomiting, salivation, and sometimes heavy sweating (Kołacz and Dobrzański 2006; von Borell et al. 2007).

Constant exposure of animals to ammonia may cause decreased yield (e.g., lower daily gains) and cannibalism, for example, in pigs. The changes observed in the respiratory tract with NH_3 concentrations of 30–100 ppm are of a functional and structural character. At this stage, changes of reversible character are still observed. Increased ammonia concentrations in the air predispose the environment to the proliferation of pathogenic bacteria in the upper respiratory tract (e.g., *Pasteurella multocida* in pigs). In turn, cattle exposure to hydrogen sulfide leads to a lack of appetite, loss of condition, milk yield decrease, subcutaneous hematoma formation, and strong acceleration of heart and breathing rates (Kołacz and Dobrzański 2006). Apart from ammonia and hydrogen sulfide, it is possible to identify about 400 other trace gases, which are generally listed as volatile substances, so-called odors (Opaliński et al. 2009). They are observed not only in the form of gases but also in the form of agglomerates molecularly linked with dusts, which are their carriers in air.

Among the methods used to reduce emission of odorous compounds from livestock buildings, the following can be distinguished: air scrubbing (Opaliński et al. 2009, 2010), biofiltration (Tymczyna et al. 2004), use of manure covers (VanderZaag et al. 2008), litter amendments, and masking agents (McCrory and Hobbs 2001; Borowski et al. 2010). Moreover, supplementation of various types of feed additives such as natural antimicrobial additives and plant-derived essential oils (Varel 2002), multistrain probiotics (Zhang and Kim 2014), or supplemental synthetic amino acids (Chavez et al. 2004) have been investigated.

5.8.5 Dusts

Irritating activity on mucous membranes, especially in the case of the respiratory tract, is also demonstrated for dusts. They may be of mineral or organic origin; however, organic dusts are predominant in livestock buildings. These include ground fodder, litter particles, and also the animals themselves—the dusts may be formed from, for example, particles of hair coat or dander cells.

Excessive dust levels are demonstrated first of all in effects on the skin, eyes, and respiratory tract. Dusts sedimented on animal skins contaminate the skin and clog the openings of sebaceous and sweat glands, leading to functional disorders. Dust mixed with sebaceous secretion and sweat, especially with an admixture of microorganisms, irritates the skin, causing inflammatory states, while eye irritation may lead to conjunctivitis. Dusts inhaled over a prolonged time period, irrespective of their character, always cause irritations of nose mucous membrane, and bronchitis, opening the pathways for conditionally pathogenic microorganisms, which leads to numerous pathogenic states of the upper and lower respiratory tract (Kołacz and Dobrzański 2006; Demanche et al. 2009).

Apart from gaseous and mechanical contamination, the air in livestock buildings also contains microorganisms, both saprophytic and pathogenic ones. On average, only between 0.02% and 5.2% of the total number of culturable aerobic bacteria were identified as Gram-negative bacteria (Zucker et al. 2000). The kind and amount of microbiological contamination observed in the air of livestock buildings constitute one of the more significant indices of sanitary–hygienic state evaluation. Profitable conditions for microorganism and parasite development and proliferation are observed in hot, humid, poorly ventilated, and rarely disinfected buildings. Pathogenic microorganisms present in air connect with dusts, water vapor, and mucus, forming a so-called bacterial aerosol in the air, which may be saprophytic, mixed, or pathogenic.

The herd size, breeding and feeding system, type of ventilation, and airflow velocity in barns were the factors significantly differentiating the concentration of total bacteria and their level in the respirable fraction. Higher concentrations of bacterial microorganisms were observed in small herds, with litter bed system, manual feed distribution, natural ventilation, and low airflow velocity (Sowiak et al. 2012). The amount of microorganisms in the air increases with the increasing density of animals in the building. In specified conditions, when infected or ill animals, or animals acting as carriers or sowers, are present in a room, a considerable amount of pathogenic microorganism forms are observed in the air. Moreover, frequent movement of animals between particular buildings or sectors leads to microorganism passaging, which may result in pathogenicity changes. Animals moving between buildings means that the effectively formed resistance against the bacterial flora of one building is insufficient and ineffective in another building. For this reason, a stage of anti-infectious prevention, taking into account the "full room–empty room" rule, is an important issue in large-scale breeding (Kołacz and Dobrzański 2006).

5.9 TOXIC HEAVY METALS

Chemical contamination is one of the criteria for evaluation of the safety of products for consumption (Andrée et al. 2010). The special role is attributed to the assessment of the level of heavy metals, which pose a danger for humans and animals. Absolutely toxic elements include lead (Pb), cadmium (Cd), mercury (Hg), and arsenic (As). These elements are characterized by a high accumulation coefficient, and the intoxication of animals maintained in industrialized regions is sometimes a controversial issue. Toxicology divides the exposures and adverse responses into acute, subchronic, and chronic ones.

Acute poisoning by meat products is rare nowadays, and if it ever happens, it is mostly driven by microorganisms (like botulism). Chronic toxic effects of many substances are of systemic character involving a number of physiological alterations in many organs or tissues (Püssa 2013). Generally, harmful effects of heavy metals on humans may include carcinogenicity, neurotoxicity, cardiovascular toxicity, acute pulmonary and renal toxicities and testicular damage, hematopoiesis toxicity (anemia), and so on. Owing to the significance of the problem, the EFSA has issued scientific reports concerning food contamination with heavy metals (Cu, As, Cd, and Pb) over the last few years. The maximum allowable levels of Pb (0.10 mg/kg fresh weight) and Cd (0.05 mg/kg fresh weight) for beef, mutton, pork, and poultry meat were determined in the Commission Regulation 629/2008/EC (European Commission 2008). Mercury, in the form of methyl mercury, is usually not observed in over normative concentrations in these kinds of products, and therefore it is standardized only for fish.

The concentration of heavy metals in animal tissues is variable and depends on numerous factors, such as the content in fodder, water, and air, the degree of their bioavailability, as well as interactions between the elements. Their accumulation in the body can lead to harmful effects over time. Most of the reports published over the last few years suggest that products of animal-origin do not pose any danger to health, since the allowable levels of heavy metals are usually not exceeded (Andrée et al. 2010). Monitoring studies of various meats (loin of pork, chicken, and lamb) demonstrate low concentrations of Pb, Cd, and As, while higher concentrations were noted in fish tissues (Perelló et al. 2008). It is however significant, that the concentration of metals such as Pb and Cd increases in the muscles, liver, and kidneys along with the age of the animal (Perelló et al. 2008).

The cooking process is only of very limited value as a means of reducing metal concentrations. The concentrations of heavy metals, for example, cadmium in animal tissues, especially the kidney, is strongly related with the cadmium levels in feedstuff. Genetic lines in the case of pigs do not affect Cd kidney concentration (Tomović et al. 2011). The highest effect is caused by the availability of heavy metals in the local agricultural environment. Differences in the content of chemical components (e.g., mycotoxins) are possible in products from organic and conventional farms. However, no differences were noted in Pb and Cd content in meat from organic and conventional farms; in both kinds of farms, the concentration of these heavy metals in meat was low (Ghidini et al. 2005).

5.10 CONCLUSIONS

The relationships among animal welfare, animal health, and food safety are strongly emphasized in the legislation of numerous countries. The conditions and systems of housing have a specific effect on animal products and their safety. Additionally, animal rearing conditions in these systems reduce stressful events, which in turn can positively affect numerous intrinsic aspects of final product quality. An increasing number of consumers link animal welfare with better product quality or product safety. Consumers expect its cost to be compensated by high-quality parameters such as freshness, tenderness, juiciness, low fat content, high nutritive value, and exquisite taste. In addition, improving an animal's welfare can positively affect other aspects of product quality (e.g., reducing the occurrence of tough or watery meat), and disease resistance. Depending on animal farming methods, there are changes in fat content, sensory value, and fatty acid profiles of meat. The standards of animal welfare in organic farms, as in conventional ones, were created for transparency in food production processes. The meat of animals from organic systems is characterized by valuable features concerning quality as well as organoleptic features (taste, marbling). Product enrichment in health-promoting, biologically active components is possible mainly via nutrition, and that procedure leads to functional products. The problem of animal health status on organic and conventional farms is discussed; however, the prevalence of parasitic diseases is higher in animals kept on organic farms. There are no significant differences in the content of harmful chemical substances (heavy metals) in both types of farms. The aspiration to improve animal welfare conditions is not inconsistent with final product quality.

REFERENCES

Alfaia, C.P.M., Alves, S.P., Martins, S.I.V., Costa, A.S.H., Fontes, C.M.G.A., Lemos, J.P.C., Bessa, R.J.B., and Prates, J.A.M. 2009. Effect of feeding system on intramuscular fatty acids and conjugated linoleic acid isomers of beef cattle, with emphasis on their nutritional value and discriminatory ability. *Food Chem.* 114:939–946.

Álvarez-Pérez, S., Blanco, J.L., Peláez, T., Astorga, R.J., Harmanus, C., Kuijper, E., and García, M.E. 2013. High prevalence of the epidemic Clostridium difficile PCR ribotype 078 in Iberian free-range pigs. *Res. Vet. Sci.* 95:358–361.

Andrée, S., Jira, W., Schwind, K-H., Wagner, H., and Schwägele, F. 2010. Chemical safety of meat and meat products. *Meat Sci.* 86:38–48.

Andrighetto, I., Gottardo, F., Andreoli, D., and Gozzi, G. 1999. Effect of type of housing on veal calf growth performance, behaviour and meat quality. *Livest. Prod. Sci.* 57:137–145.

Averós, X., Aparicio, M.A., Ferrari, P., Guy, J.H., Hubbard, C., Schmid, O., Ilieski, V., and Spoolder, H.A.M. 2013. The effect of steps to promote higher levels of farm animal welfare across the EU. Societal versus animal scientists' perceptions of animal welfare. *Animals* 3:786–807.

Barroso, F.G., Alados, C.L., and Boza, J. 2000. Social hierarchy in the domestic goat: Effect on food habits and production. *Appl. Anim. Behav. Sci.* 69:35–53.

Becker, T. 2000. Consumer perception of fresh meat quality: A framework for analysis. *Brit. Food J.* 102:158–176.

Blokhuis, H.J., Keeling, L.J., Gavinelli, A., and Serratosa, J. 2008. Animal welfare's impact on the food chain. *Trends Food Sci. Technol.* 19:79–87.

Borowski, S., Gutarowska, B., Durka, K., Korczyński, M., Opaliński, S., and Kołacz, R. 2010. Biological deodorization of organic fertilizers. *Przem. Chem.* 4:318–323. [in Polish]

Braghieri, A., Pacelli, C., Verdone, M., Girolami, A., and Napolitano, F. 2007. Effect of grazing and homeopathy on milk production and immunity of Merino derived ewes. *Small Rumin. Res.* 69:95–102.

Chavez, C., Coufal, C.D., Niemeyer, P.L., Carey, J.B., Lacey, R.E., Miller, R.K., and Beier, R.C. 2004. Impact of dietary supplemental methionine sources on sensory measurement of odor-related compounds in broiler excreta. *Poult. Sci.* 83:1655–1662.

Daley, C.A., Abbott, A., Doyle, P.S., Nader, G.A., and Larson, S. 2010. A review of fatty acid profiles and antioxidant content in grass-fed and grain-fed beef. *Nutr. J.* 10:1–12.

Demanche, A., Bønløkke, J.H., Beaulieu, M.J., Assayag, E.I., and Cormier, Y. 2009. Swine confinement buildings: Effects of airborne particles and settled dust on airway smooth muscles. *Ann. Agric. Environ. Med.* 16:233–238.

Done, S.H., Chennells, D.J., Gresham, A.C.J. et al. 2005. Clinical and pathological responses of weaned pigs to atmospheric ammonia and dust. *Vet. Rec.* 157:71–80.

Edwards, S.A. 2005. Product quality attributes associated with outdoor pig production. *Livest. Prod. Sci.* 94:5–14.

EFSA Panel on Animal Health and Welfare (AHAW). 2012. Statement on the use of animal-based measures to assess the welfare of animals. *EFSA J.* 10(6):2767. doi:10.2903/j.efsa.2012.2767.

Eurobarometer. 2005. http://ec.europa.eu/public_opinion/archives/eb/eb63/eb63_en.htm.

Eurobarometer. 2007. Attitudes of EU citizens towards animal welfare. Special Eurobarometer 270//Wave 66.1; European Commission: Brussels, Belgium. http://ec.europa.eu/public_opinion/archives/ebs/ebs_270_en.pdf.

European Commission. 2008. Commission Regulation 629/2008/EC of 2 July 2008 setting maximum levels for certain contaminants in foodstuffs. Official Journal of the European Union, L 173, 63.7.2008.

European Commission. 2012. Communication from the commission to the European Parliament, the council and the European Economic and Social Committee on the European Union Strategy for the Protection and Welfare of Animals 2012–2015. http://ec.europa.eu/food/animal/welfare/actionplan/docs/aw_strategy_19012012_en.pdf.

FAO and IFIF. 2010. Good practices for the feed industry—Implementing the Codex Alimentarius Code of Practice on Good Animal Feeding. FAO Animal Production and Health Manual No. 9. Rome.

Farm Animal Welfare Council. 1992. FAWC updates the five freedoms. *Vet. Rec.* 17:357.

Ferguson, D.M., and Warner, R.D. 2008. Have we underestimated the impact of pre-slaughter stress on meat quality in ruminants? *Meat Sci.* 8:12–19.

Fossler, C.P., Wells, S.J., Kaneene, J.B. et al. 2005. Cattle and environmental sample-level factors associated with the presence of *Salmonella* in a multi-state study of conventional and organic dairy farms. *Prev. Vet. Med.* 67:39–53.

Fraser, D. 2008. Toward a global perspective on farm animal welfare. *Appl. Anim. Behav. Sci.* 113:330–339.

Ghidini, S., Zanardi, E., Battaglia, A., Varisco, G., Ferretti, E., Campanini, G., and Chizzolini, R. 2005. Comparison of contaminant and residue levels in organic and conventional milk and meat products from Northern Italy. *Food Addit. Contam.* 22:9–14.

Gregory, N.G. 2010. How climatic changes could affect meat quality. *Food Res. Int.* 43:1866–1873.

Gruber, S.L., Tatum, J.D., Engle, T.E., Chapman, P.L., Belk, K.E., and Smith, G.C. 2010. Relationships of behavioral and physiological symptoms of pre-slaughter stress to beef LM tenderness. *J. Anim. Sci.* 88:1148–1159.

Hansen, P.J., Drost, M., Rivera, R.M., Paula-Lopes, F.F., Al-Katanani, Y.M., Krininger, C.E. III., and Chase, C.C. 2001. Adverse impact of heat stress on embryo production causes and strategies for mitigation. *Theriogenology* 55:91–103.

Hocquette, J., Botreau, R., Legrand, I., Polkinghorne, R., Pethick, D.W., Lherm, M., Picard, B., Doreau, M., and Terlouw, E.M.C. 2014. Win–win strategies for high beef quality, consumer satisfaction, and farm efficiency, low environmental impacts and improved animal welfare. *Anim. Prod. Sci.* 54:1537–1548.

Ingenbleek, P.T.M., Harvey, D.R., Ilieski, V., Immink, V.M., de Roest, K., and Schmid, O. 2013. The European market for animal-friendly products in a societal context. *Animals* 3:808–829.

Jacob, R.H., Pethick, D.W., Clark, P., D'Souza, D.N., Hopkins, D.L., and White, J. 2006. Quantifying the hydration status of lambs in relation to carcass characteristics. *Aust. J. Exp. Agric.* 46:429–437.

Kavolelis, B. 2006. Impact of animal housing systems on ammonia emission rates. *Pol. J. Environ. Stud.* 15:739–745.

Kilchsperger, R., Schmid, O., and Hecht, J. 2010. Animal welfare initiatives in Europe—Technical report on grouping method for animal welfare standards and initiatives. Deliverable No. 1.1 of EconWelfare Project. Research Institute of Organic Agriculture FiBL, Frick, Switzerland.

Knox, R.V., and Zas, S.L. 2001. Factors influencing estrus and ovulation in weaned sows as determined by transrectal ultrasound. *J. Anim. Sci.* 79:2957–2963.

Kołacz, R., and Bodak, E. 1999. Animal welfare and its assessment criteria. *Med. Weter.* 55:147–154. [in Polish]

Kołacz, R., and Bodak, E. 2002. Animal health and welfare implications. *Annals Anim. Sci.* 1:25–30.

Kołacz, R., and Dobrzański, Z. (Ed.). 2006. *Animal Hygiene and Welfare.* AR Wrocław, Poland. [in Polish]

Kołacz, R., Korniewicz, A., Dobrzański, Z., Bykowski, P., Kołacz, D., and Korniewicz, D. 2004. Effect of dietary fish and rapeseed oils on sensory and physicochemical characteristics of pigs M. Longisimus dorsi and fatty acid composition. *J. Anim. Feed Sci.* 13:143–152.

Kouba, M. 2003. Quality of organic animal products. *Livest. Prod. Sci.* 80:33–40.

Kváč, M., Kouba, M., and Vítovec, J. 2006. Age-related and housing-dependence of *Cryptosporidium* infection of calves from dairy and beef herds in South Bohemia, Czech Republic. *Vet. Parasitol.* 137:202–209.

Lowe, T.E., Gregory, N.G., Fisher, A.D., and Payne, S.R. 2002. The effects of temperature elevation and water deprivation on lamb physiology, welfare, and meat quality. *Aust. J. Agric. Res.* 53:707–714.

Lund, V. 2006. Natural living—A precondition for animal welfare in organic farming. *Livest. Sci.* 100:71–83.

Lund, V., and Algers, B. 2003. Research on animal health and welfare in organic farming—A literature review. *Livest. Prod. Sci.* 80:55–68.

Mader, T.L., Holt, S.M., Hahn, G.L., Davis, M.S., and Spiers, D.E. 2002. Feeding strategies for managing heat load in feedlot cattle. *J. Anim. Sci.* 80:2373–2382.

MAFF (Ministry of Agriculture, Fisheries and Forestry), Australia 2007. Australian Animal Welfare Strategy. http://www.daff.gov.au/animal-plant-health/welfare/aaws.

McCrory, D.F., and Hobbs, P.J. 2001. Additives to reduce ammonia and odor emissions from livestock wastes: A review. *J. Environ. Qual.* 30:345–355.

Mee, J.F., Geraghty, T., O'Neill, R., and More, S.J. 2012. Bioexclusion of diseases from dairy and beef farms: Risks of introducing infectious agents and risk reduction strategies. *Vet. J.* 194:143–150.

Mench, J. 2008. Farm animal welfare in the U.S.A.: Farming practices, research, education, regulation, and assurance programs. *Appl. Anim. Behav. Sci.* 113:298–312.

Miranda-de la Lama, G.C., Pascual-Alonso, M., Guerrero, A. et al. 2013. Influence of social dominance on production, welfare and the quality of meat from beef bulls. *Meat Sci.* 94:432–437.

Mitlohner, F.M., Galyean, M.L., and McGlone, J.J. 2002. Shade effects on performance, carcass traits, physiology, and behavior of heat-stressed feedlot heifers. *J. Anim. Sci.* 80:2043–2050.

Nardone, A., Ronchi, B., Lacetera, N., and Bernabuci, U. 2006. Climatic effects on productive traits in livestock. *Vet. Res. Commun.* 30(Suppl. 1):75–81.

Nilzen, V., Babol, J., Dutta, P.C., Lundeheim, N., Enfält, A.C., and Lundstrom, K. 2001. Free range rearing of pigs with access to pasture grazing-effect on fatty acid composition and lipid oxidation products. *Meat Sci.* 58:267–275.

Nuernberg, K., Dannenberger, D., Nuernberg, G., Ender, K., Voigt, J., Scollan, N.D., Wood, J.D., Nute, G.R., and Richardson, R.I. 2005. Effect of a grass-based and a concentrate

feeding system on meat quality characteristics and fatty acid composition of longissimus muscle in different cattle breeds. *Livest. Prod. Sci.* 94:137–147.
OIE (World Organisation for Animal Health). 2011. Terrestrial Animal Health Code. Available from http://www.oie.int/international-standard-setting/terrestrial-code.
OIE. 2013a. Organic Rules of the Office International Des Epizooties (appendices to the International Agreement for the Creation of an Office International Des Epizooties in Paris). http://www.oie.int/about-us/key-texts/basic-texts/organic-rules/.
OIE. 2013b. The OIE's Achievements in Animal Welfare. http://www.oie.int/animal-welfare/animal-welfare-key-themes.
Opaliński, S., Korczyński, M., Kołacz, R., Dobrzański, Z., and Żmuda, K. 2009. Application of selected aluminosilicates for ammonia adsorption. *Przem. Chem.* 5:540–43. [in Polish]
Opaliński, S., Korczyński, M., Szołtysik, M., Kołacz, R., Dobrzański, Z., and Gbiorczyk, W. 2010. Application of mineral sorbents to filtration of air contaminated by odorous compounds. *Chem. Eng. Trans.* 23:369–374. [in Polish]
Otter, C., O'Sullivan, S., and Ross, S. 2012. Laying the foundations for an international animal protection regime. *J. Anim. Ethics.* 2:53–72.
Paula-Lopes, F.F., Chase, C.C., Al-Katanai, Y.M. Jr, Krininger, C.E. III., Rivera, R.M., Tekin, S., Majewski, A.C., Ocon, O.M., Olson, T.A., and Hansen, P.J. 2001. Breed differences in resistance of bovine preimplantation embryos to heat shock. *Theriogenology* 55:436.
Perelló, G., Marti-Cid, R., Llobet, J.M., and Domingo, J.L. 2008. Effects of various cooking processes on the concentrations of arsenic, cadmium, mercury, and lead in foods. J. Agric. Food Chem. 56:11262–11269.
PPG. 2007. The opportunities and challenges of running a responsible UK food and drink business. The Plough to Plate Group, Business in the Community, London. http://www.bitc.org.uk/princes_programmes/rural_action/future_of_british_food_and_farming/plough_to_plate.html.
Pugliese, C., and Sirtori, F. 2012. Quality of meat and meat products produced from southern European pig breeds. *Meat Sci.* 90:511–518.
Püssa, T. 2013. Toxicological issues associated with production and processing of meat. *Meat Sci.* 95:844–853.
Realinia, C.E., Ducketta, S.K., Britob, G.W., Dalla Rizzab, M., and De Mattos, D. 2004. Effect of pasture vs. concentrate feeding with or without antioxidants on carcass characteristics, fatty acid composition, and quality of Uruguayan beef. *Meat Sci.* 66:567–577.
Rembiałkowska, E., and Badowski, M. 2012. Nutritional value of organic meat and potential human health response. In Ricke, S.C., Van Loo, E., Johnson, M.G., and O'Brian, C.A. (eds.), *Organic Meat Production and Processing*. Wiley–Blackwell, USA, pp. 239–254.
Rembiałkowska, E., and Wiśniewska, K. 2010. Meat quality from organic production. *Med. Weter.* 66:188–191. [in Polish]
Rushen, J., Butterworth, A., and Swanson, J.C. 2011. Animal behavior and well-being symposium. Farm animal welfare assurance: Science and application. *J. Anim. Sci.* 89:1219–1122.
Santos, C., Almeida, J.M., Matias, E.C., Fraqueza, M.J., Roseiro, C., and Sardina, L. 1997. Influence of lairage environmental conditions and resting time on meat quality in pigs. *Meat Sci.* 45:253–262.
Slominski, A., Zjawiony, J., Wortsman, J., Semak, I., Stewart, J., Pisarchik, A., Sweatman, T., Marcos, J., Dunbar, C., and Tuckey, R.T. 2004. A novel pathway for sequential transformation of 7-dehydrocholesterol and expression of the P450scc system in mammalian skin. *Eur. J. Biochem.* 271:4178–4188.
Sowiak, M., Bródka, K, Buczyńska, A., Cyprowski, M., Kozajda, A., Sobala, A., and Szadkowska-Stańczyk, I. 2012. An assessment of potential exposure to bioaerosols among swine farm workers with particular reference to airborne microorganisms in the respirable fraction under various breeding conditions. *Aerobiologia* 28:121–133.

Sundrum, A., Butfering, L., Henning, M., and Hoppenbrock, K.H. 2000. Effects of on-farm diets for organic pig production on performance and carcass quality. *J. Anim. Sci.* 78:1199–1205.

Tomović, V., Petrović, L.J., Tomović, M., Kevrešan, Ž., Jokanović, M., Džinić, N., and Despotović, A. 2011. Cadmium levels of kidney from 10 different pig genetic lines in Vojvodina (northern Serbia). *Food Chem.* 129:100–103.

Troy, D.J., and Kerry, J.P. 2010. Consumer perception and the role of science in the meat industry. *Meat Sci.* 86:214–226.

Truszczyński, M., and Kołacz, R. 2009. Extending the OIE activity by adding food safety and animal welfare areas. *Med. Weter.* 65:731–734. [in Polish]

Tymczyna, L., Chmielowiec-Korzeniowska, A., and Saba, L. 2004. Biological treatment of laying house air with open biofilter use. *Pol. J. Environ. Stud.* 13:425–428.

Vaarst, M., Padel, S., Hovi, M., Younie, D., and Sundrum, A. 2005. Sustaining animal health and food safety in European organic livestock farming. *Livest. Prod. Sci.* 94:61–69.

Van de Perre, V., Ceustermans, A., Leyten, J., and Geers, R. 2010. The prevalence of PSE characteristics in pork and cooked ham—Effects of season and lairage time. *Meat Sci.* 86:391–397.

VanderZaag, A.C., Gordon, R.J., Glass, V.M., and Jamieson, R.C. 2008. Floating covers to reduce gas emissions from liquid manure storages: A review. *Appl. Eng. Agric.* 24:657–671.

van de Weerd, H.A., Docking, C.M., Day, J.E.L., Avery, P.J., and Edwards, S.A. 2003. A systematic approach towards developing environmental enrichment for pigs. *Appl. Anim. Behav. Sci.* 84:101–118.

van Gastelen, S., Westerlaan, B., Houwers, D.J., and van Eerdenburg, F.J.C.M. 2011. A study on cow comfort and risk for lameness and mastitis in relation to different types of bedding materials. *J. Dairy Sci.* 94:4878–4888.

Varel, V.H. 2002. Livestock manure odor abatement with plant-derived oils and nitrogen conservation with urease inhibitors: A review. *J. Anim. Sci.* 80(E. Suppl. 2):E1–E7.

Veissier, I., Butterworth, A., Bock, B., and Roe, E. 2008. European approaches to ensure good animal welfare. *Appl. Anim. Behav. Sci.* 113:279–297.

Velarde, A., and Dalmau, A. 2012. Animal welfare assessment at slaughter in Europe: Moving from inputs to outputs. *Meat Sci.* 92:244–251.

Ventanas, S., Ventanas, J., and Ruiz, J. 2007. Sensory characteristics of Iberian dry-cured loins: Influence of crossbreeding and rearing system. *Meat Sci.* 75:211–219.

von Borell, E., Ozpinar, A., Eslinger, K.M., Schnitz, A.L., Zhao, Y., and Mitloehner, F.M. 2007. Acute and prolonged effects of ammonia on hematological variables, stress responses, performance, and behavior of nursery pigs. *J. Swine Health Prod.* 15:137–145.

Walshe, B.E., Sheehan, E.M., Delahunty, C.M., Morrissey, P.A., and Kerry, J.P. 2006. Composition, sensory and shelf life stability analyses of *Longissimus dorsi* muscle from steers reared under organic and conventional production systems. *Meat. Sci.* 73:319–325.

Warner, R.D., Ferguson, D.M., McDonagh, M.B., Channon, H.A., Cottrell, J.J., and Dunshea, F.R. 2005. Acute exercise stress and electrical stimulation influence the consumer perception of sheep meat eating quality and objective quality traits. *Aust. J. Exp. Agric.* 45:553–560.

Warren, H.E., Scollan, N.D., Enser, M., Hughes, S.I., Richardson, R.I., and Wood, J.D. 2008. Effects of breed and a concentrate or grass silage diet on beef quality in cattle of 3 ages. I: Animal performance, carcass quality and muscle fatty acid composition. *Meat Sci.* 78:256–269.

Węglarz, A. 2010. Quality of beef from semi-intensively fattened heifers and bulls. *Anim. Sci. Pap. Rep.* 28:207–218.

Whiting, T.L. 2013. Policing farm animal welfare in Federated Nations: The problem of dual federalism in Canada and the USA. *Animals* 3:1086–1122.

Wistuba, T.J., Kegley, E.B., Apple, J.K., and Rule, D.C. 2007. Feeding feedlot steers fish oil alters the fatty acid composition of adipose and muscle tissue. *Meat Sci.* 77:196–203.

WSPA. 2007. Provisional Draft UDAW. http://www.wspa.org.au/Images/Proposed_UDAW_Text%20-%20ENGLISH_tcm30-2544.pdf#false.

Zhang, Z.F., and Kim, I.H. 2014. Effects of multistrain probiotics on growth performance, apparent ileal nutrient digestibility, blood characteristics, cecal microbial shedding, and excreta odor contents in broilers. *Poult. Sci.* 93:364–370.

Zucker, B.A., Trojan, S., and Müller, W. 2000. Airborne Gram-negative bacterial flora in animal houses. *J. Vet. Med. S. B* 47:37–46.

Zurita-Herrera, P., Bermejo, J.V.D., Henríquez, A.A., Vallejo, M.E.C., and Costa, R.G. 2013. Effects of three management systems on meat quality of dairy breed goat kids. *J. Appl. Anim. Res.* 41:173–182.

6 Preslaughter Handling, Welfare of Animals, and Meat Quality

Temple Grandin

CONTENTS

6.1 INTRODUCTION

Careful, calm handling of cattle, pigs, or sheep will improve both animal welfare and help preserve meat quality. This chapter will discuss results from research studies and practical information on welfare for the stage in the animal's life when it is transported from the farm through preslaughter, handling, and restraint. The effects of stunning on meat quality will be covered in Chapter 8. The public is becoming increasingly concerned about how farm animals are treated (Rollin et al. 2011). Today, many mobile telephones are video cameras, and activities that were previously hidden from the public can now be recorded and instantly uploaded to the Internet. Everyone who is involved with harvesting animals for meat must always be aware of the importance of maintaining both good animal welfare and protecting product quality. This chapter will draw information from both the research literature and the author's extensive practical experience in several hundred slaughter plants in many countries. Its aim is to help bridge the gap between scientific research and practical applications that can be used by transporters and plant managers.

For both good animal welfare and meat quality, there are three types of problems that should be avoided. The first is loss or damage to an animal, the second is a reduction in meat quality, and the third is stress that would be detrimental to welfare. Death losses, nonambulatory animals, and bruises that occur during loading at farm, transport, and handling at the abattoir (slaughter plant) are examples of the first type of problem. Examples of loss of quality are tough meat; dark, firm, and dry (DFD) meat; and pale, soft, exudative (PSE) meat. Dark cutting, also called DFD, is a quality defect because the meat becomes darker and drier than normal. Beef will lose its bright red color and the pH rises. A major problem with DFD beef is that it has a shorter shelf life (Blixt and Borch 2002) and consumers prefer normal beef (Vilioen et al. 2002). PSE (watery) meat has a low pH. Pigs are more prone to PSE than cattle and affected pork is paler than normal and lacks the ability to hold water. At the supermarket, this results in a dry pork chop sitting in a puddle of fluid in the bottom of the package. For a more detailed description of the physiology, refer to Chapter 2.

6.2 PREVENTING TRANSPORT DEATH LOSSES AND NONAMBULATORY ANIMALS

There are five main factors associated with death losses and nonambulatory animals during handling and transport between the farm and the abattoir (harvest). They are rough handling, transport of weak, emaciated, or crippled animals, overloading trucks, heat stress, high doses of beta-agonists, and animal genetics. It is essential to keep trucks moving to prevent death losses. Heat builds up quickly in a stationary vehicle and animals may die. The author has observed that transport of weak, emaciated, or crippled animals is a major cause of downed nonambulatory animals in both cull dairy cows and sows. Another major cause of death losses and nonambulatory animals is overloading of trucks. Ritter et al. (2006) found that a pig weighing 129 kg should be loaded at 0.48 m^2/pig on per pig. Guidelines on maximum stocking density for all types of livestock can be found online at FASS (2010), animalhandling.org (2013), and EU.

6.2.1 Animals Should Be Fit for Transport

A major cause of nonambulatory (downer) animals in old dairy cows and breeding sows is the failure to send an animal to slaughter when it is still fit for transport. Schwartzkopf-Genswein et al. (2012) state that the animals that are less fit for travel have the greatest welfare and meat quality issues. The author's own observations in abattoirs indicate that poor animal condition before it left the farm was a major reason why it became nonambulatory. Emaciated, weak, lame dairy cows are often too weak to remain standing for the entire journey. Ahola et al. (2011) surveyed auctions and found that 13% of the cull dairy cows arrived in an emaciated or near emaciated state and only 4% of the beef breed cows were emaciated. Loading weak and unfit animals is most likely to occur with old breeding stock that have little economic value (Grandin 2010a). The author has observed that the four most important factors for providing acceptable animal welfare during transport are loading animals on the vehicle that are fit for transport, being careful not to overload trucks, good driving practices, where sudden stops and rapid acceleration are avoided, and quiet, careful handling during loading and unloading.

6.2.2 Genetic Factors

Genetics can also be a major factor that can contribute to death losses (Alvarez et al. 2009). Pigs that are homozygous positive for the halothane porcine stress gene can have high levels of death losses (Murray and Johnson 1998). Many U.S. and U.K. breeders in the 1980s opted for this trait because it produced lots of lean muscle. During transport, pigs that were homozygous negative had the lowest death losses, the heterozygotes were intermediate, and the homozygous positive animals had the most losses (Murray and Johnson 1998). Today, most U.S., Canadian, and Danish producers have eliminated halothane stress gene genetics, but there are some countries where it is still common. A reduction in the breeding of halothane stress gene pigs is confirmed by USDA/FSIS death loss statistics that show that death losses in the United States started to decrease after 2002 (USDA/FSIS) (Ritter et al. 2009). Pigs that have the stress gene trait will often become nonambulatory, quiver all over, and their skin may have a red splotchy appearance. If a pig with stress gene genetics becomes nonambulatory, it should be cooled by providing shade and wetting the floor. Do not douse it with cold water as the physiological shock may kill it.

6.2.3 Effect of Beta-Agonists

Another possible cause of increased transit death losses and nonambulatory animals is beta-agonist feed supplement such as ractopamine or zilpaterol. These supplements enhance lean muscle growth by expanding fast twitch fibers and may increase death losses of feedlot cattle (Montgomery et al. 2009; Baxa et al. 2010; Longeragan et al. 2014). The Elanco label on ractopamine (Paylean) states that it may increase death losses in pigs. High doses of ractopamine can make pigs more difficult to handle, increase aggression, and increase hoof cracking (Marchant-Forde et al. 2003; Poletto et al. 2009, 2010). James et al. (2013) found that pigs fed ractopamine were more susceptible to stress when they were handled roughly. The dose was 20 mg/day for four weeks. Vogel et al. (2011) found that beta-agonists were also associated with heat stress.

6.3 PREVENTING BRUISING

Bruised meat cannot be used for food in most developed countries and it must be removed from the carcass. In cattle, the muscle may be severely bruised, but on the live animal, the outside of the hide and hair may appear normal. The main causes of bruises on cattle are overloading of trucks, horns, and rough handling. Grandin (1981) and Huertas et al. (2010) found that cattle handled roughly and shocked many times with electric prods had twice as many bruises when they arrived at the abattoir. Overloading of trucks was another major cause of bruises (Eldridge and Winfield 1988; Tarrant et al. 1988, 1992). When a truck is overloaded, an animal that falls is not able to get back up and it gets trampled by other animals. There is an optimum truck loading density for cattle (Eldridge and Winfield 1988). If cattle are packed in too tightly, bruises increase, but if they are too loose, they are also more likely to fall down. Good driving practices are essential to reduce bruises as sudden stops and rapid acceleration will throw animals off balance and bruising may increase.

Horns are another major cause of bruises (Shaw et al. 1976). Tipping the ends of the horns will not reduce bruises (Ramsey et al. 1976) and cutting off horns of adult cattle is extremely stressful and this practice shortly prior to slaughter is not acceptable from an animal welfare standpoint. Breeding cattle that are polled (hornless) is one good method for reducing bruises. Angus cattle are naturally polled and many other *Bos taurus* (non-*indicus*) beef breeds have good polled genetic lines.

Bruises can occur after captive bolt stunning before the animal is bled (Meischke and Horder 1976), but after the animal is bled, it cannot be bruised. Rough handling or bumping on sharp edges during handling at the slaughter plant can cause bruises. Sharp edges with a small diameter are more likely to cause bruises than smooth rounded surfaces. To eliminate sharp edges, the track for a guillotine gate should be recessed into the wall (Figure 6.1). Cattle and sheep that pass through auctions before they arrive at the slaughter plant are likely to have more bruises compared to cattle shipped directly to slaughter (Hoffman et al. 1998; Cockram and Lee 1991). Correa et al. (2010) compared loading pigs on the farm with electric prods or nonelectric driving aids. Electric prod use increased the incidence of bruising and blood splash (Chapter 2).

6.4 EFFECT OF FINANCIAL INCENTIVES

Other factors that contribute to low percentages of animals that arrive either bruised or dead are financial incentives. When transporters have a financial incentive to deliver animals in good condition, bruising and other losses will be reduced. Holding producers and transporters financially accountable for both bruises and death losses motivates them to handle and transport animals with care (Grandin 2010a). When producers were paid based on carcass quality and bruises were deducted from their payments, bruises were significantly reduced (Grandin 1981). The worst way to pay transporters is based on kilograms of animals loaded on a truck. This method of

FIGURE 6.1 Guillotine gate with track recessed into the concrete wall to prevent bruising. It also provides sufficient back clearance to prevent animals from balking and refusing to move under it. (Photo by Temple Grandin.)

payment provides no financial incentive to reduce bruises, death losses, or nonambulatory animals.

6.5 STRESS PRINCIPLES AND MEAT QUALITY

There is a basic principle on how stress during handling and transport may affect meat quality. Long-term stresses, such as long distance transport or bulls mounting or fighting may cause problems with higher pH and DFD meat (Jones and Tong 1989; Gallo et al. 2003). Short-term stresses, such as the high use of electric prods a few minutes before slaughter, may increase meat quality problems in pigs and the incidence of tough meat in cattle (Hambrecht et al. 2005; Warner et al. 2007; Edwards et al. 2010a,b). In pigs, lactate levels measured immediately after multiple shocks were 32 mM and only 4–6 mM in carefully handled pigs (Benjamin et al. 2001). In pigs, longer-term stresses such as transport or handling stressors, which occur at the farm during loading resulted in reduced lactic acid, less PSE, and a higher pH after slaughter (Edwards et al. 2010c; Gajara et al. 2013). The physiology of the animals is complex and there are situations where these principles will not always apply. The effects of long-term stresses on beef quality are more variable. Ferguson and Warner (2008) state that in ruminants, meat quality defects can occur that are not dependent on pH. In both species, behavioral agitation shortly before slaughter raises lactate, lowers pH, and toughens meat (Warner et al. 2007; Edwards et al. 2010a; Gruber et al. 2010).

6.6 TRANSPORT AND FACTORS OUTSIDE THE PLANT THAT AFFECT MEAT QUALITY

Spending a long time in the marketing channels between the farm and the slaughter plant is detrimental to beef quality. Vogel et al. (2011) found that cull Holstein dairy cows have high pH and borderline dark cutting after they have been through an auction market. Over a long career, the author has observed that meat quality and the quantity of meat are two opposing goals in young grain-fed cattle and pigs. When animals are bred, supplemented, or implanted with hormones to produce the maximum amount of lean muscle, problems with either dark cutters, tough meat, or death losses may increase. The author observed that during the early 1990s, when U.S. pork producers genetically selected pigs for rapid weight gain and lean pork, the meat became tougher due to more lean growth. The author observed that a steak knife was required to cut it. Cattle that are fed to produce maximum yield of meat with high doses of beta-agonists will also have tougher meat (Arp et al. 2013). Aggressive implant strategies where steers are given high doses of trembolane acetate hormone implants are also more likely to produce dark cutters (Scanga et al. 1998). Zilpaterol increased dark cutters in steers (Longeragan et al. 2014). Trembolane acetate is a synthetic male hormone, which may explain its effects on dark cutters.

In many parts of the world, bulls are kept intact because they provide more kilograms of lean meat compared to steers. The disadvantage of producing bulls is tougher beef and the propensity to be dark cutters (Tarrant 1981). To prevent dark cutters, intact bulls must be transported and held at the slaughter plant lairage in groups of pen mates (Price and Tennessen 1981; Puolanne and Alto 1981; Tennessen et al. 1984). Bulls are more aggressive than steers and when strange bulls are mixed, they will often fight. If bulls fight or mount each other during transport and lairage, the percentage of animals that will become dark cutters may increase (Tarrant 1989). If a group of bulls has to be mixed, it is best to mix them 2 weeks or more *before* they leave the farm. This provides the bulls time to develop a new social order and enable their glycogen levels to recover.

The author has observed that problems with dark cutters under U.S. conditions may be sporadic. Groups of similar cattle that have been raised identically sometimes produce many dark cutters and at other times have none. Canadian researchers report a similar finding. The percentage of dark cutters averaged 2%, but a few truckloads of cattle had over 30% dark cutters (Warren et al. 2010). A good analogy for explaining to meat plant workers which groups of cattle may have more dark cutters is to consider the concept of a car running out of gas (petrol). The "gas" is glycogen and when it runs out, the animal may become a dark cutter because no "fuel" is left to produce lactic acid. As long as there is gas left in the tank, the animal is less likely to become a dark cutter. The above analogy is oversimplified, but in the practical world, it helps predict which groups of cattle would be more likely to have high percentages of dark cutters. Many different stressors can act in an additive manner to cause the animal to run out of "fuel."

Scanga et al. (1998) found that feedlot-fed steers that were pushed to produce more muscle mass with trembolane acetate were more likely to have dark cutters. The author observed that when trembolane acetate implants first became available in the United States, producers gave high doses, and many steers and heifers grew wider heads and necks and physically resembled bulls. When the abattoirs started reducing payments to producers who had high percentages of dark cutters, steers and heifers that had the physical appearance of bulls stopped arriving and dark cutters were greatly reduced. There may be a complex interaction between physiological mechanisms, because a survey of over 200,000 beef carcasses at a Western Australia plant showed that cattle with a larger eye muscle had fewer carcasses with dark cutters (McGilchrist et al. 2012). McGilchrist et al. (2011) performed another study that showed that Angus steers selected for higher muscling were less responsive to an adrenalin challenge and may be less likely to have dark cutters. Tenderness was not measured in these studies. Trembolane acetate is a synthetic male hormone and increasing muscle mass either using a male hormone implant or keeping cattle as intact bulls may increase the number of dark cutters. It is likely that there are differences in physiological mechanisms.

Animals that have greater glycogen reserves may have sufficient energy left in their muscles so they do not become dark cutters even when handling problems occur at the plant. Fluctuating temperatures may increase both DFD and PSE. South African researchers reported that PSE in pigs increased in the autumn (Gajara et al. 2013). In the southern hemisphere, the seasons are reversed and there would be fluctuating, gradually warming temperatures. Another common event that may increase dark cutters is a sudden climatic change 48 to 24 h before slaughter. Scanga et al. (1998) found that temperature fluctuations 2 days before slaughter increased dark cutting. An example would be very cold weather changing to warm or vice versa. Many different factors can contribute to the depletion of the animal's muscle energy and cause dark cutting. Nutrition can also have an effect on dark cutters. Cattle that have been fed high-energy-grain diets may be less likely to become dark cutters compared to cattle fed low-energy diets. Dark cutting beef may be reduced by feeding a high-energy diet for 2 weeks before slaughter (Immonen et al. 2000). To avoid dark cutting and a high pH, an animal must get to either the stunning or religious slaughter point before it runs out of glycogen.

6.7 GOOD LAIRAGE MANAGEMENT PRACTICES

Research clearly shows that resting pigs for 1 or 2 h in the lairage improves pork quality and reduces PSE (Milligan et al. 1998; Warriss 2003). Pigs need a minimum of 1 h of undisturbed rest time in the lairage pens. Sufficient space should be provided so all the pigs can all lie down at the same time without being on top of each other. During hot weather, showering pigs is recommended. Showering should not be used if pigs are shivering (Knowles et al. 1998). A pig lairage must have a roof or shade to protect pigs from the sun. Further recommendations and research can be found at Warriss (2003). Practical experience in large U.S. plants indicates that overnight rest in the lairage provides no additional pork quality advantages. In grain-fed beef cattle, abattoir managers in the United States schedule cattle deliveries to avoid holding cattle overnight. They do this to reduce dark cutters. Warren et al. (2010) surveyed 1363 truckloads of Canadian cattle. They reported that overnight lairage increased dark cutters. In both the United States and Canada, most of these cattle were grain fed and implanted with growth-promoting hormones. There are situations where holding the cattle and lambs for longer periods in the lairage is beneficial. A Turkish study found that lambs that had very high cortisol levels of over 117 ng/mL had darker meat when they were slaughtered within 30 min of arrival at the abattoir (Ekiz et al. 2012). Approximately, 18 h of lairage improved meat quality.

Two other studies on lambs and tame bulls that were fed concentrate feed while in lairage showed benefits of longer lairage times (Liste et al. 2011; Teke et al. 2014). The ideal lairage times for cattle and sheep may depend on many factors such as distance transported, temperament, time off feed, type of nutrition, feeding in lairage, and handling methods.

If possible, it is best for animals to be held with the same pen mates they had on the farm. If pigs have to be mixed in the lairage, the author recommends mixing large groups. When large groups of over 80–100 pigs are mixed, fighting will be reduced compared to groups of five or six pigs. The author has observed that small groups of five or six pigs in a small pen may engage in intense fighting. In a small pen, a pig that is being attacked cannot escape. Research on the farm has shown that large groups fight less when mixed (Turner et al. 2001). As stated earlier, bulls must never be mixed. They should be transported and held in the lairage with the same pen mates they had on the farm.

6.8 LOW STRESS HANDLING IS ESSENTIAL

Rough handling and animals becoming agitated during the last few minutes before slaughter may cause meat quality problems. Shocking cattle multiple times with an electric prod shortly before slaughter was associated with tougher meat and higher cortisol levels (Warner et al. 2007; Hemsworth et al. 2011). Studies have shown that cattle that became agitated while they are restrained for veterinary procedures on the farm produced more dark cutters or tougher meat (Voisinet et al. 1997; King et al. 2006; Hall et al. 2011). Cattle that became more agitated in the abattoir lairage pens when two people walked in their pens had both higher plasma lactate and higher Warner Bratzler shear force (Gruber et al. 2010). The grain-fed cattle in this study were likely to have had adequate glycogen stores when they arrived at the plant because they came from a nearby experiment station. Transport to the plant varied from 64 min to 90 min at moderate temperatures. In another study, Silveira et al. (2012) found that calm steers had higher pH. This may be due to calm steers being less likely to become stressed shortly before slaughter, which would increase lactate. Animals with an excitable temperament when they are handled on the farm are also more likely to become agitated during handling at the abattoir and have higher cortisol levels (Bourquet et al. 2010). A study with lambs indicated that high cortisol levels at slaughter was associated with tougher meat (Eriksen et al. 2013).

Studies with pigs show that handling practices during the last 5 min before slaughter have a significant effect on pork quality (Hambrecht et al. 2005; Edwards et al. 2010b). Pigs that are prodded with electric prods in the handling race or jammed together with other pigs have higher lactate and more PSE because the lactate measurement was taken within 5 min of jamming or prodding (Edwards et al. 2010a,b). Blood lactate can be measured with a simple handheld meter at bleeding and used as a measure of handling quality shortly before slaughter (Edwards et al. 2010a,b). High sound levels from pigs squealing, when they are moved to the stunner is correlated to both physiological measures of stress and poor pork quality (Warriss et al. 1994). At one abattoir, managers installed a traffic light connected to a sound decibel meter to inform the employees when squealing in the stunning area was too loud, in which case the traffic light turned red. The plant's quality assurance department measured both pork quality and sound volume and the loudness levels associated with a decline in quality were determined. The traffic light would remain green if the sound levels remained below this level.

6.9 DO ANIMALS KNOW THEY ARE GOING TO SLAUGHTER?

This is one of the most common questions asked by the public. The author has observed that both cattle and pigs behave the same way during handling on the farm and during movement to slaughter (Grandin 2007). If they knew they were going to get slaughtered, they should become more agitated at the abattoir. Cortisol levels in cattle after restraint in a cattle headgate (head bale) at the farm and after slaughter were similar (Grandin 1997; Mitchell et al. 1988; Gruber et al. 2010; Hemsworth et al. 2011). This shows that the stress levels the animals experienced at the abattoir were similar to handling on the farm. When animals enter an abattoir, they may react to being in a new environment. French researchers found that animals that become agitated on the farm during handling are the same animals that become agitated at the abattoir (Bourquet et al. 2010). They concluded that the animals react to the novel plant environment and the animal is fearful of a strange environment, and not fearful of death. Lewis et al. (2008) also found that novel alleys and ramps raised heart rate.

6.10 RESTRAINING ANIMALS FOR STUNNING OR RELIGIOUS SLAUGHTER

The use of stressful methods of restraint should be avoided. The OIE World Animal Health Organization (2008) states that the following methods of restraint should not be used for cattle, pigs, sheep, or other similar animals:

- Suspension by the leg or legs
- Electrical immobilization—It is highly aversive and must not be confused with electrical stunning that induces instantaneous insensibility. Research shows it is detrimental to animal welfare (Grandin et al. 1986; Pascoe 1986)
- Dragging conscious animals or cutting leg tendons
- Puntilla—Severing the spine to immobilize an animal

Restraint methods have definite effects on physiological measures of stress. Inversion of cattle onto their backs for over 90 s was associated with significantly higher cortisol levels compared to holding the animal in an upright position (Dunn 1990). Restraint by hanging by one leg was more stressful than restraint in a comfortable, upright position (Westervelt et al. 1976). Excessive pressure from a restraint device will cause cattle to vocalize (Grandin 2001). It has been found that high vocalization levels were associated with higher cortisol levels (Dunn 1990; Bourquet et al. 2012).

The use of a poorly designed head restraint device where the cattle had to be poked repeatedly with an electric prod to induce them to enter it resulted in high cortisol levels (Ewbank et al. 1992). Captive bolt stunning without a head holder resulted in lower cortisol levels (Ewbank et al. 1992). This is an example of a trade-off between stunning accuracy and handling stress. If a head holder is well designed and operated correctly, the cortisol levels are likely to be the same for both systems. In a well-designed system, the cattle will enter the head holder easily and no electric prods will be required. Head holders are available for both individual stunning boxes and conveyor restrainers (Grandin 2003, 2007). A head holder on a conveyor restrainer can easily handle over 200 cattle per hour. Lighting is critical to prevent balking and refusal to enter a head holder (Grandin 2007). Changing the position of a light and preventing cattle from seeing people through a head holder by installing a shield will often improve ease of entry. To reduce stress, either captive bolt stunning or religious slaughter should be performed immediately after the head is restrained.

6.11 EVALUATE THE ENTIRE HANDLING AND STUNNING SYSTEM

Another example of a trade-off is the CO_2 stunning of pigs. The CO_2 group handling systems that eliminate the single-file race have the advantage of reducing handling stress because electric prods and jamming in a single-file race is eliminated. Electric stunning provides the advantage of instantaneous insensibility, but handling of the pigs is more difficult because they have to line up in a single file. Electric prods can be totally eliminated in the group handling system that is used for CO_2 stunning. In the best single-file race systems, an electric prod may have to be used on 5% or less of the pigs (Grandin 2012). Cattle on pasture will naturally walk in a single file and single-file races work well for them. Pigs are less willing to walk in a single file and group handling is superior. Some discomfort during anesthetic induction may be a reasonable trade-off for overall best welfare. However, it is the author's opinion that gas mixtures that cause an animal to attempt to escape from the container are not acceptable (Grandin 2010b). There is a tendency to evaluate the stunning method in isolation from the handling method. The entire process should be viewed as a system.

6.12 VOCALIZATION SCORING OF CATTLE AND PIG DISTRESS

If cattle bellow or pigs squeal during handling and restraint, this is a clear indication of distress. In an abattoir, vocalization (bellow, moo, squeal) scoring should be conducted while the animal is held in the restraint device for either stunning or religious slaughter. High percentages of pigs or cattle that vocalize during restraint are an indicator of a definite problem that needs to be corrected. Squealing in pigs and bellowing and mooing in cattle during distressful events is associated with physiological measures of stress (Dunn 1990; Warriss et al. 1994; White et al. 1995; Weary et al. 1998; Hemsworth et al. 2011). Grandin (1998a, 2001) found that vocalization in cattle was associated with an obvious aversive event 99% of the time. The aversive events that caused cattle to moo or bellow were

- Prodded with an electric prod
- Excessive pressure from a restraint device
- Gates slammed on animals

- Missed captive bolt stuns
- Sharp edges that gouged the animal

Abattoirs with a well-designed and correctly operated stun box or restrainer will average 2% or less of the cattle vocalizing (Grandin 1998b, 2001, 2005). Vocalization scores of 5% or less can be easily achieved in correctly designed restraint boxes used for religious slaughter (Grandin 2006, 2012). When excessive pressure was applied by a restraint device, the percentage of cattle that vocalized in three different plants was 35%, 25%, and 23% (Grandin 1998b, 2001; Bourquet et al. 2012). Cattle held in a restrainer that holds them in an upright position for religious slaughter (slaughter without stunning) vocalized less than cattle held inverted on their backs (Dunn 1990; Velarde et al. 2014). Struggling was higher in the upright restraint compared to an inverted position (Velarde et al. 2014). Unfortunately, Velarde et al. (2014) did not differentiate between struggling that occurred before or after the loss of consciousness. Struggling is not a welfare concern if it occurs after the animal loses the ability to stand. When animals lose the ability to stand (loss of posture), they are unconscious (Benson et al. 2012) and AVMA (2013). Vocalization during religious slaughter cannot occur after the throat is cut because the larynx is disconnected from the lungs.

6.12.1 Vocalization Scoring

To score vocalization, each bovine or pig is scored as either silent or a vocalizer, and this is on a per animal basis. Vocalization scoring must *not* be used for sheep. Sheep do not vocalize when they are injured or scared. Since sheep are a defenseless prey species, they do not want to advertise to predators that they are in pain. For all species, vocalization scoring should never be conducted in the lairage when animals are not being handled. It is only measured in the stun box, religious slaughter box, or during movement of the animals into the restrainer or box. Vocalization scoring for goats needs to be developed.

6.13 BASIC HANDLING CONCEPTS TO REDUCE STRESS

6.13.1 Nonslip Flooring

Animals panic and become agitated when they start to slip. One of the most common causes of cattle refusing to stand quietly in a stun box is slipping on the floor and in this case one or more feet will do a series of rapid small slips. Stun box floors can be made nonslip with either steel rods or rubber mats.

6.13.2 Move Small Groups of Cattle and Pigs

Quiet, low-stress handling will require more walking. At the farm, moving small groups of 4–5 pigs for truck loading, reduced stress (Lewis and McGlone 2007). At the abattoir, if the crowd pen that leads to the single-file race is filled only half full, this will facilitate handling both cattle and pigs (Grandin 2007). In each facility,

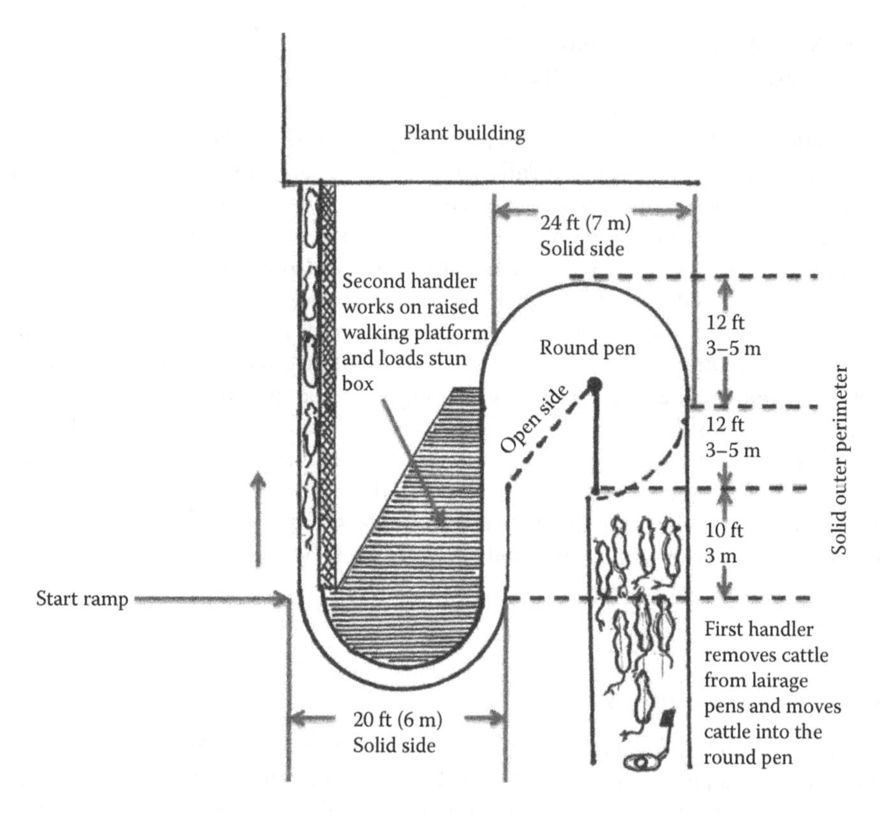

FIGURE 6.2 The handler waits until the single-file race is half empty before filling the round pen. This promotes following behavior. To facilitate cattle entry, it is essential to build a loft (3 m) straight single-file race that is connected to the round pen. (Diagram by Temple Grandin.)

management will need to determine the optimum group size. Good handling will require more walking to move small groups of animals. Handlers also need to take advantage of the natural following behavior and they should wait until the single-file race is partially empty before attempting to refill it (Figures 6.2 and 6.3).

6.13.3 Flight Zone

Handlers also need to understand the animal's flight zone (Grandin 1980; Grandin and Deesing 2008) (Figure 6.4). Tame animals have no flight zone and often they should be led instead of driven. Extensively raised animals that seldom see people will have a large flight zone. When people enter the flight zone, the animals will move away. Extensively raised cattle that are held in a single-file race may rear up if handlers get too close because they are unable to move away. The animal will often stop rearing and settle back down if the person backs up out of the flight zone. In slaughter plants, there are often lots of people, vehicles, and other activity around the handling facility. Installation of solid side panels will prevent wild cattle from seeing people and help keep them calmer. Muller et al. (2008) reported that using a solid

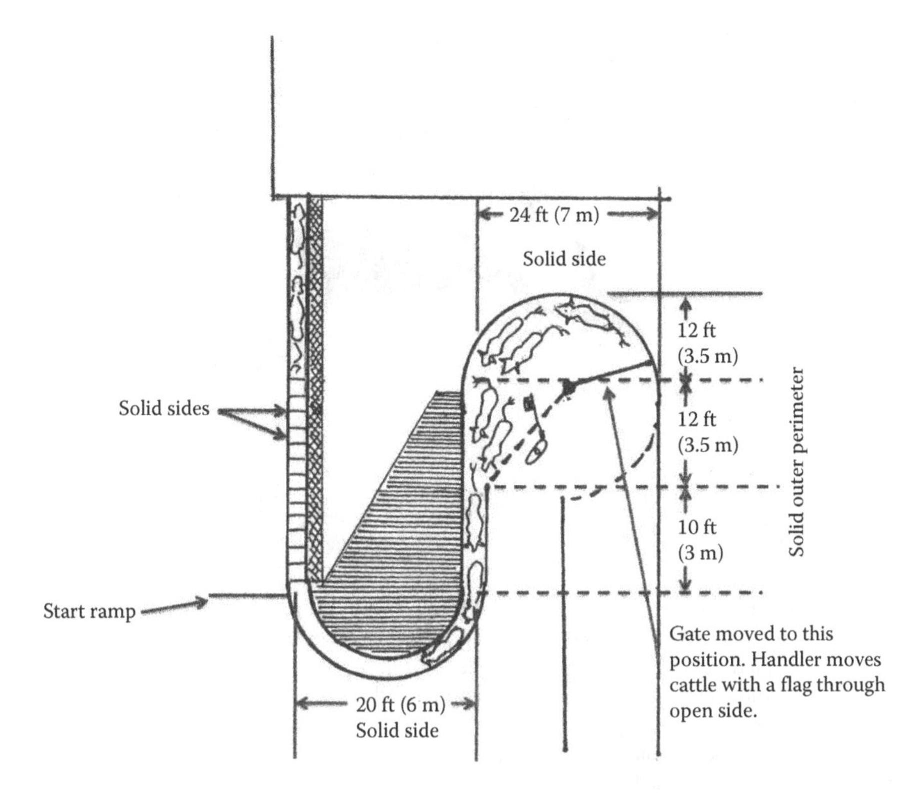

FIGURE 6.3 The cattle move easily through the full half circle round pen because it takes advantage of their natural behavior to go back to where they came from. The cattle will circle around the handler who works at the floor level through the open bar panel. Outer perimeter fences are solid to block distractions. (Diagram by Temple Grandin.)

panel to block a bovine's view of a person standing close to them reduced behaviorial reactivity of extensively raised beef steers.

6.13.4 Point of Balance

People moving cattle, pigs, or other livestock should avoid the mistake of standing at the animal's head and poking it on the rear. To move an animal forward in a single-file race, handlers should position themselves behind the point of balance at the shoulder. Animals will often walk forward when a person walks quickly by them in the opposite direction of desired movement.

6.13.5 Effects of Lighting

Lighting has a major effect on animal movement (Grandin 1996, 2001; Tanida et al. 1996; Klingimair et al. 2011). Changes in lighting can facilitate animal movement and reduce electric prod use (Grandin 2001). Pigs and cattle tend to move from a darker area to a more brightly illuminated place (Van Putten and Elshof 1978;

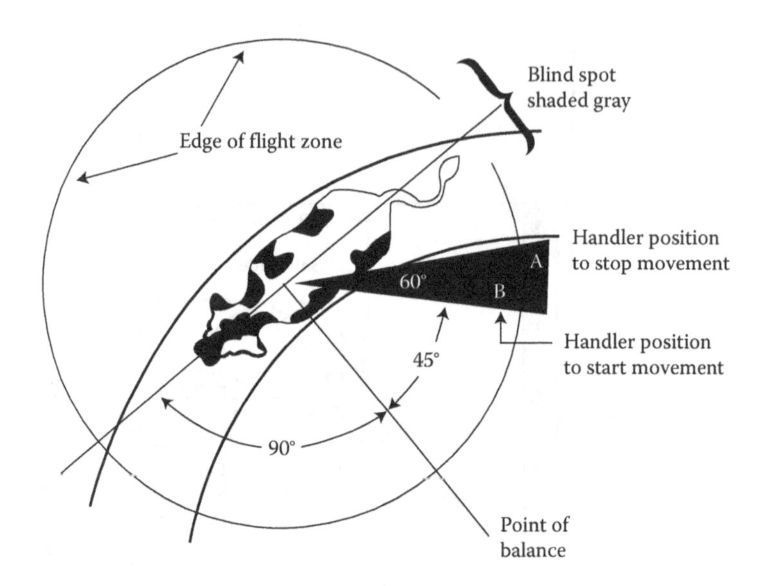

FIGURE 6.4 Flight zone diagram that shows the best handler positions and the point of balance at the shoulder. When animals are being moved through a single-file race, the handler should be located behind the point of balance. The flight zone is represented by the circle. When the handler enters the flight zone, the animal will move away. If the animal is in a race and the handler remains standing inside the flight zone, a wild animal that is not accustomed to close contact with people may struggle or rear. If this happens, the handler should back away. (Diagram by Temple Grandin, 1993.)

Grandin 1982). Pigs, cattle, and sheep will often enter a stun box more easily and balk less if a light is installed at the stun box entrance to provide indirect lighting of its interior (Grandin 2001). Reflections on wet surfaces or shiny metal can also cause animals to stop and refuse to move. Experimentation with adding lights and moving lights can be used to eliminate reflections.

6.13.6 Other Distractions That Cause Animals to Refuse to Move

Calm animals will stop and refuse to move if they see distractions that attract their attention (Table 6.1). When they see something moving, they will stop and point their eyes and ears directly at it. When they see a change in flooring, they may stop and put their heads down to look at it. Changes in contrast or differences in flooring materials will often cause animals to balk and stop. Uniformity of flooring and building materials will help facilitate movement. In several plants, the author improved animal movement by changing employee hat or coat color to reduce contrast with the walls. Animals may also stop and refuse to move if they see moving objects or people ahead of them. One advantage of a curved single-file race is that it helps prevent approaching animals from seeing people up ahead (Figures 6.4 and 6.5). People should avoid looking at approaching animals through the front of a stun box or a restrainer used for religious slaughter.

TABLE 6.1
Distractions That Cause Animals to Balk and Refuse to Move through a Facility

Stationary Distractions	Moving Distractions
• Floor changes from concrete to metal	• People visible to approaching animals
• A drain on the floor	• Moving equipment such as a conveyor
• Coat hung on a fence	• Reflections on shiny metal that move
• Hose lying on the floor	• Moving chain that is dangling in a race
• Change in floor color	• Plastic sheeting that moves
• Metal strip across an alley	• Slowly turning fan blades
• Height of stun box door too low	• Air blowing toward approaching animals

Loud noise, especially intermittent sounds, can be stressful to livestock (Talling et al. 1998) and people should not yell at animals as this is stressful (Waynert et al. 1999; Pajor et al. 2003; Hemsworth et al. 2011). Cattle have greater heart rate increases in response to the sounds of people yelling compared to sounds of gates slamming (Waynert et al. 1999). Animals of all species will often balk and refuse to move if air is blowing back toward them through a stun box door. Changing plant ventilation will often improve animal movement.

6.14 ACCLIMATING ANIMALS ON THE FARM TO BEING MOVED

Many on-farm conditions can have an effect on the ease of handling at the abattoir. The effect of flight zone has already been discussed. Pigs that have had previous

FIGURE 6.5 Curved single-file race at a large cattle plant. A curved race prevents the cattle from seeing people up ahead when they enter the race from the crowd pen. (Photo by Temple Grandin.)

experiences with being moved in alleys on the farm will be easier to move when they are handled in the future (Abbot et al. 1987; Geverink et al. 1998; Krebs and McGlone 2009). Genetic differences in temperament are less likely to affect beef quality in frequently handled *Bos taurus* cattle that are raised indoors (Turner et al. 2011). This may be due to the fact that acclimated animals are less likely to react to the novel environment.

Animals are very specific in their learning. Leiner and Fendt (2011) found that habituating a horse to tolerate a suddenly opened umbrella did not transfer to other very different novel objects such as a plastic sheet. Cattle that have a small flight zone when handled by a familiar horse and rider may have a much larger flight zone and become wild and agitated when they are moved for the first time by a person on foot (Grandin 2007; Grandin and Deesing 2008). Some of these cattle may be dangerous to handle at the abattoir. To prevent this problem, cattle from extensive ranches and properties where all handling is done on horseback need to be trained to move in and out of pens by people on foot before they arrive at the slaughter plant. Pigs will be easier to handle at the plant if people have walked through the fattening pens. Pigs differentiate between people in the alley and people in their pens. Walking in the pens on a regular basis will train the pigs to quietly get up and flow around a person moving through them and will make them easier to move at the abattoir. Brown et al. (2008) reported that walking in the pens helped to improve pork quality.

6.15 PREVENTION OF BLOOD SPLASH IN ANIMALS SLAUGHTERED WITHOUT STUNNING

Careful handling can help prevent blood splash (petechial hemorrhages) in animals that are slaughtered without stunning. The author has observed that blood spotting in the meat is especially a problem in cattle. Sheep are less likely to have blood spotting problems compared to cattle. Baldwin and Bell (1963a,b) found that sheep and cattle have differences in the anatomy of the neck. Cutting the carotids almost eliminates the entire blood supply to the brain in sheep, where this is not true for cattle. Sheep will also lose sensibility faster than cattle (Blackmore 1984). This difference may explain why cattle are more prone to blood splash than sheep. Practical experience by the author in many kosher slaughter plants has shown that immediate

TABLE 6.2
Procedures for Upright Restraint Box Operation for Religious Slaughter to Improve Cattle Welfare and Reduce Blood Splash

- Avoid the use of electric prods
- Having calm animals entering the box
- Avoid excessive pressure
- 5% or less of the cattle vocalize (moo, bellow) in the box or while entering it (Grandin, 2006, 2012)
- Cut immediately after the head is restrained
- Reduce pressure applied to the body after the cut
- Minimize the length of time that pressure is applied to the animal to restrain it

post-cut or pre-cut stunning with a captive bolt will reduce blood splash. Some plants that perform slaughter without stunning have blood splash levels in the beef loin of 20%–30%. This can be reduced by careful handling and restraint. Grandin and Regenstein (1994) and Grandin (1992, 1994) discuss in detail the correct operation of an upright restraint box. This improves animal welfare and may lower the percentage of cattle with blood splash in the loin to 2%–5% (Table 6.2). When slaughter without stunning is done with good technique, 90% or more of the cattle will collapse within 30 s (Gregory et al. 2010; Erika Voogd, personal communication).

6.16 MEASURING HANDLING AND STUNNING PRACTICES

To maintain good performance of both handling and stunning, objective scoring should be used (Grandin 1998b, 2010b, 2013). People manage the things that they measure. Measuring prevents people from reverting to old bad practices and not realizing it. The following variables should be measured:

- Percentage of animals effectively stunned with one application of the stunner. The score should be 95% or better for captive bolt, and 99% or better for correct electrode placement for electric stunning.
- Percentage of cattle or pigs rendered insensible before hoisting.
- Percentage of animals that vocalize in the stun box, restrainer, or religious slaughter box should be 5% or less.
- Percentage of animals falling should be 1% or less anywhere in the entire facility.
- Percentage moved with an electric prod should be 5% or less to get an excellent score.

Data collected in slaughter plants indicate that there are large differences between plants (Grandin 1998b; Hultgren et al. 2014).

6.16.1 Video Auditing

The advantage of assessing handling and stunning by video cameras is that it prevents the problem of people "acting good" when they know they are being watched. Two large meat companies Cargill and JBS Swift are currently using video auditing that is monitored remotely over the Internet. A third-party auditing firm provides the service. Experience has shown that this may be more effective than video cameras that are monitored exclusively by people in the plant. The most effective video auditing programs are a combination of audits conducted by plant management and video audits conducted by an outside auditing company.

6.17 CONCLUSIONS

Both animal welfare and meat quality will be improved when low stress handling methods are used. Short-term stresses such as electric prod use or jamming in the race a few minutes before slaughter are associated with tougher meat and high lactate. Multiple factors that contribute to longer-term stresses such as bulls fighting, adverse weather, or long transport times may increase DFD meat.

There are three types of problems that have an effect on both meat quality and animal welfare. They are (1) death losses, bruises, and nonambulatory animals; (2) meat quality issues such as dark cutters (DFD), PSE, and meat toughness; and (3) stressors that are detrimental to welfare such as electric prods, animals fighting, or agitated behavior. Factors associated with increased cattle bruising are overloaded trucks, horned animals, moving through auction markets and rough handling. Petechial hemorrhages in cattle kosher slaughtered without stunning in an upright restraint box can be reduced by having calm animals, reducing pressure applied by the restraint device, lowering the percentage of vocalizing animals and minimizing the time the animal is fully restrained.

REFERENCES

Abbott, T.A., Hunter, E.J., Guise, J.H., and Penny, R.H.C. 1987. The effect of experience of handling on pigs willingness to move. *App. Anim. Behav. Sci.* 54:371–375.

Ahola, J.K., Foster, H.A., Vanoverbeke, D.L. et al. 2011. Survey of quality defects in market beef and dairy cows sold through livestock auctions and markets in the Western United States, 1. Incidences rates. *J. Anim. Sci.* 89:1474–1483.

Alvarez, D., Garrido, M.D., and Banon, S. 2009. Influence of pre-slaughter process on pork quality: An overview. *Food Process Int.* 25:233–250.

American Veterinary Medical Association (AVMA). 2013. *Guidelines for the Euthanasia of Animals*, 2013 edition. Schaumburg, IL.

Arp, T.S., Howard, S.T., Woerner, D.R., Scanga, J.A., McKenna, D.R., Kolath, W.H., Chapman, P.L., Tatum, J.D., and Belk, K.E. 2013. Effect of ractopomine hydrochloride and zilpaterol hydrochloride supplementation on longissimus muscle shear force and sensory attributes of beef steers. *J. Anim. Sci.* 91:5989–5997.

Baldwin, B.A., and Bell, F.R. 1963a. The effect of temporary reduction in cephalic blood flow on EEG of sheep and calf electroencephalography. *Clin. Neurophysiol.* 15:465–473.

Baldwin, B.A., and Bell, F.R. 1963b. The anatomy of cerebral circulation of sheep and ox: The dynamic distribution of the blood supplied by the carotid and vertebral arteries to cranial regions. *J. Anat.* 97:203–215.

Baxa, T.J., Hutcheson, J.P., Miller, M.F., Brooks, J.C., Nichols, W.T., Streeter, M.N., Yates, D.A., and Johnson, B.J. 2010. Additive effects of steroidal implant and zilpaterol hydrochloride on feedlot performance, carcass characteristics, and skeletal muscle messenger ribonucleic acid abundance in finishing steers. *J. Anim. Sci.* 88:330–337.

Benjamin, M.E., Gonyou, H.W., Ivers, D.L., Richard, L.F., Jones, D.J., Wagner, J.R., Seneriz, R., and Anderson, D.B. 2001. Effect of handling method on the incidence of stress response in market swine in a model system. *J. Anim. Sci.* 79(Suppl. 1):279.

Benson, E.R., Alphen, R.L., Rankin, M.K., Capedo, M.P., Kenny, C.A., and Johnson, A.L. 2012. Evaluation of EEG-based determination of unconsciousness vs. loss of posture in broilers. *Res. Vet. Sci.* 52:960–964.

Blackmore, D.L. 1984. Difference in the behavior between sheep and cattle during slaughter. *Res. Vet. Sci.* 37:223–236.

Blixt, Y., and Borch, E. 2002. Comparison of shelf life of vacuum packed pork and beef. *Meat Sci.* 60:371–381.

Bourquet, C., Deiss, V., Gobert, M., Durand, D., Boissey, A., and Terlouw, C. 2010. Characterizing emotional reactivity of cows to understand and predict their stress reactions to slaughter procedures. *App. Anim. Behav. Sci.* 125:9–21.

Bourquet, C., Deiss, V., Tannugi, C.C., and Terlouw, E.M. 2012. Behavioral and physiological reactions of cattle in a commercial abattoir: Relationships with organizational aspects of the abattoir and animal characteristics. *Meat Sci.* 68:158–168.

Brown, J.A., Mandell, I.B., deLange, F.M., Purslow, P., Robinson, J.A., Squires, J.E.J., and Widowski, T. 2008. Relationship between a pig's temperament and handling experiences, and measures of stress at slaughter and meat quality. Allen D. Leman Swine Conference, Research Reports, www.londonswineconference.ca.

Cockram, M.S., and Lee, R.A. 1991. Some preslaughter factors affecting the occurrence of bruising in sheep. *Br. Vet. J.* 147:120–125.

Correa, J.A., Torrey, S., Devillers, N., Laforest, J.P., Gongous, H.W., and Facitano, L. 2010. Effects of different moving devices at loading on stress response and meat quality in pigs. *J. Anim. Sci.* 88:4086–4093.

Dunn, C.S. 1990. Stress reactions of cattle undergoing ritual slaughter using two methods of restraint. *Vet. Rec.* 126:522–535.

Edwards, L., Engle, T.E., Correa, J.A., Paradis, M.A., Grandin, T., and Anderson, D.B. 2010a. The relationship between exsanguination blood lactate concentration and carcass quality in slaughter pigs. *Meat Sci.* 85:435–445.

Edwards, L.N., Grandin, T., Engle, T.E., Porter, S.P., Ritter, M.J., Sosnick, A.A., and Anderson, D.B. 2010b. Use of exsanguination blood lactate to assess the quality of pre-slaughter handling. *Meat Sci.* 86:384–390.

Edwards, L.N., Grandin, T., Engle, T.E., Ritter, M.J., Sosnicki, A.A., Carlson, B.A., and Anderson, D.B. 2010c. The effects of pre-slaughter pig management from the farm with processing plant on pork quality. *Meat Sci.* 86:938–944.

Ekiz, B., Ekiz, E.E., Kocak, D., Yaleintan, H., and Yimaz, A. 2012. Effect of preslaughter management regarding transportation and time in lairage on certain stress parameters, carcass, and meat quality characteristics in Kivireik lambs. *Meat Sci.* 90:967–975.

Eldridge, G.A., and Winfield, C.G. 1988. The behavior and bruising of cattle during transport at difference space allowances. *Aust. J. Exp. Agric.* 28:695–698.

Eriksen, M.S., Rodbottan, R., Grandah, A.M., Friestad, M., Andersen, I.L., and Meidell, C.M. 2013. Mobile abattoir versus conventional slaughterhouse—Impact on stress parameters and meat quality characteristics in Norwegian lambs. *App. Anim. Behav. Sci.* 149:21–29.

Ewbank, R., Parker, M.J., and Mason, C.W. 1992. Reactions of cattle to head restraint at stunning: A practical dilemma. *Anim. Welfare* 1:55–63.

FASS 2010. *Guide for Care and Use of Agricultural Animals in Research and Teaching*, 3rd edition. Federation of Animal Science Societies, Campaign, IL.

Ferguson, D.M., and Warner, R.D. 2008. Have we underestimated the impact of preslaughter stress on meat quality in ruminants? *Meat Sci.* 803:12–19.

Gajara, C.S., Nikukwana, T.T., Marume, U., and Muchenje, J. 2013. Effects of transportation time, distance stocking density, temperature, and lairage time on the incidence of pale soft exudative (PSE) and the physic-chemical characteristics of pork, *Meat Sci.* 95:520–525.

Gallo, C., Lizondo, G., and Knowles, T.G. 2003. Effects of journey and lairage time on steers transported to slaughter in Chile. *Vet. Rec.* 152:361–364.

Geverink, N.A., Kappers, A., Van dc Burgwal, E., Lambooij, E., Blokhuis, J.H., and Wiegant, V.M. 1998. Effects of regular moving and handling on the behavioral and physiological responses of pigs to pre-slaughter treatment and consequences for meat quality. *J. Anim. Sci.* 76:2080–2085.

Grandin, T. 1980. Observations of cattle behavior applied to the design of cattle handling facilities. *App. Anim. Ethics.* 6:10–31.

Grandin, T. 1981. Bruises on southwestern feedlot cattle. *J. Anim. Sci.* 53 (Suppl. 1): 213.

Grandin, T. 1982. Pig behavior studies applied to slaughter plant design. *App. Anim. Ethics.* 9:141–151.

Grandin, T. 1992. Observations of cattle restraint devices for stunning and slaughtering. *Anim. Welfare* 1:85–91.
Grandin, T. 1993. Behavioral principles of cattle handling under extensive conditions. In Grandin, T. (ed.), *Livestock Handling and Transport*. CABI Publishing, Wallingford, Oxfordshire, UK, pp. 43–57.
Grandin, T. 1994. Euthanasia and slaughter of livestock. *J. Am. Vet. Med. Assoc.* 204:1354–1360.
Grandin, T. 1996. Factors that impede animal movement at slaughter plants. *J. Am. Vet. Med. Assoc.* 209:757–759.
Grandin, T. 1997. Assessment of stress during handling and transport. *J. Anim. Sci.* 75:249–257.
Grandin, T. 1998a. The feasibility of using vocalization scoring as an indicator of poor welfare during slaughter. *App. Anim. Behav. Sci.* 56:121–128.
Grandin, T. 1998b. Objective scoring of animal handling and stunning practices in slaughter plants. *J. Am. Vet. Med. Assoc.* 212:36–93.
Grandin, T. 2001. Cattle vocalizations are associated with handling and equipment problems in slaughter plants. *App. Anim. Behav. Sci.* 71:191–201.
Grandin, T. 2003. Transferring results from behavioral research to industry to improve animal welfare on the farm, ranch and slaughter plants. *App. Anim. Behav. Sci.* 81:216–228.
Grandin, T. 2005. Maintenance of good animal welfare standards in beef slaughter plants by use of auditing programs. *J. Am. Vet. Med. Assoc.* 226:370–373.
Grandin, T. 2006. Vocalization scoring of restraint for kosher slaughter of cattle for an Animal Welfare Audit, www.grandin.comritual/vocal.scoring.restraint.cattle.welfare.audit.html, accessed December 27, 2013.
Grandin, T. 2007. Handling and welfare of livestock in slaughter plants. In Grandin, T. (ed.), *Livestock Handling and Transport*. CABI Publishing, Wallingford, Oxfordshire, UK, pp. 329–353.
Grandin, T. 2010a. Improving livestock, poultry, and fish welfare in slaughter plants with auditing programs. In Grandin, T. (ed.), *Improving Animal Welfare: A Practical Approach*. CABI Publishing, Wallingford, Oxfordshire, UK, pp. 160–185.
Grandin, T. 2010b. Auditing animal welfare at slaughter plants. *Meat Sci.* 86:56–65.
Grandin, T. 2012. Developing measures to audit welfare of cattle and pigs at slaughter, *Anim. Welfare* 21:351–356.
Grandin, T. 2013. Making slaughter houses more humane. *Annu. Rev. Anim. Biosci. Sci.* 1:491–512.
Grandin, T., Curtis, S.E., and Widowski, T.M. 1986. Electro-immobilization versus mechanical restraint in an avoid-avoid choice test. *J. Anim. Sci.* 62:1469–1480.
Grandin, T., and Deesing, M. 2008. *Humane Livestock Handling*, Storey Publishing, North Adams, MA.
Grandin, T., and Regenstein, J.M. 1994. Religious slaughter and animal welfare: A discussion for meat scientists. *Meat Focus Int.* 3:115–123.
Gregory, N.G., Fielding, H.R., von Wenzlawowicz, M., and von Holleben, K.V. 2010. Time to collapse following slaughter without stunning of cattle. *Meat Sci.* 85:66–69.
Gruber, S.L., Tatum, T.D., Engle, T.E., Chapman, P.L., Belk, K.E., and Smith, G.C. 2010. Relationship between behavioral and physiological symptoms of pre-slaughter stress on beef longissimus muscle tenderness. *J. Anim. Sci.* 88:1148–1159.
Hall, N.L., Buchanan, D.S., Anderson, V.L., Ilse, B.R., Carlin, K.R., and Berg, E.P. 2011. Working chute behavior of feedlot cattle can be an indication of cattle temperament and beef carcass composition and quality. *Meat Sci.* 89:52–57.
Hambrecht, E., Eissen, J.I., Newman, D.J., Smith, C.H., Verstegen, M.W., and den Hartog, L.A. 2005. Preslaughter handling affects pork quality and glycolytic potential in two muscles differing in fiber type composition. *J. Anim. Sci.* 83:900–907.
Hemsworth, P.H., Rice, M., Karlen, M.G., Calleja, L., Barnett, J.L., Nash, J., and Coleman, G.J. 2011. Human-animal interactions at the abattoir, Relationships between handling and animal stress in sheep and cattle. *App. Anim. Behav. Sci.* 135:24–33.

Hoffman, D.E., Spire, M.F., Schwenke, J.R., and Unrah, J.A. 1998. Effect of sources of cattle and distance transportations to a commercial slaughter facility on carcass bruising in mature beef cow. *J. Anim. Sci.* 212:668–672.

Huertas, S.M., Gil, A.D., Piaggss, J.M., and Von Eerdenbury, F.J.C.M. 2010. Transportation of beef cattle to slaughterhouse and how this relates to animal welfare carcass bruising in an extensive production system. *Anim. Welfare* 19:281–285.

Hultgren, J., Wiberg, S., Berg, C., Cvek, K., and Kolstrup, L. 2014. Cattle behaviours and stockperson interactions related to improved animal welfare at Swedish slaughter plants. *App. Anim. Behav. Sci.* 152:23–37.

Immonen, K., Ruusunen, M., Hissa, K., and Puolanne, E. 2000. Bovine muscle glycogen concentration in relation to finishing slaughter and ultimate pH. *Meat Sci.* 55:25–31.

James, B.W., Tokach, M.D., Goodhand, R.D., Nelssen, J.L., Dritz, S.S., Owen, K.O., Woodworth, J.C., and Sulane, R.L. 2013. Effects of dietary L-carnatine and ractopomine HCL on the metabolic response in finishing pigs. *J. Anim. Sci.* 91:4426–4439.

Jones, S.D.M., and Tong, A.W. 1989. Factors influencing the commercial incidence of dark cutting beef. *Can. J. Anim. Sci.* 69:649–647.

King, D.A., Sciehle-Pfeiffer, C.E., Randel, R.D., Welsh, T.H., Oliphant, R.A., Baird, E.W., Curley, K.P., Vann, R.C., Hale, D.S., and Savell, J.W. 2006. Influence of animal temperament and stress responsiveness on the carcass quality and beef tenderness of feedlot cattle. *Meat Sci.* 74:546–556.

Klingimair, K., Steven, K.B., and Gregory, N.C. 2011. Luminare glare in indoor handling facilities. *Anim. Welfare* 20:263–269.

Knowles, T.G., Brown, S.N., Edwards, J.E., and Warriss, P.D. 1998. Ambient temperature below which pigs should not be continuously showered in lairages. *Vet. Rec.* 143:575–578.

Krebs, N., and McGlone, J.J. 2009. Effects of exposing pigs to moving and odors in a simulated slaughter chute. *App. Anim. Behav. Sci.* 116:179–185.

Leiner, L., and Fendt, M. 2011. Behavioral fear and heartrate response of horses after exposure to novel objects effects of habituation. *App. Anim. Behav. Sci.* 131:104–109.

Lewis, C.R.G., Hulbert, L.E., and McGlone, J.J. 2008. Novelty causes elevated heart rate and immune changes in pigs exposed to handling alleys and ramps. *Live. Sci.* 116:338–341.

Lewis, G.R.G., and McGlone, J.J. 2007. Moving finishing pigs in different group sizes: Cardiovascular responses, time and ease of handling. *Live. Sci.* 107:85–90.

Liste, G., Miranda de la Lama, G.C., Campo, M.M., Villarrael, M., Muela, E., and Maria, G.A. 2011. Effect of lairage on lamb welfare and meat quality. *Anim. Prod. Sci.* 51:952–958.

Longeragan, G.H., Thomson, D.V., and Scott, H.M. 2014. Increased mortality of cattle administered the B-adrenergic agonists ractopomine hydrochloride and zilpaterol hydrochloride. *PLOS ONE* 9(3):e91177. doi:10.1371/journal/pone0091177 (accessed March 13, 2014).

Marchant-Forde, J.M., Lay, D.C., Pajor, J.A., Rickert, B.T., and Schinckel, A.P. 2003. The effects of ractopomine on the behavior and physiology of finishing pigs. *J. Anim. Sci.* 81:416–422.

McGilchrist, P., Alston, C.L., Gardner, G.E., Thomson, K.L., and Pethick, D.W. 2012. Beef carcasses with larger eye muscle area, lower ossification scores and improved nutrition have a lower incidence of dark cutting. *Meat Sci.* 92:474–480.

McGilchrist, P., Pethick, D.W., Bonny, S.P.E., Greenwood, P.L., and Gardner, G.E. 2011. Beef cattle selected for increased muscularity have a reduced muscle response and increased adipose tissue response to adrenaline. *Animals* 5:875–884.

Meischke, H.R.C., and Horder, J.C. 1976. A knocking box effect on bruising in cattle. *Food Tech. Aust.* 28:369–371.

Milligan, S.D., Ramsey, C.B., Miller, M.F., Kaster, C.S., and Thompson, L.D. 1998. Resting of pigs and hot-fat trimming and accelerated chilling of carcasses to improve meat quality. *J. Anim. Sci.* 76:74–86.

Mitchell, G., Hattingh, J., and Ganhao, M. 1988. Stress in cattle assessed after handling, transport, and slaughter. *Vet. Rec.* 123:201–205.

Montgomery, J.L., Krehbiel, C.R., Cranston, J.J., Yates, D.A., Hutcheson, J.P., Nichols, W.T., Streeter, M.N., Swingle, R.S., and Montgomery, T.H. 2009. Effects of dietary zilpaterol hydrochloride on feedlot performance and carcass characteristics of beef steers fed with and without monensin and tylosin. *J. Anim. Sci.* 87:1013–1023.

Muller, R., Schartzkopf-Genswein, K., Shah, M.S., and Vonkeyerlingk, M.S. 2008. Effect of neck injection and handling visibility on behavioral reactivity of beef steers. *J. Anim. Sci.* 86:1215–1222.

Murray, A.C., and Johnson, C.P. 1998. Importance of the halothane gene on muscle quality and preslaughter death in Western Canadian pigs. *Can. J. Anim. Sci.* 78:543–548.

OIE 2008. *Terrestrial Animal Health Code: Guidelines for the Slaughter of Animals for Human Consumption, Organization Mundial de la Sante Animals.* World Organization and Animal Health, Paris, http://www.oie.int/eng/en_index.htm.

Pajor, E.A., Rushen, J., and dePassille, A.M.B. 2003. Dairy cattle's choice of handling treatments in a Y-maze. *App. Anim. Behav. Sci.* 80:93–107.

Pascoe, P.J. 1986. Humaneness of electro-immobilization unit for cattle. *Am. J. Vet. Res.* 10:2252–2256.

Poletto, R., Cheng, H.W., Neisel, R.L., Garner, J.P., Richert, B.T., and Marchant-Forde, J.N. 2010. Aggressiveness and brain amine concentrations in dominant and subordinate finishing pigs fed the β-adrenoreceptor agonist ractopomine. *J. Anim. Sci.* 88:310–320.

Poletto, R., Rostagno, M.H., Richert, B.T., and Marchant-Forde, J.N. 2009. Effects of a 'step up' ractopomine feeding program, sex and social rank on growth performance, foot lesions and enterobacteriaccae shedding in finishing pigs. *J. Anim. Sci.* 87:304–311.

Price, M.A., and Tennessen, T. 1981. Preslaughter management and dark cutting in carcasses of young bulls. *Can. J. Anim. Sci.* 61:205–208.

Puolanne, E., and Aalto, H. 1981. The incidence of dark cutting beef in young bulls in Finland, In Hood, D.E. and Tarrant, P.V. (eds.), *The Problem of Dark Cutting Beef.* Martinus Nijhoff, The Hague, the Netherlands, pp. 462–475.

Ramsey, W.R., Meischke, H.R.C., and Anderson, B. 1976. The effect of tipping of horns and interruptions of the journey on bruising in cattle. *Aust. Vet. J.* 52:285–286.

Ritter, M.J., Ellis, M., Berry, N.L., and Curtis, S.E. 2009. Review: Transport losses in market weight pigs, 1: A review of definitions, incidence and economic impact. *Prof. Anim. Sci.* 25:404–414.

Ritter, M.J., Ellis, M., Brinkman, J., Decker, J.M., Keffaber, K.K., Kocher, M.E., Peterson, B.A., Schlipf, J.M., and Wolter, B.F. 2006. Effect of floor space during transport of market weight pigs on the incidence of transport losses at the packing plant and the relationships between transport conditions and losses. *J. Anim. Sci.* 84:2856–2864.

Rollin, B.E., Broom, D.M., Fraser, D., Golab, G., Arnot, C., and Shapiro, P. 2011. In Pond, W.G., Bazer, F.W., and Rollin, B.E. (eds.), *Animal Welfare in Animal Agriculture: Husbandry and Stewardship in Animal Production.* Taylor & Francis Group LLC, Boca Raton, FL, pp. 75–120.

Scanga, J.A., Belk, K.E., Tatum, J.D., Grandin, T., and Smith, G.C. 1998. Factors contributing to the incidence of dark cutting beef. *J. Anim. Sci.* 76:2040–2047.

Schwartzkopf-Genswein, K.S., Faucitano, L., Dadgar, S., Shand, P., Gonzalez, L.A., and Crowe, T.G. 2012. Road transport of cattle, swine, and poultry in North America and its impact on animal welfare, carcass, and meat quality: A review. *Meat Sci.* 92:227–243.

Shaw, F.D., Baxter, R.I., and Ramsay, W.R. 1976. The contribution of horned cattle to carcass bruising. *Vet. Rec.* 98:255–257.

Silveira, I.D.B., Fischer, V., Farinatte, L.H.E., Dari, J.R., Filho, C.A., Glasenapp De Menezes, L.F. 2012. Relationship between temperament with performance and meat quality of feedlot steers with predominantly Charolais or Nellore breed. *Revista Brasileira de Zootechnia* 41(6):1468–1476.

Talling, J.C., Waran, N.K., Wathes, C.M., and Lines, J.A. 1998. Sound avoidance by domestic pig depends on characteristics of the signal. *App. Anim. Behav. Sci.* 58:255–266.

Tanida, H., Miura, A., Tanaka, T., and Yoshimoto, T. 1996. Behavioral responses of piglets to darkness and shadows. *App. Anim. Behav. Sci.* 49:173–183.

Tarrant, P.V. 1981. The occurrence, causes and economic consequences of dark cutting beef—A survey of current information. In Hood, D.E., and Tarrant, P.V. (eds.), *The Problem of Dark Cutting in Beef*, Springer, Martinus Nijhoff Publishers, The Hague, the Netherlands, pp. 3–36.

Tarrant, P.V. 1989. Animal behavior and environment in the dark cutting condition in beef: A review. *Irish J. Food Sci. Technol.* 13:1–21.

Tarrant, P.V., Kelly, F.J., and Harrington, D. 1988. The effect of stocking density during 4-hour transport to slaughter, on behaviour, blood constituents and carcass bruising in Friesian steers. *Meat Sci.* 24:209–222.

Tarrant, P.V., Kenny, F.J., Harrington, D., and Murphy, M. 1992. Long distance transportation of steers to slaughter: Effect of stocking density on physiology, behaviour and carcass quality. *Live. Prod. Sci.* 30:223–238.

Tekes, B., Akdag, F., Ekiz, B., and Ugurlu, M. 2014. Effects of difficult lairage times after long distance transportation on carcass and meat quality characteristics of Hungarian Simmental bulls. *Meat Sci.* 96:224–229.

Tennessen, T., Price, M.A., and Berg, R.T. 1984. Comparative responses of bulls and steers to transportation. *Can. J. Anim. Sci.* 64:333–338.

Turner, S.F., Horgan, G.W., and Edwards, S.A. 2001. Effect of social group size on aggressive behavior between unacquainted pigs. *App. Anim. Behav. Sci.* 74:203–215.

Turner, S.P., Navajas, E.A., Hyslop, J.J., Ross, D.W., Richardson, R.I., Prieto, N., Bell, M., Jack, M.L., and Roehe, R. 2011. Association between response to handling and growth and meat quality in frequently handled *Bos taurus* beef cattle. *J. Anim. Sci.* 89:4239–4248.

Van Putten, G., and Elshof, W.J. 1978. Observations of the effects of transport on the well being and lean quality of slaughter pigs. *Anim. Reg. Stud.* 1:247–271.

Velarde, A., Rodriguez, P., Dalmau, A. et al. 2014. Religious slaughter: Evaluation of current practices in selected countries. *Meat Sci.* 96:278–287.

Vilioen, H.F., deKock, H.L., and Webb, E.C. 2002. Consumer acceptability of dark firm and dry (DFD) and normal pH steaks. *Meat Sci.* 61:181–185.

Vogel, K.D., Claus, J.R., Grandin, T., Vetzel, G.R., and Schaefer, D.M. 2011. Effect of water and feed withdrawal and health status on blood and serum components, body weight loss, and meat and carcass characteristics of Holstein slaughter cows. *J. Anim. Sci.* 89:538–548.

Voisinet, B.D., Grandin, T., O'Connor, S.F., Tatum, J.D., and Deesing, M.J. 1997. *Bos indicus* cross feedlot cattle with excitable temperaments have tougher meat and a higher incidence of borderline dark cutters. *Meat Sci.* 46:367–377.

Warner, R.D., Ferguson, D.M., Cottrell, J.J., and Knee, B.W. 2007. Acute stress induced by preslaughter use of electric prodders causes tougher meat. *Aust. J. Exp. Agric.* 47:782–788.

Warren, L.A., Mandell, I.B., and Bateman, K.G. 2010. Road transport conditions of slaughter cattle: Effects on the prevalence of dark firm and dry beef. *Can. J. Anim. Sci.* 90:471–482.

Waynert, D.E., Stookey, J.M., Schartzkopf-Genswein, J.M., Watts, C.S., and Waltz, C.S. 1999. Response of beef cattle to noise during handling. *App. Anim. Behav. Sci.* 62:27–42.

Warriss, P.D. 2003. Optimal lairage times and conditions for slaughter pigs: A review. *Vet. Rec.*153:170–176.

Warriss, P.D., Brown, S., and Adams, S.J.M. 1994. Relationship between subjective and objective assessment of stress at slaughter and meat quality in pigs. *Meat Sci.* 38:329–340.

Weary, D.M., Braithwaite, L.A., and Fraser, D. 1998. Vocal response to pain in piglets. *App. Anim. Behav. Sci.* 56:161–172.

Westervelt, R.G., Kinsman, D., Prince, R.P., and Giger, W. 1976. Physiological stress measurement during slaughter of calves and lambs. *J. Anim. Sci.* 42:831–834.

White, R.G., DeShazer, J.A., Tressler, C.J., Borcher, G.M., Davey, S., Waninge, A., Parkhurst, A.M., Milanuk, M.J., and Clems, E.T. 1995. Vocalizations and physiological response of pigs during castration with and without anesthetic. *J. Anim. Sci.* 73:381–386.

7 Stress Reactivity, Stress at Slaughter, and Meat Quality

Claudia E.M. Terlouw

CONTENTS

7.1 INTRODUCTION

7.1.1 ANIMAL STRESS AND WELFARE

Throughout rearing and during the preslaughter period, animals may be confronted with stress-inducing situations. Stress has often been described in terms of the state of the animal when it is incapable of adapting, behaviorally and physiologically, to environmental or physical challenges (Fraser et al. 1975; Broom 1987). While it is important to understand the impact of physical challenges on the physiological and behavioral adaptive capacity of the animal, it is also important to take into account the emotions that animals may experience (Dawkins 1980; Duncan 1996; Dantzer 2002; Désiré et al. 2002). Studies on the behavior, physiology, and anatomy of the brain have shown that animals are capable of experiencing negative and positive emotions (Paul et al. 2005; Boissy et al. 2007). Particularly, the limbic system, known for its involvement in emotions in humans, also exists in the brain of nonhuman mammals (LeDoux 2000). The limbic system is involved in many behavioral reactions to stressful situations in mammals (Damasio 1998; Panksepp 2005). Thus, while environmental and physical challenges put a strain on the animal to adapt to the situation, they are also likely to cause negative emotions in the animal. The latter aspect is central in the context of animal welfare matters. In this chapter, it is considered that a farm animal is stressed if it experiences negative emotions (Terlouw 2005; Veissier and Boissy 2007). To understand the impact of emotional stress on meat quality, the physical and physiological adaptive responses to challenges must be taken into account.

7.1.2 Stress at Slaughter

The preslaughter period may be stressful for many reasons (Terlouw et al. 2008; Grandin 2013). Before leaving for the abattoir, animals may be gathered on a loading platform or in a pen to facilitate subsequent loading. Food may be withheld to avoid travel sickness, for convenience or for financial reasons. Subsequent loading, unloading, and transport conditions depend on the facilities on the farm and at the abattoir, and on the layout of the truck, the driving style, and the distance traveled (Cockram et al. 2004). After unloading, animals may wait for several hours, often overnight, in the lairage area in the abattoir. At the final stages of the process, the animal will be driven to the stunning box or area, where it will be bled, usually after being rendered unconscious by a stunning procedure. All these different slaughter stages may be associated with stress for the animals. The stressors may be distinguished in different categories. They may have a physical origin, such as food deprivation, fatigue, or pain, although these forms of stress probably also have an emotional component (Horswill et al. 1990; Danziger 2006). Other stressors have an emotional origin, such as unfamiliarity, human presence, and disturbance of the social group (Terlouw et al. 2012). Sudden events may also be a cause of stress (Grandin 1999). Thus, during the slaughter process, animals are subjected to many potential stressors, simultaneously and successively, in a series of different environmental contexts.

7.1.3 Stress and Meat Quality

It has been long known that slaughter conditions may have a significant impact on meat quality. The underlying mechanisms are well described (Bendall 1973; Hambrecht et al. 2005): Following slaughter, biochemical reactions continue, but as blood is no longer circulating, there is a cessation of the supply of oxygen and nutrients. Consequently, glycogen locally stored in the muscle is catabolized anaerobically, and again due to the absence of blood circulation, the products of the reactions, protons and lactate, accumulate in the muscle, causing the pH to decline. This decline is initially fast, then slows and stabilizes at a value called ultimate pH, reached approximately 24 h postmortem. The extent of pH decline depends strongly on muscular glycogen reserves before slaughter (see Chapter 4). Increased activity during the hours preceding slaughter reduces muscle glycogen reserves and may result in high ultimate pH, which is the cause of dark, firm, dry (DFD) meat (see Chapter 2). For example, earlier studies found that in bulls, aggressive behavior caused an increase in the production of DFD meat (Tarrant et al. 1992). A higher ultimate pH (>6.2) is associated with increased meat tenderness and juiciness, but also with unpleasant texture, flavor, and taste (Dransfield 1981). Generally, at higher ultimate pH values, the meat is darker. Meat with relatively high ultimate pH (generally above 6.0) has some very negative characteristics, in addition to its dark color, it is generally difficult to keep due to facilitated bacterial development.

While the magnitude of pH decline depends mainly on muscle glycogen reserves, the rate of pH decline depends on muscle metabolic activity immediately before slaughter (Bendall 1973; Hambrecht et al. 2005). If metabolic activity is high before slaughter, this will continue after the death of the animal. For example, treadmill exercise immediately before slaughter resulted in a faster pH decline, causing the production of pale, soft, exudative (PSE) meat (Rosenvold and Andersen 2003). A fast early postmortem pH decline causes the denaturation of muscle proteins, due to the association of low pH and relatively high muscle temperature (Scheffler and Gerrard 2007). The effects of the latter variables on the rate and extent of protein denaturation, oxidation and proteolysis, lipid oxidation, color characteristics, water holding capacity, and sensory aspects of meat are well known (Wismer-Pedersen 1987; Huff-Lonergan et al. 2002; Huff-Lonergan and Lonergan 2005; Rosenvold and Andersen 2003; Moeller et al. 2010; also Chapter 3). Increased metabolic activity just before slaughter is particularly noticeable in pigs and fowl because these species have high proportions of glycolytic fibers in their muscles.

Although the above phenomena, increased ultimate pH or increased rate of pH decline, involve the status of muscle energy reserves and energy metabolism, they are also strongly linked to the stress status of the animal. Emotional stress often increases the activity of the sympathetic nervous system, resulting in an increased heart rate and the secretion of catecholamines (adrenaline and noradrenaline) in the blood. This increased sympathetic activity allows the animal to physically react to the stress-inducing stimulus, for example, to flee or to fight. Of specific interest in the present context is that catecholamines stimulate glycogen breakdown, particularly in exercising muscle, which is generally the case during slaughter (Lacourt and Tarrant 1985; Fernandez et al. 1994a; Febbraio et al. 1998). Exercise further increases muscle temperature (Henckel et al. 2000; Rosenvold and Andersen 2003), which in turn enhances muscle glycogenolysis rate (Febbraio et al. 1996; Starkie et al. 1999). In the postmortem muscle too, pH decline is faster at higher temperatures (Monin 1973; Astruc et al. 2002). In addition, the more the animal is stressed, the more difficult it is to handle, leading to further physical activity and emotional stress, while fear reaction of the animals increases risks in terms of security for both operators and animals (Grandin 2013).

The association of increased activity and hormonal changes due to stress during the preslaughter period may thus stimulate glycogen breakdown. This may lead to low muscle glycogen reserves and increase the risk of the production of meat with increased ultimate pH or, if it occurs just before slaughter, to an enhanced rate of metabolism and thus to a faster early postmortem pH and slower temperature decline (Ferguson and Warner 2008). Stress may further influence the tenderness of the meat. While the relationship between pH decline and meat tenderness is well established for the meat of pigs and fowl, the relationship between pH decline and tenderness in cattle is less straightforward (Hwang and Thompson 2001; Boudjellal et al. 2008). Despite the absence of a clear, linear relationship between pH decline and tenderness, several studies indicate that stress before slaughter may influence tenderness of beef. In one study, cows were slaughtered using two standardized procedures, one designed to either avoid stress as much as possible, the other to induce a certain amount of physical effort and emotional stress. The cows slaughtered with the latter, additional stress procedure, produced significantly tougher *Longissimus* meat than cows with the minimal stress procedure. These effects were not directly related to rate or amplitude of pH decline (Bourguet et al. 2010; Terlouw et al. 2015). Warner et al. (2007) found that the use of electric prodders on cattle was associated with lower sensory ratings, including tenderness, by consumers. Gruber et al. (2010) showed a relationship between elevated blood lactate concentration at bleeding and reduced beef tenderness also suggesting that acute stress just before slaughter can negatively influence tenderness.

In conclusion, even moderate stress during slaughter may influence postmortem pH and temperature decline and sensory qualities of meat, including beef tenderness. As indicated above, the stress status of an animal is related to its emotional state. Hence, the animal's stress status depends on the way the animal evaluates its environment. Following is a discussion of how we can try to understand more precisely the causes of stress at slaughter.

7.2 PREDISPOSITION FOR HIGH OR LOW STRESS REACTIVITY

7.2.1 Stress Reactivity and Genetic Background

The effects of the presence of the allele "halothane sensitivity" on stress reactions have been extensively described (see Chapter 10). This allele influences essentially the cellular response to stress, but not the way animals evaluate their environment (Terlouw et al. 2001). Different breeds or lines, noncarriers of the halothane gene, may vary in the way they react, behaviorally and physiologically, to various challenges. This is illustrated by a study comparing Duroc and Large White pigs (Terlouw and Rybarczyk 2008). Each pig was submitted to two tests: (i) exposure to an unfamiliar object and (ii) human exposure. Durocs touched the person significantly more often than Large Whites. As Durocs were more active during this test than Large White pigs, they also had higher heart rates. This breed difference was specific for the motivation to touch man, because no differences were found in the unfamiliar object test. The results show that the two breeds evaluated the presence of a person differently.

Similar results have been found in cattle. For example, beef cattle had a greater flight distance when they were approached by a human than dairy cattle reared under the same conditions (Murphey et al. 1981). Significant differences in reactivity to handling and other challenges were also observed between different beef cattle breeds (Fordyce et al. 1988; Voisinet et al. 1997; Gauly et al. 2001). When young Angus, Blond d'Aquitaine, and Limousin bulls were compared in a human exposure test and a surprise test (sudden opening of an umbrella; Figure 7.2), the breeds differed in more than 10 behaviors during the stress reactivity tests. Blond d'Aquitaine bulls were more reactive than Angus, while Limousins had mostly intermediate levels. For example, Blond d'Aquitaine bulls showed more escape attempts and startle responses in the surprise test, and more vigilance in the presence of the immobile human, than Angus bulls (Bourguet et al. 2015). This is consistent with results from a study in an abattoir, where Blond d'Aquitaine bulls were more reactive than Charolais bulls (Bourguet et al. 2011).

Studies on other species have also shown a genetic basis for stress reactivity. For example, divergent lines of Leghorn chickens selected for high and low feather pecking also differed in reactivity to restraint in terms of increase in cortisol (Kjaer and Guémené 2009). Similarly, trout of different selection lines showed different reactivity to manual restraint, as measured by cortisol (Lefèvre et al. 2008).

When comparing physiological stress status between different cattle breeds at slaughter, several studies found different urinary catecholamines and cortisol levels, at least when expressed with respect to urinary creatinine levels (Ndlovu et al. 2008; Muchenje et al. 2009; O'Neill et al. 2012; Bourguet et al. 2015). These higher levels of catecholamine levels were associated with darker meat (Muchenje et al. 2009). Chickens of a French-label line showed increased behavioral reactivity, in terms of wing flapping on the shackle line, than a fast-growing standard line and these increased levels of wing flapping were associated with a faster rate of pH decline (Debut et al. 2003). Increased behavioral and increased physiological reactivity are not always associated. For example, at slaughter, young bulls appeared more behaviorally reactive while heifers had higher cortisol levels (Bourguet et al. 2011; Probst et al. 2014).

Not all studies show breed-related differences in stress status at slaughter. For example, the Duroc and Large White pigs, which have been shown to have different stress reactivity during rearing, did not differ in their physiological stress responses at slaughter carried out shortly after the reactivity tests (Terlouw and Rybarczyk 2008). Similarly, different goat breeds showed similar increases in stress status at slaughter (Ergul Ekiz and Yalcintan 2013).

7.2.2 Stress Reactivity and Prior Experience

The way animals react to stressful situations also depends on prior experience. Several studies showed that repeated positive handling increases the willingness of pigs to be approached by or to approach a human and repeated negative handling decreases this willingness (Gonyou et al. 1986; Paterson and Pearce 1992; Tanida et al. 1995; Hemsworth et al. 1996; Terlouw and Porcher 2005). The effect of prior experience may be rather long-lasting as shown by an experiment with cattle. Calves were reared during the first 3 months of life outdoors with little human contact or indoors with frequent human contact. After 4 months, calves were reared together. At 8 months of age, indoor-reared calves were easier to handle and less aggressive toward humans (Boivin et al. 1994).

Positive experiences during rearing may reduce reactivity, including at slaughter. Pigs reared in an enriched environment, with larger pens and straw bedding, were less active than controls and showed, in contrast to controls, no cortisol response to transport (De Jong et al. 2000). They were also less active during lairage. Pigs reared outdoors or in enriched environments were less active and had lower heart rates during unfamiliar object and isolation tests. At slaughter, they fought less and had lower levels of physiological indicators of stress than pigs reared conventionally (Olsson et al. 1999; Klont et al. 2001; Barton-Gade 2008; Terlouw et al. 2009;

Foury et al. 2011). It is believed that rearing at high density or under barren conditions may disturb the normal development of social behavior or increase fearfulness, explaining the increased aggression during mixing (Schouten 1986; Olsson et al. 1999; O'Connell et al. 2004). Other studies found in contrast that pigs reared under standard intensive conditions were easier to load, but showed a stronger increase in salivary cortisol during transport and lairage than pigs reared in larger pens or in an enriched environment (Geverink et al. 1999; De Jong et al. 2000; Klont et al. 2001; Chaloupkova et al. 2007).

Work on cattle found similar results. Several studies showed that the rearing system and the type and frequency of contacts with the stockperson influence the fear reactions to various situations during the preslaughter period. For example, bulls bred by farmers believing that bulls are sensitive to aspects of their environment showed less cortisol increase at slaughter compared to farmers having a more distant attitude toward their animals (Mounier et al. 2006). However, these bulls were more difficult to load, possibly due to their greater familiarity with humans and thus they were less reactive to human approach and therefore more difficult to drive forward (Mounier et al. 2008). Similarly, pigs less fearful of humans were more difficult to drive and these pigs received more negative interventions by abattoir personnel (Hemsworth et al. 2002; Terlouw et al. 2005). Keeping the animals in their rearing group during slaughter may, in contrast, facilitate the slaughter procedure because of group cohesion. For example, bull calves kept in their social group during the preslaughter period had a smaller increase in cortisol levels in blood collected at bleeding (Mounier et al. 2006). This effect was even more pronounced when the animals had been together throughout their rearing period (Mounier et al. 2006).

Various studies have found that the rearing history of the animal may influence muscle glycogen breakdown during the preslaughter period. Repeated negative handling during rearing, by refusing physical contact with pigs, associated with the presence of the negative handler at slaughter, resulted in lower glycogen levels immediately before slaughter (Terlouw et al. 2005). Calves that had been in contact with humans during rearing had higher muscle glycogen content at slaughter (Lensink et al. 2000a). Pigs negatively handled on-farm had lower glycogen levels 5 and 40 min postmortem (D'Souza et al. 1998). Effects may also be observed in terms of postmortem muscle metabolism. Muscles from calves reared by farmers with a negative attitude presented a faster rate of pH decline in the early postmortem period (Lensink et al. 2001). These observations suggest that negative experience with humans during rearing may enhance reactivity to the slaughter procedure, possibly due to increased fear of humans, resulting in increased muscle glycogen catabolism before and after slaughter. Another study shows that bulls that had not been mixed during finishing had a lower ultimate pH, probably because during transfer to slaughter, they were calmer and glycogen stores were not depleted (Mounier et al. 2006). The effects of prior experience may have consequences on meat quality aspects. Calves that have received positive and frequent contact from their breeder were easier to handle and had lower heart rates during loading and produced redder meat compared to calves not having received such contact (Lensink et al. 2000a,b, 2001). Similarly, higher levels of fighting among pigs reared indoors were correlated with lower levels of glycogen content at slaughter, higher ultimate pH, and decreased water holding capacity, compared to outdoor pigs (Terlouw et al. 2009). Pigs reared under standard intensive conditions produced meat with lower water holding capacity and higher ultimate pH than pigs reared in an enriched environment (Klont et al. 2001; Foury et al. 2011), although an absence of differences (Geverink et al. 1999) or opposite results have also been reported (Lambooij et al. 2004).

Overall, results show that prior experience may influence reactivity to humans and to other aspects of the slaughter procedures. The consequences in terms of stress and meat quality are variable. For example, animals with positive experiences with humans may be calmer and easier to drive and produce meat of better quality. However, lack of fear of humans may also lead to increased difficulties during driving and or handling. These variable results may be explained by differences in the preslaughter and slaughter context. For example, if a calm animal is transferred to a group of other calm animals and with a skilled operator, the stress levels of this animal will probably be relatively low. A potentially calm animal may in contrast be relatively more stressed if it is transferred by a less skilled operator or surrounded by excitable animals if it is not used to this situation.

7.3 PREDICTING DIFFERENCES BETWEEN INDIVIDUALS IN MEAT QUALITY

7.3.1 Differences in Stress Reactions at Slaughter and in Meat Quality

Differences in stress reactivity exist not only between animals of different breeds or rearing systems, but also between individuals, even from the same breed or line, and rearing system. These individual differences in stress responses at slaughter may explain differences in the rate or extent of pH decline between animals from a similar genetic and rearing background, slaughtered under similar conditions. For example, in various experiments on pigs, the amount of fighting and ultimate pH were positively correlated, explaining up to 67% of the variability (r^2) in the ultimate pH of thigh muscles (Terlouw et al. 2005, 2009; Terlouw and Rybarczyk 2008). In practical terms, this means that where mixing results in fighting, ultimate pH, particularly of thigh muscles, will increase proportionally to the amount of fighting. The often strong effect of fighting on pH decline may be explained by the combination of physical effort and the physiological changes that fighting involves. The amount of fighting was correlated with the degree of carcass damage (Terlouw et al. 2009) and amount of fighting and carcass damage score were correlated with an increase in plasma catecholamines (noradrenaline and adrenaline), cortisol, glucose, lactate, and free fatty acid levels (Warriss and Brown 1985; Fernandez et al. 1994b). The increases in lactate and glucose were at least partly due to the effects of catecholamines, as plasma glucose and lactate were positively correlated with plasma catecholamine levels (Fernandez et al. 1994b). Possibly, the pH of thigh muscles are more sensitive to the effects of fighting because they are more involved in the physical effort of fighting, but the pH decline and/or water holding capacity of other muscles, like the *Longissimus*, may also be affected (Terlouw and Rybarczyk 2008; Terlouw et al. 2009).

An increased heart rate just before slaughter or increased urinary levels of catecholamine were correlated with faster early postmortem pH decline in pigs, cows, and bulls, explaining up to 52% in the variation in early pH between animals (Terlouw and Rybarczyk 2008; Bourguet et al. 2010, 2015). High heart rate is generally associated with high catecholamine levels and fast ante- and postmortem metabolism leading to faster pH decline. In another experiment, pigs presenting higher urinary levels of catecholamines also had higher ultimate pH (Foury et al. 2011). In this experiment, increased urinary catecholamine levels were further positively correlated with skin damage, suggesting that pigs with high catecholamine levels had fought more. Thus, high physical activity during the preslaughter period associated with the high catecholamine levels of these pigs led to a depletion of glycogen stores and increased ultimate pH (Foury et al. 2011). Similar results were found for cattle. In an experiment, about 22,000 cattle were studied at a South African abattoir during the winter season. In this experiment, the percentage of dark cutting beef (pH ≥ 5.8) increased exponentially with increased bruising scores. It was further shown that gender and feedlot where the animal was reared also played a role, as at similar bruising scores, the percentage of dark cutting meat was lower for heifers and animals reared at feedlot A (Figure 7.1; Viljoen 2011).

The differences in stress reactivity between individuals indicate that the animal's stress status depends on its different evaluation of the situation in terms of the threat it represents for them. It is therefore important, when studying stress in animals, to take into account the animal's point of view, as will be demonstrated below.

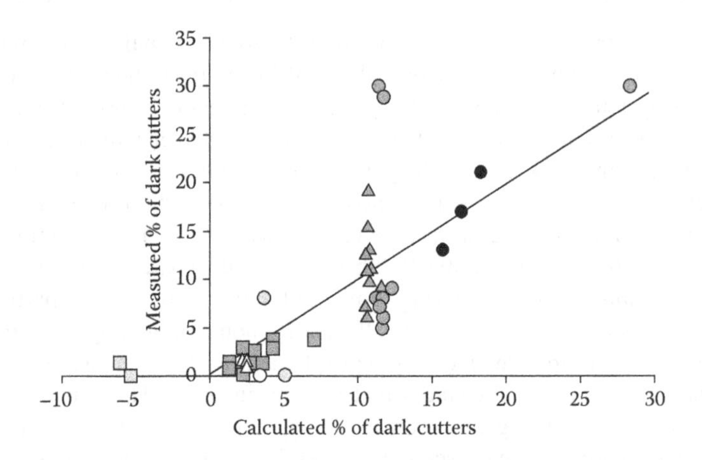

FIGURE 7.1 Correlation between calculated and measured % of dark cutters (adjusted $r = 0.73$; $p < 0.001$). Calculated % of dark cutters was based on the following best-fitting linear equation: $12.1 - 0.4 * \%\text{bruise} + 0.05 * (\%\text{bruise})^2 - 8.3 * \text{heifer} - 9.7 * \text{feedlot A}$. This equation indicates that risk of dark cutting increases exponentially with increasing percentages of bruised animals, but is lower for heifers or for animals reared in feedlot A. Squares, triangles, and circles represent animals from feedlot A, C, and D, respectively. Light, gray, and dark colored symbols represent heifers, steers, and bulls, respectively. Each symbol represents the mean of a load of between 36 and 191 cattle. The regression analysis was based on average values of cattle loads. Feedlot B was excluded because in contrast to the other feedlots, initial exploratory analyses found no relationships between % of bruised cattle and % of dark cutters. (Data from Viljoen 2009. http://repository.up.ac.za/bitstream/handle/2263/26081/dissertation.pdf?sequence=1)

7.3.2 Predictability of Stress Reactions at Slaughter and Resulting Meat Quality

Generally, individuals show a certain consistency in their way to react to stress. Consistency in stress reactivity has been shown for different species and at different ages (pigs, cattle, laboratory animals, humans: Bohus et al. 1987; Hessing et al. 1994; Boissy and Bouissou 1995). For example, gilts, that left their home pen relatively easily compared to their group members were easier to move through a corridor and were less reactive to humans or to restraint in a nose sling (Lawrence et al. 1991). Similarly, heifers that were more reluctant to explore an unfamiliar object compared to their experimental counterparts, were also more reluctant to feed near a fear-inducing stimulus and spent more time with their head upright in an unfamiliar environment (Boissy and Bouissou 1995). A similar study has shown consistent differences in stress reactivity in sheep (Deiss et al. 2009).

Given the consistency in stress reactivity of animals, various studies have attempted to predict stress reactions at slaughter on aspects of meat quality, based on individual differences in stress responses observed during rearing. One experiment found that in the early postmortem period, the muscles of pigs that were more attracted by humans in a reactivity test during rearing were cooler, had a higher pH, and contained less lactate compared with pigs less attracted by humans. It suggests that the former pigs reacted less strongly to the slaughter procedure immediately before slaughter, or maybe more specifically less strongly to human presence during this period. Consequently, physical and metabolic muscle activity was lower, resulting in slower glycogen breakdown (higher pH) and lower muscle temperature (Terlouw and Rybarczyk 2008). Similarly, cows that showed more fear responses to an unfamiliar environment in a reactivity test compared to other group members, had warmer muscles and lower pH in the early postmortem period. These cows had further, during slaughter, higher levels of noradrenaline in their urine, and higher heart rates, indicating that cows that are fearful of unfamiliarity were more stressed during the slaughter period, compared to their less fearful counterparts (Bourguet et al. 2010). In another experiment, the reactivity of young bulls of different beef breeds to the sudden opening of an umbrella was tested a few weeks before slaughter (Figure 7.2). Bulls of different beef breeds that showed higher heart rates immediately after the opening of the umbrella had lower pH 40 min postmortem, and thus a faster early postmortem pH decline, in the *Longissimus* muscle. This suggests that in these bulls, the presence of sudden events during the slaughter procedure contributed to the slaughter stress and that the latter has influenced postmortem muscle metabolism (Bourguet et al. 2015). Overall, results show that it is possible to identify before slaughter those individuals that are likely to be more reactive to the slaughter process and that if this reactivity is not taken into account by adapting slaughter procedures, there may be consequences for the quality of their meat.

Differences in ultimate pH between individuals may also be explained by individual differences in the way animals react to the slaughter procedure. For example, even when reared in the same environment, pigs differ in their tendency to engage in aggressive interactions during mixing. It is possible to identify these pigs before slaughter. Pigs fighting more for food (Terlouw et al. 2005) and during dyadic encounters (Erhard et al. 1997) in tests carried out during the rearing period fought more during mixing before

FIGURE 7.2 Young Aberdeen Angus bull in the surprise test, just before (a) and just after (b) the sudden opening of the umbrella. (Adapted from Bourguet, C. et al. 2015. *Appl. Anim. Behav. Sci.* 164:41–55; photo copyright Inra—Florent Giffard.)

slaughter. For reasons that remain to be elucidated, several studies found that pigs that explore an unfamiliar object more during a reactivity test are also more aggressive during mixing or in other circumstances (Lawrence et al. 1991; Olsson et al. 1999; O'Connell et al. 2004). Lambs those were relatively reactive to social separation or to novelty also showed increased stress reactions at slaughter as indicated by higher preslaughter cortisol levels and ultimate pH compared to their less reactive counterparts (Deiss et al. 2009).

Several studies in cattle found that reactivity to stress measured during rearing may also help to predict future tenderness of the meat. Warner Bratzler shear force values were greater in steers with an excitable temperament when compared to calmer steers (Voisinet et al. 1997; King 2005). Cows those were more reactive to human presence and social isolation produced tougher meat (Terlouw et al. 2015). In an experiment on calves, significant relationships were found between flight speeds measured during rearing and meat quality aspects, including color and tenderness, though these relationships showed inconsistencies across years (Burrow et al. 1999). Fordyce et al. (1988) rated 410 bulls and cows for their stress reactivity in two tests, both in the visible presence of humans. One test scored the amount of movement of each animal when placed individually in a crush; the other, reactivity in terms of degree of movement when introduced with other animals in a small circular yard. For each test, animals were categorized in one of the five classes according to their score. This study found that bulls and cows with higher crush or yard scores had on average more bruises at slaughter (Figure 7.3) probably due to their greater reactivity to stress in general, or to humans, in particular. It was found that bruising and gender explained 77% of the variability between animals in mechanical measurements of tenderness (Warner Bratzler peak force: WBPF), averaged by category (Figure 7.4). The impact of gender was about 1.5 times stronger than the impact of bruising, as indicated by the standardized coefficients. Overall, the results indicate that more reactive cattle have a greater risk to have bruises and consequently, to produce less tender meat.

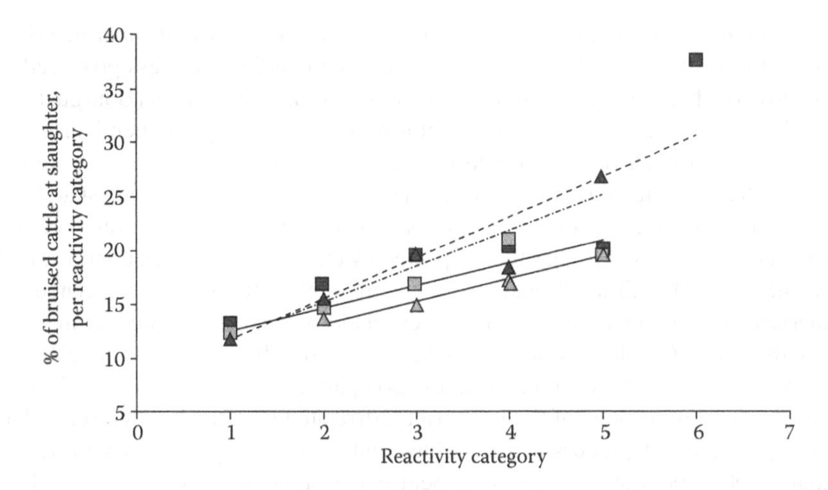

FIGURE 7.3 Correlations between reactivity (crush: squares and yard: triangles) scores during rearing and % of bruised cattle at slaughter (per reactivity category). Average % of bruised cattle increased coherently and linearly with increasing crush or yard scores. Dark and light symbols represent bulls and cows, respectively. Crush: $r = 0.84$ and 0.93 for bulls and cows, respectively. Yard: $r = 0.94$ and 0.99 for bulls and cows, respectively. $p < 0.05$ for all r-values. Each symbol represents the mean of each score category. The regression analysis was also based on average values of score categories. (Data from Fordyce, G. et al. 1988. *Aust. J. Exp. Agric.* 28:689–693.)

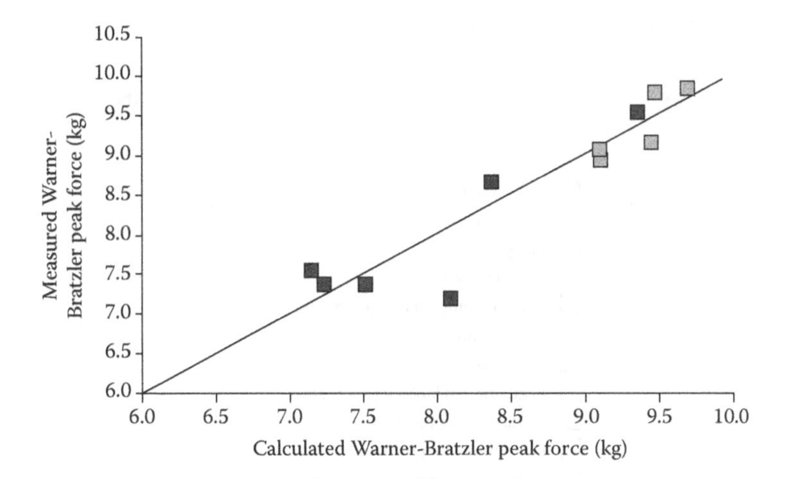

FIGURE 7.4 Correlation between calculated and measured Warner-Bratzler Peak Force (WBPF). Calculated WBPF was based on the following best-fitting linear equation: 43.3 + 0.28 * crush score – 5.94 * pH 24 h – 1.97 * bull. This equation indicates that risk of high WBPF increases linearly with increased crush scores and decreases linearly with higher pH at 24 h pm, and that WBPF values are lower for bulls. Gray and dark colored symbols indicate bulls and cows, respectively. Each symbol represents the mean of each score category. The regression analysis was also based on average values of score categories. (Data from Fordyce, G. et al. 1988. *Aust. J. Exp. Agric.* 28:689–693.)

Counteracting mechanisms can also be observed. In the experiment mentioned above (Bourguet et al. 2010), cows slaughtered with additional stress produced significantly tougher meat than cows slaughtered with minimal stress (Bourguet et al. 2010; Terlouw et al. 2015). To start exploring the underlying biochemical mechanisms, the expression of about 3000 genes in the *Longissimus* muscle sampled 40 min after slaughter was studied for each individual cow. In the group slaughtered with minimal stress, the amount of genes of which the expression was correlated with tenderness was not higher than expected by chance (i.e., 5%). In contrast, in the cows slaughtered with additional stress, the amount of transcripts correlated with tenderness was significantly higher than expected by chance (Terlouw et al. 2015). An online Gene Ontology tool was used to determine the biological processes that were represented by the correlated genes. Compared to the complete set of studied genes, greater proportions of the transcripts correlated with tenderness were related to the generation of precursor metabolites and energy: respiratory chain activity, oxidative phosphorylation, and citric acid cycle processes (Figure 7.5). All correlations were positive. These data indicate that several, apparently counteracting

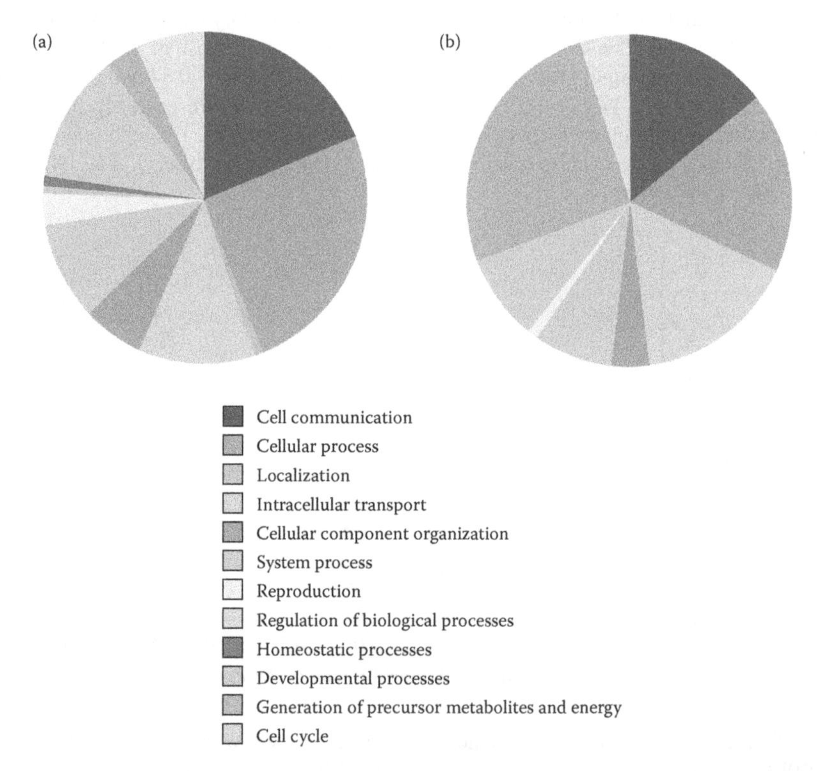

FIGURE 7.5 Distribution of the biological processes of the complete set of genes studied (a) and of the genes positively correlated with tenderness for cows slaughtered with additional stress (b). Only processes under or overrepresented in the latter group with respect to the complete set are shown, that is, 40% of the genes were excluded. Processes related to intracellular transport and the generation of precursor metabolites and energy were overrepresented in genes positively correlated with tenderness. (Data from I. Cassar-Malek, personal communication.)

mechanisms, underlie tenderness: stress reduces tenderness, but the negative effects of stress on tenderness were limited in cows showing a relatively high expression of genes involved in oxidative metabolic activity (Terlouw et al. 2015). In coherence with this, other recent studies show that the relationships between biomarkers and tenderness may be reversed, depending on the metabolic and contractile properties of the muscle and on the stress level (Picard et al. 2014; Terlouw et al. 2015). These counteracting or opposite effects may explain the difficulties that scientists have encountered to find stable and universal protein markers of the propensity of cattle to produce tender meat (Koohmaraie 1996; Guillemin et al. 2012; Picard et al. 2014).

Overall, results indicate that animals differ consistently in their stress reactivity, that is, animals differ consistently in the way they evaluate the situation in terms of the threat it represents for them. This stress reactivity depends to a great extent on their genetic background and their prior experiences. The determination of the animal's stress reactivity may help predict the quality of the meat it may produce. The findings show that relatively low stress reactivity may be a predictor of the likelihood that an animal will produce meat of good quality. However, as indicated earlier, very unreactive animals may present difficulties during handling.

7.4 CONCLUSIONS

In conclusion, during the slaughter period, animals are subjected to many potential stressors, simultaneously and successively, in different environmental contexts. These stressors may be of physical or emotional origin. They cause negative emotional responses, and are therefore an ethical issue. The emotional response underlies physiological responses, including the release of certain hormones and incites the animal to respond behaviorally, leading to an increased physical effort. The more the animal is stressed, the more difficult it is to handle, leading to further physical activity and emotional stress. Stress reactivity at slaughter can be predicted based on stress reactions in specifically designed tests carried out during rearing. Prior experience and genetic background have a significant impact on stress reactivity, including at slaughter.

The combined effects of physiological changes due to emotional stress and metabolic changes due to physical effort may lead to changes in meat quality. These reactions to stressful aspects of the slaughter procedures may result in relatively low muscle glycogen reserves at the moment of slaughter, which may cause relatively high ultimate pH. They may also lead to an acceleration of muscle metabolism at the moment of slaughter, which continues after death and may lead to a faster postmortem pH decline. Increased stress at slaughter may also be associated with alterations in sensory quality, including a reduction of tenderness, in cattle and other species. In the case of beef, this does not appear directly related to the kinetics of the pH decline. The biochemical mechanisms underlying the effects of stress on tenderness seem complex, and partly mutually counteracting.

Finally, selecting an animal for very low stress reactivity will not be beneficial. Animals that are very unreactive may be difficult to drive, which is unwanted, both for animal welfare reasons, and economical reasons. Overall, results indicate that adapting abattoir procedures must take into account the animal's point of view, which will be beneficial for animal welfare questions and meat quality, and also for security matters of stock workers and animals.

REFERENCES

Astruc, T., Talmant, A., Fernandez, X., and Monin, G. 2002. Temperature and catecholamine effects on metabolism of perfused isolated rabbit muscle. *Meat Sci.* 60:287–293.

Barton-Gade, P. 2008. Effect of rearing system and mixing at loading on transport and lairage behaviour and meat quality: Comparison of outdoor and conventionally raised pigs. *Animal* 2:902–911.

Bendall, J.R. 1973. Post-mortem changes in muscle. In *The Structure*, part 2 (Bourne GH, Eds.). Academic Press, London, pp. 244–230.

Bohus, B., Benus, R.F., Fokkema, D.S., Koolhaas, J.M., Nyakas, C., Van Oortmerssen, G.A., Prins, A.J.A., De Ruiter, A.J.H., Scheurink, A.J.W., and Steffens, A.B. 1987. Neuroendocrine states and behavioral and physiological stress responses. In *Progress in Brain Research,* V.M.W.E.R. de Kloet and D.d. Wied (Eds.), Vol. 72, Elsevier, Amsterdam/New York/Oxford, pp. 57–70.

Boissy, A., Arnould, C., Chaillou, E. et al. 2007. Emotions and cognition: A new approach to animal welfare. *Anim. Welfare* 16:37–43.

Boissy, A., and Bouissou, M.F. 1995. Assessment of individual differences in behavioural reactions of heifers exposed to various fear-eliciting situations. *Appl. Anim. Behav. Sci.* 46:17–31.

Boivin, X., Leneindre, P., Garel, J.P., and Chupin, J.M. 1994. Influence of breed and rearing management on cattle reactions during human handling. *Appl. Anim. Behav. Sci.* 39:115–122.

Boudjellal, A., Becila, S., Coulis, G., Herrera-Mendez, C.H., Aubry, L., Lepetit, J., Harhoura, K., Sentandreu, M.A., Ait-Amar, H., and Ouali, A. 2008. Is the pH drop profile curvilinear and either monophasic or polyphasic? Consequences on the ultimate bovine meat texture. *Afr. J. Agric. Res.* 3:195–204.

Bourguet, C., Deiss, V., Boissy, A., and Terlouw, E.M.C. 2015. Young Blond d'Aquitaine, Angus and Limousin bulls differ in emotional reactivity: Relationships with animal traits, stress reactions at slaughter and post-mortem muscle metabolism. *Appl. Anim. Behav. Sci.* 164:41–55.

Bourguet, C., Deiss, V., Gobert, M., Durand, D., Boissy, A., and Terlouw, E.M.C. 2010. Characterising the emotional reactivity of cows to understand and predict their stress reactions to the slaughter procedure. *Appl. Anim. Behav. Sci.* 125:9–21.

Bourguet, C., Deiss, V., Tannugi, C.C., and Terlouw, E.M.C. 2011. Behavioural and physiological reactions of cattle in a commercial abattoir: Relationships with organisational aspects of the abattoir and animal characteristics. *Meat Sci.* 88:158–168.

Broom, D.M. 1987. Applications of neurobiological studies to farm animal welfare. In *Biology of Stress in Farm Animals: An Integrated Approach.* P.R. Wiepkema and P.W.M. Van Adrichem (Eds.), Vol. 42, Springer, the Netherlands, pp. 101–110.

Burrow, H.M., Shorthose, W.R., and Stark, J.L. 1999. Relationships between temperament and carcass and meat quality attributes of tropical beef cattle. *Proc. Assoc. Advmt. Anim. Breed. Genet.* 13:227–230.

Chaloupkova, H., Illmann, G., Neuhauserova, K., Tomanek, M., and Valis, L. 2007. Preweaning housing effects on behavior and physiological measures in pigs during the suckling and fattening periods. *J. Anim. Sci.* 85:1741–1749.

Cockram, M.S., Baxter, E.M., Smith, L.A., Bell, S., Howard, C.M., Prescott, R.J., and Mitchell, M.A. 2004. Effect of driver behaviour, driving events and road type on the stability and resting behaviour of sheep in transit. *Anim. Sci.* 79:165–176.

Damasio, A.R. 1998. Emotion in the perspective of an integrated nervous system. *Brain Res. Rev.* 26:83–86.
Dantzer, R. 2002. Can farm animal welfare be understood without taking into account the issues of emotion and cognition? *J. Anim. Sci.* 80:E1–E9.
Danziger, N. 2006. Mise au point bases neurologiques de l'affect douloureux. *Rev. Neurol.* 162:395–399.
Dawkins, M.S. 1980. *Animal Suffering: The Science of Animal Welfare*. Chapman & Hall, London.
De Jong, I.C., Prelle, I.T., Van de Burgwal, J.A., Lambooij, E., Korte, S.M., Blokhuis, H.J., and Koolhaas, J.M. 2000. Effects of rearing conditions on behavioural and physiological responses of pigs to preslaughter handling and mixing at transport. *Can. J. Anim. Sci.* 80:451–458.
Debut, M., Berri, C., Baeza, E., Sellier, N., Arnould, C., Guemene, D., Jehl, N., Boutten, B., Jego, Y., Beaumont, C., and Le Bihan-Duval, E. 2003. Variation of chicken technological meat quality in relation to genotype and preslaughter stress conditions. *Poultry Sci.* 82:1829–1838.
Deiss, V., Temple, D., Ligout, S., Racine, C., Bouix, J., Terlouw, C., and Boissy, A. 2009. Can emotional reactivity predict stress responses at slaughter in sheep? *Appl. Anim. Behav. Sci.*119:193–202.
Désiré, L., Boissy, A., and Veissier, I. 2002. Emotions in farm animals: A new approach to animal welfare in applied ethology. *Behav. Proc.* 60:165–180.
Dransfield, E. 1981. Eating quality of DFD beef. In *The Problem of Dark-Cutting in Beef*, D.E. Hood and P.V. Tarrant (Eds.), Vol. 10, Springer, the Netherlands, pp. 344–361.
D'Souza, D.N., Warner, R.D., Dunshea, F.R., and Leury, B.J. 1998. Effect of on-farm and preslaughter handling of pigs on meat quality. *Aust. J. Agric. Res.* 49:1021–1025.
Duncan, I.J.H. 1996. Animal welfare defined in terms of feelings. *Acta Agric. Scan. A-An.* 29–35.
Ergul Ekiz, E., and Yalcintan, H. 2013. Comparison of certain haematological and biochemical parameters regarding pre-slaughter stress in Saanen, Maltese, Gokceada and hair goat kids. *J. Fac. Vet. Med. Istanbul Univ.* 39:189–196.
Erhard, H.W., Mendl, M., and Ashley, D.D. 1997. Individual aggressiveness of pigs can be measured and used to reduce aggression after mixing. *Appl. Anim. Behav. Sci.* 54:137–151.
Febbraio, M.A., Lambert, D.L., Starkie, R.L., Proietto, J., and Hargreaves, M. 1998. Effect of epinephrine in trained men. *J. Appl. Physiol.* 84:465–470.
Febbraio, M.A., Snow, R.J., Stathis, C.G., Hargreaves, M., and Carey, M.F. 1996. Blunting the rise in body temperature reduces muscle glycogenolysis during exercise in humans. *Exp. Physiol.* 81:685–693.
Ferguson, D.M., and Warner, R.D. 2008. Have we underestimated the impact of pre-slaughter stress on meat quality in ruminants? *Meat Sci.* 80(1):12–19.
Fernandez, X., Meunier-Salaün, M.C., and Ecolan, P. 1994a. Glycogen depletion according to muscle and fibre types in response to dyadic encounters in pigs (*Sus scrofa domesticus*)—Relationships with plasma epinephrine and aggressive behaviour. *Comp. Biochem. Physiol.* 4:869–879.
Fernandez, X., Meunier-Salaun, M.C., and Mormede, P. 1994b. Agonistic behavior, plasma stress hormones, and metabolites in response to dyadic encounters in domestic pigs: Interrelationships and effect of dominance status. *Physiol. Behav.* 56:841–847.
Fordyce, G., Whythes, J.R., Shorthose, W.R., Underwood, D.W., and Shepherd, R.K. 1988. Cattle temperaments in extensive beef herds in Northern Queensland 2. Effect of temperament on carcass and meat quality. *Aust. J. Exp. Agric.* 28:689–693.
Foury, A., Lebret, B., Chevillon, P., Vautier, A., Terlouw, C., and Mormede, P. 2011. Alternative rearing systems in pigs: Consequences on stress indicators at slaughter and meat quality. *Animal* 5:1620–1625.

Fraser, D., Ritchie, J.S.D., and Faser, A.F. 1975. The term "stress" in a veterinary context. *Brit. Vet. J.* 131:653–662.
Gauly, M., Mathiak, H., Hoffmann, K., Kraus, M., and Erhardt, G. 2001. Estimating genetic variability in temperamental traits in German Angus and Simmental cattle. *Appl. Anim. Behav. Sci.* 74:109–119.
Geverink, N.A., Jong, I.C., Lambooij, E., Blokhuis, H.J., and Wiegant, V.M. 1999. Influence of housing conditions on responses of pigs to preslaugter treatment and meat quality. *J. Anim. Sci.* 79:285–291.
Gonyou, H.W., Hemsworth, P.H., and Barnett, J.L. 1986. Effects of frequent interactions with humans on growing pigs. *Appl. Anim. Behav. Sci.* 16:269–278.
Grandin, T. 1999. Safe handling of large animals. *Occup. Med.* 14:195–212.
Grandin, T. 2013. *Recommended Animal Handling Guidelines and Audit Guide: A Systematic Approach to Animal Welfare*. AMI Foundation.
Gruber, S.L., Tatum, J.D., Engle, T.E., Chapman, P.L., Belk, K.E., and Smith, G.C. 2010. Relationships of behavioral and physiological symptoms of preslaughter stress to beef longissimus muscle tenderness. *J. Anim. Sci.* 88:1148–1159.
Guillemin, N.P., Jurie, C., Renand, G., Hocquette, J.F., Micol, D., Lepetit, J., and Picard, B. 2012. Different phenotypic and proteomic markers explain variability of beef tenderness across muscles. *Int. J. Biol.* 4:26–38.
Hambrecht, E., Eissen, J.J., Newman, D.J., Smits, C.H.M., Verstegen, M.W.A., and Den Hartog, L.A. 2005. Preslaughter handling effects on pork quality and glycolytic potential in two muscles differing in fiber type composition. *J. Anim. Sci.* 83:900–907.
Hemsworth, P.H., Barnett, J.L., Hofmeyr, C., Coleman, G.J., Dowling, S., and Boyce, J. 2002. The effects of fear of humans and pre-slaughter handling on the meat quality of pigs. *Aust. J. Agric. Res.* 53:493–501.
Hemsworth, P.H., Verge, J., and Coleman, G.J. 1996. Conditioned approach-avoidance responses to humans: The ability of pigs to associate feeding and aversive social experiences in the presence of humans with humans. *Appl. Anim. Behav. Sci.* 50:71–82.
Henckel, P., Karlsson, A., Oksbjerg, N., and Petersen, J.S. 2000. Control of post mortem pH decrease in pig muscles: Experimental design and testing of animal models. *Meat Sci.* 55:131–138.
Hessing, M.J.C., Hagelsø, A.M., Schouten, W.G.P., Wiepkema, P.R., and Van Beek, J.A.M. 1994. Individual behavioral and physiological strategies in pigs. *Physiol. Behav.* 55:39–46.
Horswill, C.A., Hickner, R.C., Scott, J.R., Costill, D.L., and Gould, D. 1990. Weight loss, dietary carbohydrate modifications, and high intensity, physical performance. *Med. Sci. Sports Exerc.* 22:470–476.
Huff-Lonergan, E., Baas, T.J., Malek, M., Dekkers, J.C.M., Prusa, K., and Rothschild, M.F. 2002. Correlations among selected pork quality traits. *J. Anim. Sci.* 80:617–627.
Huff-Lonergan, E.L., and Lonergan, S.M. 2005. Mechanisms of water-holding capacity of meat: The role of postmortem biochemical and structural changes. *Meat Sci.* 71:194–204.
Hwang, I.H., and Thompson, J.M. 2001. The interaction between pH and temperature decline early postmortem on the calpain system and objective tenderness in electrically stimulated beef longissimus dorsi muscle. *Meat Sci.* 58:167–174.
King, D.A. 2005. *Evaluation of the Relationship between Animal Temperament and Stress Responsiveness to M. Longissimus lumborum Tenderness in Feedlot Cattle*. PhD dissertation, University of Texas, USA.
Kjaer, J.B., and Guémené, D. 2009. Adrenal reactivity in lines of domestic fowl selected on feather pecking behavior. *Physiol. Behav.* 96:370–373.
Klont, R.E., Hulsegge, B., Hoving-Bolink, A.H., Gerritzen, M.A., Kurt, E., Winkelman-Goedhart, H.A., De Jong, I.C., and Kranen, R.W. 2001. Relationships between behavioral and meat quality characteristics of pigs raised under barren and enriched housing conditions. *J. Anim. Sci.* 79:2835–2843.

Koohmaraie, M. 1996. Biochemical factors regulating the toughening and tenderization processes of meat: Meat for the Consumer 42nd International Congress of Meat Science and Technology. *Meat Sci.* 43:193–201.

Lacourt, A., and Tarrant, P.V. 1985. Glycogen depletion patterns in myofibres of cattle during stress. *Meat Sci.* 15:85–100.

Lambooij, E., Hulsegge, B., Klont, R.E., Winkelman-Goedhart, H.A., Reimert, H.G.M., and Kranen, R.W. 2004. Effects of housing conditions of slaughter pigs on some post mortem muscle metabolites and pork quality characteristics. *Meat Sci.* 66:855–862.

Lawrence, A.B., Terlouw, E.M.C., and Illius, A.W. 1991. Individual differences in behavioural responses of pigs exposed to non-social and social challenges. *Appl. Anim. Behav. Sci.* 30:73–86.

LeDoux, J.E. 2000. Emotion circuits in the brain. *Annu. Rev. Neurosci.* 23:155–184.

Lefèvre, F., Cos, I., Pottinger, T.G., and Bugeon, J. 2008. Sélection génétique sur la réponse au stress et stress à l'abattage: Conséquences sur la qualité de la chair chez la truite arc-en-ciel. *Viandes Prod. Carnés*, Hors-série: 177–178.

Lensink, B.J., Boivin, X., Pradel, P., Le Neindre, P., and Veissier, I. 2000a. Reducing veal calves' reactivity to people by providing additional human contact. *J. Anim. Sci.* 78:1213–1218.

Lensink, B.J., Fernandez, X., Boivin, X., Pradel, P., Le Neindre, P., and Veissier, I. 2000b. The impact of gentle contacts on ease of handling, welfare, and growth of calves and on quality of veal meat. *J. Anim. Sci.* 78:1219–1226.

Lensink, B.J., Fernandez, X., Cozzi, G., Florand, L., and Veissier, I. 2001. The influence of farmers' behavior on calves' reactions to transport and quality of veal meat. *J. Anim. Sci.* 79:642–652.

Moeller, S.J., Miller, R.K., Aldredge, T.L., Logan, K.E., Edwards, K.K., Zerby, H.N., Boggess, M., Box-Steffensmeier, J.M., and Stahl, C.A. 2010. Trained sensory perception of pork eating quality as affected by fresh and cooked pork quality attributes and end-point cooked temperature. *Meat Sci.* 85:96–103.

Monin, G. 1973. Réaction a l'électronarcose et glycogénolyse post mortem chez le porc. *Ann. Zootech.* 22:73–81.

Mounier, L., Colson, S., Roux, M., Dubroeucq, H., Boissy, A., and Veissier, I. 2008. Positive attitudes of farmers and pen-group conservation reduce adverse reactions of bulls during transfer for slaughter. *Animal* 2:894–901.

Mounier, L., Dubroeucq, H., Andanson, S., and Veissier, I. 2006. Variations in meat pH of beef bulls in relation to conditions of transfer to slaughter and previous history of the animals. *J. Anim. Sci.* 84:1567–1576.

Muchenje, V., Dzama, K., Chimonyo, M., Strydom, P.E., and Raats, J.G. 2009. Relationship between pre-slaughter stress responsiveness and beef quality in three cattle breeds. *Meat Sci.* 81:653–657.

Murphey, R.M., Moura Duarte, F.A., and Torres Penedo, M.C. 1981. Responses of cattle to humans in open spaces: Breed comparisons and approach-avoidance relationships. *Behav. Genet.* 11:37–48.

Ndlovu, T., Chimonyo, M., Okoh, A.I., and Muchenje, V. 2008. A comparison of stress hormone concentrations at slaughter in Nguni, Bonsmara and Angus steers. *Afr. J. Agric. Res.* 3:96–100.

O'Connell, N.E., Beattie, V.E., and Moss, B.W. 2004. Influence of social status on the welfare of growing pigs housed in barren and enriched environments. *Anim. Welfare* 13:425–431.

Olsson, I.A.S., De Jonge, F.H., Schuurman, T., and Helmond, F.A. 1999. Poor rearing conditions and social stress in pigs: Repeated social challenge and the effect on behavioural and physiological responses to stressors. *Behav. Proc.* 46:201–215.

O'Neill, H.A., Webb, E.C., Frylinck, L., and Strydom, P. 2012. Urinary catecholamine concentrations in three beef breeds at slaughter. *S. Afr. J. Anim. Sci.* 42:545–549.

Panksepp, J. 2005. Affective consciousness: Core emotional feelings in animals and humans: Neurobiology of animal consciousness. *Conscious. Cogn.* 14:30–80.

Paterson, A.M., and Pearce, G.P. 1992. Growth, response to humans and corticosteroïds in male pigs housed individually and subjected to pleasant, unpleasant or minimal handling during rearing. *Appl. Anim. Behav. Sci.* 34:315–328.

Paul, E.S., Harding, E.J., and Mendl, M. 2005. Measuring emotional processes in animals: The utility of a cognitive approach. *Neurosci. Biobehav. Rev.* 29:469–491.

Picard, B., Gagaoua, M., Micol, D., Cassar-Malek, I., Hocquette, J.F., and Terlouw, C.E. 2014. Inverse relationships between biomarkers and beef tenderness according to contractile and metabolic properties of the muscle. *J. Agric. Food Chem.* 62(40):9808–9818.

Probst, J.K., Spengler Neff, A., Hillmann, E., Kreuzer, M., Koch-Mathis, M., and Leiber, F. 2014. Relationship between stress-related exsanguination blood variables, vocalisation, and stressors imposed on cattle between lairage and stunning box under conventional abattoir conditions. *Livest. Sci.* 164:154–158.

Rosenvold, K., and Andersen, H.J. 2003. Factors of significance for pork quality—A review. *Meat Sci.* 64:219–237.

Scheffler, T.L., and Gerrard, D.E. 2007. Mechanisms controlling pork quality development: The biochemistry controlling postmortem energy metabolism: 53rd International Congress of Meat Science and Technology (53rd ICoMST). *Meat Sci.* 77:7–16.

Schouten, W. 1986. Rearing conditions affect behaviour in later life. PhD thesis, Groningen University, Netherlands.

Starkie, R.L., Hargreaves, M., Lambert, D.L., Proietto, J., and Febbraio, M.A. 1999. Effect of temperature on muscle metabolism during submaximal exercise in humans. *Exp. Physiol.* 84:775–784.

Tanida, H., Miura, A., Tanaka, T., and Yoshimoto, T. 1995. Behavioral response to humans in individually handled weanling pigs. *Appl. Anim. Behav. Sci.* 42:249–259.

Tarrant, P.V., Kenny, F.J., Harrington, D., and Murphy, M. 1992. Long distance transportation of steers to slaughter: Effect of stocking density on physiology, behaviour and carcass quality. *Livest. Prod. Sci.* 30:223–238.

Terlouw, C., Ludriks, A., Schouten, W., Vaessens, S., Fernandez, X., Andanson, S., and Père, M.C. 2001. Prédominance de l'allèle de sensibilité à l'halothane. *Viandes Prod. Carnés* 22:1–10.

Terlouw, E.M.C. 2005. Stress reactions at slaughter and meat quality in pigs: Genetic background and prior experience: A brief review of recent findings. *Livest. Prod. Sci.* 94:125–135.

Terlouw, E.M.C., Arnould, C., Auperin, B., Berri, C., Le Bihan-Duval, E., Deiss, V., Lefèvre, F., Lensink, B.J., and Mounier, L. 2008. Pre-slaughter conditions, animal stress and welfare: Current status and possible future research. *Animal* 2:1501–1517.

Terlouw, E.M.C., Berne, A., and Astruc, T. 2009. Effect of rearing and slaughter conditions on behaviour, physiology and meat quality of Large White and Duroc-sired pigs. *Livest. Sci.* 122:199–213.

Terlouw, E.M.C., Bourguet, C., and Deiss, V. 2012. Stress at slaughter in cattle: Role of reactivity profile and environmental factors. *Anim. Welfare* 21:43–49.

Terlouw, E.M.C., Cassar-Malek, I., Picard, B., Bourguet, C., Deiss, V., Arnould, C., Berri, C., Duval, E., Lefèvre, F., and Lebret, B. 2015. Stress en élevage et à l'abattage: Impacts sur les qualités des viandes. *INRA Prod. Anim.* 28(2):169–182.

Terlouw, E.M.C., and Porcher, J. 2005. Repeated handling of pigs during rearing. I. Refusal of contact by the handler and reactivity to familiar and unfamiliar humans. *J. Anim. Sci.* 83:1653–1663.

Terlouw, E.M.C., Porcher, J., and Fernandez, X. 2005. Repeated handling of pigs during rearing. II. Effect of reactivity to humans on aggression during mixing and on meat quality. *J. Anim. Sci.* 83:1664–1672.

Terlouw, E.M.C., and Rybarczyk, P. 2008. Explaining and predicting differences in meat quality through stress reactions at slaughter: The case of Large White and Duroc pigs. *Meat Sci.* 79:795–805.

Veissier, I., and Boissy, A. 2007. Stress and welfare: Two complementary concepts that are intrinsically related to the animal's point of view. *Physiol. Behav.* 92:429–433.

Viljoen, H.F. 2011. *Meat Quality of Dark Cutting Cattle*. Master dissertation, University of Pretoria, South Africa (Retrieved March 25, 2014).

Voisinet, B.D., Grandin, T., O'Connor, S.F., Tatum, J.D., and Deesing, M.J. 1997. Bos Indicus-cross feedlot cattle with excitable temperaments have tougher meat and a higher incidence of borderline dark cutters. *Meat Sci.* 46(4):367–377.

Warner, R.D., Ferguson, D.M., Cottrell, J.J., and Knee, B.W. 2007. Acute stress induced by the preslaughter use of electric prodders causes tougher beef meat. *Aust. J. Exp. Agric.* 47:782–788.

Warriss, P.D., and Brown, S.N. 1985. The physiological responses to fighting in pigs and the consequences for meat quality. *J. Sci. Food Agric.* 36:87–92.

Wismer-Pedersen, J. 1987. Water. In Price, J.F. and Schweigert, B.S. (eds.), *The Science of Meat and Meat Products*, 3rd edition. Food and Nutrition Press, Westport, CT, pp. 141–154.

8 Slaughter-Line Operations and Their Effects on Meat Quality

Wiesław Przybylski, Joe M. Regenstein, and Andrzej Zybert

CONTENTS

8.1 INTRODUCTION

Issues with the meat quality of slaughtered animals have been discussed since the pioneering work of Ludvigsen (1953) more than 60 years ago. This researcher studied the problems of water holding capacity (WHC), and the color and textural properties of meat as affected by genetic and environmental factors. These problems have economic consequences for the whole supply chain associated with the production of meat. Calculations for the marketing of Polish pork showed that the economic losses associated with the occurrence of meat defects (pale, soft, and exudative [PSE], dark, firm, and dry [DFD]) during meat distribution as well as those that develop during processing and product packaging constitute about 2.4% of the value of the pig industry (Pospiech et al. 1998). It is estimated that the factors that determine meat quality are about 30% genetic and 70% environmental (Przybylski et al. 2011). About 40% of meat defects are due to events occurring in abattoirs. Among them, key issues are animal stunning, time between stunning and bleeding, bleeding position, method of scalding, and postslaughter carcass chilling (Przybylski et al. 2011).

Ante-mortem handling and slaughter are strongly influenced by the behavioral and physiological status of animals, which is reflected in the final meat quality. Physical activity and prolonged stress before slaughter decrease the muscle glycogen reserves, contributing to a higher ultimate pH in meat (Terlouw 2005). Physical activity and the stress before slaughter have been associated with a faster pH decline (Terlouw 2005; see Chapters 6 and 7). Any stress-induced changes before slaughter have been related to the release of adrenaline into the blood stream, again affecting meat quality. Adrenaline induces glycogen breakdown and postmortem pH decline and raises muscle temperature (Monin 2003). Slaughter procedures are likely to also have a very strong effect on ante- and postmortem metabolism in muscle tissue. The effects of slaughter conditions on behavior, physiology, and subsequent meat quality are variable and depend on factors ranging from the genetic background to the handling of animals before slaughter (Terlouw 2005). So, the quality of the meat obtained from slaughtered animals is an indicator that reflects the stress before slaughter, and during slaughter, which is related to animal welfare (see Chapters 6 and 7).

8.2 GENERAL CHARACTERISTICS OF MECHANICAL, ELECTRICAL, AND GAS ANESTHESIA STUNNING METHODS

Slaughter may take place on the farm or in an abattoir. On-farm slaughter is the traditional way to obtain food, often by killing the animals in the fall so that they provide food over the winter (Figure 8.1). Currently, in most western countries, this type of slaughter is prohibited if the meat is to be sold and the same applies in Central and Eastern Europe. Nevertheless, in some countries, such slaughtering must be reported to a local veterinarian and the meat is still allowed for personal consumption. This

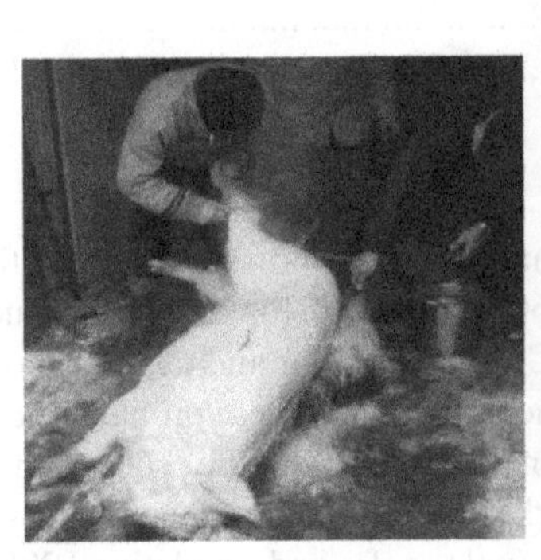

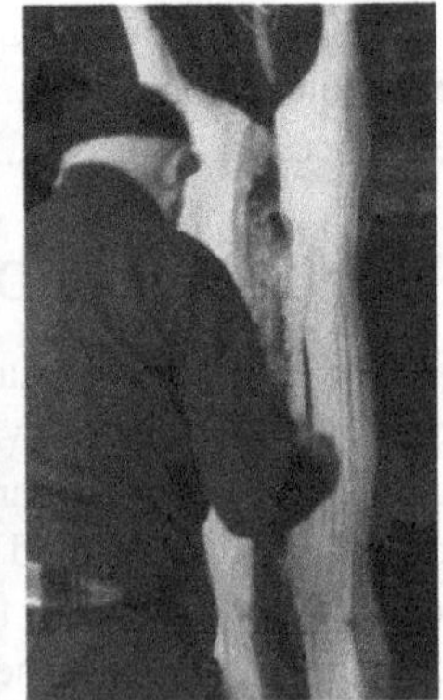

FIGURE 8.1 Slaughter of a pig on the farm. (Photo: W. Przybylski.)

method is likely to give the best meat quality as the animals are less stressed. This is due to the very limited ante-mortem handling of the animals (transport, loading, unloading, etc.) and any handling that is required is often by people and in facilities that are familiar to the animal (Eriksen et al. 2013).

In each country, the procedure for the commercial slaughter of animals is strictly defined by law. For example, in the European Union, the slaughter of animals is regulated by Council Regulations (EC) No 1099/2009 dated September 24, 2009 for the protection of animals at the time of killing. In meat processing plants, three types of stunning prior to bleeding are used: mechanical, electrical, and controlled atmosphere (gas anesthesia). The aim of stunning is to cause a loss of consciousness and feeling in animals prior to sticking and bleeding. Religious slaughter uses the bleeding step to cause rapid unconsciousness and will be described in Section 8.6.

Mechanical or percussive stunning (PS) can be divided into two broad types: penetrative and nonpenetrative (concussive) stunning (Shaw 2004). The penetrative mechanical stunning uses a captive bolt pistol that fires a steel bolt through the animal's skull into the brain. These stunners are normally powered pneumatically or by using blank cartridges. The concussive stunners have a convex steel "mushroom head" plate at the end that impinges onto the head, cracking the skull without penetrating it. Loss of consciousness with this method of stunning is probably the result of a concussion and oscillations inside the skull due to the impact of the stunner (Shaw 2004). This method may be reversible, so the bleeding must be done quickly. The shooting position, for example, in cattle, for frontal stunning has as the place of impact the intersection of two lines from the base of each horn to the eye on the opposite side of the head (Figure 8.2a). According to Shaw (2004), mechanical stunning

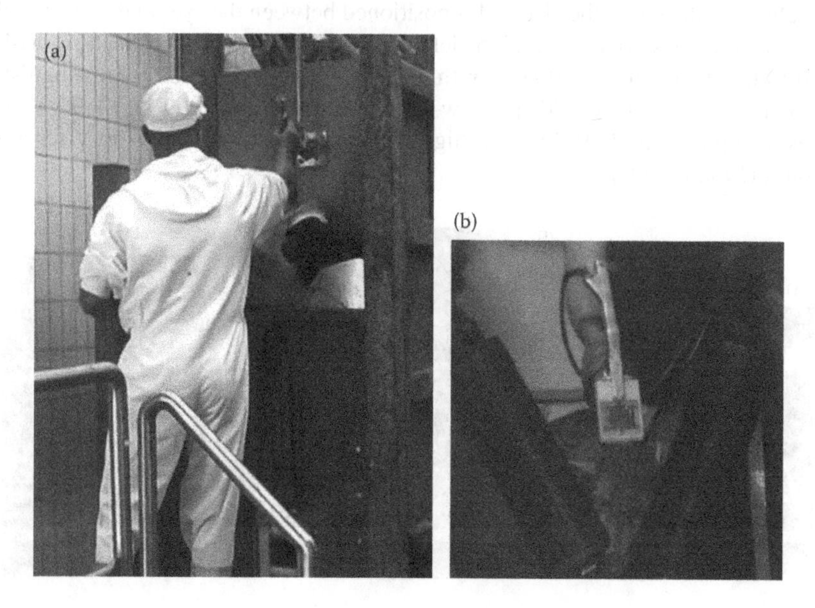

FIGURE 8.2 Stunning of cattle and sheep. (a) Head stunning of cattle with a captive bolt. (b) Electrical head stunning of sheep. (From (a) EURO-PAN. With permission. (b) E.S. Toohey. With permission.)

is humane when it is correctly done with the appropriate equipment. However, the equipment requires consistent maintenance and the workers must be trained to use it properly. Mis-stunning leads to serious animal welfare problems and extreme stress in animals that have reduced meat quality. Current standards (e.g., American Meat Institute [AMI]) allow for up to 5% mis-stunning (Regenstein 2014, unpublished data).

Electrical stunning (ES) is induced by an electrical current passing through the brain, resulting in a brain status similar to a grand mal epileptic-type seizure (Terlouw et al. 2008; Zivotofsky and Strous 2012). The epileptic process is characterized by rapid and extreme depolarization of the membrane potential that induces unconsciousness and leads to the loss of normal neuronal function (Lambooij 2004; Zivotofsky and Strous 2012). Properly conducted ES leads to a rapid loss of consciousness in animals and is regarded as a humane and generally acceptable method (Zivotofsky and Strous 2012). However, again, if done improperly, it can lead to a state of electric shock without actual loss of the ability to feel pain (Zivotofsky and Strous 2012). In small plants, animals are stunned using hand-held tongs (low voltage) or automatically (high voltage) using a V-type restrainer (Figure 8.3a and b). The stunning tongs are applied on both sides of the head between the eyes and ears, or on the back of the head (head-only method); or on the head and withers or loin back (head-to-back method). The former can be a reversible stun while the latter is an irreversible killing stun. In abattoirs where a large number of animals are slaughtered, several automatic ES methods are available with the application of high voltage. The animals are individually moved using "V" or "Midas" restrainers (Figure 8.3b and c). At the end of the restrainer, the animals are automatically brought in contact with the electrodes positioned between the eyes and ears or with a third electrode positioned behind the left shoulder, chest, or brisket (head-to-brisket; head-to-heart) (Lambooij 2004). For this reason, the two methods are distinguished by their electric voltage with the low-voltage method generally at 70–90 V with times of application of 10–15 s and high-voltage method with 250–700 V for a few seconds (Monin 2003).

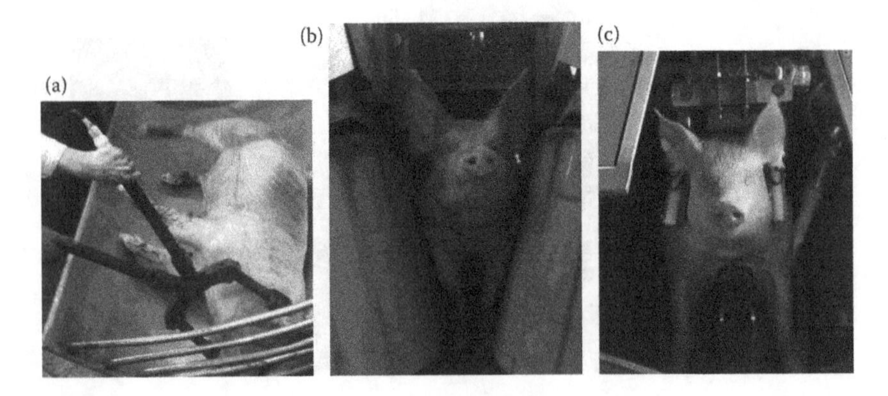

FIGURE 8.3 Electrical stunning of pigs in a slaughterhouse. (a) By hand using tongs. (b) Restrainer "V." (c) Central rail "Midas" automated device with transporter to immobilize. ((a) Photo: W. Przybylski; (b) photo: W. Przybylski; and (c) EURO-PAN. With permission.)

TABLE 8.1
Recommended Minimum Current for Electrical Stunning of Slaughter Animals

Category and Animal Type	Minimum Current (A)		Source
	EU Regulation	Literature	
Cattle		1.2–1.5	Lambooij (2004); Hopkins (2010); Zivotofsky and Strous (2012); Nakyinsinge et al. (2013); Hopkins (2014)
Calves	1.25	0.5–1.5	
Cows	1.28	1.2–3.5	
Steers	1.28	1.5–2.5	
Sheep	1.00	1.00	
Lambs	1.00	>0.5–0.6	
Goat	1.00	0.7–1.0	
Pigs	1.30	1.30	

For worker safety reasons, the latter methods require the use of automatic stunners. Currently, the EU is considering raising the minimum voltage for the low-voltage method. The recommended minimum current for ES of slaughter animals according to EU regulations and described in the literature is presented in Table 8.1. The minimum current is a function of the electrical impedance of the body and is necessary for the occurrence of unconsciousness of animals (Lambooij 2004; Zivotofsky and Strous 2012). Malaysia is a Muslim country that permits mild ES. It suggests slightly different and higher values for the current: 0.5–1.5 A for 2 s for calves, 1.5–2.5 A for 2–3 s for steers, 2–3 A for 2.5–3.5 s for cows, and 2.5–3.5 A for 3–4 s for bulls (Nakyinsige et al. 2013). Most methods of ES use alternating current (AC) at a frequency of 50 Hz with a sinusoidal waveform (Lambooij 2004). ES is also considered a humane method when done right, although again the issue of mis-stuns leads to problems.

The third method of stunning slaughter animals, most commonly used to stun pigs and poultry, is based on the use of gas mixtures (gas anesthesia, pharmacological method) containing CO_2 and air, A, or N_2 (Raj 2004). A CO_2 concentration above 70% with an exposure for 90 s is recommended and most commonly used (Monin 2003; Raj 2004; Grandin 2010). In this method, loss of consciousness occurs through the inhibition of neurons by CO_2 due to the pH falling below 7.1 in the cerebrospinal fluid (CSF). This is probably coupled with an excessive release of γ-aminobutyric acid in the brain, which is the major inhibitory amino acid neurotransmitter (Raj 2004). It should be noted that the inhalation of CO_2 does not cause a reduction of O_2 levels in the blood and there is no anoxia. The inhalation of A and/or N_2 induces anoxia of the brain and unconsciousness (Raj 2004). These solutions (using a mixture containing a low concentration of CO_2 in A and/or N_2) are also considered better than high concentrations of carbon dioxide (Raj 2004). It should be noted that stunning with high concentrations of CO_2 is also stressful for animals (Raj 2004; Gregory 2005; Terlouw et al. 2008; Grandin 2010), but in comparison with other methods, it is often recommended as being less stressful for animals (Gregory 2005;

Terlouw et al. 2008). This method also improves carcass and meat quality in the case of pigs and poultry (Raj 2004; Terlouw et al. 2008).

8.2.1 Effect of Stunning Methods on Meat Quality of Slaughter Animals

8.2.1.1 Pigs Stunning Methods and Pork Quality

Currently, pigs are usually stunned using electrical current or gas anesthesia. Captive bolt stunning of pigs is very stressful and should be discouraged (Overstreet et al. 1975; Althen et al. 1977). Burson et al. (1983) demonstrated that pigs stunned with a captive bolt had a higher incidence of blood splash than those where electrical or CO_2 stunning was used. In the study of Bertram et al. (2002), when a captive bolt pistol was used on pigs, this caused higher drip loss in comparison to the other methods of stunning (CO_2, electrical). Of all the farm animals, pigs are the most susceptible to stress during preslaughter handling and stunning (Raj 2004).

In smaller abattoirs, the most commonly used ES equipment is hand-held tongs applied with low voltage and amperage, while larger plants generally use a high-voltage automatic system. Studies on the influence of voltage on meat quality most often show that pigs stunned by high voltage have an increased incidence of PSE meat (from 220 to 430 V) or more extensive glycolysis, resulting in a low ultimate pH (400 vs. 700 V) (Gregory 2008; Przybylski et al. 2011). Many studies have shown that ES may result in acceleration of postmortem glycolysis, resulting in a rapid pH decline postslaughter (Velarde et al. 2000; Channon et al. 2003). Although the study of Fauciano et al. (1998) showed higher incidences of PSE meat in pigs stunned using low voltage (110 vs. 200 V), the effect was attributed to poor handling in lairage and a delayed stun-to-stick interval and also to the low voltage, resulting in insufficient stunning. The research of Terlouw and Rybarczyk (2008) indicated that the reaction to slaughter stress also depends on the breed of pig. In that study, slaughter conditions influenced ultimate pH and drip loss more with Large White pigs than with Durocs. This was explained by the high preslaughter heart rate and higher early postmortem metabolism observed in Large Whites. Stunning with high voltage can also cause an increase in the occurrence of ecchymosis and petechial hemorrhages (Channon et al. 2003; Monin 2003).

Another disadvantage of high-voltage and automatic ES is the necessity to use a single-line restrainer. There are two types of restrainers commonly used, either two slightly inclined conveyer belts ("V" restrainer; see Figure 8.3b) or a central rail (Midas) (Figure 8.3c) (Terlouw et al. 2008). Studies on a comparison of these two types of restrainers have demonstrated that the "V" restrainer is more stressful for pigs, causes a sharp rise in heart rate, and increases the risk of PSE defects (Chevillon 2001). In some abattoirs using automatic ES, a third electrode is applied on the back or chest to cause cardiac fibrillation or arrest, to avoid animals regaining consciousness (Terlouw et al. 2008). In a study by Vogel et al. (2011) using head/heart ES with both voltage and amperage control (313 V, 2.3 A, 3 s application of stunning wand behind the ears followed by 3 s application of stunning wand to the ventral region of the ribcage directly caudal to the junction of the humerus and scapula while the stunned pig was in lateral recumbency), this was demonstrated to be a better method in comparison to head-only stunning (313 V, 2.3 A, 6 s). The head/heart method reduces the incidence of signs of return to sensibility without significant effects on meat quality and also less risk of ecchymosis. A study by Channon et al. (2002) indicated that manual ES of pigs using head-only tongs (1.3 A for 4 s) reduced the percentage drip loss, reduced muscle lightness, and reduced the rate of muscle pH decline compared with pigs manually and electrically stunned using head-to-brisket stunning (1.3 A for 4 s). In another study, head-to-back manual ES (1.3 A for 4 s) increased the incidence of PSE defects in the *Longissimus thoracis et lumborum* and *Biceps femoris* muscles and the incidence of fractures in comparison to head-only stunning (1.3 A for 4 s) (Channon et al. 2003). Channon et al. (2003) also showed a significant effect of current level and the time of stunning on pork quality. In this study, the effect of different current levels (0.9, 1.3, and 2.0 A) was evaluated as was time of application (4 and 19 s). The results showed that higher currents and longer times increased the incidence of PSE defects in the *Longissimus thoracis et lumborum* and *Biceps femoris* muscles and ecchymosis in the shoulder.

In Europe, a CO_2 method for pig stunning is actually the most popular in large abattoirs. A number of studies have shown that stunning pigs with CO_2 compared with ES leads to a lower incidence of bone fractures, ecchymosis, petechial hemorrhages, and PSE, and also improves meat quality (Channon et al. 2000, 2002, 2003; Velarde et al. 2000, 2001; Monin 2003; Becerril-Herrera et al. 2009; Przybylski et al. 2011). The studies of Channon et al. (2000) and Velarde et al. (2001) demonstrated significant improvements in pork quality even in pigs carrying the halothane gene. CO_2 in comparison to ES reduces the incidence of PSE defects, ecchymosis, and petechiae in the meat of pigs with the Nn ($RYR1^T RYR1^C$) genotype. Nevertheless, the studies of Hambrecht et al. (2003, 2004a,b) suggested that CO_2 stunning is not a guarantee of better meat quality in terms of color and water holding properties. These discrepancies and differences may result from the application of different concentrations of CO_2 and various exposure times of animals to the anesthesia. There is also a broad debate about the humaneness of this method of stunning at high concentrations of CO_2 (Monin 2003; Terlouw et al. 2008; Grandin 2010). Raj and Gregory (1995) showed that food-deprived pigs refuse to enter an area containing 90% CO_2 to obtain a food reward. Generally, abattoirs use 70%–90% CO_2. The minimum concentration that is usually recommended is 80% (according to EC Council Regulations No 1099/2009) and at such a concentration, pigs lose their standing posture after about 22 s (Gregory 2008; Terlouw et al. 2008). Further research has shown that 90% CO_2 causes less physical stress, improves meat quality, and is more effective, but it can also cause behavioral aversion mentioned above (Nowak et al. 2007; Gregory 2008; Grandin 2010; Antosik et al. 2011). The study by Nowak et al. (2007) found that stunning with 80% or 90% CO_2 is acceptable for animal welfare only in combination with a longer exposure time of 100 s and shorter stun-to-stick times of 25–35 s. In the case of 90% CO_2, the delay time (40–50 s) to stick is also acceptable in terms of animal welfare. In conclusion, better meat quality is obtained from stunning with 90% CO_2 and 100 s exposure time and a stun-to-stick interval of up to about 40–50 s.

8.2.1.2 Cattle Stunning Methods and Beef Quality

The most popular method used in many countries for adult cattle and calf stunning is the captive bolt, although in some countries, several abattoirs use an automatic ES system (Önenç and Kaya 2004; Gregory 2005; Piotrowski et al. 2006; Terlouw et al. 2008). The effectiveness of the first method on rate and duration of unconsciousness depends on a number of factors such as the correct immobilization of the animal, the correct choice of cartridge, the way the device is applied to the skull, giving the bolt an appropriate direction of movement, and the technical efficiency of the shooting device (Piotrowski et al. 2006; Gregory et al. 2007; Terlouw et al. 2008). The data show that only about 30% of the beef plants in the United States could stun 95% of the cattle correctly with one shot from a captive bolt (Grandin 2010). The shooting position on the head is located at the cross-over point between imaginary lines drawn between the base of the horns and the opposite eyes (Figure 8.2a). The penetrative captive bolt stunning method results in a higher level of central nervous system tissue contamination than nonpenetrating sledgehammer stunning (Gregory 2005; Lim et al. 2007). This contamination increases the risk of the spread of bovine spongiform encephalopathy (BSE) prions from the brain and spinal cord to edible tissue and the possibility of human infection to a new variant of Creutzfeldt–Jakob disease (Lim et al. 2007; Hopkins 2014). For this reason, it is better to use a percussive, nonpenetrative mushroom-headed, captive bolt device. But these are harder to use correctly, require head constraint, and have a higher degree of mis-stun. According to Hopkins (2014), the following criteria must be achieved for an effective stun using a captive bolt: collapse of the animal, no corneal reflex, eye balls not rotated, and absence of normal rhythmic breathing. Research has shown that for the effective stun of bulls, more than one shot is required in comparison to steers or heifers (Gregory et al. 2007). However, a very strong stroke during stunning could influence the level of released stress hormones and have negative effects on meat quality (Hopkins 2014).

ES may also be used with cattle. For young calves and adult cattle immobilized in a restraint system, manual stunning can be applied (Terlouw et al. 2008). For less calm animals or in abattoirs with a large number of animals, an automatic system should be used. In this system, two electrodes are applied to the head (e.g., 280 V for 4 s) and then a third electrode in the cardiac region that induces cardiac ventricular fibrillation and cardiac arrest (the total time is then 20 s). ES also has some disadvantages. For example, it may induce tonic/clonic seizures, making it difficult to correctly stick and bleed (Wotton et al. 2000).

Research has shown that cattle stunning methods may affect meat quality (Önenç and Kaya 2004; Gregory 2005; Piotrowski et al. 2006; Terlouw et al. 2008). The results of Piotrowski et al. (2006) showed that ES (three-electrode device with two main electrodes applied to the head—350 V, 2.5 A, 50 Hz, and after 4 s, the third electrode is applied to the chest 2–2.5 A—20 s) of both bulls and heifers significantly increased the extent of pH decline in the *Longissimus thoracis et lumborum* muscle shortly after slaughter and also 24 h postmortem in comparison to animals stunned mechanically using the radical apparatus (penetrating method). These differences were explained as being due to the partial electrostimulation of carcasses during the stunning process. Better results for degree of bleeding were also observed in animals stunned electrically.

Önenç and Kaya (2004) used different stunning methods to study beef quality. Young Holstein Friesian bulls were stunned with either ES (400 V, 1.5 A, 10 s, tongs applied on both sides of the head, behind the ears) or PS using a nonpenetrative stunner, mushroom-headed captive bolt, while the controls were not stunned (NS). Animals from the ES and PS groups had higher muscle glycogen concentrations compared to NS animals. The rate of muscle pH decline was significantly faster in PS compared with NS and ES. There were no significant differences in ultimate pH, WHC (filter paper press method), or color measures between the groups. Cooking loss was higher in stunned groups compared to the control group. Sensory quality of beef (odor, flavor, and tenderness) was better or not different in the percussive captive bolt-stunned animals in comparison to ES-stunned animals. However, these differences are difficult to explain. The authors concluded that percussive captive bolt stunning of cattle improved meat quality compared with cattle electrically stunned using head-only tongs and those nonstunned under approved Turkish slaughter procedures. Bourguet et al. (2011) showed that conventionally stunned cattle with a penetrating captive bolt had a higher pH 1 h after bleeding in the *Longissimus thoracis et lumborum* muscle than traditional halal-slaughtered cattle. But these differences could also be attributed to other factors such as unloading and traceability procedures, time spent in abattoir, time spent in the slaughter corridor, electrical prods during entrance in the box, and so on. This study showed that differences in slaughter procedures may influence postmortem muscle metabolism and that breed and gender influence behavioral and/or physiological reactions.

8.2.1.3 Sheep Stunning Methods and Sheep Meat Quality

Sheep can be stunned by application of mechanical, electrical, and CO_2 anesthesia methods. The electrical methods are considered the most economical and effective for sheep stunning. The most common method uses a pair of tongs (or electrodes) placed on both sides of the head (Bórnez et al. 2009a) (Figure 8.2b). Head-only ES in sheep can lead to serious poststunning convulsions in the animal and can increase the rate of postmortem muscle glycolysis (Velarde et al. 2003). Moreover, unadvisable (e.g., incorrect parameters of electrical current) ES can cause cardiac dysfunction, circulatory arrest, convulsions, and fractures (Kirton et al. 1981; Gregory 2005). Some authors have reported that ES caused ecchymosis (blood splash) in muscle and petechial hemorrhages (speckle) in connective tissue and muscle fascia (Lister et al. 1981; Gregory 2005). It is often necessary to use pointed wet electrodes to penetrate the wool and to establish good contact with the skin (Terlouw et al. 2008). Factors favoring the occurrence of these defects can be preslaughter stress (i.e., hot weather) and ineffective stunning that leads to higher blood pressure. Others have indicated that the occurrence of petechial hemorrhages increases with the length of time between stunning and bleeding (Kirton et al. 1980/1981). Sticking very quickly after stunning minimizes the incidence of hemorrhages (Kirton et al. 1978). The study of Velarde et al. (2003) suggested that meat quality and the incidence of hemorrhages are unaffected by head-only ES of light lambs of the Ripollesa breed, for 3 s at a constant voltage of 250 V, 50 Hz sinusoidal AC. This study also showed that blood lost relative to body weight was significantly higher in electrically stunned lambs in comparison to nonstunned (4.6% vs. 4.3%) (Velarde et al. 2003).

Another study on ES (125 V, 10 s, electrodes applied on both sides of the head behind the ears) of lambs of the Manchega breed did not show any significant effect on the decrease in pH, WHC, shear force, and color parameters in comparison to nonstunned animals (religious slaughter) (Vergara and Gallego 2000).

A number of recent studies have shown that stunning with CO_2 could be an alternative stunning method for lambs (Bórnez et al. 2009a,b; 2010; Vergara et al. 2009). These studies focused on evaluating the effectiveness of four stunning treatments using different CO_2 concentrations (80% and 90%) and exposure times (60 and 90 s) in comparison to ES (110 V, 50 Hz, 5 s, plate electrodes applied on both sides of the head behind the ears) of light (12.8 kg live weight, 30 days old) lambs of the Manchego breed. For those animals correctly stunned, a concentration of 90% CO_2 for 60 s was considered most effective (100% correctly stunned lambs). Nevertheless, this approach seems to be too slow for high-throughput chains slaughtering large numbers of sheep. Other studies that compared different methods of gas-stunning on meat quality indicated that high CO_2 concentrations (90%) could be considered as the best method (no differences between exposure times of 60 or 90 s were found) as the meat was more stable with aging time in terms of color and WHC (Bórnez et al. 2009b). Vergara et al. (2009) using the same methods found no differences in color, WHC, and drip loss between methods in Manchega lambs slaughtered at 25 kg live weight (70 days old). Only meat from the group with 80% CO_2 for 90 s appeared to be less tender (higher shear force). The authors mentioned that the effects of gas stunning methods on meat quality depended on lamb age, because in lighter and younger lambs (in a study of Bórnez et al. 2009b), less tender meat was reported in electrically stunned animals.

8.3 IMMOBILIZATION AFTER STUNNING

Very often, after stunning, the animals begin to move, convulse, and kick. This is a risk to staff safety because it increases the risk of injury by knives and other equipment being kicked, that is, during hoisting on the overhead rail and during initial skinning. There are four methods used to reduce animal movement after stunning that can improve worker safety (Gregory 2004). These are pithing (destruction of the integrity of the spinal cord by using a flexible pithing rod inserted into the brain through the hole created by the captive bolt), severing the spinal cord, spinal discharge, and electroimmobilization (Gregory 2004). The first three methods are rarely used, for example, pithing is currently prohibited because it is considered inhumane and also increases the risk of BSE contamination (Gregory 2004). Some abattoirs use electroimmobilization for reducing carcass convulsions and kicking (Gregory 2004; Hopkins 2014). Most frequently, this method is used for cattle and sometimes for sheep and in these cases, the electric current flows through the body after stunning (Figure 8.4). The carcasses of animals are usually placed in a horizontal position and

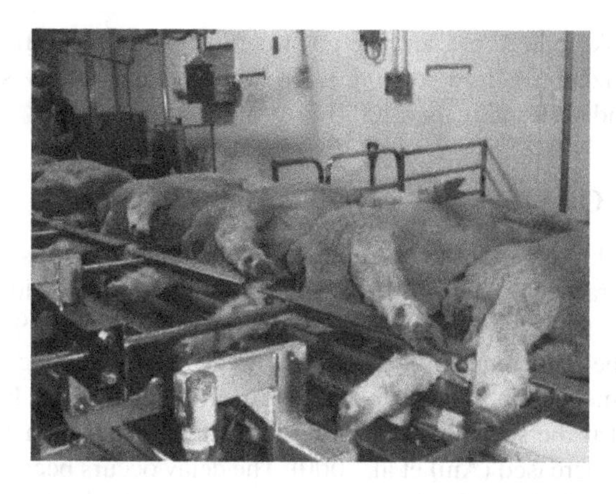

FIGURE 8.4 Immobilization of sheep immediately post-exsanguination. (From E.S. Toohey. With permission.)

treated with an electric current using varying parameters (Table 8.2). For example, in New Zealand, in case of cattle stunned using the head-only method, electroimmobilization reduces animal movement and allows safe dressing of the carcass (Devine et al. 1986; Hopkins 2014). Recently, a new approach was applied in Australian sheep (Figure 8.4) and cattle. In the case of cattle, after stunning, the animal is placed on a V-bed (Table 8.2), which is lined with electrodes. This method significantly reduces animal movement after stunning, allowing safe shackling and exsanguination, thus reducing the risk of injuries and improving worker safety. Evidence indicates that this application does not have any detrimental effect on meat quality, particularly pH (Hopkins 2014). In the case of sheep, the research of Toohey and Hopkins (2007) also showed a positive effect of immobilization on reducing animal movement immediately after exsanguination. The study showed no significant effect of high-frequency immobilization on meat quality traits such as sarcomere length, drip loss,

TABLE 8.2
Parameters of Electrical Current Applied for Immobilization of Slaughter Animals

Slaughter Animal Species	Parameters of Electrical Current	Source
Cattle	20 s after stunning, 80 V, 14.3 Hz, 300 mA for 30–37 s via electrodes connected to the nose and anus	Hopkins (2014)
Cattle	Constant current, 2000 Hz, 1–2 A, pulse with 0.15 ms applied for up to 15 s	
Sheep	Exposure time 25–35 s, 2000 Hz, 400 V, 9 A over 7 animals, pulse width of 150 μs	Toohey and Hopkins (2007)
Cattle and sheep	Application less than 30 s, 80–100 V DC at 15 pulses per second with a 5-ms pulse duration	Gregory (2004)

shear force, cooking loss, pH, and the color parameters of meat. In some countries, electroimmobilization before sticking is prohibited, since poorly stunned animals can feel pain and suffer (Gregory 2004).

8.4 EFFECT OF STICKING AND BLEEDING

Sticking with a knife is done after stunning to open blood vessels. This is followed by blood loss, brain hypoxia, and the death of the animal. It should be done as soon as possible after stunning to reduce blood splash (Burson et al. 1983; Bolton 2004) and to reduce the occurrence of other defects in the meat. The time from stunning to sticking recommended in practice for pigs is generally about 20 s. It has been shown in research that if the stunning-to-sticking interval is longer than 15 s, the chances of recovery are increased (Anil et al. 2000). The delay occurs because of the need to shackle the stunned pigs and hoist it onto an overhead rail prior to sticking in case of exsanguination in the vertical position. The time to lose brain responsiveness following effective sticking should on average be 18 s (Anil et al. 2000, quoted by Wotton and Gregory 1986). Bleeding is most often done with pigs hanging vertically by one leg or in the normal horizontal position. Orientation of the carcasses during bleeding influences its rate and efficiency, and muscle hemorrhages (Gregory 2005). Bleeding of pigs in the vertical position leads to convulsions followed by PSE in the ham from the shackled leg side (Pisula and Florowski 2011), especially with late sticking or if the sticking is incorrectly done. As a result of late sticking, some pigs have been found to be showing signs of recovery (return of rhythmic breathing) during the bleed out (Anil et al. 2000).

The study of Anil et al. (2000) in abattoirs with low (120 V) and high (200 V) voltage stunning of pigs showed that less than half of the sticking operations were regarded as trouble-free. The rest were found to be either delayed or the pigs were difficult to handle because of carcass convulsions. The authors observed short (4.5 cm) and long (11.2 cm) sticking cuts. The long sticking cuts resulted in a more rapid rate of exsanguination of pigs than with short cuts. The results of this study indicated that following head-only ES, a relatively long thoracic cut should result in a more humane slaughter. Consequently, this can also lead to better meat quality because it provides better welfare for slaughter pigs.

According to Velarde et al. (2000), in sheep, the length of time between stunning and bleeding affects the occurrence of hemorrhages. Sticking very quickly after stunning minimizes their occurrence (Kirton et al. 1978). Another study showed that delayed bleeding increased the amount of residual blood in beef carcasses and increased carcass weight and heart activity. The additional residual blood could increase protein, Fe content, and yield of meat (Vimini et al. 1983a,b), but on the other hand may cause the meat to be darker. Hopkins et al. (2006) have demonstrated that the use of both a thoracic stick (severing the vena cava and the aorta) and application of electrical current (10 or 14 Hz) to lamb carcasses immediately after exsanguination will increase the amount of collectable blood and this has the potential to both reduce the biological oxygen demand of abattoir waste and provide additional income for those abattoirs that process blood. The additional benefit of this solution is the production of lighter and redder meat.

8.5 EFFECT OF SCALDING, DEHAIRING, SINGEING, AND DEHIDING

When pig carcasses are not skinned, the hairs are removed by using two different techniques. Scalding is used in many countries (Bolton 2004; Pisula and Pospiech 2011). Traditionally, this is achieved by immersing the carcasses in hot water in a scald tank or using steam in a vertical scalder. High temperature causes the denaturation of proteins and facilitates the removal of hair. The temperature of the water should be 60–70°C for 5–10 min. The combination of time and temperature depends on pig breed and season. Extended scalding can increase carcass temperature. Gardner et al. (2006) showed that extending scalding from 5 to 8 min caused an increase in ultimate pH in *Semimembranosus* and *Biceps femoris* muscles and Warner–Bratzler shear force of broiled loin chops, measured using a texture analyzer. The scalding bath is often followed by mechanical agitation to remove the hair and this may also account for quality decreases. Troeger and Woltersdorf (1986) reported that thermal and mechanical processes associated with scalding and hair removal accelerated biochemical reactions in the muscle and can influence protein denaturation and also water-holding properties of meat.

Scalding also improves carcass hygiene by reducing bacterial numbers and the incidence of pathogens such as *Salmonella*. Pearce et al. (2004) showed a reduction in the incidence of surface *Salmonella* from 31% to 1% during scalding for approximately 8 min using a scald tank at a temperature of about 61°C.

Other abattoirs use singeing to remove pigs' hair. The pigs' bodies go through a series of gas burners before they are scraped and brushed in the usual way (Figure 8.5). This technique has the reputation among meat processors and butchers of producing

FIGURE 8.5 Gas singeing of pig carcasses. (Photo: W. Przybylski.)

pork with a better meat quality and longer shelf life than the meat from scalded carcasses (Monin et al. 1995). Chilling loss was reported to be slightly higher in scalded carcasses (Monin et al. 1995) and singed carcasses showed a slightly higher ultimate pH in the *Semimembranosus* and *Adductor femoris*, and lower reflectance and higher WHC (measured by a imbibition time method) in the *Biceps femoris*. The overall meat quality was significantly higher in singed carcasses. In other studies on the production of dry-cured hams from pigs dehaired by scalding or singeing, it has been shown that scalded hams lose more weight than the singed ones during processing (Monin et al. 1997). The salt content was higher in scalded hams; they were saltier (sensory assessment) and had more pronounced aromas of dry ham, rancidity, and hazelnut, and less aroma of fresh meat (Monin et al. 1997).

In some countries, the skins are removed from the pig without scalding or immediately after scalding. Scalding at temperatures below 60°C gives skins of better quality for the manufacturing of leather (Mowafy and Cassens 1975). Nevertheless, leather produced from unscalded pigs is thicker and has greater tensile strength than that produced from skins that have undergone the scalding process (Bolton 2004). Maribo et al. (1998) compared the meat quality of pig carcasses that had been scalded, singed, or dehided without hair removal and then fast chilled. The results of this work show that dehided carcasses had the lowest drip loss, internal reflection values, and darkest meat color, but also had the toughest meat, compared to the scalded carcasses. The authors suggested that the results showed the importance of reducing the temperature as early as possible postmortem to reduce protein denaturation and improve meat quality (drip loss, meat color), but this also increased the risk of toughening when chilling conditions are too fast. Dehiding can be recommended as a possible method of improving meat quality. To eliminate the risk of tough meat, an aging process must be applied.

8.6 RELIGIOUS SLAUGHTER

The Muslim and Jewish religions are based on a series of laws. For them, the consumption of food is regulated by a detailed set of laws that are part of a larger set of laws that influence all aspects of their life. To be able to understand the impact of these laws on the quality of meats, it is necessary to understand the details of a few of these laws to see how these affect the slaughter of animals. For Muslims, the major concern with respect to meat is whether the religious slaughter itself is considered acceptable. For Jews, that is only a small part of the process of preparing meat from acceptable animals for consumption. The commercially relevant kosher animals include beef, sheep, and goat, although water buffalo, bison, deer, and elk may also be slaughtered regularly as kosher. Muslims accept all of the previously mentioned animals and may also slaughter camel and rabbit, and some Muslims accept horse. A comprehensive review of both kosher and halal for readers in the broader context of the food industry can be found in Regenstein et al. (2003).

The kosher (kashrus) dietary laws determine which foods are "fit or proper" for consumption by Jewish consumers and are established in the first five books of Hebrew scripture (Torah). The Muslim laws that determine which foods are "lawful" are established in the Quran. Both kosher and halal meat slaughter and processing requires methods that are different from the normative activities of the meat industry in the western world.

8.6.1 Allowed Animals

For kosher, ruminants with split hoofs that chew their cud, the traditional domestic birds, and fish with fins and removable scales, and a few jumping insects (locust) are generally permitted while all other animals are excluded. For Muslims, the pig is uniquely prohibited, while animals like the camel and rabbit are permitted. Generally, the acceptable animals for halal are broadly animals that would be classified as essentially herbivores.

Judaism has many animal welfare laws dealing with the importance of not harming an animal (Tzar Balay Chayim) and these laws are reflected in the importance of properly slaughtering an animal so as to minimize pain. It is also important that slaughter be done with respect for the animal. The Muslim community also takes animal welfare extremely seriously and at slaughter requires that no animal see another animal being slaughtered nor should they see the knife being sharpened. They also are concerned that the animals not be fed any filth, such as animal by-products, particularly prior to slaughter when under the control of farmers.

8.6.2 Prohibition of Blood

Blood as the life fluid is prohibited in both religions. Thus, in addition to animal welfare considerations, the goal is to assure that the slaughter of the animals leads to the removal of as much blood as possible; thus, the emphasis on the animal being alive with their heart pumping at the time of slaughter. Ruminants and fowl must be slaughtered according to Jewish law by a specially trained religious slaughterman (the shochet) using a special knife designed specifically for the purpose of slaughter (the chalef). The knife must be extremely sharp and have a very straight blade that is at least twice the diameter of the neck of the animal to be slaughtered. This is now being recognized as an important consideration with respect to religious slaughter of animals and is slowly moving into western regulatory language. A great deal of training goes into making sure the knife is razor-sharp and absolutely free of nicks. The knife is checked by running its entire working blade along a finger nail both before and after each slaughter of ruminants.

The shochet will rapidly cut the jugular veins and the carotid arteries along with the esophagus and trachea without burrowing, tearing, or ripping the animal. "This process when done properly leads to a rapid death of the animal, a sharp cut is also known to be less painful" (Grandin and Regenstein 1994). Before the slaughter occurs, the shochet will quickly check the neck of the animal to be sure it is clean so that there is nothing on the neck that could harm the knife. If there is a problem, the neck of the animal needs to be washed and this slows the process. During the actual slaughter, the following five kosher laws must be observed:

No Pausing (Shehiyyah): Can be multiple continuous strokes. (Muslim requirements are similar.)

No Pressure (Derasah): Concern that the head falls back on the knife.

No Burrowing (Haladah): The knife has to be doing its job by cutting.

No Deviating (Hagrama): There is a correct area for cutting. Work by Dr. Temple Grandin (personal communication to Dr. Regenstein) suggests that the upper limit of this allowed range is the best for overall animal welfare. The area for cutting the neck under Muslim laws is similar. The problem is that a higher cut than permitted may cut the larynx and the bones that are in it which may harm the knife.

No Tearing (Ikkur): If the neck is stretched too tight, tearing may occur before the cutting. If it is too loose, then pressure on the knife may occur. It is for this reason that Dr. Grandin has developed a special head holder for religious slaughter that is designed so the head holder is away from the cutting area on the neck, gives the right amount of tension, and allows the eyes to be observed as the eyes are the last organ in the head to become insensible.

The shochet checks the chalef before and after the slaughter of each animal. If any problem occurs with the knife, the animal becomes treife, that is, not kosher. The shochet also checks the cut on the animal's neck after each slaughter to make sure it was done correctly.

All adult Muslims with normal mental health are permitted to slaughter animals. They are taught to use a sharp knife. The traits of this knife and its sharpness have not been specified to date. All of the preliminary steps to slaughter must be optimized to ensure that animals will be "unstressed" at the time of slaughter. This is important for all slaughter systems to ideally have calm animals so one has a higher quality meat. Various systems and restraint equipment are available to help improve the religious slaughter of animals.

For over 10 years, the AMI slaughter standards have called for an upright religious slaughter of animals, using one of the many restraining devices available for this purpose. For some groups within the Jewish and Muslim communities, however, upright slaughter may be unacceptable. Upside-down slaughter is preferred as better reflecting traditional slaughter and/or is felt to be better in assuring that the rules of religious slaughter of animals are not violated. The AMI standards have been modified to account for this need, although the long-term goal is to move to upright religious slaughter.

In all cases, almost all of the animal welfare and quality requirements are not directly related to meeting religious requirements. These requirements can and should be met so as to lead to meat that is equal in quality to that of meat slaughtered using secular methods. For both humane and safety reasons, plants that conduct religious slaughter of animals need to install and properly maintain modern restraining equipment to hold the animal in place and ideally to hold the head in place to prevent head movement during slaughter. Having the animal's feet off the ground with a comfortably supported animal seems to have a calming effect on the animal and therefore is highly recommended (personal observations of Dr. Regenstein). Regardless of the system used for the religious slaughter, the animals must be allowed to become unconscious and bleed out and be completely insensible before any other slaughter procedures are performed for quality, humane, and religious reasons.

Prior to undertaking a period of slaughter, the shochet says a prayer. The Muslim slaughter man, on the other hand, will say a prayer over each animal. In both cases, the animal is not stunned prior to slaughter, which is controversial in the western world where the requirement for stunning has become a matter of secular faith. Many people simply cannot believe that a well-done religious slaughter may actually be better than current stunning systems where the current AMI standard for secular slaughter permits 5% of the animals to be mis-stunned on the first try and the actual figures in the best slaughter houses in the United States tend to be between 1% and 2% (Temple Grandin, personal communication). Having an animal mis-stunned will lead to an animal under very high stress, which is not good animal welfare and will have a negative effect on meat quality. If the slaughter is done in accordance with Jewish or Muslim law and with the highest standards of modern animal handling practices, the animal will die without showing any signs of stress (Grandin and Regenstein 1994) and the number of failures from an animal welfare point of view should be quite low although more work is always needed to lower it further.

It is often difficult to separate the impact of preslaughter handling from those aspects directly related to religious slaughter itself with respect to animal welfare or meat quality. There are two important issues that need to be considered when evaluating different forms of religious slaughter of animals. They are the stressfulness of each restraint method and the potential pain perception during the incision, which may depend on the sharpness of the knife and the absence of nicks until unconsciousness (loss of posture), but not to insensibility (loss of the eye reflex). Poorly designed systems, whether for religious or conventional slaughter, can cause great stress. Many stress problems are caused by rough handling and by excessive use of electric prods. The very best mechanical systems will cause distress if operated by abusive, uncaring people. This will affect the quality of the meat as discussed elsewhere in this chapter. In Europe, there has been much concern about the stressfulness of restraint devices used for both conventional slaughter (where the bovid is stunned) and religious slaughter. Ewbank et al. (1992) found that cattle restrained in a poorly designed head holder (i.e., where over 30 s was required to drive the animal into the holder), had higher cortisol levels than cattle stunned with their heads free. Cattle will voluntarily place their heads in a well-designed head restraint device that is properly operated by a trained operator (Grandin 1992). Tume and Shaw (1992) reported a very low cortisol level, about 15 ng/mL, in cattle during stunning and slaughter. Their measurements were made with cattle held in a head restraint (personal communication via Dr. Grandin who spoke to Shaw in 1993). Cortisol levels during on-farm restraint of extensively reared cattle range from 25 to 63 ng/mL (Mitchell et al. 1988; Zavy et al. 1992). Thus, some of the treatments given to animals on the farm were more stressful than slaughter!

For ritual slaughter (and for captive bolt stunning with a nonpenetrating stunner), devices to restrain the body are strongly recommended. Animals remain calmer in head restraint devices when the body is also restrained. Stunning or slaughter must occur within 10 s after the head is restrained (Grandin and Regenstein 1994). The variable of reactions to the incision must be separated from the variable of the time required for the animal to become completely insensible. Recordings of EEG or evoked potentials measure the time required for the animal to lose consciousness.

They are not measurements of pain. Careful observations of the animal's behavioral reactions to the cut are one of the best ways to determine if cutting the throat without prior stunning is perceived as painful by the animal. Observations of over 3000 cattle and formula-fed veal calves were made in three different U.S. kosher slaughter plants. The plants had state-of-the-art upright restraint systems. The systems have been described in detail by Dr. Grandin (1988, 1991, 1992, 1993, 1994). The cattle were held in either a modified American Society for the Prevention of Cruelty to Animals (ASPCA) pen or a double rail (center track) conveyor restrainer. Very little pressure was applied to the animals by the rear pusher gate in the ASPCA pen and head holders were equipped with pressure-limiting devices. The animals were handled gently and calmly and blood on the equipment did not appear to upset the cattle. They voluntarily entered the box when the rear gate was opened and some cattle licked the blood. In all three restraint systems, the animals had little or no reaction to the throat cut and there was a slight flinch when the blade first touched the throat. This flinch was much less vigorous than an animal's reaction to an ear-tag punch and there was no further reaction as the cut proceeded and both carotids were severed in all animals. Some animals in the modified ASPCA pen were held so loosely by the head holder and the rear pusher gate that they could have easily pulled away from the knife. These animals made no attempt to pull away. In all three slaughter plants, there was almost no visible reaction of the animal's body or legs during the throat cut. Body and leg movements can be easily observed in the double rail restrainer because it lacks a pusher gate and very little pressure is applied to the body. Body reactions during the throat cut were much fewer than the body reactions and squirming that occurred during testing of various chin lifts and forehead hold-down brackets. Testing of a new chin lift required deep, prolonged invasion of the animal's flight zone by a person. Penetration of the flight zone of an extensively raised animal by people will cause the animal to attempt to move away (Grandin 1993). The throat cut caused a much smaller reaction than penetration of the flight zone. It appears that the animal is not aware that its throat has been cut. Bager et al. (1992) reported a similar observation with calves. Further observations of 20 Holstein, Angus, and Charolais bulls indicated that they did not react to the cut. The bulls were held in a comfortable head restraint with all body restraints released. They stood still during the cut and did not resist head restraint. After the cut, the chin lift was lowered, the animal either immediately collapsed or it looked around like a normal alert animal. Within 5–60 s, the animals went into a hypoxic spasm and sensibility appeared to be lost. Calm animals had almost no spasms and excited cattle had very vigorous spasms, which may contribute to blood splash. Calm cattle collapsed more quickly and appeared to have a more rapid onset of insensibility (Temple Grandin, personal communication). Munk et al. (1976) reported similar observations with respect to the onset of spasms. The spasms were similar to the hypoxic spasms that occur when cattle become unconscious in a V-shaped stanchion due to pressure on the lower neck. Observations in feed-yards during handling for routine husbandry procedures indicate that pressure on the carotid arteries and surrounding areas of the neck can kill cattle within 30 s (Grandin and Regenstein 1994).

The details spelled out in Jewish law concerning the design of the knife and the cutting method appear to be important in preventing the animal from reacting to the cut. The fact that the knife is razor sharp and free of nicks may be the critical factors in getting the reactions detailed above. As previously mentioned, the cut must be made continuously without hesitation or delay. It is also prohibited for the incision to close back over the knife during the cut. This is called "covering." The prohibition against covering appears to be important in reducing the animal's reaction to the cut. Further observations of kosher slaughter conducted in a poorly designed head holder, that is, one that allowed the incision to close back over the knife during the cut, resulted in vigorous reactions from the cattle during the cut. The animals kicked violently, twisted sideways, and shook the restraining device. Cattle that entered the poorly designed head holder in an already excited, agitated state had a more vigorous reaction to the throat cut than calm animals. These observations indicated that head holding devices must be designed so that the incision is held open during and immediately after the cut. Occasionally, a very wild, agitated animal went into a spasm [that] resembled an epileptic seizure immediately after the cut. This almost never occurred in calm cattle.

The issue of time to unconsciousness is also an important one. Scientific researchers agree that sheep lose consciousness within 2–15 s after both carotid arteries are cut (Nangeroni and Kennett 1963; Blackmore 1984; Gregory and Wotton 1984). Studies with cattle and calves indicate that most animals lose consciousness rapidly. However, some animals may have a period of prolonged sensibility (Blackmore 1984; Daly et al. 1988) that lasts for over a minute. Other studies with bovines also indicate that the time required for them to become unconscious is more variable than for sheep and goats (Munk et al. 1976; Gregory and Wotton 1984). The differences between cattle and sheep can be explained by differences in the anatomy of their blood vessels, that is, bovines have additional small blood vessels in the back of their head that are not cut during the slaughter process. Observations of both calf and cattle slaughter indicate that problems with prolonged consciousness can be corrected (Temple Grandin, personal communication). When a shochet uses a rapid cutting stroke, 95% of the calves collapse almost immediately (Grandin 1987). When a slower, less decisive stroke is used, there is an increased incidence of prolonged sensibility. Approximately 30% of the calves cut with a slow knife stroke had a righting reflex and retained the ability to walk for up to 30 s (Temple Grandin, personal communication).

Gregory (1988) provided a possible explanation for the delayed onset of unconsciousness. A slow knife stroke may be more likely to stretch the arteries and induce an occlusion. Rapid loss of consciousness will occur more readily if the cut is made as close to the jawbone as religious law will permit, and the head holder is partially loosened, but still in place immediately after the cut. The chin lift should, however, remain up so the cut is still open. Excessive pressure applied to the chest by the rear pusher gate will slow bleed out. Gentle operation of the restrainer is essential. Observations indicate that calm cattle lose consciousness more rapidly and they are less likely to have contracted occluded blood vessels, that is, blood splash. Calm cattle will usually collapse within 10–15 s. Recently, the time to unconsciousness (drop to the ground) in a glatt (a higher standard for kosher meat based on lung inspection, see below) kosher plant in North America was found to be less than 10 s for 34 of 36 cattle (Temple Grandin, personal communication).

Captive bolt and electrical stunning will induce instantaneous insensibility when they are properly applied. However, improper application can result in significant stress. All stunning methods trigger a massive secretion of epinephrine (Warrington 1974; Van der Wal 1978). This outpouring of epinephrine is greater than the secretion that would be triggered by an environmental stressor or a restraint method. Since the animal is expected to be unconscious, it does not feel the stress. One can definitely conclude that improperly applied stunning methods would be much more stressful than kosher slaughter with the long straight razor sharp nick-free knife. Kilgour (1978), one of the pioneers in animal welfare research, came to a similar conclusion on stunning and slaughter. In some religious slaughter plants, animal welfare is compromised when animals are pulled out of the restraint box before they have lost consciousness. Observations clearly indicate that disturbance of the incision or allowing the cut edges to touch causes the animal to react strongly. Dragging the cut incision of a conscious animal against the bottom of the head-opening device is likely to cause pain. Animals must remain in the restraint device with the head holder and body restraint loosened until they collapse. The belly lift must remain up during bleed-out to prevent bumping of the incision against the head opening when the animal collapses.

Most reports suggest that the difference in blood loss between religious and nonreligious slaughter are insignificant and this is neither a Jewish nor a Muslim issue and is probably not a factor in meat quality. Observations in many plants indicate that slaughter without stunning requires greater management attention to the procedures than stunning to maintain good welfare and high meat quality. Religious slaughter is a procedure that can be greatly improved by the use of a total quality management (TQM) approach to continual incremental improvements in the process. In plants with existing upright restraint equipment, significant improvements in animal welfare and reductions in petechial hemorrhages can be made by making the following changes: (1) training of employees in gentle calm cattle handling, (2) modifying the restrainer as per the specifications in this chapter, (3) eliminating distractions that make animals balk, and (4) careful attention to the exact cutting method. Kosher-slaughtered animals are subsequently inspected for visible internal organ defects by rabbinically trained inspectors called a "bodek." Muslims after slaughter turn the animal back over to the secular authorities. If the bodek finds that an animal has a defect, the animal is deemed unacceptable and becomes treife and is not usable for kosher. There is no "trimming" of the defective portions as is generally permitted under secular law.

Consumer desire for more stringent kosher meat inspection requirements in the United States, especially when slaughter began to go to higher speeds, has led to the development of a standard for kosher meat, mainly with respect to cattle, that meets a stricter inspection of the mammalian animal's lungs. This inspection is done in two parts. Initially during the initial evisceration of the animal, the heart and lungs are left in the animal and the bodek does an internal inspection to look for any lung adhesions. Meat with less than two such adhesions (sirchas; see Figure 8.6) is referred to as "glatt kosher." Some Jewish groups require a total absence of adhesions even in adult large animals, that is, cattle. Such meat is referred to as "Beit Yosef" meat. Note that young (e.g., veal calves) and small (e.g., all sheep and goats) red meat animals

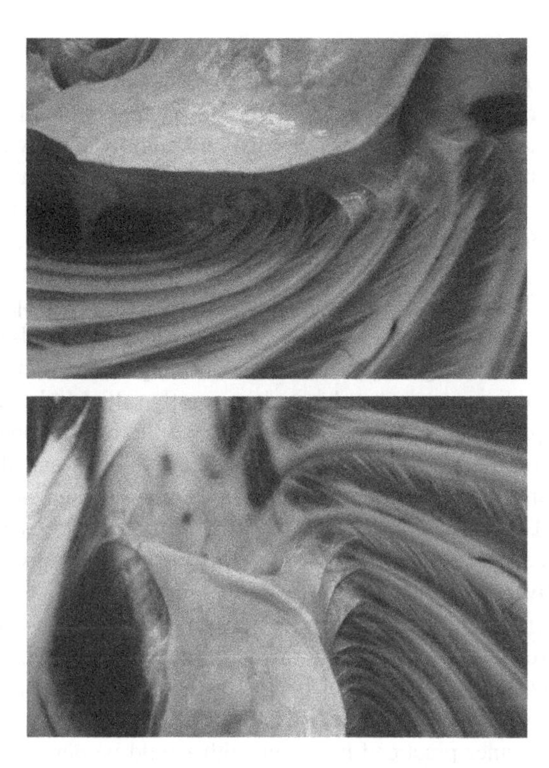

FIGURE 8.6 Sirchas (lung adhesions). (Pictures courtesy of Judy Moses, Spirit of Humane, Boyceville, WI, www.spiritofhumane.com.)

must always be without any adhesions. There has been a recent report that suggests that animals with lung adhesions grow slower, that is, have a lower feed conversion and take longer to reach an equal weight compared to healthy animals. The impact on meat quality has not been examined. Preliminary research using ultrasound to look at the lungs of living animals suggests that many animals with sirchas could be directed away from kosher slaughter (Regenstein, Stouffer, Wanner, and Pufpaff, unpublished research). The same methodology could be used to determine the management practices that lead to problems and allow for better animal health protocols to be developed, which may positively impact meat quality.

In modern times, to decrease the amount of blood splash (capillary rupture in the meat, particularly in the hindquarter), a postslaughter nonpenetrating captive bolt stun is often used with non-glatt kosher meat. This also allows the animal to be hung on the shackle more rapidly. Meat must be further prepared by properly removing certain veins, arteries, prohibited fats, blood, and the sciatic nerve. The parts of the veins and arteries removed are high in blood, the prohibited fats are those mainly in the belly cavity that were used for sacrifices in ancient times, and the sciatic nerve commemorates Jacob's struggle with the angel. This process is called "nikkur" in Hebrew and "treiboring" in Yiddish. The person who is specifically trained to do this is called a "menacker." In practical terms, this means that only the front quarter cuts of kosher red meat are used in the United States and most western countries because it is difficult to remove the sciatic nerve and the hindquarter cuts would be in small pieces. This means that the lower-quality (less tender) front quarter cuts are used in the modern kosher home, thus, there is an interest in the higher end marketing grades, for example, prime and choice, to make up for the lower quality of the cuts used. To further remove the prohibited blood, red meat and poultry must then be soaked and salted (melicha) within 72 h of slaughter. The soaking is done for a half hour in cool water followed by salting for one hour with all surfaces, including cut surfaces, being covered with ample amounts of salt. The salted meat is then rinsed well three times. The presence of the salt is generally believed to speed up rancidity development and this meat may not freeze well for extended periods of time with the penetration of the salt being less than 0.5 cm. Cooking removes most of this salt in expelled liquid and because most red meat is currently salted and soaked as primals, very few of the actual retail cuts contain salt. Once the meat is properly koshered, any remaining "red-liquid" is no longer considered "blood" and the meat can be used without further concern for these issues. Because of its high blood content, liver cannot be soaked and salted, but must instead be broiled until it is at least over half cooked using special equipment reserved for this purpose. The liver is then rinsed, after which the liver can be used in any way the user wishes.

There has been much discussion and controversy among Muslim consumers as well as Islamic scholars over the issue of the permissibility of consuming meat of animals killed by the Ahl-al-Kitab or people of the book, meaning, among certain other faith communities, Jews and Christians. The issue focuses on whether meat prepared in the manner practiced by either faith would be permitted for Muslims. In the Quran, this issue is presented only once in Sura V, verse 5, in the following words: "This day all good things are made lawful for you. The food of those who have received the Scripture is lawful for you, and your food is lawful for them." The majority of Islamic scholars are of the opinion that the food of the Ahl-al-Kitab must meet the criteria established for halal and for wholesome food including proper slaughter of animals. They believe that the following verse establishes a strict requirement for Muslims "And eat not of that whereupon Allah's name hath not been mentioned, for lo! It is abomination." [Quran VI:121]. In recent years, some members of the Orthodox rabbinate have ruled that the saying of the Muslim takbir, that is, the blessing "Allah is great (Bismillah Allah Aqaba)" in Arabic by the shochet is permitted.

8.7 CHILLING OF CARCASSES

The storage of meat at low temperatures is the most common method used through the centuries to prolong its keeping quality. Originally, natural caves or holes dug in the ground, where the temperature was relatively low even in the warm seasons of the year were used by early humans to store food that they had caught by hunting or fishing. Later, cold water from streams and ice gathered in the winter season from lakes were used to keep the meat at low temperatures (Leighton and Douglas 1910; Klettner 1996). Toussaint-Samat in *A History of Food* (2009) describes that in ancient times fish from the Rhine and the Baltic were surrounded with ice, insulated with furs, and sent off to markets in Rome.

Commercial meat freezing based on mechanical removal of heat from freshly slaughtered animals has been in use since the second half of the nineteenth century (Lawrie 2006; James and James 2009; Toussaint-Samat 2009). Today, chilling of carcasses is the most important step in the cold-chain for preserving meat. According to North and Lovatt (2012), chilling is the process of cooling meat while meat remains above its freezing temperature. The main purpose of carcass chilling is to increase the safety of meat by reducing the growth of pathogens and extending the storage life of meat (James 1996; James and James 2009). Moreover, some scientific reports (Jones et al. 1987; Long and Tarrant 1990; Wal 1997) show that rapid air (using blowers) or spray chilling when compared with conventional cold air systems decreased evaporate weight loss (from 2% to 3% to 1% for bovine and porcine carcasses), slowed down the rate of postmortem metabolic changes and shortened the carcass chilling time leading to the possibility of earlier meat fabrication after slaughter. European regulations currently require that carcasses must reach an internal temperature below 7°C before transport or cutting (James 1996). Recently, the EFSA (European Food Safety Authority) has suggested that these regulations need to be revisited to put more emphasis on the control of the surface temperature, which is where more pathogens are found. Improving all of these factors is of economic benefit to the meat industry.

The rate of cooling meat also has other implications besides its effects on microbiology, weight loss or financial benefits connected with a shorter chilling time. The chilling also influences the rate of postmortem metabolism in muscle with beneficial effects on the rate and range of postmortem muscle glycolysis, the rate of pH decline and drip (or purge) loss (Kerth et al. 2001; Zybert et al. 2007; Shackelford et al. 2012; Xu et al. 2012).

Huff-Lonergan and Page (2001) and James and James (2009) describe how the four basic mechanisms that are responsible for removing heat from carcasses work—convection, conduction, evaporation/condensation, and radiation. Radiation requires large temperature differences between the surface of the meat being cooled and that of surrounding surfaces to achieve heat flow. Evaporative cooling is the result of moisture migration toward the surface where evaporation occurs. In practice, meat cooling typically involves a combination of conduction and convection, where internal conduction carries the heat to the surface and then the convection (natural or forced) carries that heat away from the surface. In this process, good physical contact between the meat to be cooled and the cooling medium is required (Huff-Lonergan and Page 2001; James and James 2009; Warris 2010). The most important and most common mechanism used is convection. Natural convection (also called free convection) occurs naturally during the cooling process, but in this situation, the carcass slowly moves from a hot to a cold state. Forced convection using mechanical methods cools a carcass by blowing chilled air across the surface, which increases the rate of heat removal (Huff-Lonergan and Page 2001; James and James 2009; Warris 2010).

There are many factors that influence the chilling rate such as size, shape, and fatness of the carcass, surface area of the carcass that is available for heat flow and the distance between carcasses suspended in the cooling room, initial carcass temperature, the temperature difference between the surface of the carcass and the cooling medium (air or water), and the speed with which the cooling medium circulates over the warm carcasses (Aalhus et al. 2001; Lawrie 2006; James and James 2009).

In conventional chilling systems, the carcasses are held in chilling rooms at 1–4°C with air velocity of 0.1–0.8 m/s for 24 h (Klettner 1996). In some alternative commercial systems two stages are used, where the air temperature and/or air velocity differ. Initially carcasses are chilled in a prechiller that works at a higher rate of heat exchange (lower temperature and/or higher air velocity), then the carcasses are transferred to the main chiller, which has a lower rate of heat exchange (usually with temperatures at 1–4°C) to carry out the final cooling (Klettner 1996; James and James 2009; Warris 2010). In the literature, this type of accelerated chilling system may be referred to as rapid, ultra-rapid, very fast (VFC), or blast chilling (Aalhus et al. 2002; Juarez et al. 2009; Li et al. 2012), but as Savell et al. (2005) states, there is no "consistent definition used by the authors in describing systems." These accelerated air chilling systems use temperatures from –20°C to –35°C with an air velocity 3–5 m/s for 1–3 h and a subsequent conventional chilling regime (Klettner 1996; Springer et al. 2003; Zhang et al. 2007; Tomović et al. 2013). VFC is usually defined as the achievement of a temperature of 0°C within 4 h postmortem (Honikel 1998) or 0°C within 5 h, while according to Joseph (1996), it is –1°C within 5 h postmortem. Some European meat processors also use a three-phase chilling system where carcasses are conveyed through refrigerated rooms with different temperatures (–15°C, –10°C, and –5°C) and air speeds of 2–3 m/s in the process of cooling (Hambrecht et al. 2004a, b). The main purpose of accelerated chilling systems is not only to reduce chilling time but also to decrease evaporative weight loss (Long and Tarrant 1990; Jones et al. 1993; Wal 1997).

Both in North America and in Europe, spray chilling of beef, lamb, and poultry is becoming a more common practice. Spray chilling is based on intermittent or continuous spraying of carcasses with water during early stages of cooling of hot sides followed by a drying period prior to load-out (Park et al. 2007; Shackelford et al. 2012). Spray chilling lowers the temperature of carcasses more quickly than conventional chilling, but not as quickly as accelerated (blast) chilling (Huff-Lonergan and Page 2001; Shackelford et al. 2012). Moreover, superficial muscles like the *Longissimus* are cooled more rapidly than interior muscles like the *Semimembranosus* (Hambrecht et al. 2004a, b). Jones et al. (1987) showed that spray-chilled pork carcasses compared with conventionally chilled sides had significantly lower temperatures in the loins (10.4°C vs. 11.7°C) and hams (18.9°C vs. 20.4°C) at 6 h postmortem. Milligan et al. (1998) reported that blast-chilled carcasses (at –32°C for 100 min with air flow of 2 m/s) compared to conventionally chilled sides had ham temperatures 6°C (at 2.5 h postmortem) to 10°C (after 5.5 h) lower. Shackelford et al. (2012) noted that blast-chilled pork sides compared with spray-chilled carcasses had about 11–12°C lower temperatures in the *Longissimus* muscle in the period between 2 and 7 h after slaughter. Also, Hambrecht et al. (2004a,b) reported that rapidly chilled pork carcasses (in a three-phase chilling tunnel) compared to conventionally chilled sides had temperatures about 10°C lower at 2.5 h postmortem and about 5°C at 6.5 h after slaughter. Moreover, the carcasses reached 10°C after approximately 5–6.5 h postmortem using a blast or three-phase chilling tunnel (Figure 8.7). With the use of conventional or spray chilling, such temperatures were achieved 7–8 h after slaughter (Milligan et al. 1998; Kerth et al. 2001; Josell et al. 2003; Hambrecht et al. 2004a,b; Zybert et al. 2009; Shackelford et al. 2012).

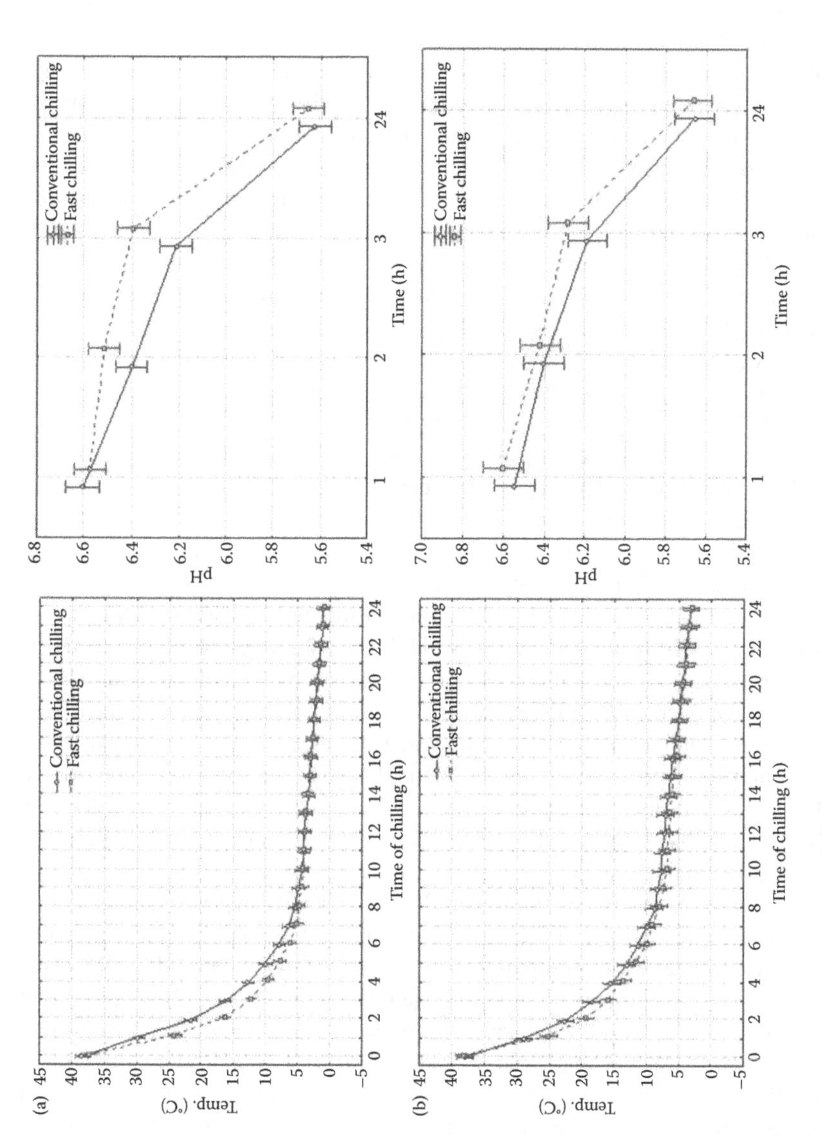

FIGURE 8.7 Temperature and pH profiles at different chilling systems for Duroc (a) and (Landrace × Yorkshire) × Hampshire (b) pigs. (Based on data shown by Zybert, A. et al. 2007. In *Proceedings of the 53rd ICoMST*, Beijing, China, pp. 293–294; Zybert, A. et al. 2009. In *Proceedings of the 55th ICoMST*, Copenhagen, Denmark, pp. 204–207.)

8.7.1 Impact of Chilling on pH Decline

Glycolysis is a fundamental biochemical process in the postmortem conversion of muscle to meat and simultaneously a key factor for meat quality (see Chapter 4). In postmortem muscle, where the circulation is stopped, pyruvic acid—the product of glycolysis—is converted during anaerobic respiration to lactic acid and H^+ with the reduction carried out by NADH. This results in the decline of pH. Usually, glycolysis ceases when all the glycogen is depleted. It has been suggested that several factors stop postmortem glycolysis, including the muscle glycogen content at death, an adenosine monophosphate deficiency and inactivation of glycolytic enzymes by pH (Pearson and Young 1989) (see Chapter 4). Zybert et al. (2009) showed that fast chilled (in a three-phase chilling tunnel) carcasses of (Landrace × Yorkshire) × Hampshire crossbreeds had significantly lower (about 9 µmol/g wet tissue) lactate content at 24 and 48 h after slaughter than conventionally chilled sides.

One significant postmortem change in muscle due to anaerobic metabolism of glycogen is the lowering of muscle pH. Typical pattern of meat acidification is not linear with time postmortem (Przybylski et al. 1994; Warris 2010) and normally pH declines gradually from approximately 7.0–7.2 in living muscle to 5.3–5.7 at 24–48 h after slaughter. There is a strong negative correlation between glycolytic potential (or glycogen content) and ultimate pH as reported by van Laack (2001) and Zybert et al. (2008) (−0.61 and −0.45, respectively). Moreover, in fast chilled carcasses, Zybert et al. (2008) reported a higher negative correlation between glycolytic potential measured 45 min postmortem and pH measured at 24 h compared to conventionally chilled (Landrace × Yorkshire) × Duroc sides.

In general, chilling influences the pH decline (faster chilling usually slows the fall of pH while slow chilling results in a faster pH drop), but the results published in the literature can vary in terms of the effect of chilling systems. Milligan et al. (1998) showed that accelerated chilling decreased the rate of pH decline in pork *Longissimus* muscle at 4.5, 5.5, and 24 h after slaughter, which Josell et al. (2003) confirmed at 5 and 7 h postmortem. Springer et al. (2003) found that blast chilling when compared to conventional chilling slowed the pH decline at 2.5 and 4.5 h, but showed no effect at 24 h. Zhang et al. (2007) compared a blast chilling system to conventional chilling and observed that within the first 5 h postmortem, a lower rate of pH decline was observed in muscles from the blast chilling treatment. Rosenvold and Borup (2011) noted that at 35 min after slaughter, the pH of the *Longissimus* muscle was significantly higher in the stepwise chilled carcasses compared to those conventionally chilled. Therkildsen et al. (2012) proved the same at 45 min after slaughter. Xu et al. (2012) observed that the rate of pH decline in pork *Longissimus* muscle between 3 and 12 h postmortem was significantly slower in rapidly chilled sides compared to conventionally chilled carcasses. By contrast, Zybert et al. (2009) found no difference in the pH of (Land race × Yorkshire) × Hampshire meat measured at 2, 3, and 24 h after slaughter, but Zybert et al. (2013) showed that there was a higher pH in the *Longissimus* muscle of Duroc pigs at 2, 3, 24, and 48 h postmortem, when the fast chilling (three-phase chilling tunnel) was compared with conventional chilling (Figure 8.7).

For beef, Trevisani et al. (1998) and Van Moeseke et al. (2001) showed lower early postmortem pH falls for very fast chilled carcasses although there were no statistical differences at 24 h after slaughter. Also, Li et al. (2012) observed that very fast chilled carcasses compared to slow chilled sides had significantly higher pH in the *Longissimus lumborum* muscle at 5 and 10 h after slaughter. Aalhus et al. (2002) measured blast (at –20°C) and conventionally (at –2°C) chilled beef sides and showed that blast-chilled carcasses had a significantly higher pH at 3, 5, 7, and 10 h postmortem compared with conventionally chilled sides. Li et al. (2006) compared rapid chilling to a conventional system and observed that pre-rigor rapid chilling without electrical stimulation had no impact on the rate of pH decline of carcasses ($\Delta pH = pH_1 - pH_3$), but decreased the rate of pH decline of electrically stimulated sides. After the relatively fast fall of the pH within the first 3 h postmortem, chilling had no influence on the rate of pH decline in the *Longissimus* muscle of Chinese Yellow crossbred bulls at 5, 7, 9, 11, and 24 h after slaughter. Gariepy et al. (1995) found that the pH of the loins from blast-chilled carcasses was slightly higher at 24 h and 6 days postmortem. On the contrary, several studies showed no effect of accelerated (blast) chilling on the pH of pork, beef, or lamb meat (Table 8.3).

Some studies have shown that the use of different chilling systems for rapid temperature decline can be an effective method to reduce the incidence of PSE in meat. Savell et al. (2005) in their review paper described investigations by Borchert and Briskey (1964) who found that liquid nitrogen chilling prevents the formation of PSE in meat without affecting the pH at 24 h, while Reagan and Honikel (1985) showed that chilling pork to internal temperatures of 20–25°C within 2–3 h postmortem can reduce PSE in meat. Kerth et al. (2001) observed a reduction in the incidence of PSE in loins (from 38% to 17%) and hams (from 32% to 10%) for stress-sensitive pigs when the sides were cooled using accelerated chilling. Park et al. (2007) compared spray and rapid chilling to a conventional chilling system and observed that the spray and rapid chilled carcasses had about 22% lower incidence of PSE meat than conventionally chilled sides. On the contrary, Hambrecht et al. (2004a,b) concluded that rapid air chilling cannot prevent the appearance of inferior pork quality caused by high preslaughter stress. Previously, Offer (1991) stated that in extreme cases, carcasses with a very rapid pH decline may have completed glycolysis before entering the cooler. Also, some authors (Long and Tarrant 1990; Milligan et al. 1998; Zybert et al. 2009) stated that accelerated air chilling affects pH mainly in the period between 3 and 4 h postmortem.

8.7.2 Impact of Chilling on Color

The color of meat is probably the most important quality indicator governing its purchase because consumers use proper coloration as a sign of freshness and wholesomeness (Grunert et al. 2004). The color of the meat is determined by several factors such as pigment concentration and its oxidation, intramuscular fat content, lipid oxidation, and the muscle structure (it is the fiber composition and the spatial relationships that determine the scattering of light) (Mancini and Hunt 2005). Also, the ultimate pH and the rate of pH decline together with temperature strongly affect the color of meat. James and James (2009) described that the red color from oxymyoglobin in meat is more stable at lower temperatures because the rate of pigment oxidation to metmyoglobin decreases. Moreover, at normal ultimate pH values

TABLE 8.3
Results from Selected Studies Concerning the Impact of Various Air Chilling Systems on Meat Quality Attributes

Chilling Parameters	Traits				Source
	pH	Colour	Drip Loss	Tenderness	
(AC) −32°C (100 min) airflow 2m/s followed by 2°C **vs. (CC)** 2°C to 24 h	+ †	+	O	–	Milligan et al. (1998)
(AC) −32°C for					Springer et al. (2003)
60 min	–	+	O	O	
90 min	+††	+	O	O	
120 min	+	+	O	O	
150 min. airflow 2 m/s followed by 2°C (air flow 1.2 m/s) **vs. (CC)** 2°C (air flow 1.2 m/s)	+†	+	O	O	
(AC) −30°C (3 h) airflow 5 m/s followed by 2°C **vs. (CC)** 2°C to 24 h (0.5 m/s)	+†††	O	O	O	Tomovic et al. (2013)
(AC) −25°C (1 h) airflow 2.5 m/s followed by 2°C **vs. (CC)** 2°C to 24 h	O	O	+	O	Juarez et al. (2009)
(AC) −23°C (70 min) airflow 3 m/s + 10°C (6 h) followed by 4°C **vs (AC)** −23°C (76 min) airflow 3 m/s followed by 4°C 2°C to 24 h	+††††	O	O	+	Rosenvold and Borup (2011)
(AC) −20°C (3.5 h) airflow 1.5 m/s followed by 4°C **vs. (CC)** 4°C to 24 h (0.2 m/s)	O	X	X	+	Mc Geehin et al. (1999)
(AC) −20°C (30 min) followed by 4°C **vs. (CC)** 4°C to 24 h	+	+	+	O	Xu et al. (2012)
−15°C (50 min), 5°C (5 h), −2°C (6 h) and 5°C until cutting **vs.** 5°C (5 h), −2°C (6 h) and 5°C until cutting	+	O	O	O	Josell et al. (2003)
(AC) −15°C (15 min), −10°C (38 min) −1°C (38 min), airflow 2 m/s followed by 4°C (0.5 m/s) **vs. (CC)** 3–5°C (31 min) airflow 3m/s followed by 4°C (1.5 m/s)	+	O	O	X	Hambrecht et al. (2004a)
(AC) −14°C (2 h) airflow 3 m/s followed by 4°C (0.5 m/s) **vs. (CC)** 4°C to 24 h (0.5 m/s)	+†††††	X	X	O	Zhu et al. (2011)

Note: AC, accelerated chilling; CC, conventional chilling; +, positive effect (improve); –, negative effect (deterioration); O, no influence; X, not investigated, statistically confirmed: †, at 4.5, 5.5 h post mortem; ††, at 3.5, 4.5 and 5.5 h post mortem; †††, at 8 h post mortem, ††††, at 35 min and 8 h post mortem; †††††, at 3, 5, 7, 9, and 11 h post mortem.

in meat, the fibers hold less water and the O_2-consuming enzymes are less active, resulting in a brighter appearance of meat. On the other hand, if low pH is achieved at higher temperatures, then some partial denaturation of myosin and sarcoplasmic proteins may occur, thereby, increasing the light scattering observed. A number of studies (Jones et al. 1993; Milligan et al. 1998; Kerth et al. 2001; Springer et al. 2003) reported darker muscle color after fast (blast) chilling, although Feldhusen et al. (1995) noted that after 4 h of chilling, the musculature of sprayed ham became lighter; however, after 20 h, there were no significant differences in the color values. Also, Zybert et al. (2013) found lighter *Longissimus* muscle color in fast chilled sides compared with conventionally chilled sides. On the other hand, others have shown no effect of cooling rate on meat color (Table 8.3).

Beef and lamb contain more myoglobin than pork. There are some reports that rapid chilling in beef resulted in a darker meat color (Aalhus et al. 2001, 2002). Aalhus et al. (2002) showed that at 3, 5, 7, and 10 h postmortem very fast chilled beef carcasses had darker meat compared to the brighter meat of conventionally chilled sides, but after 6 days, no differences in color parameters were observed. Also, Aalhus et al. (2001), Trevisani et al. (1998), and Van Moeseke et al. (2001) found a slightly darker color of beef when the sides were chilled rapidly. Gariepy et al. (1995) showed no statistical differences in beef meat color for carcasses that were blast or conventionally chilled, while Boakye and Mittal (1996) found that beef meat lightness (L^*) was increased by a faster chilling rate.

8.7.3 Impact of Chilling on Water Content

Drip loss is also a very important factor strongly connected with consumers' perception of meat quality. It also has a detrimental impact on the appearance of fresh meat cuts during retail sale and impacts on the sensory quality of meat (Purslow et al. 2001; Huff-Lonergan and Lonergan 2005). The loss of water from meat is controlled by several structural and biochemical mechanisms such as the muscle tissue temperature postmortem, myofibrillar shrinkage and contraction, shrinkage of the myofilament lattice postmortem caused by the pH fall, and structural changes at the fiber bundle level that lead to an increase of extracellular space and/or myosin denaturation (Honikel et al. 1986; Huff-Lonergan and Lonergan 2005). The normal ultimate pH is very close to the isoelectric point of the major proteins in muscle especially myosin (5.3–5.4). Huff-Lonergan and Lonergan (2005) describe this as the point at which the electric charges on the amino and carboxyl groups of the proteins are approximately equal. The areas with the opposite charges are attracted to each other, which results in a reduction in the water content held by the proteins. Additionally, when repulsion forces are diminished, the space between the thick and thin filaments is reduced. This in turn results in the loss of space within the myofibril for water, which is released into the extra-myofibrillar spaces (Offer et al. 1989; Purslow et al. 2001).

On the other hand, a rapid pH decline in the first hour postmortem when the muscle is still warm causes the denaturation of muscle proteins and cellular membrane destruction, which leads to increased water loss from meat (Honikel et al. 1986). Spray chilling has a very little effect on the drip loss (Long and Tarrant 1990). Some studies have shown that blast or accelerated air chilling can improve WHC or decrease drip loss (Zybert et al. 2008; Shackelford et al. 2012; Therkildsen et al. 2012). However, others have shown no effect of accelerated air chilling on WHC or drip loss (Table 8.3). Jones et al. (1993) also reported that blast chilling did not decrease protein denaturation. However, Honikel (2004) stated that PSE-prone muscles must be chilled rapidly and early to reduce high drip loss because achieving a temperature of 34°C within about 45 min after slaughter improves the WHC. Beef tends to lose similar amounts or less water than pork (Chambaz et al. 2003).

Aalhus et al. (2001) and Li et al. (2006) stated that a greater loss in the *Longissimus lumborum* muscle in conventionally chilled sides was observed compared to those that were blast chilled. Aalhus et al. (2002) showed that blast chilling (–35°C) had no influence on drip loss at 3 and 5 h postmortem. However, extended blast chilling (7 and 10 h after slaughter) significantly increased drip. In contrast, Li et al. (2012) observed that very fast chilled carcasses compared to sides chilled at 7°C and 14°C had significantly higher purge loss.

8.7.4 Impact of Chilling on Tenderness

Chilling has an influence on the tenderness of meat, if it is carried out too rapidly, when the meat is still in pre-rigor condition. After death, but before the onset of rigor mortis, contraction of myofibrils is possible when ATP is present in sufficient concentration and Ca^{2+} ions are released from the sarcoplasmic reticulum (or the mitochondria). According to Honikel et al. (1983), contraction can happen if muscles are chilled rapidly (cold shortening) or very slowly (rigor shortening). Cold shortening was defined by Locker and Hagyard (1963) as a rapid decline in muscle temperature below 14°C before the onset of rigor mortis (minimum shortening occurred in the temperature region 14–19°C). Some studies showed that the relationship between pH value and temperature at the onset of rigor mortis are very important factors responsible for the development of cold shortening (Hannula and Puolanne 2004). Susceptibility to cold shortening varies between species and between muscle types. Generally, red muscles are more susceptible to cold shortening than white muscles (Bendall 1973; Lawrie 2006). Also, species such as pigs and chickens are less prone to cold shortening than beef. According to the literature data, cold-shortened meat appears when the temperature of pre-rigor muscles is significantly reduced when muscles still contain ATP and the pH value is high. Cold shortening can be prevented by moderate chilling or acceleration of glycolysis through the use of electrical stimulation. Bendall (1973) reported that muscles with a temperature below 10°C and pH above 6.2 are more susceptible to cold shortening than muscles with a temperature above 16°C. Olsson et al. (1994) stated that in meat with a high pH and a temperature lower than 7°C, cold shortening might occur. Honikel et al. (1983) suggested that to keep shortening to a minimum, the temperature at pH values above 6.0 must be above 18°C; below pH 6.0 but before the onset of rigor mortis (pH 5.7–5.9), the temperature must be around 12–18°C, while after the onset of rigor, the temperature can drop below 10°C.

Although several studies have shown that accelerated (blast) chilling can decrease meat tenderness (Jeremiah et al. 1992; Jones et al. 1993; Jacob et al. 2012), several studies have found no differences in tenderness when fast chilling systems were compared with conventional systems (Table 8.3). However, Aalhus et al. (2001), Li et al. (2006), and Zhang et al. (2007) reported that blast chilling in combination with electrical stimulation can be recommended to industry to improve meat tenderness.

8.8 CONCLUSIONS

Slaughter procedures are likely to have a larger effect on postmortem metabolism in muscle tissue. The effects of slaughter conditions on behavior, physiology, and subsequent meat quality are variable. The quality of the meat obtained from slaughtered animals is an indicator that reflects stress before slaughter, and during slaughter, which is related to animal welfare. About 40% of meat defects are due to events occurring in the abattoirs. Among them, key issues are animal stunning methods, time between stunning and bleeding, bleeding position, method of scalding, and postslaughter carcasses chilling. In meat processing plants, three types of specific stunning prior to bleeding are permitted and used: mechanical, electrical, and controlled atmosphere (gas anesthesia). The aim of stunning is to cause a loss of consciousness and feeling in animals prior to sticking and bleeding. Religious slaughter uses the bleeding step to cause rapid unconsciousness. All described methods, if they are correctly performed, and with the appropriate equipment, are humane. Mis-stunning leads to serious animal welfare problems and extreme stress in animals that reduces meat quality. Through research, these methods of stunning and religious slaughter of animals are continuously being improved with a view to also providing better animal welfare. The research shows that respect for animal welfare also presents the opportunity to improve meat quality. So, in the future, much emphasis should be focused on the improvement of animal slaughter technology. As stated by a European Community regulation, "The protection of animals at the time of slaughter or killing is a matter of public concern that affects consumer attitudes toward agricultural products. In addition, improving the protection of animals at the time of slaughter contributes to higher meat quality and indirectly has a positive impact on occupational safety in slaughterhouses."

REFERENCES

Aalhus, J.L., Janz, J.A.M., Tong, A.K.W., Jones, S.D.M., and Robertson, W.M. 2001. The influence of chilling rate and fat cover on beef quality. *Can. J. Anim. Sci.* 81:321–330.

Aalhus, J.L., Robertson, W.M., Dugan, M.E.R., and Best, D.R. 2002. Very fast chilling of beef carcasses. *Can. J. Anim. Sci.* 82:56–67.

Althen, T.G., Ono, K., and Topel, D.G. 1977. Effect of stress susceptibility or stunning method on catecholamine levels in swine. *J. Anim. Sci.* 44:985–989.

Anil, M.H., Whittington, P.E., and McKinstry, J.L. 2000. The effect of the sticking method on the welfare of slaughter pigs. *Meat Sci.* 55:315–319.

Antosik, K., Koćwin-Podsiadła, M., and Goławski, A. 2011. Effect of different CO_2 concentration on the stunning effect of pigs and selected quality traits of their meat—A short report. *Pol. J. Food Nutr. Sci.* 61:69–72.

Bager, F., Braggins, T.J., Devine, C.E., Graafhus, A.E., Mellor, D.J., Taener, A., and Upsdell, M.P. 1992. Onset of insensibility in calves: Effects of electropletic seizure and exsanguinations on the spontaneous electrocortical activity and indices of cerebral metabolism. *Res. Vet. Sci.* 52:162–173.

Becerril-Herrera, M., Alonso-Spilsbury, M., Lemus-Flores, C., Guerrero-Legarreta, I., Olmos-Hernández, A., Ramirez-Necoechea, R., and Mota-Rojas, D. 2009. CO_2 stunning may compromise swine welfare compared with electrical stunning. *Meat Sci.* 81:233–237.

Bendall, J.R. 1973. The biochemistry of rigor mortis and cold contracture. In *Proceedings of the 19th European Meeting of Meat Research Workers*, Paris, pp. 1–27.

Bertram, H.C., Stødkilde-Jørgensen, H., Karlsson, A.H., and Andersen, H.J. 2002. Post mortem energy metabolism and meat quality of porcine *M. longissimus dorsi* as influenced by stunning method—A ^{31}P NMR spectroscopic study. *Meat Sci.* 62:113–119.

Blackmore, D.K. 1984. Differences in the behaviour of sheep and calves during slaughter. *Res. Vet. Sci.* 37:223–226.

Boakye, K., and Mittal, G.S. 1996. Changes in colour of beef m. *Longissimus dorsi* muscle during ageing. *Meat Sci.* 42(3):347–354.

Bolton, D.J. 2004. Slaughter-line operation/pigs. In Jensen, W.K., Devine, C., and Dikeman, M. (eds.), *Encyclopedia of Meat Science*. Elsevier Ltd., Oxford, UK, pp. 1243–1249.

Borchert, L.L., and Briskey, E.J. 1964. Prevention of pale, soft, exudative porcine muscle through partial freezing with liquid nitrogen postmortem. *J. Food Sci.* 29:203–209.

Bórnez, R., Linares, M.B., and Vergara, H. 2009a. Systems stunning with CO_2 gas on Manchego light lambs: Physiologic responses and stunning effectiveness. *Meat Sci.* 82:133–138.

Bórnez, R., Linares, M.B., and Vergara, H. 2009b. Effect of stunning with different carbon dioxide conterntrations and exposure times on suckling lamb meat quality. *Meat Sci.* 81:493–498.

Bórnez, R., Linares, M.B., and Vergara, H. 2010. Physiological responses of Manchega suckling lambs: Effect of stunning with different CO_2 concentrations and exposure times. *Meat Sci.* 85:319–324.

Bourguet, C., Deiss, V., Tannugi, C.C., and Terlouw, E.M.C. 2011. Behavioural and physiological reactions of cattle in a commercial abattoir: Relationships with organizational aspects of the abattoir and animal characteristics. *Meat Sci.* 88:158–168.

Burson, D.E., Hunt, M.C., Schafer, D.E., Beckwith, D., and Garrison, J.R. 1983. Effect of stunning method and time interval from stunning to exsanguination on blood splashing in pork. *J. Anim. Sci.* 57:918–921.

Chambaz, A., Scheeder, M.R.L., Kreuzer, M., and Dufey, P.A. 2003. Meat quality of Angus, Simmental, Charolais and Limousin steers compared at the same intramuscular fat content. *Meat Sci.* 63:491–500.

Channon, H.A., Payne, A.M., and Warner, R.D. 2000. Halothane genotype, pre-slaughter handling and stunning method all influence pork quality. *Meat Sci.* 56:291–299.

Channon, H.A., Payne, A.M., and Warner, R.D. 2002. Comparison of CO_2 stunning with manual electrical stunning (50 Hz) of pigs on carcass and meat quality. *Meat Sci.* 60:63–68.

Channon, H.A., Payne, A.M., and Warner, R.D. 2003. Effect of stun duration and current level applied during head to back and head only electrical stunning of pigs on pork quality compared with pigs stunned with CO_2. *Meat Sci.* 65:1325–1333.

Chevillon, P. 2001. Opération de pré-abattage et d'anesthésie. La reduction des stress améliore le bien-être des porcs. *Viandes Prod. Carnés* 22:95–102.

Council Regulations (EC) No 1099/2009 of 24 September 2009 on the protection of animals at the time of killing. *Official Journal of the European Union*, L 303/pp. 1–30.

Daly, C.C., Kallweit, E., and Ellendorf, F. 1988. Cortical function in cattle during slaughter: Conventional captive bolt stunning followed by exsanguinations compared to shechita slaughter. *Vet. Rec.* 122:325–329.

Devine, C.E., Tavener, A., Gilbert, K.V., and Day, A.M. 1986. Electroencephalographic studies of adult cattle associated with electrical stunning, throat cutting and carcass electro-immoblilization. *New Zealand Veterinary Journal* 34:210–213.

Eriksen, M.S., Rødbotten, R., Grøndahl, A.M., Friestad, M., Andersen, I.L., and Mejdell, C.M. 2013. Mobile abattoir versus conventional slaughterhouse—Impact on stress parameters and meat quality characteristics in Norwegian lambs. *Appl. Anim. Behav. Sci.* 149:21–29.

Ewbank, R., Parker, M.J., and Mason, C.W. 1992. Reactions of cattle to head restraint at stunning: A practical dilemma. *Anim. Welfare* 1:55–63.

Fauciano, L., Marquardt, L., Oliveira, M.S., Sebastiany Coelho, H., and Terra, N.N. 1998. The effect of two handling and slaughter systems on skin damage, meat acidification and colour in pigs. *Meat Sci.* 50:13–19.

Feldhusen, F., Kirschner, T., Koch, R., Giese, W., and Wenzel, S. 1995. Influence on meat colour of spray-chilling the surface of pig carcasses, *Meat Sci.* 40:245–251.

Gardner, M.A., Huff-Lonergan, E., Rowe, L.J., Schultz-Kaster, C.M., and Lonergan, S.M. 2006. Influence of harvest processes on pork loin and ham quality. *J. Anim. Sci.* 84:178–184.

Gariepy, C., Delaquis, P.J., Aalhus, J.L., Robertson, M., Leblanc, C., and Rodrigue, N. 1995. Functionality of high and low voltage electrically stimulated beef chilled under moderate and rapid chilling systems. *Meat Sci.* 39:301–310.

Grandin, T. 1987. High speed double rail restrainer for stunning or ritual slaughter. *International Congress of Meat Science and Technology*, Helsinki, Finland, pp. 102–104.

Grandin, T. 1988. Double rail restrainer for livestock handling. *J. Agric. Eng. Res.* 41:327–338.

Grandin, T. 1991. Double rail restrainer for handling beef cattle. Technical paper 915004. Am. Soc. Agric. Eng., St. Joseph, MI.

Grandin, T. 1992. Observations of cattle restraint devices for stunning and slaughtering. *Anim. Welfare* 1:85–91.

Grandin, T. 1993. Management commitment to incremental improvements greatly improves livestock handling. *Meat Focus* Oct:450–453.

Grandin, T. 1994. Euthanasia and slaughter of livestock. *J. Am. Vet. Med. Assoc.* 204:1354–1360.

Grandin, T. 2010. Auditing animal welfare at slaughter plants. *Meat Sci.* 86:56–65.

Grandin, T., and Regenstein, J.M. 1994. Religious slaughter and animal welfare: A discussion for meat scientists. *Meat Focus Int.* 3:115–123.

Gregory, G., and Wotton, S.D. 1984. Time of loss of brain responsiveness following exsanguinations in calves. *Res. Vet. Sci.* 37:141–143.

Gregory, N.G. 1988. Humane slaughter. In *Proceedings of the 34th ICoMST, Workshop on Stunning Livestock*, Brisbane, Australia.

Gregory, N.G. 2004. Pithing and immobilization. In Jensen, W.K., Devine, C., and Dikeman, M. (eds.), *Encyclopedia of Meat Science*. Elsevier Ltd., Oxford, UK, pp. 1353–1354.

Gregory, N.G. 2005. Recent concerns about stunning and slaughter—A review. *Meat Sci.* 70:481–491.

Gregory, N.G. 2008. Animal welfare at markets and during transport and slaughter. *Meat Sci.* 80:2–11.

Gregory, N.G., Lee, C.J., and Widdicombe, J.P. 2007. Depth of concussion in cattle shot by penetrating captive bolt. *Meat Sci.* 77:499–503.

Grunert, K.G., Bredahl, L., and Brunsø, K. 2004. Consumer perception of meat quality and implications for product development in the meat sector—A review. *Meat Sci.* 66:259–272.

Hambrecht, E., Eissen, J.J., de Klein, W.J.H., Ducro, B.J., Smits, C.H.M., Verstegen, M.W.A., and den Hartog, L.A. 2004a. Rapid chilling cannot prevent inferior pork quality caused by high preslaughter stress. *J. Anim. Sci.* 82:551–556.

Hambrecht, E., Eissen, J.J., Nooijen, R.I.J., Ducro, B.J., Smits, C.H.M., den Hartog, L.A., and Verstegen, M.W.A. 2004b. Preslaughter stress and muscle energy largely determine pork quality at two commercial processing plants. *J. Anim. Sci.* 82:1401–1409.

Hambrecht, E., Eissen, J.J., and Verstegen, M.W.A. 2003. Effect of processing plant on pork quality. *Meat Sci.* 64:125–131.

Hannula, T., and Puolanne, E. 2004. The effect of cooling rate on beef tenderness: The significance of pH at 7°C. *Meat Sci.* 67:403–408.

Honikel, K.O. 1998. Conclusion and executive remarks. In Dransfield, E. and Roncales, P. (eds.), *Very Fast Chilling in Beef. 1. Pre-mortem and the Chilling Process. University of Bristol Press*, Bristol, UK, pp. 153–157.

Honikel, K.O. 2004. Water-holding capacity of meat. In te Pas, M.F.W., Everts, M.E., and Haagsman, H.P. (eds.), *Muscle Development of Livestock Animals: Physiology, Genetics and Meat Quality*. CABI Publishing, Oxfordshire, UK, pp. 389–400.

Honikel, K.O., Kim, C.J., Hamm, R., and Roncales, P. 1986. Sarcomere shortening of prerigor muscles and its influence on drip loss. *Meat Sci.* 16:267–282.

Honikel, K.O., Roncales, P., and Hamm, R. 1983. The influence of temperature on shortening and rigor onset in beef muscles. *Meat Sci.* 8:221–241.

Hopkins, D.L. 2010. Processing of sheep and sheep meat. In Cottle, D.J. (ed.), *International Sheep and Wool Handbook*. Nottingham University Press, UK, pp. 691–710.

Hopkins, D.L. 2014. Beef processing and carcass and meat quality. In Cottle, D.J. and Kahn, L.P. (eds.), *Beef Cattle Production and Trade*. CSIRO Publishing, Melbourne, pp. 17–46.

Hopkins, D.L., Shaw, F.D., Baud, S., and Walker, P.J. 2006. Electrical currents applied to lamb carcases—Effects on blood release and meat quality. *Aust. J. Exp. Agric.* 46:885–889.

Huff-Lonergan, E., and Lonergan, S.M. 2005. Mechanisms of water-holding capacity of meat: The role of postmortem biochemical and structural changes. *Meat Sci.* 71:194–204.

Huff-Lonergan, E., and Page, J. 2001. The role of carcass chilling in the development of pork quality. In *Fact Sheet*, Originally published as National Pork Producers Council, Pork Quality, American Meat Science Association, Des Moines, IA, USA, pp. 1–8. http://old.pork.org/filelibrary/factsheets/pigfactsheets/newfactsheets/12-03-02g.pdf (accessed on 05.2015).

Jacob, R., Rosenvold, K., North, M., Kemp, R., Warner, R., and Geesink, G. 2012. Rapid tenderisation of lamb *M. longissimus* with very fast chilling depends on rapidly achieving sub-zero temperatures. *Meat Sci.* 92:16–23.

James, S. 1996. The chill chain "from carcass to consumer". *Meat Sci.* 43:203–216.

James, S.J., and James, C. 2009. Chilling and freezing of meat and its effect on meat quality. In Kerry, J.P. and Ledward, D. (eds.), *Improving the Sensory and Nutritional Quality of Fresh Meat*. CRC Press, Boca Raton, FL, pp. 539–560.

Jeremiah, L.E., Jones, S.D.M., Kruger, B., Tong, A.K.W., and Gibson, R. 1992. The effects of gender and blast-chilling time and temperature on cooking properties and palatability of pork longissimus muscle. *Can. J. Anim. Sci.* 72:501–506.

Jones, S.D.M., Jeremiah, L.E., and Robertson, W.M. 1993. The effects of spray and blast-chilling on carcass shrinkage and pork muscle quality. *Meat Sci.* 34:351–362.

Jones, S.D.M., Murry, A.C., and Robertson, W.M. 1987. The effect of spray chilling pork carcasses on the shrinkage and quality of pork. *Can. I. Food Sci. Tech. J.* 21:102–105.

Josell, A., von Seth, G., and Tornberg, E. 2003. Sensory and meat quality traits of pork in relation to post-slaughter treatment and RN genotype. *Meat Sci.* 66:113–124.

Joseph, R.L. 1996. Very fast chilling of beef and tenderness—A report from an EU concerted action. *Meat Sci.* 43(S):217–227.

Juarez, M., Caine, W.R., Larsen, I.L., Robertson, W.M., Dugan, M.E.R., and Aalhus, J.L. 2009. Enhancing pork loin quality attributes through genotype, chilling method and ageing time. *Meat Sci.* 24:447–453.

Kerth, C.R., Carr, M.A., Ramsey, C.B., Brooks, J.C., Johnson, R.C., Cannon, J.E., and Miller, M.F. 2001. Vitamin–mineral supplementation and accelerated chilling effects on quality of pork from pigs that are monomutant or noncarriers of the halothane gene. *J. Anim. Sci.* 79:2346–2355.

Kilgour, R. 1978. The application of animal behavior and the humane care of farm animals. *J. Anim. Sci.* 46:1479–1486.

Kirton, A.H., Bishop, W.H., and Mullord, M.M. 1978. Relationship between time of stunning and time throat cutting and their effect on blood pressure and blood splash in lambs. *Meat Sci.* 2:199–206.

Kirton, A.H., Frazerhurst, L.F., Woods, E.G., and Chrystall, B.B. 1981. Effect of electrical stunning method and cardiac arrest on bleeding efficiency residual blood and blood splash in lambs. *Meat Sci.* 5:347–353.

Klettner, P.G. 1996. Kühlen und Gefrieren von Schlachttierkörpern. *Fleischwirtschaft*, 76:679–687.

Lambooij, E. 2004. Stunning—Electrical. In Jensen, W.K., Devine, C., and Dikeman, M. (eds.), *Encyclopedia of Meat Science*. Elsevier Ltd., Oxford, UK, pp. 1342–1348.

Lawrie, R.A. 2006. The storage and preservation of meat: I Temperature control. In Lawrie, R.A. and Ledward, D.A. (eds.), *Lawrie's Meat Science*. 7th edition. CRC Press, Cambridge, UK.

Leighton, G.R., and Douglas, L.M. 1910. *The Meat Industry and Meat Inspection*. The Educational Book Company, London.

Li, C.B., Chen, Y.J., Xu, X.L., Huang, M., Hu, T.J., and Zhou, G.H. 2006. Effects of low-voltage electrical stimulation and rapid chilling on meat quality characteristics of Chinese yellow crossbred bulls. *Meat Sci.* 72:9–17.

Li, K., Zhang, Y., Mao, Y., Cornforth, D., Dong, P., Wang, R., Zhu, H., and Luo, X. 2012. Effect of very fast chilling and aging time on ultra-structure and meat quality. Characteristics of Chinese yellow cattle *M. Longissimus lumborum*. *Meat Sci.* 92:795–804.

Lim, D.G., Erwanto, Y., and Lee, M. 2007. Comparison of stunning methods in the dissemination of central nervous system tissue on the beef carcass surface. *Meat Sci.* 75:622–627.

Lister, D., Gregory, N.G., and Warris, P.D. 1981. Stress in meat animals. In Lawrie, R. (ed.), *Developments in Meat Science*, Vol. 2. Applied Science Publishers, London, pp. 61–92.

Locker, R.H., and Hagyard, C.J. 1963. A cold shortening effect in beef muscles. *J. Sci. Food Agric.* 14:787–793.

Long, V.P., and Tarrant, P.V. 1990. The effect of pre-slaughter showering and post-slaughter rapid chilling on meat quality in intact pork sides. *Meat Sci.* 27:181–195.

Ludvigsen, J. 1953. "Muscular degeneration" in hogs (preliminary report). In *Proceedings of the 15th IVC Stockholm I*, Sweden, p. 602.

Mancini, R.A., and Hunt, M.C. 2005. Current research in meat color. *Meat Sci.* 71:100–121.

Maribo, H., Olsen, E.V., Barton-Gade, P., and Møller, A.J. 1998. Comparison of dehiding versus scalding and singeing: Effect on temperature, pH and meat quality in pigs. *Meat Sci.* 50:175–189.

Milligan, S.D., Ramsey, C.B., Miller, M.F., Kaster, C.S., and Thompson, L.D. 1998. Resting of pigs and hot-fat trimming and accelerated chilling of carcasses to improve pork quality. *J. Anim. Sci.* 76:74–86.

Mitchell, G., Hahingh, J., and Ganhao, M. 1988. Stress in cattle assessed after handling, transport and slaughter. *Vet. Rec.* 123:201–205.

Monin, G. 2003. Abattage des porcs et qualities des carcasses et des viands. *INRA Prod. Anim.* 16:251–262.

Monin, G., Marinova, P., Talmant, A., Martin, J.F., Cornet, M., Lanore, D., and Grasso, F. 1997. Chemical and structural changes in dry-cured hams (bayonne hams) during processing and effects of the dehairing technique. *Meat Sci.* 47:29–47.

Monin, G., Talmant, A., Aillery, P., and Collas, G. 1995. Effects on carcass weight and meat quality of pigs dehaired by scalding or singeing post-mortem. *Meat Sci.* 39:247–254.

Mowafy, M., and Cassens, R.G. 1975. Comparative study on different scalding methods and their effect on the quality of pig skin. *J. Anim. Sci.* 41:1291–1297.

Munk, M.L., Munk, E., and Levinger, I.M. 1976. *Shechita: Religious and Historical Research on the Jewish Method of Slaughter and Medical Aspects of Shechita*. Feldheim Distributors, Jerusalem.

Nakyinsige, K., Che Man, Y.B., Aghwan, Z.A., Zulkifli, I., Goh, Y.M., Bakar, F.A., Al-Kahtani, H.A., and Sazili, A.Q. 2013. Stunning and animal welfare from Islamic and scientific perspectives. *Meat Sci.* 95:352–361.

Nangeroni, L.L., and Kennett, P.D. 1963. *An Electroencephalographic Study of the Effect of Shechita Slaughter on Cortical Function of Ruminants*. Unpublished report. Department of Physiology, NY State Veterinary College, Cornell University, Ithaca, NY.

North, M.F., and Lovatt, S.J. 2012. Chilling and freezing meat. In Hui, Y.H. (ed.), *Handbook of Meat and Meat Processing*. CRC Press, Boca Raton, FL, pp. 357–380.

Nowak, B., Mueffling, T.V., and Hartung, J. 2007. Effect of different carbon dioxide concentrations and exposure times in stunning of slaughter pigs: Impact on animal welfare and meat quality. *Meat Sci.* 75:290–298.

Offer, G. 1991. Modelling of the formation pale, soft, exudative meat: Effects of chilling regime and rate and extent of glycolysis. *Meat Sci.* 30:157–184.

Offer, G., Knight, P., Jeacocke, R., Almond, R., Cousins, T., Elsey, J., Parsons, N., Sharp, A., Starr, R., and Purslow, P. 1989. The structural basis of water holding, appearance and toughness of meat and meat products. *Food Microstruc.* 8:151–170.

Olsson, U., Herzman, C., and Tornberg, E. 1994. The influence of low temperature, type of muscle and electrical stimulation on the course of rigor mortis, ageing and tenderness on beef muscles. *Meat Sci.* 37:115–131.

Önenç, A., and Kaya, A. 2004. The effects electrical stunning and percussive captive bolt stunning on meat quality of cattle processed by Turkish slaughter procedures. *Meat Sci.* 66:809–815.

Overstreet, J.W., Marple, D.N., Huffman, D.L., and Nachreiner, R.F. 1975. Effect of stunning methods on porcine muscle glycolysis. *J. Anim. Sci.* 41:1014–1020.

Park, B.Y., Kim, J.H., Cho, S.H. et al. 2007. Evidence of significant effects of stunning and chilling methods on PSE incidences. *Asian Austral. J. Anim. Sci.* 20:257–262.

Pearce, R.A., Bolton, D.J., Sheridian, J.J., McDowell, D.A., Blair, I.S., and Harrington, D. 2004. Studies to determine the critical control points in pork slaughter hazard analysis and critical control point systems. *Int. J. Fd. Micro.* 90:331–339.

Pearson, A.M., and Young, R.B. 1989. Post mortem changes during conversion of muscle to meat. In Young, R.B. (ed.), *Muscle and Meat Biochemistry*. Academic Press, San Diego, CA.

Piotrowski, E., Borzuta, K., Lisiak, D., Kien, S., and Grześkowiak, E. 2006. The effect of different methods of pre-slaughter stunning on meat quality in cattle. *Anim. Sci. Pap. Rep.* 24:223–229.

Pisula, A., and Florowski, T. 2011. Slaughterhouse production. In Pospiech, E. and Pisula, A. (eds.), *Meat—Basis of Science and Technology*. SGGW-Warsaw Life Science University Press, Warsaw, pp. 52–119.

Pisula, A., and Pospiech, E. 2011. *Meat—Basis of Science and Technology*. SGGW-Warsaw Life Science University Press, Warsaw, pp. 1–520.

Pospiech, E., Borzuta, K., Łyczyński, A., and Półkarz, W. 1998. Meat defects and their economic importance. *Pol. J. Food Nutr. Sci.* 7:7–20.

Przybylski, W., Bareja, M., Boruszewska, K., and Jaworska, D. 2011. Effect of slaughter technology on the quality of pork meat. In Borkowski, S. and Rosak-Szyrocka, J. (eds.), *Quality Improvement*. TRIPSOFT, Trnava, pp. 41–57.

Przybylski, W., Vernin, P., and Monin, G. 1994. Relationship between glycolytic potential and ultimate pH in bovine, porcine and ovine muscles. *J. Muscle Foods* 5:245–255.

Purslow, P.P., Shäfer, A., Kristensen, L. et al. 2001. Water-holding of pork: Understanding the mechanisms. In *Proceedings of the 54th Reciprocal Meat Conference*, Savoy, IL, American Meat Science Association. pp. 134–142.

Raj, A.B.M. 2004. Stunning—CO_2 and other gases. In Jensen, W.K., Devine, C., Dikeman, M. (eds.), *Encyclopedia of Meat Science*. Elsevier Ltd., Oxford, UK, pp. 1348–1353.

Raj, A.M., and Gregory, N.G. 1995. Welfare implications of gas stunning of pigs. *Anim. Welfare* 4:273–280.

Reagan, J.O., and Honikel, K.O. 1985. Weight loss and sensory attributes of temperature conditioned and electrically stimulated hot processed pork. *J. Food Sci.* 50:1568–1570.

Regenstein, J.M., Chaudry, M.M., and Regenstein, C.E. 2003. The kosher and halal food laws. *Comp. Rev. Food Sci. Food Safety* 2:111–127.

Rosenvold, K., and Borup, U. 2011. Stepwise chilling adapted to commercial conditions. Improving tenderness of pork without compromising water-holding capacity. *Acta Agric. Scan. A* 61:121–127.

Savell, J.W., Mueller, S.L., and Baird, B.E. 2005. The chilling of carcasses. *Meat Sci.* 70:449–459.

Shackelford, S.D., King, D.A., and Wheeler, T.L. 2012. Chilling rate effects on pork loin tenderness in commercial processing plants. *J. Anim. Sci.* 90:2842–2849.

Shaw, F.D. 2004. Stunning—mechanical. In Jensen, W.K., Devine, C., and Dikeman, M. (eds.), *Encyclopedia of Meat Science*. Elsevier Ltd., Oxford, UK, pp. 1336–1342.

Springer, M.P., Carr, M.A., Ramsey, C.B., and Miller, M.F. 2003. Accelerated chilling of carcasses to improve pork quality. *J. Anim. Sci.* 81:1464–1472.

Terlouw, C. 2005. Stress reaction at slaughter and meat quality in pigs: Genetic background and prior experience. A brief review of recent findings. *Livest. Prod. Sci.* 94:125–135.

Terlouw, C., and Rybarczyk, P. 2008. Explaining and predicting differences in meat quality through stress reaction at slaughter: The case of Large White and Duroc pigs. *Meat Sci.* 79:795–805.

Terlouw, E.M.C., Arnould, C., Auperin, B., Berri, C., Le Bihan-Duval, E., Deiss, V., Lefèvre, F., Lensink, B.J., and Mounier, L. 2008. Pre-slaughter conditions, animal stress and welfare: Current status and possible future research. *Animal* 2(10):1501–1517. doi: 10.1017/S1751731108002723.

Therkildsen, M., Kristensen, L., Kyed, S., and Oksbjerg, N. 2012. Improving meat quality of organic pork through post mortem handling of carcasses: An innovative approach. *Meat Sci.* 91:108–115.

Tomović, V.M., Jokanović, M.R., Petrović, L.S., Tomović, M.S., Tasić, T.A., Ikonić, P.M., Šumić, Z.M., Šojić, B.V., Škaljac, S.B., and Šošo, M.M. 2013. Sensory, physical and chemical characteristics of cooked ham manufactured from rapidly chilled and earlier deboned M. semimembranosus. *Meat Sci.* 93:46–52.

Toohey, E.S., and Hopkins, D.L. 2007. Does high frequency immobilisation of sheep post-death affect meat quality? In *Proceedings of the 67th New Zealand Society of Animal Production*,Wanaka, New Zealand, pp. 420–425.

Toussaint-Samat, M. 2009. *A History of Food*. Blackwell Publishing Ltd., Malden, MA.

Trevisani, M., Claeys, E., and Demeyer, D. 1998. Does freezing overcome the problem of cold shortening in VFC beef? In Dransfield, E. and Ronchalés, P. (eds.), *Very Fast Chilling in Beef, 2. Muscle to Meat*. University of Bristol Press, Bristol, UK, pp. 105–121.

Troeger, K., and Woltersdorf, W. 1986. Influence of scalding and dehairing during pig slaughtering on meat quality. *Fleischwitschaft* 66:893–897.

Tume, R.K., and Shaw, F.D. 1992. Beta endorphin and cortisol concentration in plasma of blood samples collected during exsanguination of cattle. *Meat Sci.* 31:211–217.

Van der Wal, P.G. 1978. Chemical and physiological aspects of pig stunning in relation to meat quality: A review. *Meat Sci.* 2:19–30.

van Laack, R.L.J.M. 2001. Metabolic factors influencing ultimate pH. In *Proceedings of the 54th Reciprocal Meat Conference*, Indianapolis, IN; American Meat Science Association. pp. 158–160.

Van Moeseke, W., De Smet, S., Claeys, E., and Demeyer, D. 2001. Very fast chilling of beef: Effects on meat quality. *Meat Sci.* 59:31–37.

Velarde, A., Gispert, M., Diestre, A., and Manteca, X. 2003. Effect of electrical stunning on meat and carcass quality in lambs. *Meat Sci.* 63:35–38.

Velarde, A., Gispert, M., Faucitano, L., Alonso, P., Manteca., X., and Diestre, A. 2001. Effect of the stunning procedure and the halothane genotype on meat quality and incidence of haemorrhages in pigs. *Meat Sci.* 58:313–319.

Velarde, A., Gispert, M., Faucitano, L., Manteca., X., and Diestre, A. 2000. The effect of stunning method on the incidence of PSE meat and haemorrhages in pork carcasses. *Meat Sci.* 55:309–314.

Vergara, H., Bórnez, R., and Linares, M.B. 2009. CO_2 stunning procedure on Manchego light lambs: Effect on meat quality. *Meat Sci.* 83:517–522.

Vergara, H., and Gallego, L. 2000. Effect of electrical stunning on meat quality of lamb. *Meat Sci.* 56:345–349.

Vimini, R.J., Field, R.A., Riley, M.L., and Varnell, T.R. 1983a. Effect of delayed bleeding after captive bolt stunning on heart activity and blood removal in beef cattle. *J. Anim. Sci.* 57:628–631.

Vimini, R.J., Field, R.A., Riley, M.L., Williams, J.C., Miller, G.J., and Kruggel, W.G. 1983b. Influence of delayed bleeding after stunning on beef muscle characteristics. *J. Anim. Sci.* 56:608–615.

Vogel, K.D., Badtram, G., Claus, J.R., Grandin, T., Turpin, S., Weyker, R.E., and Voogd, E. 2011. Head-only followed by cardiac arrest electrical stunning is an effective alternative to head-only electrical stunning in pigs. *J. Anim. Sci.* 89:1412–1418.

Wal, P.G. 1997. Kuehlung von Schweineschlachtkoerpern und deren Auswirkung auf die Fleischqualitaet. *Fleischwirschaft* 77:9:769.

Warrington, R. 1974. Electrical stunning: A review of the literature. *Vet. Bull.* 44:617–633.

Warris, P.D. 2010. *Meat Science: An Introductory Text*, 2nd edition. CABI Publishing, Wallingford, UK.

Wotton, S.B., and Gregory, N.G. 1986. Pig slaughtering procedures: Time to loss of brain responsiveness after exsanguination or cardiac arrest. *Res. Vet. Sci.* 40:148–151.

Wotton, S.B., Gregory, N.G., Whittington, P.E., and Parkman, I.D. 2000. Electrical stunning of cattle. *Vet. Rec.* 147:681–684.

Xu, Y., Huang, J.C., Huang, M., Xu, B.C., and Zhou, G.H. 2012. The effects of different chilling methods on meat quality and calpain activity of pork muscle *Longissimus Dorsi*. *J. Food Sci.* 71:C27–C32.

Zavy, M.T., Juniewicz, P.E., Phillips, W.A., and Von Tungeln, D.L. 1992. Effect of initial restraint, eaning and transport stress on baseline ACTH stimulated cortisol response in beef calves of different genotypes. *Am. J. Vet. Res.* 53:551–557.

Zhang, W.H., Peng, Z.Q., Zhou, G.H., Xu, X.L., and Wu, J.Q. 2007. Effects of low voltage electrical stimulation and chilling methods on quality traits of pork m. longissimus lumborum. *J. Muscle Foods* 18:109–119.

Zhu, L., Gao, S., and Luo, X. 2011. Rapid chilling has no detrimental effect on the tenderness of low-voltage electrically stimulated *M. longissimus* in Chinese bulls. *Meat Sci.* 88:597–601.

Zivotofsky, A.Z., and Strous, R.D. 2012. A perspective on the electrical stunning of animals: Are there lessons to be learned from human electro-convulsive therapy (ECT)? *Meat Sci.* 90:956–961.

Zybert, A., Krzęcio, E., Sieczkowska, H., Antosik, K., Podsiadły, W., and Koćwin-Podsiadła, M. 2008. Relationship between glycolytic potential and some physico-chemical and functional traits of *Longissimus lumborum* muscle including chilling method of carcasses. *Sci. Ann. Polish Soc. Anim. Prod.* 4:301–309 (In Polish with English abstract).

Zybert, A., Krzęcio, E., Sieczkowska, H., Podsiadły, W., and Przybylski, W. 2007. The influence of chilling method on glycolytic changes and pork meat quality. In *Proceedings of the 53rd ICoMST*, Beijing, China, pp. 293–294.

Zybert, A., Miszczuk, B., Koćwin-Podsiadła, M., Krzęcio, E., Sieczkowska, H., and Antosik, K. 2009. The influence of carcass chilling on glycolytic changes and pork meat quality. In *Proceedings of the 55th ICoMST*, Copenhagen, Denmark, pp. 204–207.

Zybert, A., Sieczkowska, H., Antosik, K., Krzęcio-Nieczyporuk, E., Adamczyk, G., and Koćwin-Podsiadła, M. 2013. Relationship between glycolytic potential and meat quality of durocs' fatteners including chilling system of carcasses. *Ann. Anim. Sci.* 13:645–654.

9 Breeding Strategies for Improving Meat Quality

Suzanne I. Mortimer and Wiesław Przybylski

CONTENTS

9.1 INTRODUCTION

In the meat-producing livestock species, substantial genetic gains in production (e.g., growth, lean yield, carcass composition, and feed efficiency) and reproduction traits (e.g., survival and litter size) continue to be delivered by breeding programs. In further developing these breeding programs in cattle, pigs, and sheep over recent decades, effort has been directed toward gathering the information necessary to also include the genetic improvement of meat quality traits in breeding programs, either to influence profitability or to address emerging consumer preferences for better quality and healthier meat products. As for any meat production trait, the major strategies that can be used for genetic improvement of meat quality traits are (1) selection between breeds, where a genetically superior breed is substituted for another breed; (2) selection within breeds, where genetically superior animals within a breed are chosen to be parents; and (3) crossbreeding, where animals derived from two or more breeds are mated (Simm 1998). Selection between breeds exploits the large additive genetic differences between breeds that may exist in traits that are to be improved and allows the opportunity to have a better performing population as the basis of a flock's or herd's breeding program that implements within-breed selection. Variation between breeds has been identified for many meat quality traits of meat-producing livestock breeds, as reviewed for pigs by Ciobanu et al. (2011) and Chapter 12 of this volume; for cattle by Marshall (1999), Burrow et al. (2001), and Chapter 11 of this volume; and for sheep by Thompson and Ball (1997), Bishop and Karamichou (2009), and Chapter 13 of this volume. With breeds continuing to change due to selection and/or as novel breeds come under consideration to form the parental sources for breeding programs, breed comparison studies continue to be undertaken and reported.

Consequently, for this chapter, the focus will be on the breeding strategies of selection within breeds and crossbreeding. To implement within-breed selection, key steps that allow incorporation of meat quality traits in a breeding program are deciding which traits are to be improved and deciding what measurements to record and select to improve those traits. Information that can aid these decisions can be derived from the relationships between meat quality traits and the traits that are being improved by a breeding program for meat production. These relationships are crucial to understand, as they indicate where antagonistic and beneficial associations exist between traits. This chapter summarizes these relationships for meat quality traits of cattle, pigs, and sheep; then breeding strategies that can include the genetic improvement of meat quality traits are summarized. Mechanisms for phenotyping of meat quality measures are considered. Live animal techniques for measuring meat quality traits are discussed. A discussion of the evidence for additive genetic variation in meat quality traits within breeds across livestock species is not included in this chapter, as it is well established that estimates of heritability are generally low to moderate for subjective (sensory) assessments of eating quality traits and moderate to high for objective (technological) assessments of meat quality (Thompson and Ball 1997; Marshall 1999; Burrow et al. 2001; Bishop and Karamichou 2009; Simm et al. 2009; Ciobanu et al. 2011; Warner et al. 2011; see Chapters 12 and 13 of this volume). The use of major gene effects on meat quality traits in breeding programs is discussed in Chapter 10 of this volume.

9.2 IMPACTS OF MEAT PRODUCTION BREEDING PROGRAMS ON MEAT QUALITY

The first step in the design of an efficient breeding program is definition of its breeding objective or breeding goal. During this step, the appropriate measurements on animals or traits to be improved through selection are identified, as well as the direction of desired improvement for those traits and the relative economic value of a change in each trait that defines its emphasis relative to other traits within the breeding program. Additionally, a clearly defined breeding objective is critical to the effectiveness of any genetic improvement strategy. It enables potential selection criteria, the measurements that are recorded and selected to achieve improvement in the breeding objective, to be evaluated for their accuracy in improving breeding objective traits, and then identifying the appropriate traits for use in selection strategies. As this information has been widely available for production traits of meat production systems, breeding programs have been developed and successfully contributed to improvements in quantity and carcass quality of meat products from livestock. Over time, substantial rates of genetic improvement in growth and carcass composition traits, that is, lean growth, have been achieved from breeding programs for meat production from cattle (e.g., Barwick and Henzell 2005; Amer et al. 2007), pigs (e.g., Chen et al. 2002; Fix et al. 2010), and sheep (e.g., Amer et al. 2007; Swan et al. 2009; Young and Amer 2009). While it has been relatively straightforward in relation to growth and carcass traits of meat production systems to identify which traits are to be improved, the direction of improvement of each trait, and its economic importance, there has been often insufficient information in the past to allow inclusion of meat quality traits into livestock breeding programs.

As discussed by Simm et al. (2009), the availability of information to enable meat quality traits to be a part of livestock breeding programs has been influenced by many issues: (1) the ability to clearly define the meat quality trait to be included in the breeding objective, such that the meat quality trait is relevant for the livestock breed of interest and its production and marketing systems, as well as relevant to consumer preferences; (2) in many instances for livestock industries, the existence of poor or nonexistent price signals for meat quality traits from consumers to farmers that do not provide a financial value for breeders and stimulate them to incorporate these traits in their breeding programs; (3) the requirement for a clear understanding of the nature of the improvement required in the meat quality trait to meet performance targets; for example, optimum performance levels of 4%–5% intramuscular fat (IMF) for Australian lamb have been found to be needed to achieve consumer satisfaction for palatability (Hopkins et al. 2006); and (4) the availability of cost-effective and easy-to-measure methods to accurately assess meat quality traits on slaughter and live animals, as consumer taste panel assessments of eating quality traits and objective measurements of meat quality traits from meat samples are usually expensive and laborious to obtain. However, live animal predictors of meat quality traits would allow a selection of animals to include information on their own performance, plus performances of their relatives (increasing the accuracy of selection) and for selection to occur at an early age (reducing the generation interval). As preslaughter management (nutrition levels, particularly during the finishing phase) and animal growth rates (Warner et al. 2010) and postslaughter processing conditions (Ferguson et al. 2001; Pearce et al. 2010) are known to influence the expression of meat quality traits, it is important to measure meat quality traits under as controlled production and processing conditions as possible, with those conditions reflecting best practice for the production and marketing of meat products in line with consumer preferences.

9.2.1 Relationships between Beef Production and Meat Quality Traits

Several studies have estimated genetic correlations between live animal and meat quality traits in cattle. The early studies, reviewed by Marshall (1999) and Burrow et al. (2001), tended to show few unfavorable genetic correlations between growth and meat quality traits, though these estimates were based on relatively small samples of animals and potentially varied in their control of pre- and postslaughter conditions. Genetic correlation estimates from larger, more recent studies in Australian beef herds reported by Reverter et al. (2003), based on a temperate and a tropically adapted breed, and Wolcott et al. (2009), based on two tropical breeds, found that selection for improved growth rate would lead to genetic improvement in a range of meat quality traits, including sensory tenderness score and shear force value. Genetic correlation estimates were generally low to moderate in magnitude. It should be noted that earlier Johnston et al. (2003) had concluded that the scope for genetic improvement of meat quality traits in tropically adapted breeds was greater than in temperate breeds, due to moderate heritabilities and phenotypic variances, generally favorable genetic relationships among traits and little influence of genotype by environment interactions on the meat quality traits. After reviewing published genetic relationships between growth and meat quality, Marshall (1999) had previously concluded that unfavorable associations generally should be of little concern.

These Australian studies also found that selection to reduce fatness in the live animal would tend to be not associated with unfavorable responses in meat quality traits, with genetic improvement likely for tenderness and loin muscle lightness in tropically adapted breeds, though responses would be influenced by the time of fatness assessment and differences in genotype. These genetic relationships tended to be positive and also low to moderate in size. While genetic correlation estimates between ultrasound eye muscle area and meat quality traits in a temperate breed tended to be less than ±0.20 in size, the genetic correlations among these traits tended to be more moderate in the tropical breeds; consequently, selection for increased ultrasound eye muscle area tended to result in unfavorable changes of darker meat, increased shear force values, and lower IMF (Reverter et al. 2003; Wolcott et al. 2009). In support of the weaker genetic relationships between muscling and meat quality traits in temperate breeds, Cafe et al. (2012) reported from a study of divergent Angus selection lines that there was no effect of selection for muscling score on a range of meat quality traits, including shear force value, meat color, ultimate pH, cooking loss percentage, and sensory eating quality traits (tenderness, juiciness, satisfaction, overall liking). At the same time, retail meat yield percentage was increased following selection for increased muscling score (Cafe et al. 2014). While selection for growth in Pirenaica cattle had no effect on carcass and meat quality traits (Altarriba et al. 2005), Boukha et al. (2011) concluded that Piedmontese cattle with superior growth potential would have inferior carcass conformation (as described by EUS scores derived from the EU linear grading system and rearranged into numerical scores) and paler meat with lower tenderness and water holding capacity based on genetic correlation estimates.

Of the carcass and meat quality attributes of beef, the genetic relationships between retail beef yield, fat thickness, and IMF or marbling score are considered important, as these traits are a major influence on carcass price in many markets (Burrow et al. 2001). Marbling scores are also of interest, as tenderness in beef grading schemes is often predicted by marbling score. For cooked beef products, consumer evaluations have identified tenderness as the most significant attribute in determining the level of satisfaction (e.g., Egan et al. 2001). Retail beef yield percentage has moderate to strong unfavorable genetic correlations with IMF (range of estimates −0.19 to −0.60 as reviewed by Marshall (1999) and Burrow et al. (2001) and reported by Reverter et al. (2003) and Wolcott et al. (2009)). The genetic relationships of carcass weight with IMF are variable in sign, but low. The genetic relationship of retail beef yield with marbling score is negative, but low (range of estimates of −01.4 to −0.25: Koots et al. 1994; Marshall 1999; Wolcott et al. 2009). Genetic correlations between retail beef yield and shear force value generally range between ±0.20 (Burrow et al. 2001; Reverter et al. 2003; Wolcott et al. 2009), whereas the genetic relationship of carcass weight with shear force value tends to be generally favorable (Marshall 1999; Burrow et al. 2001; Reverter et al. 2003; Wolcott et al. 2009). For both relationships, the estimates were associated with large standard errors. The genetic correlation between IMF and shear force value in cattle has indicated that selection for increased IMF will reduce shear force value and improve tenderness, though the size of this relationship has varied between studies and breeds (Burrow et al. 2001; Reverter et al. 2003; Wolcott et al. 2009).

The few studies that have estimated genetic correlations between female reproduction traits and meat quality traits in steers reported that selection for improved marbling would yield no unfavorable responses in female reproduction in tropically adapted cattle (Wolcott et al. 2014) and an earlier age at first calving in Wagyu cattle (Oyama et al. 1996). However, female reproduction in tropically adapted breeds was found to have an unfavorable genetic relationship with shear force value (less tender), though the genetic correlation estimates were associated with high standard errors (Wolcott et al. 2014). In Wagyu cattle, there was no genetic relationship found between early female reproduction and shear force value (Oyama et al. 2004). The temperament of tropically adapted breeds, as measured by a flight time test, has been shown to have consistently negative genetic correlations with shear force value (Reverter et al. 2003; Kadel et al. 2006; Wolcott et al. 2009), indicating that animals of genetically superior temperament produce more tender meat. Genetic correlation estimates reported by Wolcott et al. (2009) suggest that selection for improved residual feed intake in tropically adapted cattle breeds would result in no unfavorable responses in tenderness, cooking loss, lightness of the meat, or IMF.

9.2.2 Relationships between Pork Production and Meat Quality Traits

The reduction of pork production costs has largely been the focus of pig breeding programs in the past, with selection aimed at improving litter size, lean meat percentage, weight gain, and feed conversion ratio. Nonetheless, meat quality traits had been included among the breeding objective and selection criteria traits of breeding programs of several pig breeding organizations during the mid-1970s (Knap et al. 2002). Breeding programs more widely now have started to include meat quality traits among breeding objective traits, either due to the high economic value of these traits for some commercial lines or before any impact on profitability has occurred, as driven by expectations of consumer preferences (Knap 2014). Many studies have provided estimates of genetic correlations involving meat quality and pork production traits for the development of these breeding programs. As reviewed by Ciobanu et al. (2011), average genetic correlation estimates of carcass leanness were low and unfavorable with ultimate pH, color reflectance, and water holding capacity;

moderate and unfavorable with IMF; and unfavorable with sensory eating quality traits (range of average genetic correlations of −0.18 for juiciness to −0.48 for overall acceptability).

Several estimates are available of genetic correlations between growth and meat quality traits for pigs. Estimates for average daily weight gain (ADG) with meat quality traits reported by Miar et al. (2014) suggested that selection for ADG would have no detrimental effects on meat quality traits, results that were consistent with those of de Vries et al. (1994) and Hermesch et al. (2000a). In contrast, van Wijk et al. (2005) reported strong, unfavorable genetic correlations of ADG with pork quality traits, with selection for ADG expected to result in lower pH, reduced water holding capacity, and paler meat color, while Suzuki et al. (2005) reported that selection for ADG was expected to increase tenderness and IMF and lighten meat color. Miar et al. (2014) have reported that selection for leaner carcasses using ultrasound backfat thickness would not alter pork quality traits, except for cooking loss. However, the earlier report of Hermesch et al. (2000a) concluded that selection for increased leanness would reduce IMF (a finding that was consistent with genetic correlation estimates from earlier studies reviewed by those authors) and pH 45 min after slaughter, as well as increasing the incidence of pale, soft, and exudative (PSE) meat. Suzuki et al. (2005) also identified adverse impacts from selection for leanness of meat being less tender and darker in color. A moderate, negative genetic correlation of ultrasound loin depth with ultimate pH was observed by Miar et al. (2014), while they concluded that selection for increased ultrasound IMF would result in meat of higher pH and more yellow color in the *m. gluteus medius* but of lower cooking loss. Long-term selection for increased leanness over 20 years has been found to reduce IMF, objectively measured tenderness and pork flavor of loins (Schwab et al. 2006).

Moderate to high and unfavorable genetic correlations of feed conversion ratio with pH (45 min after slaughter), meat lightness, and drip loss percentage reported by Hermesch et al. (2000a) have indicated that selection to improve feed conversion ratio will result in meat of inferior quality and indicative of PSE. These findings were consistent with findings of earlier studies reviewed by those authors, but not the later, smaller study of Miar et al. (2014) where lightness of the loin had a moderate, positive correlation with the feed conversion ratio. Consistent with the results for feed conversion ratio, selection for low residual feed intake (more feed efficient animals) was found to result in lower IMF and lower ultimate pH (Lefaucheur et al. 2011) and increased lightness and drip loss of the loin (Lefaucheur et al. 2011; Faure et al. 2013) and meat quality traits of ham muscles were adversely affected (Faure et al. 2013). Also, Hoque et al. (2009) found more efficient animals genetically produced darker meat that had greater losses during cooking.

Genetic correlations between reproduction and meat quality traits have been found to be generally small across studies, with no clear relationships evident. Overall, IMF was associated with lower litter birth weight and lower average piglet weight at birth (Hermesch et al. 2000b) and pH and meat lightness favorably associated with age at first farrowing and piglet loss during suckling, but unfavorably associated with total piglets born (Serenius et al. 2004). Carcass leanness was unfavorably associated and ultimate pH was favorably associated with litter size components (Rosendo et al. 2010).

9.2.3 Relationships between Sheep Meat Production and Meat Quality Traits

Few reported estimates are available of genetic correlations between meat quality traits, production traits (reviewed by Mortimer et al. 2010), and reproduction traits of sheep. Mortimer et al. (2010) reported unfavorable correlations between IMF and weaning weight (−0.19 ± 0.11) and between shear force value and weaning weight (0.45 ± 0.15), though de Hollander et al. (2014) reported positive correlations in terminal sheep breeds between weaning weight and IMF (0.12) and negative correlations between weaning weight and shear force value (−0.13). In contrast to the large negative genetic correlations between meat color measures and reproductive traits reported by Afolayan et al. (2008), Safari et al. (2008) reported that the genetic correlations between meat color measures and reproductive traits were small and generally positive. Brien et al. (2013) reported unfavorable genetic correlations of lamb survival with lean meat yield (−0.33) and shear force value (0.27). Behavioral reactivity traits recorded at weaning of lambs (flight time and agitation tests) were found to have weak and generally nonsignificant genetic correlations with meat quality traits, though the estimates suggested that meat from more reactive animals would be of slightly improved eating quality, tenderness, and color (Dodd et al. 2014). From a small study of a Norwegian sheep breed, Lorentzen and Vangen (2012) reported unfavorable correlations between weight of dissected fat and IMF (0.62 ± 0.34) and ultimate pH of the *m. semimembranosus* with the percentage of lean weight (−0.65 ± 0.27). In a larger study of an Australian sheep population, Mortimer et al. (2014) found genetic correlations between hot carcass weight and the meat quality traits to be generally small and favorable, as was the case for live animal assessments of body weight, muscle, and subcutaneous fat (Mortimer et al. 2010). The need for more precise estimates of genetic correlations between meat quality traits and other traits currently included in sheep meat breeding programs is being addressed in several countries through the analyses of data recorded in resource flocks, such as the Information Nucleus (Mortimer et al. 2014) in Australia and progeny testing flocks such as the Beef + Lamb New Zealand Genetics Central Progeny Test (Beef + Lamb New Zealand Genetics 2014).

Associations between estimated breeding values (EBVs) for various live animal production traits and the meat quality in their progeny also assist our understanding of the implications of selection for production traits on meat quality. Several studies have evaluated these associations (Hopkins et al. 2005), following work that showed individual sires can affect tenderness (Hopkins and Fogarty 1998). Table 9.1 summarizes statistically significant results from several Australian studies reporting the impact of sire EBVs for growth, muscling and fat on tenderness, IMF, and other aspects of eating quality. The associations of consumer-assessed tenderness score with EBVs for growth, fatness, and muscling are consistent with available small genetic correlation estimates for shear force value for these traits (Table 9.1), though there was no impact of any of these EBVs on shear force value of either the m. *longissimus* or m. *semimembranosus* muscles in electrically stimulated carcasses (Hopkins et al. 2005, 2007a). Published genetic correlations with shear force value are small for muscling (0.15 ± 0.17) and fatness (−0.10 ± 0.16), though relationships with body weight are inconsistent (0.30 ± 0.15: Mortimer et al. 2010; −0.43: de Hollander et al. 2014).

TABLE 9.1
Summary of the Significant Regression Coefficients for Meat Quality Traits on Sire Estimated Breeding Values

Trait	PWWT (kg)	PEMD (mm)	PFAT (mm)	YEMD (mm)	YFAT (mm)	Significance Level	Background Information	Reference
Tenderness (score)		−1.91 ± 0.59				0.05	Consumer panel (m. *longissimus*)	Hopkins et al. (2005)
			1.66 ± 0.76			0.05	Consumer panel (m. *longissimus*)	Hopkins et al. (2007b)
	−0.019 ± 0.004[a]					0.01	Shear force (m. *semimembranosus* low growth)	Hopkins et al. (2007b)
Eating quality (score 0–100)								
Juiciness		−1.59 ± 0.57				0.05		Hopkins et al. (2005)
Overall liking		−1.32 ± 0.58				0.05		Hopkins et al. (2005)
Overall liking		−3.6				0.01	Consumer panel (m. *longissimus*)	Pannier et al. (2014a)
Overall liking		−3.6				0.01	Consumer panel (m. *semimembranosus*)	Pannier et al. (2014a)
Flavor liking		−1.37 ± 0.49				0.05		Hopkins et al. (2005)
Flavor liking			1.30 ± 0.60			0.05		Hopkins et al. (2007b)
Flavor liking		−3.1				0.01	Consumer panel (m. *longissimus*)	Pannier et al. (2014a)

(Continued)

TABLE 9.1 (*Continued*)
Summary of the Significant Regression Coefficients for Meat Quality Traits on Sire Estimated Breeding Values

Trait	PWWT (kg)	PEMD (mm)	PFAT (mm)	YEMD (mm)	YFAT (mm)	Significance Level	Background Information	Reference
Flavor liking		−3.1				0.01	Consumer panel (m. *semimembranosus*)	Pannier et al. (2014a)
Tenderness		−5.3	3.6			0.05	Consumer panel (m. *longissimus*)	Pannier et al. (2014a)
Tenderness	−4.0 to −5.3[b]	−5.3				0.05	Consumer panel (m. *semimembranosus*)	Pannier et al. (2014a)
Intramuscular fat		−0.11 ± 0.06				0.05	For m. *longissimus*	Hopkins et al. (2005)
			0.041 ± 0.015			0.05	For m. *longissimus*	Hopkins et al. (2007b)
			Not given			0.05	For m. *longissimus*	Pethick et al. (2010)
		−0.56	0.84[c]				For m. *longissimus*	Pannier et al. (2014c)
pH				0.012		0.05	For m. *longissimus*	Hopkins et al. (2007a)
					0.027	0.05	For m. *semitendinosus*	Hopkins et al. (2007a)
Lightness (L*)	0.15 ± 0.06					0.05		Hopkins et al. (2005)
Redness (a*)				−0.14 ± 0.06		0.05		Hopkins et al. (2007a)

(*Continued*)

TABLE 9.1 (*Continued*)

Summary of the Significant Regression Coefficients for Meat Quality Traits on Sire Estimated Breeding Values

Trait	PWWT (kg)	PEMD (mm)	PFAT (mm)	YEMD (mm)	YFAT (mm)	Significance Level	Background Information	Reference
Meat R630/R580 after display[d]		0.27				0.05	For m. *longissimus*	Calnan et al. (2014)
Iron (mg/100 g)			0.11			0.01	For m. *longissimus*	Pannier et al. (2014b)
Zinc (mg/100 g)		−0.05					For m. *longissimus*	Pannier et al. (2010)

Source: Adapted from Hopkins, D.L., Fogarty, N.M., and Mortimer, S.I. 2011. *Small Rum. Res.* 101:160–172.

Note: PWWT, post weaning weight; PEMD, post weaning muscle depth; PFAT, post weaning fat depth; YEMD, yearling muscle depth; YFAT, yearling fat depth.

[a] Log scale.

[b] Range across maternal, terminal, and Merino sire breeds.

[c] Terminal sire breeds only.

[d] The ratio of reflectance of light at wavelengths 630 and 580 nm (R630/R580) after simulated retail display.

Other eating quality traits declined as the sire EBV for muscling increased (Hopkins et al. 2005; Pannier et al. 2014a). This negative relationship could be partly explained by the reduced IMF as marbling is positively related to eating quality traits (Karamichou et al. 2006). There is a positive relationship between IMF and sire EBV for fat, which is consistent with a small positive genetic relationship (Mortimer et al. 2010) and responses in IMF reported in divergent lines selected for carcass lean content (Karamichou et al. 2006). Despite the high error around the regression coefficients for the relationships between sire EBVs and sensory traits, the results suggest that selection for increased muscling may impact adversely on sensory traits (Table 9.1). Flavor score increased as the sire EBV for fat increased, though no association was found by Karamichou et al. (2006). The positive effect of fat EBV on sensory traits must be balanced by the negative effect on lean meat production (Hopkins et al. 2007c), emphasizing the need for EBVs to be used in an index. Gardner et al. (2006) found no association between EBVs and pH among a subsample of lambs from the Hopkins et al. (2005) study, which is consistent with the generally small genetic correlations reported (Ingham et al. 2007; Greeff et al. 2008; Payne et al. 2009; Mortimer et al. 2010). Zinc levels in the m. *longissimus* reduced as sire EBV for muscle increased (Table 9.1), whereas iron levels increased as the sire EBV for fat increased. No relationship between omega-3 fatty acids and sire EBVs has been reported.

In general, the published estimates of genetic relationships between meat production traits and meat quality traits in cattle, sheep, and pigs are low to moderate in size, often favorable, and appear likely to be sufficiently small to be managed by design of appropriate breeding objectives and multitrait selection indexes. Where relationships are stronger and unfavorable between meat production traits and a meat quality trait, all traits would need to be considered in the breeding objective and simultaneous genetic improvement of the traits should be possible, though improvement would be slower compared to selection for a breeding objective with fewer traits. Although genetic relationships among the meat quality traits were not reviewed in this chapter, the likely genetic responses in other meat quality traits also will need to be considered when developing breeding programs that aim to improve a meat quality trait. For example, Mortimer et al. (2014) reported genetic parameters for IMF and shear force value recorded in lamb samples, suggesting that while improving either of these traits should result in favorable responses in fresh meat color measures and most other meat quality traits, unfavorable changes in the color stability of lamb during retail display were predicted to occur to a lesser extent. In support of the conclusion of Simm et al. (2009), inclusion of meat quality traits in breeding programs will also require more substantial knowledge of the genetic correlations of meat quality traits with meat production and reproduction traits, as well as other traits of importance such as disease resistance traits.

9.3 BREEDING STRATEGIES FOR IMPROVEMENT OF MEAT QUALITY

9.3.1 Selection Indexes

Incorporation of meat quality traits into industry breeding programs requires the development of an appropriate breeding objective that includes meat quality and the production traits of importance to the meat production and marketing system of interest (Bishop and Karamichou 2009), as well as identifying the selection criterion traits. A selection index enables the prediction of an animal's genetic merit for an aggregate breeding objective, or EBV, from a range of information sources, including information on different traits measured on an animal and information measured on the same trait on the animal and its relatives. The selection criterion traits (measured traits) are combined into a linear index for which the coefficients of the selection criterion traits maximize the response in the breeding objective. Importantly, a selection index enables the traits among the selection criterion to be different to the traits that are in the breeding objective, whereas for other methods of multitrait selection, such as independent culling levels, this is not possible. Having used selection index methodology to derive the optimal index that achieves maximum response in the breeding objective, responses following multitrait selection in the breeding objective and its component traits can be predicted. This is useful as it allows the comparison of selection responses from different breeding programs, as well as evaluation of different sources of information or selection strategies, in contributing to response in the overall breeding objective for a breeding program. For a full description of the methodology for development of a selection index, see Cameron (1997).

A selection index has been shown to be the most efficient method of combining traits to maximize response from a breeding program and its objective. Selection indexes are commonly used in breeding programs. For example, for pigs, selection indexes for specific lines are calculated following genetic evaluations to predict EBVs, conducted by commercial pig breeding companies that often operate across countries as well as by the national improvement programs that are conducted in some countries (Dekkers et al. 2011). For beef-producing cattle in Australia, breed societies for several of the larger breeds provide indexes that are appropriate for a range of production systems and markets, based on EBVs predicted by the BREEDPLAN evaluation software (Swan et al. 2012), while for the Australian sheep industry, a range of terminal sire, maternal, and dual-purpose indexes have been developed by Sheep Genetics, the organization that supports genetic evaluation nationally, for a range of lamb production systems, with EBVs predicted by the OVIS software (Swan et al. 2012). The statistical procedure, best linear unbiased prediction (BLUP), has been used widely by genetic evaluation systems to predict EBVs from pedigree and performance records, while simultaneously estimating the effects of environmental influences on the performance records. As applied in genetic evaluation systems of the livestock industries of many other countries, large multitrait animal model BLUP procedures have been the basis of genetic evaluations conducted for the Australian beef cattle and sheep industries (Swan et al. 2012).

BLUP procedures have provided a means of optimally combining information from different sources and in the Australian industries this has allowed information for genetic evaluation of meat production traits to be sourced usually from measurements of early growth and live animal ultrasound scanning of carcass traits in beef herds and early growth and live animal ultrasound scanning of muscle and fat levels in sheep flocks. For Australian herds and flocks, the substantial genetic improvement achieved in lean growth from application of BLUP procedures and selection indexes has been reported by Barwick and Henzell (2005) and Swan et al. (2009). An example of the value of the procedures used by genetic evaluation systems to predict EBVs and their enhancement over time has been described by Johnston (2007). It was only after an EBV for IMF had become available to Angus breeders following the introduction of ultrasound scanning for IMF, the subsequent development of an EBV and enhancements to BREEDPLAN in the late 1990s that a positive genetic trend in IMF was apparent, whereas previously the genetic trend was negligible for the Angus breed.

As direct measurements of meat quality traits are expensive and require slaughter of the animals, Simm et al. (2009) previously suggested that a two-stage selection strategy could reduce costs and achieve increased genetic gains from breeding programs through live assessment of indirect predictors of the meat quality trait on all selection candidates followed by direct measurement of a subset of animals. The strategy consists of a first stage of selection, which is based on an initial screening of the selection candidates using measurements of traits that are straightforward to record, but perhaps are less accurate in predicting the trait(s) of interest. A second stage of selection then is based on testing a proportion of the selection candidates using a more expensive method that is more accurate in predicting the trait(s) of interest. X-ray computed tomography (CT) scanning can be used to accurately predict carcass composition and muscle characteristics of live animals (see Section 9.5.1), but it is an expensive method. Consequently, it has been considered as a likely trait to be included in a two-stage selection strategy. In considering studies in the United Kingdom and New Zealand, Lambe et al. (2008) concluded that a two-stage selection strategy, where the first stage screening of selection candidates included body weight and ultrasound measures of muscle and fat depths followed by a second-stage screening of 15%–20% of rams on CT scans of fat and muscle traits, was the most cost-effective means of using CT scanning to increase genetic gains in carcass quality as well as maternal and growth traits in many sheep breeding programs.

More recently, the availability of genomic information, (for more details, see Chapter 10 of this volume for example), is providing substantial, increased opportunities for increasing the rates of genetic gains from breeding programs. Its availability is enabling traits such as meat quality traits to be included among the breeding objective traits, through the prediction of EBVs, and as part of selection indexes. In Australia, this has followed the genotyping of large numbers of animals in both resource herds and flocks and recording of phenotypes for a wide range of traits, including meat quality traits (Swan et al. 2012). In the genetic evaluation systems for the Australian beef cattle and sheep industries, pedigree and performance data are now being combined with genomic information to predict EBVs; for example, EBVs are now provided to sheep breeders for IMF and shear force value, as well as lean meat yield. The procedures being implemented to incorporate genomic information into these genetic evaluation systems are described in detail by Swan et al. (2012) and Swan et al. (2014), while Dodds et al. (2014) has described the application of genomic selection in New Zealand dual-purpose sheep. The application of genomic selection in the genetic evaluation systems for the pig and beef cattle industries is discussed by van Eenennaam et al. (2014), who noted that some commercial pig breeding companies had implemented genomic selection as part of their breeding programs, whereas there has been lesser adoption of genomic selection in the beef industry. In order to support the development of accurate genomic predictions for the application of genomic selection in breeding programs of the cattle and sheep industries, Swan et al. (2012) highlighted the need for continued investment to develop and maintain both training populations (where animals are genotyped and phenotyped to provide data to estimate genomic EBVs of unphenotyped animals based on their genotype) and validation populations that record a wide range of phenotypes on genotyped animals.

9.3.2 Crossbreeding

Reasons for crossbreeding of different breeds or lines as a strategy for genetic improvement have been discussed by Simm (1998). In relation to meat production, these reasons include crossing of breeds or specialized sire and dam lines that are of high genetic merit in different traits to increase the efficiency of a production system; creation of a composite or synthetic breed; grading up to a new breed or strain; exploitation of heterosis; into an existing breed or line, introduction of a single gene for a favorable trait; and from two more extreme breeds, production of animals of more intermediate performance. Given that differences in meat quality traits across breeds have been identified in pigs, cattle, and sheep (see Section 9.1), crossbreeding in meat-producing livestock is an option as a selection strategy for genetic improvement of meat quality traits. In the pig industry, crossbreeding of specialized sire and dam lines is used very extensively to produce crossbred commercial pigs (Dekkers et al. 2011), to take advantage, for example, of the improved early survival of piglets, production and carcass and meat quality traits of the sire lines, and the reproductive performance of the dam lines. Similarly, the terminal crossbreeding systems that are currently used in the sheep industries of many countries (Australia, New Zealand, and the United Kingdom) could be used to improve meat quality traits. Marshall (1999) and Burrow et al. (2001) both concluded that the significant differences in meat quality attributes of beef between breeds could be exploited by crossbreeding systems based on the selection of appropriate breeds. Nonetheless, the effects of heterosis on meat quality traits have been found to be relatively small where these effects have been examined in meat-producing livestock (e.g., Burrow et al. 2001; Ciobanu et al. 2011).

9.4 PHENOTYPING OF MEAT QUALITY TRAITS

Phenotyping, or the measurement and recording of performance in a trait, of animals for meat quality traits is critical for traits to be improved through any breeding strategy. With phenotype-based selection, the number of records available on an animal for the traits that are included in a multitrait selection index, including records for traits from its relatives, is one factor that will determine the accuracy of EBVs. Consequently, the design of a breeding program will require decisions to be made on which traits to record and which animals to record those traits on, within the family structure in place. As records for meat quality traits for a potential candidate for selection are not available as it has not been slaughtered, the accuracy of selection for these traits will depend on availability of meat quality records directly collected from sibs, ancestors, or progeny and indirectly recorded on the selection candidate itself through predictor traits for meat quality. Increases in the generation interval of the breeding program will be a consequence of testing sibs or progeny, but selection accuracy is likely to be greater from sib or progeny testing than from records for a predictor trait. Genetic improvement in the meat quality traits will then be altered, as the expected annual response to selection is a function of the selection accuracy (higher accuracy, larger responses), the selection intensity (more intense selection, larger responses), the generation interval (shorter generation interval arising from selection at an early age, larger responses), and the genetic variation for a trait.

Performance records from selection candidates themselves, their close relatives, and ancestors are usually used in breeding programs in most meat-producing livestock. In relation to meat quality traits, only a few predictor traits recorded on live animals are available to provide performance records to use in designing breeding programs (e.g., real-time ultrasound scanning of IMF), so that often there is less phenotypic information (compared to other economically important traits) recorded for genetic evaluation and selection purposes. However, other technologies for predicting meat quality traits in live animals are under development (see Section 9.5). Also, studies are being undertaken to develop high-volume, rapid, and low-cost technologies to test large numbers of animals for meat quality traits on carcasses in abattoirs and on muscle samples in testing laboratories for performance recording purposes, as well as collection of data for estimation of genetic parameters (e.g., drip loss in pork: Gjerlaug-Enger et al. 2010; fatty acid composition of pork: Gjerlaug-Enger et al. 2011). Potentially, the development of these technologies will result in increases in the range of phenotypes for meat quality traits and, over time, the number of records for these phenotypes for individual animals. This will expand the range of meat quality traits that may be considered as breeding objective traits and enable identification of appropriate selection criteria traits for use in breeding programs.

Merging records sourced from abattoirs (carcass data) and a number of industry organizations involved in the national cattle tracing system in the United Kingdom (pedigree data), Wall et al. (2013) have recently shown that it is feasible to estimate genetic parameters for carcass traits of cattle from the merged data set. In the future, this approach may also be applicable in the cattle industries of other countries, or perhaps in other livestock species, and for a wider range of carcass traits recorded in abattoirs, including meat quality traits. In the Australian beef and lamb industries, this approach may become feasible through enhancements to the web-based application, Livestock Data Link (Meat & Livestock Australia 2014), which is being developed to provide both carcass feedback information, derived from slaughter data of the National Livestock Identification Scheme, and tools to assist producers meet market specifications (A.J. Ball, personal communication). Potentially, these data could be augmented by linkage to the feedback information on eating quality and pricing differentials (price premiums, consumer willingness-to-pay) for individual cuts that could come from Meat Standards Australia (MSA), the voluntary beef and lamb grading system aimed at describing and predicting the eating quality of individual cuts in the carcass (Griffith et al. 2009).

As discussed by Dekkers and van der Werf (2014), the availability of genomic selection, based on EBVs predicted from molecular information, has not diminished the importance of performance recording and the numbers of records collected. Accurate genomic prediction of EBVs, in the first instance, still requires large amounts of data from training populations. Continued phenotype recording is then required to maintain the accuracy of genomic predictions. Overall, extensive phenotyping will still be needed to underpin genetic improvement programs. Modifications to existing performance recording programs are very likely to be needed to collect data useful for genomic prediction, as these designs become less efficient in providing information for designing breeding programs. With genomic selection, increased flexibility should occur that allows investments in performance recording to be optimized and to prioritize testing on animals for difficult-to-measure traits, such as meat quality traits, in specialized herds and flocks (Dekkers and van der Werf 2014). The Information Nucleus program of the Australian Sheep CRC (van der Werf et al. 2010) and the Beef Information Nucleus projects funded by Meat & Livestock Australia (Banks 2013) are examples of these specialized data collection flocks and herds that have been established to collect data on carcass and meat quality traits from Australian sheep and cattle. Case studies of the application of genomics in industry herds and flocks are more fully discussed by van Eenennaam et al. (2014).

9.5 LIVE ANIMAL MEASUREMENTS TO PREDICT MEAT QUALITY

The development of molecular genetics and the discovery of major genes affecting meat quality in pigs (e.g., RYR1 and RN) and other species (e.g., CAST in cattle: see Chapter 10 of this volume) offer new opportunities to improve these traits, though traditional breeding tools will still have some role in the genetic improvement of meat quality traits. However, the capability to measure or predict meat quality traits in the live animal offers the opportunity not only to manipulate these traits by avoiding deleterious effects appearing in the population but also to aid breeding decisions through the availability of phenotypes for those traits on selection candidates. This information would be in addition to the information recorded following slaughter of their relatives and should be available at an early age, as well as being easier, less expensive, and less time consuming to record.

9.5.1 Current Techniques

9.5.1.1 Invasive Methods with Application as Biochemical Markers

The pioneer work on the evaluation of meat quality attributes using biopsy samples taken from the m. *longissimus* of live pigs was conducted by Schöberlein (1976). The method has been applied by many scientists (Lahucky et al. 1982; Von Lengerken et al. 1988, 1991, 1993; Talmant et al. 1989; Le Roy et al. 1994; Cheah et al. 1993, 1995, 1998). These studies led to the development of several methods that can be used in practice to determine the level of Ca^{2+}, pH, R1 (IMP/ATP ratio), level of juice in muscle and muscle glycogen level (glycolytic potential).

As reported by Cheah et al. (1995, 1998), faulty meat could be detected using the biopsy approach in stress-resistant pigs. This *in vivo* approach allowed for the identification of pigs with a predisposition to producing PSE and reddish-pink, soft, and exudative (RSE) meat in stress-resistant and heterozygous pigs (NN and Nn of halothane genotypes). The method was based on measurement of pH, fluid (drip loss from centrifuged muscle biopsy samples) and Ca^{2+} after muscle was incubated at 39°C for 45 min. Lahucky et al. (2002) studied muscle energy metabolism in muscle biopsy samples by measuring phosphorus levels using nuclear magnetic resonance (^{31}P NMR). The authors studied pH and changes in inorganic phosphate (Pi), phosphocreatine (PCr), and adenosine triphosphate (ATP). This study showed a significant relationship ($r = 0.4–0.6$) between PCr and pH measured in biopsy samples, as well as relationships between the rate of postmortem muscle metabolism (pH) and other meat quality traits. However, these methods have not been applied in practice for breeding and selection purposes. Of these methods, only the determination of muscle glycolytic potential (GP) has been applied and tested in practice (Naveau 1994; Le Roy et al. 1994, 1998; Larzul et al. 1999a,b). GP is an indicator of the glycogen level in a living animal as proposed by Monin and Sellier (1985) and it is calculated as a total of the principal compounds susceptible to conversion to lactate. In many studies, it has been shown that GP has a marked effect on meat quality and there are significant genetic and phenotypic relationships between GP and meat quality traits (Larzul et al. 1998a,b; Monin et al. 2003) (Table 9.2). It is known that the major gene called RN has a strong effect on GP and numerous meat quality traits (Le Roy et al. 1995). This effect has been used to identify RN^- allele carriers. As was shown by Naveau (1994), the application of GP in selection (measured in biopsy samples) proved to be an efficient method for eliminating the RN^- gene from pig population in the PenArLan breeding company.

Przybylski (1998) elaborated on a method to identify pigs with a predisposition to produce PSE and acid meat after slaughter that can be applied in populations of pigs free from $RYR1^T$ and RN^- alleles. The heritability of GP is high in pig populations where the RN^- gene is segregated and lower in others such as the Large White breed (Table 9.2). Selection over six generations for a reduction of GP in the *m. longissimus* in an experimental herd of Large White pigs resulted in a correlated response in backfat thickness, with an increase in adiposity (Larzul et al. 1995) and a decrease of the carcass lean-to-fat ratio (Larzul et al. 1999b). Recent studies of Closter et al. (2011) confirmed these results as they showed that selection for the elimination of the RN^- allele from Danish Hampshire pigs reduced genetic gain in lean meat percentage while improving meat quality traits. The study of Larzul et al. (1998b) showed that GP estimated *in vivo* could possibly provide a selection criterion for pig meat technological quality. However, the biopsy technique cannot be used in practice now due to animal welfare laws.

There have also been studies of the relationships between metabolites in blood serum and meat quality traits. A number of reports (Choe et al. 2009; Edwards et al. 2010; Choe and Kim 2014) have produced evidence of a significant relationship of blood lactate, glucose, and serum cortisol levels in pigs with meat quality traits. Also in pigs, Yu et al. (2007) and Przybylski et al. (2009) observed an association between the plasma lipoprotein profile (cholesterol fractions and triglycerides) and IMF level. However, these results should be considered preliminary and require further research.

TABLE 9.2
Heritability and Phenotypic (r_P) and Genetic (r_G) Correlations of Glycolytic Potential and Intramuscular Fat in Pigs and Marbling Score in Cattle (Measured *In Vivo*) with Meat Quality Traits, Growth Rate, and Carcass Traits

Species	Trait	Heritability		r_P	r_G
Pigs	Muscle glycolytic potential	0.25–0.90			
			Growth rate	0.00 to −0.03	−0.03 to 0.07
			Backfat thickness	−0.07 to −0.26	−0.10 to −0.60
			pH (LD)	0.10	0.05
			pH_u (LD, SM, GS, BF, AF)	−0.40	−0.50
			Reflectance (LD, GS, BF)	0.20	0.30
			Water holding capacity (LD, GS, BF)	−0.10	−0.20
			Napole yield (LD, SM) or technological yield of cooking ham	−0.40	−0.70
	Intramuscular fat	0.69			
			Back fat thickness	0.17	0.13
			M. longissimus thickness	−0.14	−0.46
Cattle	Marbling score	0.15–0.55			
			Fat thickness	−0.02 to 0.05	0.05 to 0.83
			M. longissimus area	0.07	−0.15 to 0.17

Source: Adapted from Larzul, C. et al. 1995. *Journées Rech. Porcine en France* 27:171–174; Larzul, C. et al. 1998a. *INRA Prod. Anim.* 11:183–197; Le Roy, P. et al. 1994. *Journées Rech. Porcine en France* 26:311–314; Reverter, A. et al. 2000. *J. Anim.* Sci. 78:1786–1795; Crews Jr., D.H. et al. 2003. *J. Anim.* Sci. 81:1427–1433; Lee, D.H., and Kim, H.C. 2004. *Asian-Aust. J. Anim.* Sci. 17:6–12; Lee, D.H., Choudhary, V., and Lee, G.H. 2006. *Asian-Aust. J. Anim. Sci.* 19:468–474; Maignel, L., Daigle, J.P., and Sullivan, B. 2009. *Journées Rech. Porcine en France* 41:13–18.

Note: LD, *m. longissimus dorsi*; SM, *m. semimembranosus*; GS, *m. gluteus superficialis*; BF, *m. biceps femoris*; AF, *m. adductor femoris*.

9.5.1.2 Noninvasive Physical Methods

Ultrasound technology has been applied to measure meat traits in live animals for a number of years with a focus on fat depth, muscle depth, and marbling or IMF (Park et al. 1994a,b; Newcom et al. 2002; Maignel et al. 2009). The estimated correlation between predicted IMF and carcass IMF is about 0.60. For this reason, in many countries, it seems appropriate to include this trait in the estimation of breeding values and include it in selection indexes (Tyra and Żak 2013). Application of *in vivo* methods has accelerated genetic improvement, particularly in terms of the lean meat content of carcasses. Maignel et al. (2009) confirmed the results of Newcom et al. (2002) and obtained a correlation between IMF measures *in vivo* and chemical analysis of 0.55, with a standard error of prediction of 0.71%. The estimated heritability of IMF predicted *in vivo* was 0.69 and low, positive phenotypic and genetic correlations with backfat thickness and negative correlations with the depth of the *m. longissimus* were observed (Table 9.2). Przybylski et al. (2010a) showed that the accuracy of this method may be increased by including the level of triglycerides in blood plasma. Maignel et al. (2010) suggested that selection for higher IMF was associated with higher proportions of saturated and monounsaturated fatty acids in the loin, lower proportions of polyunsaturated fatty acids, and slightly lower collagen content.

In cattle, many studies have confirmed the scope to use real-time ultrasound technology to measure IMF *in vivo* and include this trait in selection indexes (Reverter et al. 2000; Crews et al. 2003; Lee and Kim 2004; Lee et al. 2006). The heritability of IMF measured by ultrasound is considered moderate and the trait is positively and strongly correlated with fat thickness, but weakly with the area of the *m. longissimus* (Table 9.2). The trait is measured and used in the genetic evaluation systems and breeding programs of many countries.

In recent years, X-ray CT is being used in commercial sheep breeding programs in some countries to estimate carcass quality (Lambe et al. 2008; Jay et al. 2014). The traits that have been examined include IMF and flavor, juiciness, and overall palatability estimated on the basis of IMF (Karamichou et al. 2006; Lambe et al. 2008, 2009; Macfarlane et al. 2009). According to Lambe et al. (2008), predictors derived from CT were correlated with IMF, but this correlation was influenced by breed and was around 0.6 in Texel lambs and 0.4–0.5 in Scottish Blackface lambs. Macfarlane et al. (2009) found that IMF levels in Charolais, Suffolk, and Texel breeds were predicted with moderate accuracy by using CT technology. In the study of Lambe et al. (2009), residual correlations between meat eating quality and predictor CT traits were low to moderate in size (about ±0.4) and were dependent on the breed and the level of lean. Improved models for prediction of IMF in loins of Texel lambs have been reported by Clelland et al. (2014) that show promise for being used to generate data for estimation of genetic parameters.

A new noninvasive technique, bioelectrical bioimpedance analysis (BIA), has been demonstrated to be an accurate means of body composition evaluation that is applicable in live subjects (Avril et al. 2013). This method has long been used to predict body composition in humans. In comparison to other methods such as ultrasound or CT scanning, the equipment cost is lower and the method shows greater precision for total fat (Avril et al. 2013). BIA has shown successful results in the prediction of tissue composition in cattle and sheep (Berg et al. 1996). The degree of conductance of an applied electric current varies because of the impedance (resistance and reactance), which varies between tissues. Przybylski et al. (2010b) showed a significant relationship between predictors derived from BIA analysis and pH (r = 0.48) in live pigs.

9.5.2 Techniques in Development

Monin (1998) showed that ^{31}P-NMR spectroscopy allows the measurement of intracellular pH and energy metabolism indicators in live piglets. However, the results so far published using noninvasive NMR in live animals were obtained on piglets of less than 25 kg, which limits the practical application. A small, promising study in pigs by Ravn ct al. (2008) has shown that parameters recorded by evoked noninvasive surface electromyography (SEMG) analysis can be associated with meat quality traits (shear force value). Evoked SEMG monitors electrical signals generated in muscle fibers following external stimulation through electrodes applied at the skin surface, with the signals being related to muscle fiber composition and structure. Although Weschenfelder et al. (2013) reported that infrared thermographic image maps of the temperature of the skin surface of the eye area (ocular infrared thermography) of pigs had low correlations with blood lactate levels and meat quality traits; these authors did recommend that the image capture techniques be further developed and evaluated under conditions where there is greater variation in pork quality attributes. In Duroc and Large White-based terminal sire lines of pigs, Hermesch and Jones (2012) investigated the value of hematological traits as selection criterion for pork quality and iron content of pork, where a HemoCue Hb 2011 analyzer (a device used in human medicine) was used to rapidly measure hemoglobin levels in blood collected from pigs on-farm. Though they considered their heritability estimate (0.18) to be low for hemoglobin level in blood collected from pigs at 21 weeks of age, the significant positive genetic correlations with meat redness and iron content suggested that on-farm measures of blood hemoglobin levels could be used as a selection criterion for the improvement of these meat quality traits. A number of different biophysical methods for meat quality assessment have been described by Damez and Clejron (2008) that could be applied in the future.

9.6 CONCLUSIONS

The design of appropriate breeding objectives and multitrait selection indexes will allow meat quality traits to be included in breeding programs of meat-producing livestock, given that the available estimates of genetic relationships between meat production traits and meat quality traits in cattle, sheep, and pigs are low to moderate in size and often favorable. With the arrival of genomic selection, the potential for this to occur is even greater, as it provides a means for meat quality traits to be included in breeding objectives of selection programs and among the selection criterion traits of selection indexes. As a result, EBVs of increased accuracy would be possible for some meat quality traits, while for other meat quality traits EBVs would become available for the first time. Investment in extensive phenotyping will be required to support genetic evaluation systems and breeding programs, particularly to ensure the maintenance of the accuracy of genomic predictions of EBVs for meat quality, as well as for other traits of importance in breeding programs. This will be assisted by continued research to develop measurements in live animals as predictors of meat quality.

REFERENCES

Afolayan, R.A., Fogarty, N.M., Gilmour, A.R., Ingham, V.M., Gaunt, G.M., and Cummins, L.J. 2008. Genetic correlations between reproduction of crossbred ewes and the growth and carcass performance of their progeny. *Small Rum. Res.* 80:73–79.

Altarriba, J., Varona, L., Moreno, C., Yagüe, G., and Sañudo, C. 2005. Consequences of selection for growth on carcass and meat quality in Pirenaica cattle. *Livest. Prod. Sci.* 95:103–114.

Amer, P.R., Nieuhwhof, G.J., Pollott, G.E., Roughsedge, T., Conington, J., and Simm, G. 2007. Industry benefits from recent genetic progress in sheep and beef populations. *Animal* 1:1414–1426.

Avril, D.H., Lallo, C., and Mlambo, V. 2013. The application of bioelectrical impedance analysis in live tropical hair sheep as a predictor of body composition upon slaughter. *Trop. Anim. Health Prod.* 45:1803–1808.

Banks, R.G. 2013. Progress in implementation of a Beef Information Nucleus portfolio in the Australian beef industry. *Proc. Assoc. Advmt. Anim. Breed. Genet.* 20:399–402.

Barwick, S.A., and Henzell, A.L. 2005. Development successes and issues for the future in deriving and applying selection indexes for beef breeding. *Aust. J. Exp. Agric.* 45:923–933.

Beef + Lamb New Zealand Genetics. 2014. Central progeny test results 2013–2014. http://www.beeflambnz.com/Documents/Farm/Central%20Progeny%20Test%20Results%20 2013-14.pdf.

Berg, E.P., Neary, M.K., Forrest, J.C., Thomas, D.L., and Kaufmann, R.G. 1996. Assessment of lamb carcass composition from live animal measurement of bioelectrical impedance or ultrasonic tissue depths. *J. Anim. Sci.* 74:2672–2678.

Bishop, S.C., and Karamichou, E. 2009. Genetic and genomic approaches to improving sheep meat quality. In Kerry, J.P. and Ledward, D.A. (eds.), *Improving the Sensory and Nutritional Quality of Fresh Meat*. Woodhead Publishing, Cambridge, UK, pp. 249–263.

Boukha, A., Bonfatti, V., Cecchinato, A., Albera, A., Gallo, L., Carnier, P., and Bittante, G. 2011. Genetic parameters of carcass and meat quality traits of double muscled Piemontese cattle. *Meat Sci.* 89:84–90.

Brien, F.D., Rutley, D.L., Mortimer, S.I., and Hocking Edwards, J.E. 2013. Genetic relationships between lamb survival and meat traits. *Proc. Assoc. Advmt. Anim. Breed. Genet.* 20:237–240.

Burrow, H.M., Moore, S.S., Johnston, D.J., Barendse, W., and Bindon, B.M. 2001. Quantitative and molecular genetic influences on properties of beef: A review. *Aust. J. Exp. Agric.* 41:893–919.

Cafe, L.M., McKiernan, W.A., and Robinson, D.L. 2014. Selection for increased muscling improved feed efficiency and carcass characteristics of Angus steers. *Anim. Prod. Sci.* 54:1412–1416.

Cafe, L.M., McKiernan, W.A., Robinson, D.L., and Walmsley, B.J. 2012. Additional measurements on muscle line cattle. Final Report B.BFG.0049. Meat & Livestock Australia Limited, North Sydney, pp. 1–45.

Calnan, H.B., Jacob, R.H., Pethick, D.W., and Gardner, G.E. 2014. Factors affecting the colour of lamb meat from the *longissimus* muscle during display: The influence of muscle weight and muscle oxidative capacity. *Meat Sci.* 96:1049–1057.

Cameron, N.D. 1997. *Selection Indices and Prediction of Genetic Merit in Animal Breeding*. CAB International, Wallingford, Oxon, UK.

Cheah, K.S., Cheah, A.M., and Just, A. 1998. Identification and characterization of pigs prone to producing RSE (reddish-pink, soft and exudative) meat in normal pigs. *Meat Sci.* 48:249–255.

Cheah, K.S., Cheah, A.M., and Krausgrill, D.I. 1995. Variations in meat quality in live halothane heterozygotes identified by biopsy samples of *m. longissimus dorsi*. *Meat Sci.* 39:293–300.

Cheah, K.S., Cheah, A.M., Lahucky, R., Mojto, J., and Kovac, L. 1993. Prediction of meat quality in live pigs using stress-susceptible and stress-resistant animals. *Meat Sci.* 34:179–189.

Chen, P., Baas, T.J., Mabry, J.W., Dekkers, J.C.M., and Koehler, K.J. 2002. Genetic parameters and trends for lean growth rate and its components in U.S. Yorkshire, Duroc, Hampshire, and Landrace pigs. *J. Anim. Sci.* 80:2062–2070.

Choe, J.H., Choi, Y.M., Lee, S.H., Nam, Y.J., Jung, Y.C., Park, H.C., Kim, Y.Y., and Kim, B.C. 2009. The relation of blood glucose level to muscle fiber characteristics and pork quality traits. *Meat Sci.* 83:62–67.

Choe, J.H., and Kim, B.C. 2014. Association of blood glucose, blood lactate, serum cortisol levels, muscle metabolites, muscle fiber type composition, and pork quality traits. *Meat Sci.* 97:137–142.

Ciobanu, D.C., Lonergan, S.M., and Huff-Lonergan, E.J. 2011. Genetics of meat quality and carcass traits. In Rothschild, M.F. and Ruvinsky, A. (eds.), *The Genetics of the Pig*. 2nd edition. CAB International, Wallingford, Oxfordshire, UK, pp. 355–389.

Clelland, N., Bünger, L., McLean, K.A., Conington, J., Maltin, C., Knott, S., and Lambe, N.R. 2014. Prediction of intramuscular fat levels in Texel lamb loins using X-ray computed tomography scanning. *Meat Sci.* 98:263–271.

Closter, A.M., Guldbrandtsen, B., Henryon, M., Nielsen, B., and Berg, P. 2011. Consequences of elimination of the Rendement Napole allele from Danish Hampshire. *J. Anim. Breed. Genet.* 128:192–200.

Crews Jr., D.H., Pollak, E.J., Weaber, R.L., Quaas, R.L., and Lipsey, R.J. 2003. Genetic parameters for carcass traits and their live animal indicators in Simmental cattle. *J. Anim. Sci.* 81:1427–1433.

Damez, J.-L., and Clejron, S. 2008. Meat quality assessment using biophysical methods related to meat structure. *Meat Sci.* 80:132–149.

de Hollander, C.A., Moghaddar, N., Kelman, K.R., Gardner, G.E., and van der Werf, J.H.J. 2014. Is variation in growth trajectories genetically correlated with meat quality traits in Australian terminal lambs? *Proceedings of the 10th World Congress on Genetics Applied to Livestock Production*, pp. 1–3. https://asas.org/docs/default-source/wcgalp-proceedings-oral/346_paper_9018_manuscript_365_0.pdf?sfvrsn=2.

de Vries, A.G., van der Wal, P.G., Long, T., Eikelenboom, G., and Merks, J.W.M. 1994. Genetic parameters of pork quality and production traits in Yorkshire populations. *Livest. Prod. Sci.* 40:277–289.

Dekkers, J.C.M., Mathur, P.K., and Knol, E.F. 2011. Genetic improvement of the pig. In Rothschild, M.F. and Ruvinsky, A. (eds.), *The Genetics of the Pig*. 2nd edition. CAB International, Wallingford, Oxfordshire, UK, pp. 390–425.

Dekkers, J.C.M., and van der Werf, J.H.J. 2014. Breeding goals and phenotyping programs for multi-trait improvement in the genomics era. *Proceedings of the 10th World Congress on Genetics Applied to Livestock Production*, pp. 1–6. https://asas.org/docs/default-source/wcgalp-proceedings-oral/008_paper_10296_manuscript_1312_0.pdf?sfvrsn=2

Dodd, C.L., Hocking Edwards, J.E., Hazel, S.J., and Pitchford, W.S. 2014. Flight speed and agitation in weaned lambs: Genetic and non-genetic effects and relationships with carcass quality. *Livest. Sci.* 160:2–20.

Dodds, K.G., Auvray, B., Lee, M., Newmann, S.A.-N., and McEwan, J.C. 2014. Genomic selection in New Zealand dual purpose sheep. *Proceedings of the 10th World Congress on Genetics Applied to Livestock Production*, pp. 1–6. https://asas.org/docs/default-source/wcgalp-proceedings-oral/333_paper_10352_manuscript_1331_0.pdf?sfvrsn=2

Edwards, L.N., Engle, T.E., Correa, J.A., Paradis, M.A., Grandin, T., and Anderson, D.B. 2010. The relationship between exsanguination blood lactate concentration and carcass quality in slaughter pigs. *Meat Sci.* 85:435–440.

Egan, A.F., Ferguson, D.M., and Thompson, J.M. 2001. Consumer sensory requirements for beef and their implications for the Australian beef industry. *Aust. J. Exp. Agric.* 41:855–859.

Faure, J., Lefaucheur, L., Bonhomme, N., Ecolan, P., Meteau, K., Metayer Coustard, S., Kouba, M., Gilbert, H., and Lebret, B. 2013. Consequences of divergent selection for residual feed intake in pigs on muscle energy metabolism and meat quality. *Meat Sci.* 93:37–45

Ferguson, D.M., Bruce, H.L., Thompson, J.M., Egan, A.F., Perry, D., and Shorthose, W. 2001. Factors affecting beef palatability—Farmgate to chilled carcass. *Aust. J. Exp. Agric.* 41:879–891.

Fix, J.S., Cassady, J.P., van Heugten, E., Hanson, D.J., and See, M.T. 2010. Differences in lean growth performance of pigs sampled from 1980 and 2005 commercial swine fed 1980 and 2005 representative feeding programs. *Livest. Sci.* 128:108–114.

Gardner, G.E., Pethick, D.W., Greenwood, P.L., and Hegarty, R.S. 2006. The effect of genotype and plane of nutrition on the rate of pH decline in lamb carcasses and the expression of metabolic enzymatic markers. *Aust. J. Agric. Res.* 57:661–670.

Gjerlaug-Enger, E., Aass, L., Ødegård, J., Kongsro, J., and Vangen, O. 2011. Genetic parameters of fat quality in pigs measured by near-infrared spectroscopy. *Animal* 5:1495–1505.

Gjerlaug-Enger, E., Aass, L., Ødegård, J., and Vangen, O. 2010. Genetic parameters of meat quality traits in two pig breeds measured by rapid methods. *Animal* 4:1832–1843.

Greeff, J.C., Safari, E., Fogarty, N.M., Hopkins, D.L., Brien, F.D., Atkins, K.D., Mortimer, S.I., and van der Werf, J.H.J. 2008. Genetic parameters for carcass and meat quality traits and their relationships to liveweight and wool production in hogget Merino rams. *J. Anim. Breed. Genet.* 125:205–215.

Griffith, G., Rodgers, H., Thompson, J., and Dart, C. 2009. The aggregate economic benefits to 2007/08 from the adoption of Meat Standards Australia. *Australas. Agribus. Rev.* 17: Paper 5.

Hermesch, S., and Jones, R.M. 2012. Genetic parameters for haemoglobin levels in pigs and iron content in pork. *Animal* 6:1904–1912.

Hermesch, S., Luxford, B.G., and Graser, H.-U. 2000a. Genetic parameters for lean meat yield, meat quality, reproduction and feed efficiency traits for Australian pigs. 2. Genetic relationships between production, carcass and meat quality traits. *Livest. Prod. Sci.* 65:249–259.

Hermesch, S., Luxford, B.G., and Graser, H.-U. 2000b. Genetic parameters for lean meat yield, meat quality, reproduction and feed efficiency traits for Australian pigs. 3. Genetic parameters for reproduction traits and genetic correlations with production, carcase and meat quality traits. *Livest. Prod. Sci.* 65:261–270.

Hopkins, D.L., and Fogarty, N.M. 1998. Diverse lamb genotypes. 2. Meat pH, colour and tenderness. *Meat Sci.* 49:477–488.

Hopkins, D.L., Fogarty, N.M., and Mortimer, S.I. 2011. Genetic related effects on sheep meat quality—A review. *Small Rum. Res.* 101:160–172.

Hopkins, D.L., Hegarty, R.S., and Farrell, T.C. 2005. Relationship between sire estimated breeding values and the meat and eating quality of meat from their progeny grown on two planes of nutrition. *Aust. J. Exp. Agric.* 45:525–533.

Hopkins, D.L., Hegarty, R.S., Walker, P.J., and Pethick, D.W. 2006. Relationship between animal age, intramuscular fat, cooking loss, pH, shear force and eating quality of aged meat from sheep. *Aust. J. Exp. Agric.* 46:879–884.

Hopkins, D.L., Stanley, D.F., Martin, L.C., Ponnampalam, E.N., and van de Ven, R. 2007c. Sire and growth path effects on sheep meat production. 1. Growth and carcass characteristics. *Aust. J. Exp. Agric.* 47:1208–1218.

Hopkins, D.L., Stanley, D.F., Martin, L.C., Toohey, E.S., and Gilmour, A.R. 2007a. Genotype and age effects on sheep meat production. 3. Meat quality. *Aust. J. Exp. Agric.* 47:1155–1164.

Hopkins, D.L., Stanley, D.F., Toohey, E.S., Gardner, G.E., Pethick, D.W., and van de Ven, R. 2007b. Sire and growth path effects on sheep meat production. 2. Meat and eating quality. *Aust. J. Exp. Agric.* 47:1219–1228.

Hoque, M.A., Katoh, K., and Suzuki, K. 2009. Genetic associations of residual feed intake with serum insulin-like growth factor-I and leptin concentrations, meat quality, and carcass cross sectional fat area ratios in Duroc pigs. *J. Anim. Sci.* 87:3069–3075.

Ingham, V.M., Fogarty, N.M., Gilmour, A.R., Afolayan, R.A., Cummins, L.J., Gaunt, G.M., Stafford, J., and Hocking Edwards, J.E. 2007. Genetic evaluation of crossbred lamb production 4. Genetic parameters for first-cross animal performance. *Aust. J. Agric. Res.* 58:839–846.

Jay, N.P., van de Ven, R.J., and Hopkins, D.L. 2014. Comparison of rankings for lean meat based on results from a CT scanner and a video image analysis system. *Meat Sci.* 98:316–320.

Johnston, D.J. 2007. Genetic trends in Australian beef cattle—Making real progress. *Proc. Assoc. Advmt. Anim. Breed Genet.* 17:8–15.

Johnston, D.J., Reverter, A., Ferguson, D.M., Thompson, J.M., and Burrow, H.M. 2003. Genetic and phenotypic characterisation of animal, carcass and meat quality traits for temperate and tropically adapted beef breeds. 3. Meat quality traits. *Aust. J. Agric. Res.* 54:135–147.

Kadel, M.J., Johnston, D.J., Burrow, H.M., Graser, H.-U., and Ferguson, D.M. 2006. Genetics of flight time and other measures of temperament and their value as selection criteria for improving meat quality traits in tropically adapted breeds of beef cattle. *Aust. J. Agric. Res.* 57:1029–1035.

Karamichou, E., Richardson, R.I., Nute, G.R., McLean, K.A., and Bishop, S.C. 2006. Genetic analyses of carcass composition, as assessed by X-ray computer tomography, and meat quality traits in Scottish Blackface sheep. *Anim. Sci.* 82:151–162.

Knap, P.W. 2014. Pig breeding goals in competitive markets. *Proceedings of the 10th World Congress on Genetics Applied to Livestock Production*, pp. 1–6. https://asas.org/docs/default-source/wcgalp-proceedings-oral/007_paper_8901_manuscript_518_0.pdf?sfvrsn=2.

Knap, P.W., Sosnicki, A.A., Klont, R.E., and Lacoste, A. 2002. Simultaneous improvement of meat quality and growth and carcass traits in pigs. *Proceedings of the 7th World Congress on Genetics Applied to Livestock Production*, Communication 11-07, pp. 1–8.

Koots, K.R., Gibson, J.P., and Wilton, J.W. 1994. Analyses of published genetic parameter estimates for beef production traits. 2. Phenotypic and genetic correlations. *Anim. Breed. Abstr.* 62:825–853.

Lahucky, R., Baulain, U., Henning, M., Demo, P., Krska, P., and Liptaj T. 2002. *In vitro* ^{31}P NMR studies on biopsy skeletal muscle samples compared with meat quality of normal and heterozygous malignant hyperthermia pigs. *Meat Sci.* 61:233–241.

Lahucky, R., Fischer, K., and Augustini, C. 1982. Zur Vorhersage der Fleischbeschaffenheit am lebenden Schwein mit Hilfe Schuβbiopsie [Predicting meat quality in the live pig by shot biopsy]. *Fleischwirtsch.* 62:1323–1326.

Lambe, N.R., Navajas, E.A., Fisher, A.V., Simm, G., Roehe, R., and Bünger, L. 2009. Prediction of lamb meat eating quality in two divergent breeds using various live animal and carcass measurements. *Meat Sci.* 83:366–375.

Lambe, N.R., Navajas, E.A., Schofield, C.P., Fisher, A.V., Simm, G., Roehe, R., and Bünger, L. 2008. The use of various live animal measurements to predict carcass and meat quality in two divergent lamb breeds. *Meat Sci.* 80:1138–1149.

Larzul, C., Le Roy, P., Gogué, J., Talmant, A., Jacquet, B., Lefaucheur, L., Ecolan, P., Sellier, P., and Monin, G. 1999a. Selection for reduced muscle glycolytic potential in large white pigs. II. Correlated responses in meat quality and muscle compositional traits. *Genet. Sel. Evol.* 31:61–76.

Larzul, C., Le Roy, P., Gogué, J., Talmant, A., Monin, G., and Sellier, P. 1999b. Selection for reduced muscle glycolytic potential in large white pigs. III. Correlated responses in growth rate, carcass composition and reproductive traits. *Genet. Sel. Evol.* 31:149–161.

Larzul, C., Le Roy, P., Monin, G., and Sellier, P. 1998a. Variabilité génétique du potentiel glycolytique du muscle chez le porc [Genetic variability of muscle glycolytic potential in pigs]. *INRA Prod. Anim.* 11:183–197.

Larzul, C., Le Roy, P., Sellier, P., Jacquet, B., Gogué, J., Talmant, A., Vernin, P., and Monin, G. 1998b. Le potentiel glycolytique du muscle mesuré sur le porc vivant: un nouveau critère de sélection pour la qualité de la viande? [The muscle glycolytic potential measured on live pigs: a new selection criterion for meat quality?]. *Journées Rech. Porcine en France* 30:81–85.

Larzul, C., Le Roy, P., Gogué, J., Talmant, A., Vernin, P., Lagant, H., Monin, G., and Sellier, P. 1995. Résultats de quatre générations de sélection sur le potentiel glycolytique musculatuire mesuré *in vivo* [Results of four generations of selection for in vivo glycolytic potential of muscle]. *Journées Rech. Porcine en France* 27:171–174.

Le Roy, P., Caritez, J.C., Billon, Y., Elsen, J.M., Talmant, A., Vernin, P., Lagant, H., Larzul, C., Monin, G., and Sellier, P. 1995. Etude de l'effet du locus RN sur les caractères de croissance et de carcasse. Premier résultats [Effect of the RN locus in growth and carcass traits: First results]. *Journées Rech. Porcine en France* 27:165–170.

Le Roy, P., Larzul, C., Gogué, J., Talmant, A., Monin, G., and Sellier, P. 1998. Selection for reduced muscle glycolytic potential in Large White pigs. I. Direct responses. *Genet. Sel. Evol.* 30:469–480.

Le Roy, P., Przybylski, W., Burlot, T., Bazin, C., Lagant, H., and Monin, G. 1994. Etude des relations entre le potentiel glycolytique du muscle et les caractères de production dans les lignées Laconie et Penshire [Study of relations between muscular glycolytic potential and production traits in the Laconie and Penshire lines]. *Journées Rech. Porcine en France* 26:311–314.

Lee, D.H., Choudhary, V., and Lee, G.H. 2006. Genetic parameter estimates for ultrasonic meat qualities in Hanwoo cows. *Asian-Aust. J. Anim. Sci.* 19:468–474.

Lee, D.H., and Kim, H.C. 2004. Genetic relationship between ultrasonic and carcass measurements for meat qualities in Korean steers. *Asian-Aust. J. Anim. Sci.* 17:6–12.

Lefaucheur, L., Lebret, B., Ecolan, P., Louveau, I., Damon, M., Prunier, A., Billon, Y., Sellier, P., and Gilbert, H. 2011. Muscle characteristics and meat quality traits are affected by divergent selection on residual feed intake in pigs. *J. Anim Sci.* 89:996–1010.

Lorentzen, T.K., and Vangen, O. 2012. Genetic and phenotypic analysis of meat quality traits in lamb and correlations to carcass composition. *Livest. Sci.* 143:201–209.

Macfarlane, J.M., Lewis, R.M., Emmans, G.C., Young, M.J., and Simm, G. 2009. Predicting tissue distribution and partitioning in terminal sire sheep using X-ray computed tomography. *J. Anim. Sci.* 87:107–118.

Maignel, L., Daigle, J.-P., and Sullivan, B. 2009. Utilisation de la technologie ultrasons pour la prediction *in vivo* du pourcentage de gras intramusculaire de la longe et perspectives d'utilisation en amelioration génétique porcine [Using ultrasound technology to predict intramuscular fat of loin in live pigs and perspectives on the use of ultrasounds in pig genetic improvement]. *Journées Rech. Porcine en France* 41:13–18.

Maignel, L., Daigle, J.-P., Wyss, S., Plourde, N., Gariepy, C., and Sullivan, B. 2010. Relations entre pourcentage de gras intramusculaire, le profile n acides gras et la teneur en collagène du muscle long dorsal chez le porc Duroc, et conséquences pour la selection sur le gras intramusculaire [Relationships between intramuscular fat content, fatty acid profile and collagen content in *Longisimus dorsi* muscle in Duroc pigs, and consequences for selection on intramuscular fat]. *Journées Rech. Porcine en France* 42:187–188.

Marshall, D.M. 1999. Genetics of meat quality. In Fries, R. and Ruvinsky, A. (eds.), *The Genetics of Cattle*. CAB International, Wallingford, Oxfordshire, UK, pp. 605–636.

Meat & Livestock Australia. 2014. LDL Update. *Feedback*, September 2014, Meat & Livestock Australia Ltd., North Sydney, Australia, p. 10.

Miar, Y., Plastow, G., Bruce, H. et al. 2014. Genetic and phenotypic correlations between performance traits with meat quality and carcass characteristics in commercial crossbred pigs. *PLoS ONE* 9:e110105. doi:10.1371/journal.pone.0110105.

Monin, G. 1998. Recent methods for predicting quality of whole meat. *Meat Sci.* 49:S231–S243.

Monin, G., Przybylski, W., and Koćwin-Podsiadła, M. 2003. Glycolytic potential as meat quality determinant. *Anim. Sci. Pap. Rep.* 21:109–120.

Monin, G., and Sellier, P. 1985. Pork of low technological quality with a normal rate of muscle pH fall in the immediate post-mortem period: The case of the Hampshire breed. *Meat Sci.* 13:49–63.

Mortimer, S.I., van der Werf, J.H.J., Jacob, R.H. et al. 2014. Genetic parameters for meat quality traits of Australian lamb meat. *Meat Sci.* 96:1016–1024.

Mortimer, S.I., van der Werf, J.H.J., Jacob, R.H. et al. 2010. Preliminary estimates of genetic parameters for carcass and meat quality traits in Australian sheep. *Anim. Prod. Sci.* 50:1135–1144.

Naveau, J. 1994. Selection programme for eliminating N and RN-genes determining the quality of pork. *IInd International Conference on the Influence of Genetic and Non-Genetic Traits on Carcass and Meat Quality*, Siedlce, November 7–8, 1994, pp. s.69–s.76.

Newcom, D.W., Baas, T.J., and Lampe, J.F. 2002. Prediction of intramuscular fat percentage in live swine using real-time ultrasound. *J. Anim. Sci.* 80:3046–3052.

Oyama, K., Katsuta, T., Anada, K., and Mukai, F. 2004. Genetic parameters for reproductive performance of breeding cows and carcass traits of fattening animals in Japanese Black (Wagyu) cattle. *Anim. Sci.* 78:195–201.

Oyama, K., Mukai, F., and Yoshimura, T. 1996. Genetic relationships among traits recorded at registry, judgement, reproductive traits of breeding females and carcass traits of fattening animals in Japanese Black cattle. *Anim. Sci. Tech. (Japan)* 67:511–518.

Pannier, L., Gardner, G.E., Pearce, K.L., McDonagh, M., Ball, A.J., Jacob, R.H., and Pethick, D.W. 2014a. Associations of sire estimated breeding values and objective meat quality measurements with sensory scores in Australian lamb. *Meat Sci.* 96:1076–1087.

Pannier, L., Pethick, D.W., Boyce, M.D., Ball, A.J., Jacob, R.H., and Gardner, G.E. 2014b. Associations of genetic and non-genetic factors with concentrations of iron and zinc in the *longissimus* muscle of lamb. *Meat Sci.* 96:1111–1119.

Pannier, L., Pethick, D.W., Geesink, G.H., Ball, A.J., Jacob, R.H., and Gardner, G.E. 2014c. Intramuscular fat in the *longissimus* muscle is reduced in lambs from sires selected for leanness. *Meat Sci.* 96:1067–1075.

Pannier, L., Ponnampalam, E.N., Gardner, G.E., Butler, K.L., Hopkins, D.L., Ball, A.J., Jacob, R.H., Pearce, K.L., and Pethick, D.W. 2010. Prime Australian lamb supplies key nutrients for human health. *Anim. Prod. Sci.* 50:1115–1122.

Park, B., Whittaker, A.D., Miller, R.K., and Bray, D.E. 1994a. Measuring intramuscular fat in beef with ultrasonic frequency analysis. *J. Anim. Sci.* 72:117–125.

Park, B., Whittaker, A.D., Miller, R.K., and Hale, D.S. 1994b. Predicting intramuscular fat in beef *longissimus* muscle from speed of sound. *J. Anim. Sci.* 72:109–116.

Payne, G.M., Campbell, A.W., Jopson, N.B., McEwan, J.C., Logan, C.M., and Muir, P.D. 2009. Genetic and phenotypic parameter estimates for growth, yield and meat quality traits in lamb. *Proc. NZ Soc. Anim. Prod.* 69:210–214.

Pearce, K.L., van de Ven, R., Mudford, C., Warner, R.D., Hocking-Edwards, J.E., Jacob, R., Pethick, D.W., and Hopkins, D.L. 2010. Case studies demonstrating the benefits on pH and temperature decline of optimising medium-voltage electrical stimulation of lamb carcasses. *Anim. Prod. Sci.* 50:1107–1114.

Pethick, D.W., Pannier, L., Gardner, G.E., Geesink, G.H., Ball, A.J., Hopkins, D.L., Jacob, R.H., Mortimer, S.I., and Pearce, K.L. 2010. Genetic and production factors that influence the content of intramuscular fat in the meat of prime lambs. In Matteo Crovetto, G. (ed.), *Energy and Protein Metabolism and Nutrition*. Wageningen Academic Publishers, The Netherlands, EAAP Publication No. 127: 673–674. Presented at the International Symposium on Energy and Protein Metabolism and Nutrition, PARMA, Italy, September 6–10, 2010.

Przybylski, W. 1998. A method for detection of pigs with predisposition to producing of PSE and acid meat. *Pol. J. Food Nutr. Sci.* 7/48:246–250.

Przybylski, W., Abbe, R., Niemyjski, S., Jaworska, D., and Olczak, E. 2010a. The analysis of possibility of application of USG technique for measuring of IMF in pigs *in vivo*. *LXXV Congress Polish Animal Science Association*, Olsztyn, Poland.

Przybylski, W., Gromadzka-Ostrowska, J., Olczak, E., Jaworska, D., Niemyjski, S., and Santé-Lhoutellier, V. 2009. Analysis of variability of plasma leptin and lipids concentration in relations to glycolytic potential, intramuscular fat and meat quality in P76 pigs. *J. Anim. Feed Sci.* 18:296–304.

Przybylski, W., Jaworska, D., Kajak-Siemaszko, K., Niemyjski, S., and Siekański, M. 2010b. The assessment of possibility of application of bioelectrical impedance analysis (BIA) for estimation of quality of pork. *LXXV Congress Polish Society Animal Science*, Olsztyn, Poland.

Ravn, L.S., Andersen, N.K., Rasmussen, M.A., Christensen, M., Edwards, S.A., Guy, J.H., Henckel, P., and Harrison, A.P. 2008. *De electricitatis catholici musculari*—Concerning the electrical properties of muscles, with emphasis on meat quality. *Meat Sci.* 80:423–430.

Reverter, A., Johnston, D.J., Ferguson, D.M., Perry, D., Goddard, M.E., Burrow, H.M., Oddy, V.H., Thompson, J.M., and Bindon, B.M. 2003. Genetic and phenotypic characterisation of animal, carcass and meat quality traits for temperate and tropically adapted beef breeds. 4. Correlations among animal, carcass and meat quality traits. *Aust. J. Agric. Res.* 54:149–158.

Reverter, A., Johnston, D.J., Graser, H.-U., Wolcott, M.L., and Upton, W.H. 2000. Genetic analyses of live-animal ultrasound and abattoir carcass traits in Australian Angus and Hereford cattle. *J. Anim. Sci.* 78:1786–1795.

Rosendo, A., Druet, T., Péry, C., and Bidanel, J.P. 2010. Correlative responses for carcass and meat quality traits to selection for ovulation rate or prenatal survival in French Large White pigs. *J. Anim. Sci.* 88:903–911.

Safari, E., Fogarty, N.M., Hopkins, D.L., Greeff, J.C., Brien, F.D., Atkins, K.D., Mortimer, S.I., Taylor, P.J., and van der Werf, J.H.J. 2008. Genetic correlations between ewe reproduction and carcass and meat quality traits in Merino sheep. *J. Anim. Breed. Genet.* 125:397–402.

Schöberlein, L. 1976. Die Schuβbiopsie—eine neue Methode zur Entnahme von Muskelproben [Gun biopsy—A new approach to muscle sampling]. *Mh. Vet. Med.* 31:457–460.

Schwab, C.R., Baas, T.J., Stalder, K.J., and Mabry, J.W. 2006. Effect of long-term selection for increased leanness on meat and eating quality traits in Duroc swine. *J. Anim. Sci.* 84:1577–1583.

Serenius, T., Sevon-Aimónen, M.-L., Kause, A., Mäntysaari, E.A., and Mäki-Tanila, A. 2004. Genetic associations of prolificacy with performance, carcass, meat quality, and leg conformation traits in the Finnish Landrace and Large White pig populations. *J. Anim. Sci.* 82:2301–2306.

Simm, G. 1998. *Genetic Improvement of Cattle and Sheep*. Farming Press, Tonbridge.

Simm, G., Lambe, N., Bünger, L., Navajas, E., and Roehe, R., 2009. Use of meat quality information in breeding programmes. In Kerry, J.P. and Ledward, D.A. (eds.), *Improving the Sensory and Nutritional Quality of Fresh Meat*. Woodhead Publishing, Cambridge, UK, pp. 264–291.

Suzuki, K., Irie, M., Kadowaki, H., Shibata, T., Kumagai, M., and Nishida, A. 2005. Genetic parameter estimates of meat quality traits in Duroc pigs selected for average daily gain, *longissimus* muscle area, backfat thickness, and intramuscular fat content. *J. Anim. Sci.* 83:2058–2065.

Swan, A.A., Brown, D.J., and Banks, R.G. 2009. Genetic progress in the Australian sheep industry. *Proc. Assoc. Advmt. Anim. Breed Genet.* 18:326–329.

Swan, A.A., Brown, D.J., Daetwylar, H.D., Hayes, B.J., Kelly, M., Moghaddar, N., and van der Werf, J.H.J. 2014. Genomic evaluations in the Australian sheep industry. *Proceedings of the 10th World Congress on Genetics Applied to Livestock Production*, pp. 1–6. https://asas.org/docs/default-source/wcgalp-proceedings-oral/334_paper_10482_manuscript_1357_0.pdf?sfvrsn=2.

Swan, A.A., Johnston, D.J., Brown, D.J., Tier, B., and Graser, H.-U. 2012. Integration of genomic information into beef cattle and sheep genetic evaluations in Australia. *Anim. Prod. Sci.* 52:126–132.

Talmant, A., Fernandez, X., Sellier, P., and Monin, G. 1989. Glycolytic potential in *Longissimus dorsi* muscle of Large White pigs, as measured after *in vivo* sampling. In *Proceedings of the 35th International Congress of Meat Science and Technology*, Copenhagen, Denmark, Communication 6.33, pp. 1129–1131.

Thompson, J.M., and Ball, A.J. 1997. Genetics of meat quality. In Piper, L. and Ruvinsky, A. (eds.), *The Genetics of Sheep*. CAB International, Wallingford, Oxfordshire, UK, pp. 523–538.

Tyra, M., and Żak, G. 2013. Analysis of the possibility of improving the indicators of pork quality through selection with particular consideration of intramuscular fat (IMF) content. *Ann. Anim. Sci.* 13:33–44.

van der Werf, J.H.J., Kinghorn, B.P., and Banks, R.G. 2010. Design and role of an information nucleus in sheep breeding programs. *Anim. Prod. Sci.* 50:998–1003.

van Eenennaam, A.L., Weigel, K.A., Young, A.E., Cleveland, M.A., and Dekkers, J.C.M. 2014. Applied animal genomics: Results from the field. *Annu. Rev. Anim. Biosci.* 2:105–139.

van Wijk, H.J., Arts, D.J.G., Matthews, J.O., Webster, M., Ducro, B.J., and Knol, E.F. 2005. Genetic parameters for carcass composition and pork quality estimated in a commercial production chain. *J. Anim. Sci.* 83:324–333.

Von Lengerken, G., Wicke, M., and Maak, S. 1993. Suitability of structural and functional traits of skeletal muscle for the genetic improvement of meat quality in pigs. *44th Annual Meeting of the European Association for Animal Production*, August 16–19, 1993, Aarhus, Denmark, Paper P1.3, p. 9.

Von Lengerken, G., Wicke, M., Maak, S., and Paulke, T. 1991. Relationship between muscle metabolism in the *Musculus longissimus* and halothane susceptibility. *Arch. Tierz. Dummerstorf* 34:553–560.

Von Lengerken, G., Wicke, M., and Schöberlein, L. 1988. Beziehungen bioptischer Kennwerte zur Belastungsempfindlichkeit und Fleischqualität des Schweines [Pork quality parameters measured in biopsy samples in relation to stress sensitivity]. *Felischwirtch.* 42:127–130.

Wall, E., Coffey, M., and Pritchard, T. 2013. Selection opportunities from using abattoir carcass data. *Proc. Assoc. Advmt. Anim. Breed Genet.* 20:253–256.

Warner, R.D., Greenwood, P.L., and Ferguson, D.M. 2011. Understanding genetic and environmental effects for assurance of meat quality. In Joo, S.-T. (ed.), *Control of Meat Quality*. Research Signpost, Kerala, India, pp. 117–145.

Warner, R.D., Jacob, R.H., Hocking Edwards, J.E., McDonagh, M., Pearce, K., Geesink, G., Kearney, G., Allingham, P., Hopkins, D.L., and Pethick, D.W. 2010. Quality of lamb meat from the Information Nucleus Flock. *Anim. Prod. Sci.* 50:1123–1134.

Weschenfelder, A.V., Saucier, L., Maldague, X., Rocha, L.M., Schaefer, A.L., and Faucitano, L. 2013. Use of infrared ocular thermography to assess physiological conditions of pigs prior to slaughter and predict pork quality variation. *Meat Sci.* 95:616–620.

Wolcott, M.L., Johnston, D.J., Barwick, S.A., Corbet, N.J., and Burrow, H.M. 2014. Genetic relationships between steer performance and female reproduction and possible impacts on whole herd productivity in two tropical beef genotypes. *Anim. Prod. Sci.* 54:85–96.

Wolcott, M.L., Johnston, D.J., Barwick, S.A., Iker, C.L., Thompson, J.M., and Burrow, H.M. 2009. Genetics of meat quality and carcass traits and the impact of tenderstretching in two tropical beef genotypes. *Anim. Prod. Sci.* 49:383–398.

Young, M.J., and Amer, P.R. 2009. Rates of genetic change in New Zealand sheep. *Proc. Assoc. Advmt. Anim. Breed. Genet.* 18:422–425.

Yu, I.T., King, Y.T., Chen, S.L., Wang, Y.H., Chang, Y.H., and Yen, H.T. 2007. Dietary conjugated linoleic acid and leucine improve pork intramuscular fat and meat quality. *J. Anim. Feed Sci.* 16:65–74.

10 Influence of Major Genes on Meat Quality

Nicola R. Lambe, Elżbieta Krzęcio-Nieczyporuk, Maria Koćwin-Podsiadła, and Lutz Bünger

CONTENTS

10.1 INTRODUCTION TO MAJOR GENES

To date, most genetic progress in quantitative traits in livestock has been made by selection on phenotypic information, or on estimates of breeding values derived from such phenotypic information, without any knowledge of the genes underlying the genetic variation. The number of these genes, and their wider properties, such as the magnitude of their effects, their frequencies, and interactions between them, have been largely ignored in this quantitative genetic approach. The genetic architecture of traits of interest has essentially been treated as a "black box," but, nevertheless, substantial rates of genetic improvement have been achieved (e.g., Dekkers and Hospital 2002; Hill 2014). However, further improvements are necessary, alongside a widening of selection goals. With advances in molecular genetic tools, the arsenal available to breeders is being revolutionized, allowing them to study the genetic makeup of individuals at the DNA level.

The development of molecular genetic techniques, as well as suitable statistical methods to analyze resulting data (reviewed by Hill 2014), has advanced substantially over the last two decades. This has resulted in the discovery of hundreds of genes, quantitative trait loci (QTL, a gene or chromosomal region significantly affecting a quantitative trait), and markers that are linked with genes or QTL (Figure 10.1), which affect important livestock production traits. This led to the establishment of the Animal QTL database (AnimalQTLdb; http://www.animalgenome.org/cgi-bin/QTLdb/index; Hu and Reecy 2007; Hu et al. 2013), which is striving to collect all publicly available mapping data for livestock animal genomes (for cattle, pigs and sheep QTL, see Table 10.1). The aim of this database is mainly to facilitate centralized recording of QTL locations and allow comparison of discoveries within and between species. In addition, new data and database tools are continually being developed to align various trait mapping data to map-based genome features, such as annotated genes. This database is linked with the NCBI gene and map viewer resources (http://www.ncbi.nlm.nih.gov/gene?term=cattle%20QTL) where more information is available.

Not all QTL/genes qualify as a major gene, which can be defined as having at least one phenotypic standard deviation (s_p) between the homozygotes for the metric trait (Hill and Knott 1990). However, QTL with large effects are rare and Hayes and Goddard (2001) found, for example, that the segregation of large QTL > 1s_p explains

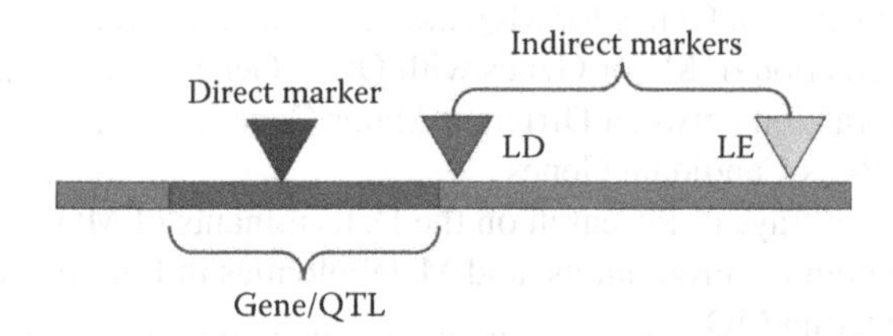

FIGURE 10.1 Diagram of the relationships between markers and the gene controlling the trait of interest. LD = linkage disequilibrium (marker likely to be inherited with the gene/QTL); LE = linkage equilibrium (alleles of the marker and gene/QTL inherited in proportions as would be expected at random from allele frequencies).

TABLE 10.1
Number of QTL, Associated Publications, and Investigated Traits Curated in the Animal QTLdb Database for Cattle, Pigs, and Sheep (January 2014)

Species	QTL	Publications	Traits
Cattle	8305	454	467
Pigs	9862	391	653
Sheep	789	90	219
Sum	18,956	935	

only about 5% of the total variance in dairy cattle, and even less for pigs. For this chapter, we will, therefore, apply a less stringent approach to "major genes" and also include QTL/genes with smaller effects than $1s_p$. Consequently, many of the QTLs curated in the AnimalQTLdb could be of relevance to the subject of this chapter. It is probably the case that the majority of genes for the traits of interest have not yet been detected. Therefore, we have detailed information on some genes, but must continue to rely on phenotypic data for the majority of genes. This situation will probably remain for the foreseeable future, until all genes have been sequenced and identified, and will not principally change with the application of genomic selection (see Sections 10.4 and 10.5). It remains the ultimate goal of complex trait dissection to identify the actual genes involved in the trait and to understand the cellular roles and functions of these genes. A synergistic approach is required, involving all sources of knowledge, comprising sequencing, genes/QTL, mapping, genome-wide association studies (GWAS), copy number variations, and phenotyping using genomics, proteomics, and metabolomics. This chapter reviews some of the identified genes/QTL that significantly influence meat quality (MQ) in major livestock species (cattle, pigs, and sheep).

10.2 MAJOR GENES AFFECTING MQ IN THE MAIN SLAUGHTER SPECIES

The number of QTL identified in Table 10.1 implies that a comprehensive review would be impossible. A selected number of well-documented QTL, and those with larger effects on MQ traits in livestock species, are discussed below. The main traits affected are summarized in Table 10.2. Although many studies have found associations between QTL and MQ traits, the results of association studies should be confirmed through independent studies before relying on this information, especially since the associations between genotype and phenotype can differ between populations or breeds (Hocquette et al. 2012.)

10.2.1 Cattle

The 8305 QTL in the cattle QTLdb (http://www.animalgenome.org/cgi-bin/QTLdb/BT/index) are based on 454 publications and represent 467 traits, of which the

TABLE 10.2
Associations between Major Genes/QTL and Meat Quality Traits

Meat Quality Trait	Species	Associated Polymorphism (Preferred Genotype)	Reference
		Glycolytic-Energetic Resources of Muscle	
Glycogen	Pigs	RN$^-$ phenotype (rn$^+$rn$^+$)	Przybylski et al. (1996), Enfalt et al. (1997), Kocwin-Podsiadla et al. (1998, 1999, 2006b)
		PRKAG3 [R200Q]	Maedus et al. (2002)
		PRKAG3 [I199V]	Ciobanu et al. (2001), Josell et al. (2003)
		PRKAG3 [T30N]	Qiao et al. (2011)
		PKM2 (CC)	Fontanesi et al. (2003)
		CAST/HinfI (AA)	Kocwin-Podsiadla et al. (2004a, 2006a), Krzecio et al. (2004, 2006)
		ACSL4 (GG)	Rusc et al. (2011)
	Cattle	MSTN	Review by Fiems (2012)
Lactate	Pigs	PRKAG3 [R200Q]	Kocwin-Podsiadla et al. (2004b, 2006e)
		PRKAG3 [T30N]	Fontanesi et al. (2008)
		PRKAG3 [I199V]	Ciobanu et al. (2001)
		PKM2 (CC)	Sieczkowska et al. (2007, 2010a)
		CAST/MspI (AA)	Kocwin-Podsiadla et al. (2004a, 2006a), Krzecio et al. (2004, 2005b)
Glycolytic potential	Pigs	RN$^-$ phenotype (rn$^+$rn$^+$)	Monin et al. (1992), Przybylski et al. (1996), Kocwin-Podsiadla et al. (2004a,b, 2006b), Carr et al. (2006)
		PRKAG3 [R200Q] (GG)	Maedus et al. (2002), Kocwin-Podsiadla et al. (2004a, 2006d)
		PRKAG3 [I199V]	Ciobanu et al. (2001)
		PKM2 (CC)	Sieczkowska et al. (2007, 2010a)
		CAST/HinfI (AA)	Kocwin-Podsiadla et al. (2004a, 2006a) Krzecio et al. (2004, 2007b)
		ACSL4 (GG)	Rusc et al. (2011)
R_1 (IMP/ATP)	Pigs	RYR1 (CC)	Kocwin-Podsiadla et al. (1995, 1998, 2004a) Przybylski et al. (1998)
		PKM2 (CC)	Sieczkowska et al. (2010a)
		CAST/Rsa (AA)	Kocwin-Podsiadla et al. (2004a, 2006a), Krzecio et al. (2004, 2005b, 2008)
		H-FABP/MspI (aa)	Kocwin-Podsiadla et al. (2004a)
		(Aa)	Sieczkowska et al. (2006a)
Proportion of glycolytic muscle fibers	Cattle	MSTN	Albrecht et al. (2006), Bellinge et al. (2005)
	Sheep	MSTN	Laville et al. (2004)
		Callipyge	Koohmaraie et al. (1995)

(*Continued*)

TABLE 10.2 (*Continued*)
Associations between Major Genes/QTL and Meat Quality Traits

Meat Quality Trait	Species	Associated Polymorphism (Preferred Genotype)	Reference
		Physical/Chemical Properties	
pH_1	Pigs	RYR1 (CC)	Kocwin-Podsiadla et al. (1995, 1998, 2004a), Przybylski et al. (1998), Otto et al. (2007)
		PRKAG3 [I199V]	Fontanesi et al. (2003, 2008)
		PKM2 (CC)	Fontanesi et al. (2008), Sieczkowska et al. (2010a)
		MYOG	Zhu and Li (2005), Jiusheng et al. (2009), Kim et al. (2009)
		CAST/HinfI	Kocwin-Podsiadla et al. (2004a, 2006a)
		CAST/RsaI	Krzecio et al. (2004, 2007b, 2008)
		H-FABP/HaeIII (Dd, dd) H-FABP/MspI (AA)	Nechtelberger et al. (2001)
pH_{24}	Pigs	RYR1 (CC)	Van den Maagdenberg et al. (2008)
		RN^- phenotype (rn^+rn^+)	Przybylski et al. (1996), Kocwin-Podsiadla et al. (2004a, 2006b), Carr et al. (2006)
		PRKAG3 [R200Q] (GG)	Kocwin-Podsiadla et al. (2004a, 2006d,e), Krzecio et al. (2008)
		PRKAG3 [I199V]	Josell et al. (2003), Qiao et al. (2011)
		PKM2 (CC)	Sieczkowska et al. (2010b)
		CAST/HinfI (AA)	Kocwin-Podsiadla et al. (2004a, 2006a)
		CAST/MspI (AA)	Krzecio et al. (2004, 2007b, 2008)
pH_{48}	Pigs	RN^- phenotype (rn^+rn^+)	Kocwin-Podsiadla et al. (2004a,b, 2006c)
		PRKAG3 [I199V]	Josell et al. (2003)
		MYOG (BB)	Kocwin-Podsiadla et al. (2004a), Krzecio et al. (2008)
		CAST/MspI (AA)	Kocwin-Podsiadla et al. (2004a, 2006a), Krzecio et al. (2004, 2005b)
		CAST /RsaI (AA)	Krzecio et al. (2008)
		ACSL4 (GG)	Rusc et al. (2011)
Electrical conductivity	Pigs	RYR1 (CC)	Kocwin-Podsiadla et al. (2004a), Van den Maagdenberg et al. (2008)
		MYOG (BB)	Kocwin-Podsiadla et al. (2004a), Krzecio et al. (2008)
		CAST/MspI (AA)	Kocwin-Podsiadla et al. (2004a, 2006a)
		CAST/Rsa (AA)	Krzecio et al. (2004, 2005b, 2008)
Meat color	Pigs	RYR1 (CC)	Kocwin-Podsiadla et al. (1993, 1995, 1998, 2004a). Otto et al. (2007)
		RN^- phenotype (rn^+rn^+)	Przybylski et al. (1996), Kocwin-Podsiadla et al. (2004a,b, 2006b), Hullberg et al. (2005)

(*Continued*)

TABLE 10.2 (*Continued*)
Associations between Major Genes/QTL and Meat Quality Traits

Meat Quality Trait	Species	Associated Polymorphism (Preferred Genotype)	Reference
		PRKAG3 [T30N]	Fontanesi et al. (2003)
		PRKAG3 [I199V]	Ciobanu et al. (2001)
		PKM2 (CC)	Sieczkowska et al. (2010a)
		MYOG	Zhu and Li (2005), Jiusheng et al. (2009)
		H-FABP/HaeIII (dd)	Kocwin-Podsiadla et al. (2004a)
		H-FABP/MspI (aa)	Sieczkowska et al. (2006a)
		(Aa)	Sieczkowska et al. (2006a)
		H-FABP/HinfI (Hh)	Nechtelberger et al. (2001)
	Cattle	MSTN	Many authors, see reviews by: Arthur (1995), Fiems (2012)
Drip loss	Pigs	RYR1 (CC)	Van den Maagdenberg et al. (2008)
		RN⁻ phenotype (rn^+rn^+)	Enfalt et al. (1997), Kocwin-Podsiadla et al. (2004a,b, 2006b)
			Bertram et al. (2000), Hullberg et al. (2005), Carr et al. (2006)
		PKM2 (CC)	Sieczkowska et al. (2010a)
		GLUT 4 (CC)	Grindflek et al. (2002), Sieczkowska et al. (2007)
		MYOG	Zhu and Li (2005)
		CAST/HinfI	Kocwin-Podsiadla et al. (2004a, 2006a), Krzecio et al. (2004, 2007b)
		CAST/RsaI	Kocwin-Podsiadla et al. (2006a)
		H-FABP/HaeIII (dd)	Kocwin-Podsiadla et al. (2004a), Antosik et al. (2006)
		H-FABP/MspI (aa)	Kocwin-Podsiadla et al. (2004a)
		(Aa)	Sieczkowska et al. (2006a)
		H-FABP/HinfI (Hh)	Nechtelberger et al. (2001)
WHC	Pigs	RYR1 (CC)	Kocwin-Podsiadla et al. (1995, 1998, 2004a)
		MYOG	Jiusheng et al. (2009)
		Technological and Culinary Usefulness	
RTN	Pigs	RN⁻ phenotype (rn^+rn^+)	Przybylski et al. (1996), Enfalt et al. (1997), Hullberg et al. (2005)
TY	Pigs	RYR1 (CC)	Kocwin-Podsiadla et al. (2004a)
		RN⁻ (rn^+rn^+)	Kocwin-Podsiadla et al. (2004a,b)
		CAST/Rsa (AA)	Kocwin-Podsiadla et al. (2004a, 2006a), Krzecio et al. (2004, 2005b, 2008)
		H-FABP/HaeIII (DD)	Kocwin-Podsiadla et al. (2004a), Sieczkowska et al. (2006a)
Cooking loss	Pigs	RN⁻ phenotype (rn^+rn^+)	Enfalt et al. (1997)

(*Continued*)

TABLE 10.2 (*Continued*)
Associations between Major Genes/QTL and Meat Quality Traits

Meat Quality Trait	Species	Associated Polymorphism (Preferred Genotype)	Reference
		PRKAG3 [R200Q](GG)	Maedus et al. (2002)
		PRKAG3 [I199V]	Josell et al. (2003)
		CAST/Hpy188I	Ciobanu et al. (2004)
		Basic Chemical Composition	
IMF content	Pigs	RN⁻ phenotype (rn^+rn^+)	Moeller et al. (2003)
		MYOG	Zhu and Li (2005)
		H-FABP/HaeIII (dd)	Gerbens et al. (1999)
		H-FABP/MspI (aa)	Sieczkowska et al. (2006a)
		H-FABP/HinfI (HH)	Gerbens et al. (1999)
			Gerbens et al. (1997, 1999)
			Urban et al. (2002)
		ACSL4 (GG)	Rusc et al. (2011)
	Cattle	MSTN	Many authors, see reviews by Arthur (1995), Bellinge et al. (2005)
		DGAT1	Thaller et al. (2003)
		TG	Thaller et al. (2003)
		FABP4	Michal et al. (2006)
		RORC	Barendse et al. (2010)
		SCD1	Taniguchi et al. (2004), Wu et al. (2012)
		IGF2	Sherman et al. (2008)
	Sheep	MSTN	Kijas et al. (2007)
		Callipyge	Koohmaraie et al. (1995)
		TM-QTL	Lambe et al. (2010b)
Protein content	Pigs	RN⁻ phenotype (rn^+rn^+)	Monin et al. (1992)
		PRKAG3 [R200Q]	Kocwin-Podsiadla et al. (2004a, 2006d,e)
		PRKAG3 [I199V]	Josell et al. (2003)
		CAST/Rsa (AA)	Kocwin-Podsiadla et al. (2004a, 2006a) Krzecio et al. (2004, 2005b)
		H-FABP/HaeIII (DD)	Kocwin-Podsiadla et al. (2004a)
		H-FABP/HinfI (Hh)	Antosik et al. (2006)
Dry matter content	Pigs	MYOG (BB)	Krzecio et al. (2008)
		H-FABP/HaeIII (DD)	Kocwin-Podsiadla et al. (2004a)
		(dd)	Sieczkowska et al. (2006a)
		H-FABP/HinfI (Hh)	Antosik et al. (2006)
Collagen content	Cattle	MSTN	Albrecht et al. (2006), Bellinge et al. (2005)

(*Continued*)

TABLE 10.2 (*Continued*)
Associations between Major Genes/QTL and Meat Quality Traits

Meat Quality Trait	Species	Associated Polymorphism (Preferred Genotype)	Reference
		Organoleptic (Sensory) Characteristics	
Meat tenderness (including mechanical tenderness)	Pigs	RYR1 (CC)	Van den Maagdenberg et al. (2008)
		RN^- phenotype (rn^+rn^+)	Przybylski et al. (1996), Enfalt et al. (1997), Kocwin-Podsiadla et al. (1998, 2004a,b, 2006b), Hullberg et al. (2005)
		CAST/HinfI (AA)	Kocwin-Podsiadla et al. (2004a, 2006a), Krzecio et al. (2004, 2007b)
		CAST/Hpy188I	Ciobanu et al. (2004)
		H-FABP/HaeIII (DD)	Kocwin-Podsiadla et al. (2004a)
	Cattle	MSTN	Review by Arthur (1995), Uytterhaegen et al. (1994)
		CAPN1	Page et al. (2002), Casas et al. (2006), Morris et al. (2006)
		CAST	Casas et al. (2006), Schenkel et al. (2006), Morris et al. (2006)
	Sheep	MSTN	Hope et al. (2013)
		Callipyge	Shackelford et al. (1997), Taylor and Koohmaraie (1998), Freking et al. (1999), Duckett et al. (1998a), Abdulkhaliq et al. (2007)
		Carwell	Hopkins and Fogarty (1998), Jopson et al. (2001)
		TM-QTL	Lambe et al. (2010a,b, 2011)
Taste/flavor	Pigs	RN^- phenotype (rn^+rn^+)	Lundstrom et al. (1994)
	Cattle	MSTN	Review by Bellinge et al. (2005), Allais et al. (2011)
Juiciness	Pigs	CAST/Hpy188I	Ciobanu et al. (2004)
	Sheep	Callipyge	Shackelford et al. (1997)

Note: R_1 (IMP/ATP)—ratio of inosine monophosphate to adenosine triphosphate; pH_1, pH_{24}, pH_{48}—pH measured at 1, 24, and 48 h after slaughter, respectively; GP—glycolytic potential; WHC—water-holding capacity; RTN—Napole yield; TY—technological yield of cured meat in terminal processing; IMF—intramuscular fat; 1—genetic groups without Duroc breed; 2—genetic groups with Duroc breed share; 3—genetic groups with heterozygous carriers of $RYR1^T$.

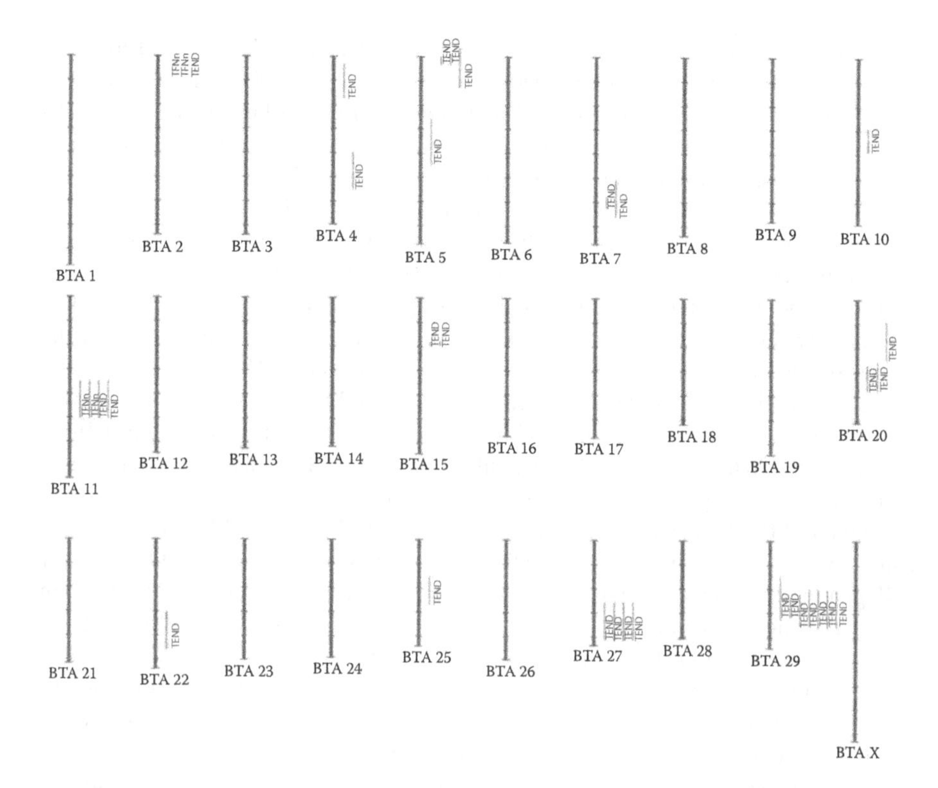

FIGURE 10.2 January 2014 screenshot from cattle QTLdb (http://www.animalgenome.org/cgi-bin/QTLdb/BT/traitmap?trait_ID=1030) showing the location of QTL/genes associated with tenderness (TEND) in cattle.

majority belong in the broad category of milk performance and milk composition, feed intake and growth, but there are also 160 affecting shear force and 154 affecting marbling score, which are two of the main MQ traits in beef. If sorted by trait type, the category "MQ" provides 1428 QTL, with 2099 curated in the trait class "meat." Figure 10.2 provides an example of different QTLs affecting tenderness that have been identified across all cattle chromosomes to date.

10.2.1.1 Myostatin (MSTN)

The growth and differentiation factor gene (GDF8; now known as myostatin, MSTN) was first shown in mice, by a knock-out study, to have a large enhancing effect on muscling (McPherron and Lee 1997). Considering this gene then as a candidate gene, studies in cattle, and later other species, identified it as highly polymorphic, with several variants associated with muscular hypertrophy, often referred to as double-muscling. In cattle, MSTN has been mapped to the distal end of chromosome 2 (Grobet et al. 1998). MSTN is responsible for repressing growth of skeletal muscles; therefore, if this gene is inactivated, or animals are treated with substances that block myostatin activity, muscular hypertrophy (enlargement of muscle fibers) and hyperplasia (increase in muscle fiber number) usually occur. Several independent mutations in MSTN (reviewed by Bellinge et al. 2005) have been found to cause this phenotype in different cattle breeds (Grobet et al. 1997, 1998) and it is believed that there may be further unidentified mutations having similar effects. For example, Grobet et al. (1998) did not find evidence that the identified MSTN mutations found to be responsible for muscular hypertrophy in other breeds were causing this observed phenotype in Limousin or Blonde d'Aquitaine cattle.

MSTN mutations also have pleiotropic effects in terms of carcass composition. Double-muscled (DM) cattle have greatly increased muscle mass, reduced fat, and reduced bone, resulting in an increased dressing percentage. The size of the effect differs between different muscles, resulting in a greater proportion of meat in the higher-priced cuts (Arthur 1995). However, other associated effects are a reduction in the size of vital organs and an increase in susceptibility to stress, as well as lower fertility, more calving difficulties, and lower calf viability (Ansay and Hanset 1979; Hanset 1991; Arthur 1995; Bellinge et al. 2005). As MSTN plays a major role in pathways controlling muscle differentiation, growth, and development, it is not surprising that MSTN variations are also associated with MQ traits. Muscles from DM cattle contain a larger proportion of white fast-twitch glycolytic muscle fibers, lower collagen (with reduced cross-linkage), and lower amounts of connective tissue than those from non-DM animals (Bellinge et al. 2005; Albrecht et al. 2006). Levels of intramuscular fat (IMF) and marbling (the visual appearance of IMF) are also lower, resulting in reduced flavor. This fat depot also contains a higher proportion of polyunsaturated fats, beneficial for human health (Bellinge et al. 2005; Allais et al. 2010). Meat from DM cattle has a lower myoglobin content, which has implications not only for oxygen transport but also for meat color, and meat from DM cattle is well documented as being paler than meat from non-DM animals. However, the meat from DM cattle is also more likely to suffer from DFD (dry, firm, dark meat) due to a high ultimate pH, reflecting fiber type, as discussed below. The likelihood of high ultimate pH is increased in DM cattle by a reduced amount of available glucose in the muscle, a high proportion of glycolytic muscle fibers (increasing the rate of glycolysis postmortem) and increased susceptibility to stress (reviewed by Fiems 2012).

Different studies have produced contradictory results of the effects of DM on tenderness, although the general conclusion from most studies is that tenderness is increased (Uytterhaegen et al. 1994). This fits with the lowered collagen content of DM meat, since collagen increases toughness. Differences between studies in the direction of this association with tenderness have been attributed to nongenetic effects, such as management and processing factors (Arthur 1995). Differences in calpain and calpastatin activities between DM and non-DM cattle have been reported, although results are inconsistent across different breeds, muscles, years, and studies (reviewed by Fiems 2012) and the role of myofibrillar tenderization in the effects on tenderness of meat from DM animals is unclear (Arthur 1995; Arthur and Hearnshaw 1995).

The reported effects of DM on muscle histology and MQ have implications for the slaughter process. The higher susceptibility to stress requires more careful preslaughter handling to avoid DFD. An increased muscle mass in the carcass, especially of heavier animals, requires consideration in terms of the chilling regime, to avoid heat-toughening (which is likely when the pH falls below 6 at temperatures >35°C).

In summary, the availability of genetic tests to detect different mutations in this gene allows the use of MSTN in genetically controlled breeding strategies. Its substantial effects on carcass value (dressing out percentage, saleable meat yield, conformation) are well documented, but adverse effects on fitness (dystocia, stress susceptibility, fertility) are also known, and its varying effects on different MQ traits need to be considered. Additionally, there is a need to know the current frequency in the breed of interest and its mode of inheritance, which has remained controversial for some time, but evidence of a partial recessive inheritance is accumulating (e.g., Grobet et al. 1997). The latter might be of use to somewhat balance positive and negative effects and suggests, in some cases, benefits from the use of heterozygous animals or the use of a genetic background with less dramatic effects.

10.2.1.2 DGAT1 (Diacylglycerol *O*-Acyltransferase 1) and Other QTL Affecting IMF

Several polymorphisms have been identified as predictors of IMF or marbling in beef. The diacylglycerol *O*-acyltransferase 1 (*DGAT1*) and the thyroglobulin (*TG*) genes are both located in the centromeric region of bovine chromosome 14 and polymorphisms in both genes have been shown to affect IMF deposition (Thaller et al. 2003). Other examples include a single-nucleotide polymorphism (SNP) in the fatty acid binding protein 4 gene (*FABP4*) that is associated with marbling score (Michal et al. 2006), an SNP in the retinoic acid receptor-related orphan receptor C (*RORC*) gene that affects marbling and IMF (Barendse et al. 2010) and an SNP in the stearoyl-CoA desaturase (*SCD1*) enzyme reported by Taniguchi et al. (2004) affecting fatty acid composition, as well as IMF and marbling score (Wu et al. 2012).

Of these polymorphisms affecting IMF, the K232A SNP on the DGAT1 gene is one of the more widely studied in different cattle breeds and populations. DGAT1 is a microsomal enzyme that catalyzes the final step of triglyceride synthesis and the lysine/alanine polymorphism on the gene encoding for this enzyme has previously been shown to affect milk fat content (Grisart et al. 2002; Winter et al. 2002). The original study by Thaller et al. (2003), which identified the association between this polymorphism and IMF in German Holstein and Charolais cattle, found a stronger association with IMF in the leg muscle (*semitendinosus*) than the loin muscle (*longissimus lumborum*). The lysine allele of the DGAT1 polymorphism (known as K) was found to be associated with greater IMF levels (as was the case for milk fat in previous studies; Winter et al. 2002), and was found to be recessive (Thaller et al. 2003). Pannier et al. (2010) found differences in allele frequencies at the K232A SNP between different breeds of cattle in Ireland, but found no evidence that this polymorphism significantly affected IMF in crossbred commercial cattle, in either the *longissimus* or *semimembranosus* muscles, although only two animals of the genotype KK were included in the crossbred study population. In a study of Spanish crossbred cattle, a significant association between the DGAT1 genotype and back fat thickness was found, but not between the DGAT1 genotype and IMF (Aviles et al. 2013), although this was measured in the *longissimus* muscle. IMF in this muscle was found to show less association with DGAT1 than IMF in the *semitendinosus* in the original study by Thaller et al. (2003). However, a study of Swedish cattle populations (Li et al. 2013) found a significant association between the DGAT1 genotype and both IMF (%) and marbling score of the *longissimus* muscle, with heterozygous (KA) animals having higher levels of fat than homozygous (AA) animals. Insufficient homozygous KK animals were available to assess the effects of carrying two copies of the allele associated with greater IMF. Other SNPs were identified on DGAT1 in Chinese cattle breeds that affected some carcass and MQ traits, including marbling score (Yuan et al. 2013), when measured in the loin region.

From the studies referenced above, allele frequencies for the K232A polymorphism differ markedly between breeds and between populations, as do the reported associations with IMF. Several of these studies conclude, therefore, that these markers have potential for use in certain cattle populations where associations with IMF have been confirmed. However, widespread use across different breeds or populations cannot be recommended.

10.2.1.3 CAPN1 (μ-Calpain Gene) and CAST (Calpastatin Gene)

Postmortem tenderization of muscle is due to enzymatic degradation of myofibrillar proteins (see Chapter 3). The μ-calpain (CAPN1) gene is known to encode μ-calpain, the protease responsible for this process, while the calpastatin (CAST) gene encodes the inhibitor of that protease (Koohmaraie 1996). Markers associated with significant effects on meat tenderness have been identified in the CAPN1 gene on bovine chromosome 29 and the CAST gene on chromosome 7 (Page et al. 2002; Casas et al. 2006; Morris et al. 2006; Schenkel et al. 2006).

Commercial marker tests for meat tenderness that use polymorphisms on the CAPN1 and CAST genes have been marketed by companies in the United States and Australia. These genetic tests (GeneSTAR Tenderness and Igenity *Tender*-GENE) were independently validated by the National Beef Cattle Evaluation Consortium in the United States on over 1000 *Bos taurus* and *Bos indicus* cattle (Van Eenennaam et al. 2007) and highly significant associations with tenderness, as measured by Warner–Bratzler shear force, were reported. The difference between the genotypes associated with the most tender and least tender meat was 9.8 N of shear force (~0.6 s_p). However, some studies using different beef cattle populations in different countries have found inconsistencies between breeds in the effects of SNPs in these genes on meat tenderness, or in which markers result in the largest effects on meat tenderness. For example, Casas et al. (2006) found significant associations between SNPs on the CAST and CAPN1 genes in two mixed-breed populations (with only *Bos taurus* or including *Bos indicus* influences), but not in a third purebred Brahman population. Allais et al. (2011) found that the effects of different markers for CAST and CAPN1 on meat tenderness in three French beef breeds were breed-specific. While effects of CAPN1 polymorphisms on tenderness were confirmed in crossbred Irish cattle (Costello et al. 2007), reported associations between CAST polymorphisms and shear force were not found in this same population (Reardon et al. 2010). The authors suggested that this could imply a different linkage disequilibrium (LD) relationship between this marker and the causative location in this population. LD is due to the proximity on the chromosome of the marker to the causative gene/QTL and determines how likely they are to be inherited together (Figure 10.1).

Associations with other MQ traits have been identified. For example, Casas et al. (2006) identified significant relationships between genotypes at SNPs in the CAST and

CAPN1 genes and flavor intensity of beef steaks. In the Irish study discussed above, where no association was found between the SNP on the CAST gene and tenderness, this same SNP was found to significantly affect loin pH and color parameters of both loin and leg muscles (Reardon et al. 2010). A study of bulls from five Swedish beef breeds (Li et al. 2013) found that meat color (after 6 days exposure to air) and marbling were significantly associated with an SNP on the CAPN1 gene (CAPN1:c.947), although this SNP did not affect water-holding capacity (WHC), pH, or other MQ traits, whereas an SNP on the CAST gene (CAST:c.155) did not affect any of the MQ traits tested.

10.2.2 Sheep

The SheepQTLdb (http://www.animalgenome.org/cgi-bin/QTLdb/OA/index; Table 10.1) contains all curated sheep QTL, amounting to 789 to date, based on 90 publications, with associations to 219 traits. This number is of course much smaller (≪10%) than for cattle and pigs, reflecting the different research focus in some countries, rather than a true "species effect." The top ovine QTL listed are related to nematode resistance, aseasonal reproduction, milk yield, carcass composition, and growth, rather than to MQ, but sorting for trait types provides 6, 5, and 2 QTL for meat color, texture, and meat eating quality (MEQ), respectively. On each of the 26 ovine autosomes, 30 QTL have been identified on average (ranging from 6 on chromosome 19–148 on chromosome 2).

10.2.2.1 Myostatin (MSTN)

A number of polymorphisms of the MSTN gene on chromosome 2 have been reported to segregate in various breeds or strains of sheep, including Belgian, NZ, UK, and Australian Texel, Poll Dorset, Lincoln, Charollais, NZ Romney, Norwegian White, and East Friesian. As described above for cattle, MSTN inhibits skeletal muscle growth, and thus lambs with certain mutations on this gene have increased muscle growth. The c.*1232 G > A mutation on MSTN was first reported in Texel sheep in New Zealand (Broad et al. 2000) and a commercial marker test (MyoMAX®) is now available (Dodds et al. 2007). The mutation affects muscle and fat traits in some Texel-sired lambs (Johnson et al. 2005b), with no effect on shear force, color, and pH of the *longissimus* and *semimembranosus* muscles (Johnson et al. 2005a). Laville et al. (2004) reported a shift toward more glycolytic muscle fibers in lambs carrying the mutation, alongside increases in muscle thickness and weight, particularly in the hind limb, but no other MQ traits were studied.

One area of concern for mutations that affect muscling is always the increased likelihood of decreasing IMF levels, which can lead to a detrimental effect on eating quality (Hopkins et al. 2007a). Kijas et al. (2007) reported a reduction in IMF% and sensory scores associated with this MSTN mutation, but no effect on shear force, although there was evidence of another QTL that affected MQ traits. An inferred reduction in IMF percentage, based on muscle density measured by CT scanning, was also reported for crossbred carriers of the mutation by Masri et al. (2011a). In a small experiment that included both homozygote and heterozygote male lambs for the mutation, it was shown that the mutation did not cause a significant reduction in IMF percentage, shear force, or compression of the *longissimus* and *semimembranosus* muscles or impact on collagen content and solubility, but did improve the sensory tenderness of the *semimembranosus* (Hope et al. 2013).

The mode of inheritance for the MSTN mutation in sheep seems to be influenced by breed and by trait, but for muscle growth traits, it is usually reported as partially recessive (Hadjipavlou et al. 2008; Masri et al. 2011b), implying that the heterozygous animals differ only moderately from homozygous wild-type animals, which has an impact on the use of this mutation in sheep breeding strategies, especially in crossbreeding schemes.

10.2.2.2 Callipyge

The callipyge effect, originally identified in U.S. sheep flocks, is manifested as a major increase in hindquarter muscling caused by muscle hypertrophy (Cockett et al. 1999). The condition is due to a mutation in the callipyge gene on chromosome 18 and is expressed in heterozygous animals with maternal imprinting, meaning that the mutation is only expressed when a single copy is inherited from the sire (termed "polar overdominance") (Cockett et al. 1996; Freking et al. 1998; Shackelford et al. 1998). Whereas animals with one paternally inherited copy have superior carcass characteristics (leaner, higher dressing percentage; Freking et al. 1998), this mutation leads to a drastic impairment of MQ. The *longissimus* muscle of callipyge lambs has much higher levels of shear force (Shackelford et al. 1997; Taylor and Koohmaraie 1998; Freking et al. 1999), even after 24 days conditioning postslaughter (Duckett et al. 1998b). Significantly reduced myofibrillar fragmentation index values measured in the loin muscle from these lambs are indicative of a decrease in protein degradation (Koohmaraie et al. 1995). It appears that postmortem proteolysis is delayed in muscle from callipyge lambs due to higher levels of calpastatin (Koohmaraie et al. 1995; Geesink and Koohmaraie 1999b). The extent of proteolysis is also reduced (Geesink and Koohmaraie 1999a), with much less structural disruption of the sarcomere (Taylor and Koohmaraie 1998). It appears the effect is manifested solely through the myofibrillar component of the *longissimus* muscle with no impact via collagen (Field et al. 1996). The IMF percentage is reduced, with no effect on the rate of pH decline (Shackelford et al. 1995), although there is a change in the proportion and size of muscle fibers with an increase in the proportion of glycolytic fibers (Koohmaraie et al. 1995). While there is increased toughness (Abdulkhaliq et al. 2007) and reduced juiciness of the *longissimus* from callipyge lambs, there is no reduction in tenderness in other muscles (Shackelford et al. 1997; Abdulkhaliq et al. 2002). Increased cooking loss in meat from callipyge lambs has been reported (Abdulkhaliq et al. 2007), although not always (Shackelford et al. 1997).

In summary, although callipyge could provide a drastic improvement in the production of lean slaughter lambs, the negative effects on MQ from callipyge do not allow its use in commercial scenarios.

10.2.2.3 Carwell (Loin-Max; LM-QTL)

An industry progeny testing program indicated that some rams from an Australian Poll Dorset stud (Carwell) produced progeny with a large positive deviation for eye muscle area (Banks et al. 1995). Subsequently, one of the four markers for the callipyge gene, TGLA 122 on chromosome 18, was found to be present in Australian animals exhibiting this increased muscling (Nicoll et al. 1998). The Carwell mutation is thought to be maternally imprinted and a marker test is now available under the trademark LoinMAX® (Dodds 2007).

The Carwell mutation shows a smaller phenotypic effect than callipyge on muscling, increasing the weight of the *longissimus* by 8%–10% (Hopkins and Fogarty 1998; Nicoll et al. 1998) and its muscle area by 4.5% (Masri et al. 2010). Carwell was found to increase shear force in the *longissimus* by 35% in male lambs that had a high preslaughter growth rate (Hopkins and Fogarty 1998), but there was no effect in female lambs under low growth rate (Hopkins et al. 1997). In a more recent study including progeny by the same Carwell sire, Hopkins et al. (2007c) found no increase in toughness of the *longissimus* or *semimembranosus* muscles or any other effects on pH or color traits. In contrast, Jopson et al. (2001) found that loins, frozen at 24 h postmortem, from carrier progeny were tougher and unacceptable, based on values of 27–30 N shear force (Hopkins et al. 2006), although the differences disappeared when the loins were aged for 6 weeks. As the lambs were processed under an accelerated conditioning procedure, this indicates the differences in shear force could be overcome by myofibrillar protein degradation (aging). While the Carwell mutation does not have such an impact on carcass and meat traits as callipyge, its effects can be modified by the production and processing environments. Unless appropriate production and processing strategies are in place, there is a need for caution when selecting sires to avoid detrimental effects of progeny on MQ.

10.2.2.4 Texel Muscling QTL (TM-QTL)

A QTL identified on chromosome 18 in Texel sheep (Walling et al. 2004) has been found to increase loin muscle dimensions and weights (by ~4%–14%) in purebred and crossbred carrier lambs (Walling et al. 2004; Macfarlane et al. 2009, 2014). The TM-QTL is not based on the same mutation as callipyge, since the callipyge mutation was not found in the Texel sires initially used to identify the TM-QTL (S.C. Bishop, personal communication). However, it is found on the same chromosome, near the callipyge region, and has also been shown to exhibit the same polar overdominant mode of inheritance (Macfarlane et al. 2010, 2014; Matika et al. 2011). It cannot be fully excluded that TM-QTL is another mutation in the same gene. In male crossbred progeny of sires carrying the TM-QTL mutation, there was increased shear force and reduced IMF of the *longissimus* muscle. There was also a significant interaction with sex, as there was no effect in female lambs (Lambe et al. 2010b). In this study, there was no effect on shear force or IMF of the *vastus lateralis*, a muscle from the hind leg. The levels of IMF reported for the *longissimus* muscle in the Lambe et al. (2010b) study were very low compared with others (e.g., Hopkins et al. 2007b,c), as they were from Texel crossbred lambs slaughtered at a fixed age of 21 weeks, regardless of live weight or condition, and were below the level for optimal eating quality (Hopkins et al. 2006). While variable times between death and electrical stimulation and variable aging periods may have impacted on the results, the greater variation for shear force in lambs carrying the mutation highlights the need for caution in the dispersal of carrier genetics.

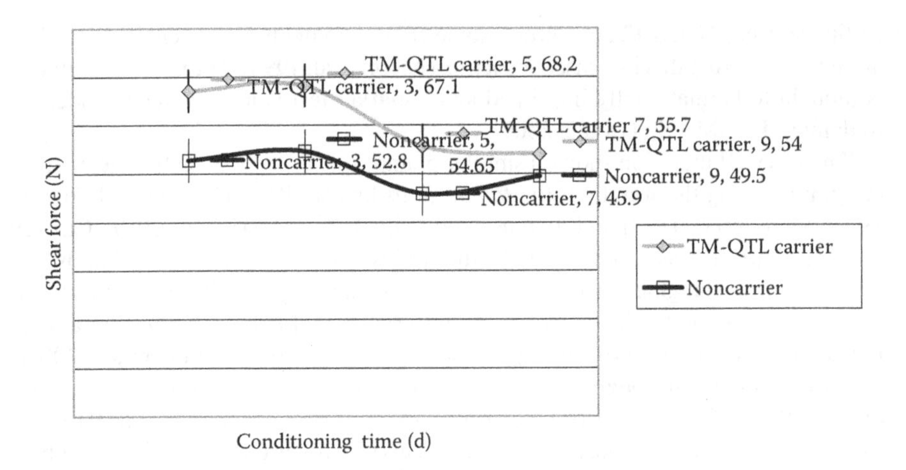

FIGURE 10.3 Mean shear force value of the *longissimus* muscle from crossbred lambs that were carriers or noncarriers of TM-QTL, after different periods of aging. (Adapted with permission from Lambe, N.R. et al. 2010a. *Meat Sci.* 85:715–720 with modification.)

In follow-up work, Lambe et al. (2010a) reported that crossbred carriers of TM-QTL had a significantly tougher *longissimus* muscle than noncarrier lambs when aged for 7 days or less (Figure 10.3) and there was a higher proportion of high shear force *longissimus* muscles among carriers than noncarriers after 3 and 5 days of aging, but not after 9 days. Sensory meat eating quality traits were not measured as part of this study and the exact relationship between shear force values and sensory assessments are difficult to assess and will vary according to the instrument used to measure shear force (Hopkins et al. 2011), and also potentially between different consumer populations. The threshold of consumer acceptability was taken as 5.5 kg force (approximately 54 N) in the study by Lambe et al. (2010a), based on previous studies using Warner–Bratzler shear force in beef and lamb (Shorthose et al. 1986; Miller et al. 2001; Platter et al. 2003; Destefanis et al. 2008), with the majority of samples (from both carriers and noncarriers) falling below this level after 7 days of aging. However, the absolute levels of shear force for all the genotypes studied would suggest unacceptable eating quality, if using the thresholds recommended by Hopkins et al. (2006). The relatively high shear force values reported by Lambe et al. (2010a) may indicate ineffective electrical stimulation. In further work by Lambe et al. (2011), Texel purebred carcasses carrying the TM-QTL were electrically stimulated and the *vastus lateralis* and *longissimus* aged for between 7 and 9 days. In this case, there was no effect of the mutation on mechanical compression (measured by two different devices), although the type of instrument used to test compression exhibited some tendency to alter the outcome. There was also no significant impact of the mutation on sensory traits, as assessed by a largely consumer panel, although the low numbers tested and the design of the taste panel tests limited the reliability of these results. Overall, the outcomes of these studies do not suggest that any detrimental effects on MQ will arise from the TM-QTL, provided minimum aging periods are applied.

10.2.3 Pigs

The pig genome mapping project has allowed localization of QTL for carcass and MQ traits important to modern consumers and processors. QTL have been found on all pig chromosomes and at the beginning of 2014 the database containing all of the identified QTL (PigQTLdb; http://www.animalgenome.org/cgi-bin/QTLdb) comprised 9862 QTL for 653 traits. Among the described QTL, the majority are related to meat and carcass quality with 1400 for texture, 358 for meat color, and 355 for pH, for example. Some of the most widely studied major genes for pig MQ are discussed below.

10.2.3.1 RYR1 (Ryanodine Receptor 1)

Localized on porcine chromosome 6, RYR1 contains two alleles: a normal allele ($RYR1^C$ previously named HAL^N) and a recessive mutant allele ($RYR1^T$ previously named HAL^n) (Fujii et al. 1991). The genotypes can be identified using a molecular marker to test pig tissues such as blood, skin, or hair (Houde et al. 1993).

Intensive breeding of pigs with the intention of increasing lean meat content has led to the occurrence of malignant hyperthermia syndrome, characterized by a rapid decrease in pH after slaughter (pH 45 min postmortem <6.0), which results in the occurrence of PSE meat in pigs (Pommier et al. 1998). The transition of C (cytosine) to T (thymine) at position 1843 in the RYR1 (ryanodine receptor 1) gene leads to a mutation that contributes to disturbed regulation of intracellular Ca^{2+} in pig skeletal muscles and leads to an alteration in Ca^{2+} homeostasis, hypermetabolism, and intense muscle contraction, resulting in malignant hyperthermia and reduced MQ (Fujii et al. 1991). The clinical symptoms are characterized by a deficit of oxygen and rapid glycolysis, accompanied by the production of lactic acid and acidosis, primarily in white muscle fibers (Fujii et al. 1991; Wendt et al. 2000). Results obtained by Laville et al. (2009) indicate that muscles of stress-susceptible (genotype TT) pigs contained fewer proteins of the oxidative metabolic pathway, fewer antioxidants, and more protein fragments. A lower abundance of small heat shock proteins and myofibrillar proteins in TT muscles may at least partly be explained by the effect of pH on their extractability.

Several studies have noted the influence of RYR1 gene polymorphisms on pH measured 35 and 45 min after slaughter; R_1 values (providing information on muscle tissue energy metabolism ratio = inosine monophosphate (IMP)/adenosine triphosphate (ATP)), which are the basis of PSE meat classification, and also on the lactate level in *longissimus lumborum* muscle (Sellier 1998; Krzecio et al. 2005a).

Pigs heterozygous for the RYR1 gene generally have a lower quality of meat. Heterozygous (CT) carriers of $RYR1^T$, compared to those resistant to stress (CC), were found to display about 1% higher drip loss from *longissimus* muscle tissue. This is explained by the more intensive glycolytic metabolism that takes place in those animals, both before and after slaughter (Kocwin-Podsiadla et al. 1995; Krzecio et al. 2005a; Otto et al. 2007).

An understanding of the PSE condition was the basis for breeding strategies aimed at eliminating the RYR1 mutation from global pig populations. It should be emphasized that the favorable effect of the $RYR1^T$ allele on lean meat content and carcass composition of TT pigs is considerably outweighed by very high frequency of PSE meat (36%–100% of carcasses). Therefore, the recessive RYR1^T allele has now been gradually, but effectively, removed from most pig breeds. The elimination of the RYR1 mutated gene from pig populations has other economic drivers, since fertility, reproduction, and daily weight gain are also significantly reduced in stress-susceptible pigs (Wendt et al. 2000). The discovery of the RYR1 recessive allele and decisions concerning elimination of the RYR1 mutation from pig breeds was the first example of practical implementation of molecular genetics in pork MQ improvement.

10.2.3.2 Rendement Napole (RN)/PRKAG3 (AMP-Activated Protein Kinase Subunit Gamma-3)

RN is a major gene in pigs (effect about 3 s_p; Le Roy et al. 1990) causing low ultimate pH and WHC in pork, which was first described by Naveau (1986) using segregation analysis of phenotypic data. The RN$^-$ allele is dominant, which implies one inherited copy from either parent can cause poor MQ. It can be identified by a Napole yield assessment, an indicator of the technological yield of ham production, or by an evaluation of the muscle glycolytic potential (GP) (Fernandez et al. 1990, 1992). This gene appears on porcine chromosome 15 in the form of two alleles: the rn$^+$ allele (normal recessive allele) and the RN$^-$ dominant allele (the allele with a negative effect on Napole yield). This gene has been found in Hampshires and Hampshire-based composite lines. The unfavorable RN$^-$ allele increases muscle glycogen level in the *longissimus* and *semimembranosus* muscles by approximately 40% to 70% (Monin et al. 1992). Meat from RN$^-$ pigs is paler than normal meat and has a low ultimate pH measured at 24 h after slaughter (<5.5), reduced WHC, lower (about 1%) protein content, and lowered yield of cured cooked hams. The RN$^-$ allele causes a large drip loss in fresh meat and also increases cooking loss (6%–9%). These combined effects result in major financial losses from carrier pigs (Monin and Sellier 1985; Fernandez et al. 1992; Przybylski et al. 1996; Enfalt et al. 1997). However, some positive effects of the mutation have been reported, with meat from RN$^-$ carriers characterized by better taste and aroma and also requiring less shear force in the slicing process (Lundstrom et al. 1994). The RN$^-$ allele also promotes high growth rate and high lean-muscle content (Enfalt et al. 1997; Kocwin-Podsiadla et al. 2000).

According to Closter et al. (2011), the preselection strategy designed by the Danish Meat Association to eliminate the RN$^-$ mutant in the Danish Hampshire population has avoided unacceptable loss of genetic gain and excessive loss of genetic variation. The design of the program has succeeded in eliminating the RN$^-$ allele without serious negative consequences either on genetic gain during the elimination program or on the level of co-ancestry in the population. The RN$^-$ mutant was found to act in either additive, dominant, recessive, or overdominant fashion, depending on the trait examined. Up to the year 2000, the phenotypic effects of the RN gene were estimated on the basis of Napole yield, then by *in vivo* or postmortem measurements of muscle GP (Fernandez et al. 1990; Lundstrom et al. 1996). Milan et al. (2000) found a mutation in the AMP-activated protein kinase subunit gamma-3 (PRKAG3) gene in codon 200 (Q200R) and showed a phenotypic effect of this substitution in the form of a reduction in the glycogen content of meat after slaughter. PRKAG3 codes for a regulatory subunit of adenosine monophosphate–activated protein kinase (AMPK), which is a heterotrimeric enzyme maintaining cellular and whole-body energy homeostasis (Andersson 2003). Following this, a new allele (V199I) in the PRKAG3 gene affecting glycogen content, ultimate pH, and color was found (Ciobanu et al. 2001). The V199I mutation had the opposite phenotypic effects to the R200Q mutation—carriers of the mutation had lower levels of muscular glycogen and therefore higher muscle pH_{24} values. Subsequent studies disclosed a number of mutations in the PRKAG3 encoding region. The Q200R mutation mainly exists in Hampshire pigs, whereas mutations of T32N, G52S, and V199I have been identified in other breeds of pigs (Milan et al. 2000; Ciobanu et al. 2001).

In subsequent years, the hypothesis that the PRKAG3 gene was responsible for increased levels of glycogen in muscle tissue was disproven and the debate on the role of this gene is still open (Andersson 2003, Fontanesi et al. 2003). No statistically significant differences between animals with AA and GG genotypes for the PRKAG3 (Q200R) gene (RN^-/RN^- and rn^+/rn^+, respectively) were found for GP level, glycogen content or protein content (typical for the RN^- phenotype) in the study of Kocwin-Podsiadla et al. (2006d). These authors showed that a high proportion (55%) of animals in a group of fatteners of genotype GG (rn^+/rn^+) had high GP (GP > 130 μmol/g with phenotype RN^-) and high muscle glycogen content. Zybert et al. (2009) found poor (6%) agreement between RN^- phenotype and PRKAG3 genotype in porkers. Also, Maedus et al. (2002) reported that the PRKAG3 mutation was not found in 27% of retail pork chop samples that had high GP values. Investigations in Italy (Fontanesi et al. 2003) also suggested other genetic influences, since high GP in some animals could not be explained by the presence of the Q200R allele. Qiao et al. (2011) excluded PRKAG3 as a causative gene for a QTL detected on porcine chromosome 15 with a strong effect on pH_{24} in *longissimus* and *semimembranosus* muscle tissue, suggesting a distinct maternally expressed QTL for pH_{24}. According to Galve et al. (2013), the decision of whether or not to use the V199I mutation of PRKAG3 as a pig selection criteria is complex and will depend on the genetic background of the selected population and/or the final use of the meat (fresh, cooked, or cured products).

10.2.3.3 PKM2 (Pyruvate Kinase Muscle 2)

Pyruvate kinase is an enzyme involved in glycolysis and catalyzes the reversible conversion of phosphenol pyruvate to pyruvate that, in anaerobic conditions, is reduced to lactic acid (Fontanesi et al. 2003; Pösö and Puolanne 2005). Pyruvate kinase appears in four forms: M1, M2, L, and R, of which forms M2 and M1 occur in muscles, heart, and brain; and L and R in liver and red blood cells, respectively (Imamura et al. 1986). M1 and M2 are coded by the gene PKM2, located on porcine chromosome 7 (Noguchi et al. 1986). A relationship between polymorphisms of PKM2 and glycogen content in the *longissimus lumborum* muscle at 45 min was reported by Fontanesi et al. (2003). Porkers with the CC genotype for PKM2, as compared to those with CT and TT genotypes, were characterized by a lower amount of glycogen and GP, lower lactic acid content and IMP/ATP ratio, darker color, and less drip loss of muscle tissue (Sieczkowska et al. 2007). Moreover, Sieczkowska et al. (2010a) observed in (Landrace × Yorkshire) × Duroc crossbreds, compared to Landrace × Yorkshire, an increased expression of the gene PKM2, closely related with an increase in IMF content, as well as a more favorable progress of glycolytic and energy metabolism during the early stage of transforming muscle tissue. Furthermore, the PKM2 gene in interaction with the GLUT4 gene (see below), which controls glucose transport into muscle cells, was found to modify IMF content in muscle tissue (Sieczkowska et al. 2007). This phenomenon was noted among Landrace pigs and was confirmed by a high phenotypic correlation ($r = 0.69$) between expression level of the PKM2 gene and IMF content in the *longissimus lumborum* muscle. This indicates the possibility of selection in the Landrace breed to breed pigs with optimal IMF content (2%–3%), as preferred by consumers (Bejerholm and Barton-Gade 1986).

10.2.3.4 GLUT4 (Insulin-Sensitive Glucose Transporter 4)

GLUT4 (also called SLC2A4), on porcine chromosome 12 (Grindflek et al. 2002), is an interesting candidate gene for MQ and carcass traits because of its role in muscle glucose metabolism and adipose tissue. GLUT4 is an isoform present exclusively in muscle and adipose cells and plays a key role in cellular glucose uptake stimulated by insulin in these cells (Abe et al. 1997). Glucose transport in cells is mediated by a family of facilitative glucose transporters that catalyze the diffusion of glucose across the plasma membrane. The SLC2A4 protein is unique among these with respect to its unusual muscle/fat tissue specific expression (Fukumoto et al. 1989; Abe et al. 1997).

There is some evidence concerning the effect of the GLUT4 gene on MQ traits. Grindflek et al. (2002) found relatively weak and inconsistent associations between GLUT4 (SLC2A4) genotypes and traits such as drip loss, color, and loin marbling, in different pig breeds and crosses from the United States and Norway. Moreover, owing to the inconsistencies in the results, they concluded that the GLUT4 (SLC2A4) polymorphism is not associated with MQ traits in the lines studied. However, Sieczkowska et al. (2007) observed that the GLUT4 gene polymorphism differentiated GP of fatteners that were simultaneously TT homozygotes at the PKM2 locus (see Section 10.3). The role of glucose transportation in metabolic pathways provides justification for the analysis of associations between GLUT4 gene polymorphism and MQ.

10.2.3.5 CAST (Calpastatin Gene)

A polymorphism at locus CAST/MspI explained considerable differentiation in MQ observed within porkers heterozygous at locus RYR1, as explained in Section 10.3 (Kocwin-Podsiadla et al. 2003). Krzecio et al. (2008) noted that the CAST AA genotype identified using RsaI endonuclease is associated with higher pH values for the *longissimus* muscle in the first 3 h after slaughter, whereas the opposite BB genotype is associated with higher pH values in the 24–144 h period postmortem.

Lindholm-Perry et al. (2009) have shown that the causative mutation in CAST associated with tenderness may affect the amount of CAST mRNA and calpastatin protein, either by regulating expression, by affecting mRNA stability or half-life, or by alternative splicing rather than causing a change in the amino acid sequence of the protein. While identification of the causative variation will provide the most reliable marker for pork tenderness, characterization of the genetic architecture of this region could provide markers that are predictive of pork tenderness.

10.2.3.6 MYOG (Myogenin)

Myogenin (MYOG), the expression of which is related to the fusion of mononucleated myoblasts into multinucleated muscle fibers, is one of the four factors of the MyoD family (known as MRF—myogenic regulatory factors) controlling myogenesis (te Pas and Visscher 1994). During the postnatal period, MYOG transcripts are accumulated principally in muscle regions dominated by slow contracting, oxidative fibers (Voytik et al. 1993). Genes of the MyoD family code transcripting factors, regulating the expression of other muscle-specific genes (Wright et al. 1989).

MYOG is one of the candidate genes responsible for muscle fiber characteristics (te Pas et al. 1999). The MYOG gene, on chromosome 9 in pigs, has been shown, in a number of studies, to be polymorphic and associated with muscle mass and growth rate (te Pas et al. 2005). However, it is unknown whether these polymorphisms are closely associated with muscle fiber and MQ characteristics (Soumillion et al. 1997; Verner et al. 2007; Krzecio et al. 2008). Associations were noted between MYOG gene polymorphisms and pH 1 h after slaughter, lightness of meat, drip loss, and WHC (Zhu and Li 2005; Jiusheng et al. 2009; Kim et al. 2009). Krzecio et al. (2008) noted that MYOG gene polymorphisms were significantly associated with the pH in meat 48 h postmortem, electrical conductivity in the muscle tissue measured at different times after slaughter, and also with dry matter content. Kim et al. (2009) suggested that diplotypes from the BspCNI and MspI sites on MYOG can be used to provide meaningful molecular markers for today's porcine breeding goals.

10.2.3.7 H-FABP (Heart Fatty Acid Binding Protein)

The search for genetic markers for IMF was initiated by Gerbens et al. (1997), who suggested that the H-FABP (heart fatty acid binding protein) gene, on porcine chromosome 6, may be responsible for this trait and that pigs with the aa/dd/HH genotype had the highest levels of IMF. Subsequent studies confirmed the effect of this gene on IMF in crossbred pig populations that included Duroc genes (Gerbens et al. 1999; Sieczkowska et al. 2006b), although these findings were not confirmed in non-Duroc pig populations (Nechtelberger et al. 2001; Sieczkowska et al. 2006b). Arnyasi et al. (2006) evaluated H-FABP as a potential candidate for the QTL and observed significant associations between H-FABP polymorphisms and IMF, which explained 30%–35% of the variation and confirmed the results of Gerbens et al. (1999). As a complex trait, IMF is likely to also be influenced by other polymorphic genes involved in lipid synthesis and fatty acid degradation.

The associations between H-FABP gene polymorphisms and pH value, meat lightness, drip loss, technological yield of cured meat, and protein content were noted in a number of studies, independent of the proportion of Duroc in the breeds/crossbreeds studied (Nechtelberger et al. 2001; Antosik et al. 2006; Sieczkowska et al. 2006a).

10.2.3.8 ACSL4 (Long-Chain Acyl-CoA Synthetase 4)

Localized on porcine chromosome 10, close to a QTL for IMF (Perez-Enciso et al. 2002) and a QTL affecting growth in Iberian × Landrace crossbreds (Perez-Enciso et al. 2005), the gene coding for long-chain acyl-CoA synthetase (ACSL4) plays an essential role in lipid biosynthesis and fatty acid degradation (Mercade et al. 2006). In mammals, ACSL catalyzes the initial step in cellular long-chain fatty acid metabolism. Mercade et al. (2006) reported that the ACSL4 gene is expressed in many different tissues, for example, liver, heart, lung, spleen, stomach, brain, skeletal muscle, backfat, uterus, ovary, and testis. In the same study, Mercade et al. (2006) also identified 10 polymorphisms and 2 haplotypes, by comparative sequencing in 12 pigs from six different pig breeds. Analysis performed by Rusc et al. (2011) demonstrated that the G allele of the ACSL4 gene affected IMF content (GG pigs had the highest IMF of 2.47%), but also some other meat parameters (water and glycogen content, GP, and pH) in the *longissimus* muscle of crossbred fatteners. Further studies should be performed to reveal whether this influence originates from different stabilities of mRNA and ASCL4 protein yield and activity, or from unknown loci physically linked with the ACSL4 gene. Figure 10.4 shows the glycolytic-energetic changes that occur postmortem relating to ultimate pork MQ, as controlled by major gene polymorphisms.

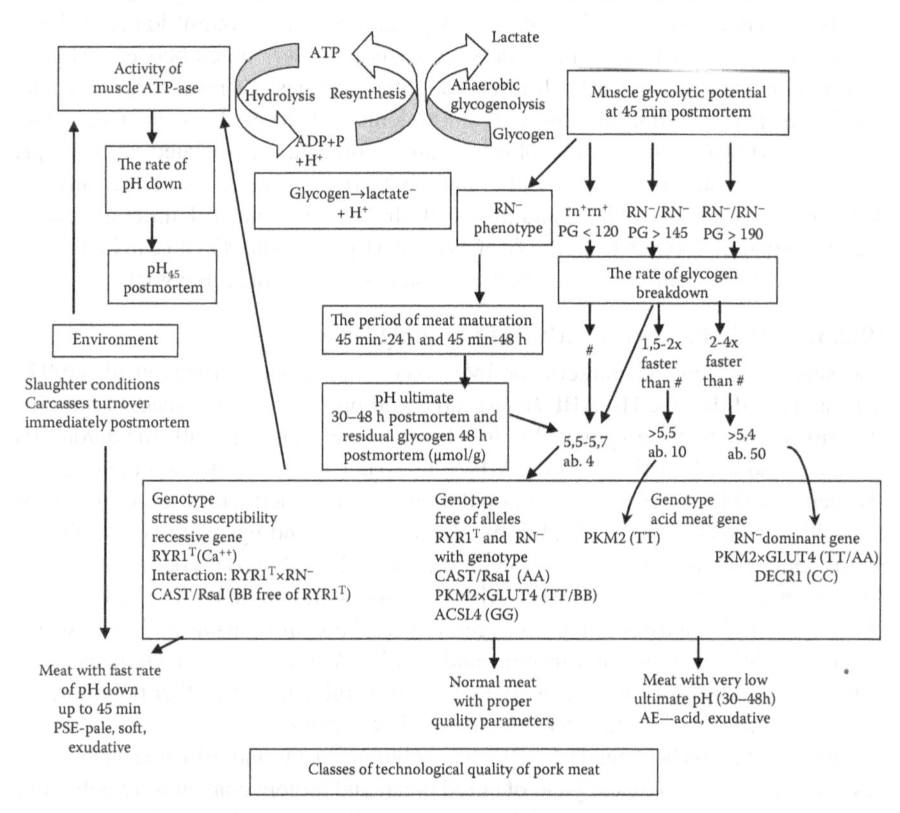

FIGURE 10.4 Glycolytic-energetic changes postmortem relating to pork meat quality, as controlled by major gene polymorphism. Gene symbols: RYR1—ryanodine receptor gene; RN—Rendement Napole gene; CAST—calpastatin gene; PKM2—pyruvate kinase muscle gene; GLUT4—insulin-sensitive glucose transporter gene; PRKAG3—AMP-activated protein kinase γ3 gene; ACSL4—long-chain acyl-CoA synthetase gene; DECR1—2,4-dienoyl CoA reductase 1 gene. *DFD—dark, firm, dry faulty meat was not shown on the scheme because of environmental conditioning. (Adapted with permission from Kocwin-Podsiadla, M. et al. 2009. *The Genomics of Cattle and Pig*. Poznań University of Life Sciences Press, Poznań, Poland. With modification.)

The associations between polymorphisms of the genes described above and MQ traits are summarized in Table 10.2. The results presented strongly indicate multifactorial genetic conditioning of MQ and highlight that further analysis of gene interaction effects on these traits would be beneficial.

10.3 INTERACTION BETWEEN GENES FOR MQ

10.3.1 Interaction of Major Genes with Other Genes

Epistasis is when the expression of a gene depends on one or more modifier genes in the genetic background of the organism. Consequently, epistatic mutations have different effects in combination than individually and genes having a major effect on a phenotype in one population may show little phenotypic expression in a different breed or population (Le Rouzic et al. 2008; Williams 2008). Therefore, the purported effects of major genes or QTL must be validated in different breeds and environments (e.g., Van Eenennaam et al. 2007). Studies in South Devon cattle on the MSTN mutation, responsible for the double-muscling of Belgian Blue cattle, led the authors to conclude that strong selection in breeds such as the Belgian Blue for the DM phenotype may have also resulted in extensive LD for alleles in other genes controlling conformation and muscling traits, together with the GDF8 deletion. However, in the South Devon cattle, which carry the same MSTN mutation, selection has been for good conformation, rather than double-muscling. Thus, in the South Devon breed, the combination of supporting alleles of the modifier genes across all the loci involved is not found, and only moderate hypermuscularity has been observed. Identification of modifier genes and the relative importance of MSTN will be vital to the understanding of the genetic control of muscle growth and conformation in cattle (Smith et al. 2000).

10.3.2 Interaction between Different Major Genes

Despite the increasing economic importance of MQ, and the number of QTL and positional candidate genes already known, the practical use of marker-assisted selection is limited by the lack of knowledge of the number and interaction of genes affecting MQ characteristics (Dekkers 2004; Rothschild et al. 2007). Gene-by-environment or gene-by-gene interactions deserve to be tested and the complexity of these associations highlights the need for complementary genetic and functional approaches to reduce the list of candidate genes and variants (Georges 2011). The effects of gene interactions on MQ traits are not fully understood and only few publications seem to exist.

In pigs, Le Roy et al. (2000) observed that the interaction between genotypes at *loci* RYR1 and RN may be detrimental, due to what is known as the "snow ball" effect—when the effect of one mutation is strengthened by the effect of the second, or may be beneficial, due to the "leveling effect"—when the effect of one mutation is leveled out by the effect of the second. The authors stated that for pH at 1 h postmortem the unfavorable effect of allele $RYR1^T$ was stronger in the carriers of allele RN^- than in animals not having this allele ("snow ball" effect), while for the Napole (RTN) technological yield, the unfavorable effect of gene RN^- was less pronounced in the case of stress-susceptible (RYR1 genotype TT) than stress-resistant (CC) animals. Przybylski et al. (2000), in analogous studies, demonstrated an interaction between the RYR1 and RN^- genes only in relation to RTN (Rendement Technology Yield)—one trait out of five examined. The RN^- gene had a significantly larger effect on the RTN value than the $RYR1^T$ gene, although more pronounced changes were observed when the genotype of crossbred pigs contained both unfavorable genes (RTN around 5%–7% lower).

There was also an interaction between RN^- and CAST genotypes in pigs. Animals with AA or AB genotypes at the CAST/MspI locus differed in some MQ traits depending on their RN^- genotype, with noncarriers of the RN^- gene having around 0.07–0.08 units higher pH_{48} and lower drip loss at 144 h postmortem (about 1.9 pp) (Kocwin-Podsiadla et al. 2003). The analysis of interactions between genotypes at *loci* CAST and RYR1 found that, among animals with the favorable RYR1 genotype (resistant to stress), there were still animals with exudative meat and this was largely explained by their CAST/Hinfl genotype, with those of genotype AB showing exudation. Moreover, animals heterozygous for the RYR1 gene, with a BB genotype at locus CAST/HinfI, were characterized by an almost doubled drip loss effect compared to AB heterozygotes at the same locus (7.80% vs. 4.04%) (Kocwin-Podsiadla et al. 2003). Even more favorable results were obtained, in terms of the variation in the rate of drip loss in RYR1 heterozygotes, when considering the interaction with CAST/MspI. This interaction explained considerable differentiation in MQ among porkers that were carriers of the RYR1 gene, probably due to different susceptibility of the calpastatin enzyme (a calpain inhibitor) to the concentration and changes in the level of Ca^{2+} released in the cell and conditioned by allele $RYR1^T$ (Krzecio et al. 2005a).

Krzecio et al. (2008) noted a statistically significant interaction between CAST/RsaI and MYOG genotypes for pH 48 h postmortem, electrical conductivity in the muscle tissue measured 3 and 24 h postmortem, and technological yield indicator of cured meat in terminal processing. The highest pH value (5.54) in the *longissimus* muscle, measured 48 h postmortem, occurred in animals with genotype AB at the MYOG locus, which were simultaneously BB homozygotes at the CAST/RsaI locus. Within the group of pigs with the same MYOG genotype (i.e., AB), the technological yield was dependent on the animal's CAST/RsaI genotype. However, no interactions were observed between genotypes at the CAST/MspI and MYOG loci, in relation to MQ traits, including chemical composition (Krzecio et al. 2007a).

There are statistically significant interactions between PKM2 and CAST/HinfI genotypes for pH, electrical conductivity (EC_2), and WHC (Zybert et al. 2007). Pigs with AA genotype at the CAST/HinfI locus were differentiated by PKM2 genotypes for these traits. Homozygotes with AA genotype at the CAST/HinfI locus and TT genotype at the PKM2 locus, in comparison to AA/CC animals (CAST/PKM2 genotypes, respectively), had a higher (0.22) pH_{144} value and lower EC_2 (0.9 ms). Regarding WHC, animals with AB and BB CAST/HinfI genotypes were differentiated by the PKM2 genotype. The results of Zybert et al. (2007) suggest that the disadvantageous effect on MQ of the TT genotype of the PKM2 gene may be suppressed if fatteners are simultaneously AA homozygotes at the CAST/HinfI locus.

An interaction between GLUT4 and PKM2 genotypes was indicated by the presence of two groups of fattening pigs with the same GP level, but strongly differentiated by drip loss and basic chemical composition of the *longissimus* muscle (Sieczkowska et al. 2007). The GLUT4 gene polymorphism differentiated GP of pigs that were TT homozygotes at the PKM2 locus. It was observed that groups of animals with TT/AA and TT/BB genotypes at loci PKM2 and GLUT4, respectively, had the same GP level, whereas the group with genotype TT/AA had significantly higher drip loss (3.64%), higher total protein content (0.65%), and about 0.5% lower IMF in comparison to the TT/BB group (Sieczkowska et al. 2007). Pork MQ traits significantly influenced by genotype-by-genotype interactions are shown in Table 10.3.

Of the major genes or QTL affecting MQ traits in sheep that have been discussed above, there is limited information on their potential interactions. The callipyge, LoinMAX, and TM-QTL are all located on chromosome 18 and it has not been conclusively ruled out that these could be different mutations on the same gene, or potentially the same causative mutation in the case of LoinMAX versus TM-QTL. The MSTN (MyoMAX) mutation is located on another chromosome (chromosome 2) and could potentially interact with the effects of these other genes, although little evidence could be found in the literature. A study of crossbred Texel × Welsh Mountain lambs found that the interaction between MyoMAX and TM-QTL was not significant ($P > 0.05$) for a range of carcass composition traits measured on the live animal by ultrasound scanning, or the carcass by video image analysis (VIA) (Masri et al. 2011b), with the one exception of the weight of the leg primal joint, as predicted by VIA.

In cattle, interaction between the CAPN1 and CAST genes has been investigated, since the biological mechanisms of these genes are heavily dependent—CAPN1 encoding for the protease responsible for myofibrillar protein degradation postmortem and CAST encoding for the inhibitor of that protease. Casas et al. (2006) found a statistically significant interaction between markers on these two genes for Warner–Bratzler shear force in a population of mixed crossbred cattle (*Bos taurus* with *Bos indicus* influence). However, these authors questioned the validity of this result, given the limited sample size and the fact that a significant interaction was not detected in another population of *Bos taurus* cattle in which there were more animals with the genotypes of interest. They suggested that an additive action of these two genes on mean shear force may be more likely across cattle populations (Casas et al. 2006). However, significant epistasis was found between SNPs on the CAPN1 and CAST genes using DNA from a variety of cattle breeds and crosses (*Bos indicus* and *Bos taurus*) by Morris et al. (2006) and Barendse et al. (2007), with reranking of some genotypes, in terms of their effects on meat tenderness, when considered in combination with marker genotypes from the other gene. Although epistasis was not found in all breeds studied by Barendse et al. (2007), or between all pairs of SNPs in each breed, the majority did show epistasis, of which additive × dominance components were most commonly identified.

Other examples of interactions between major genes for carcass and MQ have been reported. Additive and nonadditive genetic effects and epistatic interactions were investigated between markers in casein (CSN1S1) and thyroglobulin (TG) genes (Bennett et al. 2013), which have previously been associated with fat distribution in cattle. Adjusted fat thickness showed a significant dominance association

TABLE 10.3
Effects of Gene Interactions on Pork Meat Quality Traits

Gene	Meat Quality Traits	Reference
	Interaction between RYR1 Genotype and	
RN^-	Napole yield (RTN), pH_1; glycogen content; yield of loins in curing and smoking	Le Roy et al. (2000), Przybylski et al. (2000), Kocwin-Podsiadla et al. (2004a)
CAST/HinfI	DL_{48}; WHC	Kocwin-Podsiadla and Kuryl (2003), Kocwin-Podsiadla et al. (2003)
CAST/MspI	pH_{24}; DL_{48}	Kocwin-Podsiadla and Kuryl (2003), Krzecio et al. (2005b)
CAST/RsaI	pH_{45}; DL_{48}	Kocwin-Podsiadla et al. (2003), Krzecio et al. (2005b)
H-FABP	Water content; protein content	Kocwin-Podsiadla et al. (2004a)
MYOG	pH_{24}	Kocwin-Podsiadla et al. (2004a)
	Interaction between RN^- Genotype and	
CAST/HinfI	Glycogen content; pH_{24}; water content	Kocwin-Podsiadla et al. (2003, 2004a)
CAST/MspI	pH_{48}; DL_{144}	Kocwin-Podsiadla et al. (2004a, 2006c)
MYOG	Protein content	Kocwin-Podsiadla et al. (2004a)
	Interaction between H-FABP Genotype and	
CAST/MspI	Protein content	Kocwin-Podsiadla et al. (2004a)
CAST/RsaI	Protein content	Kocwin-Podsiadla et al. (2004a)
	Interaction between MYOG Genotype and	
CAST/RsaI	EC_3; EC_{24}; pH_{48}; TY	Krzecio et al. (2008)
CAST/MspI	No significant	Krzecio et al. (2007b)
	Interaction between PKM2 Genotype and	
CAST/HinfI	pH_{144}; EC_{120}; WHC	Zybert et al. (2007)
GLUT4	GP; glycogen content; DL_{96}; IMF content; protein content	Sieczkowska et al. (2007)

DL—drip loss; EC—electrical conductivity; GP—glycolytic potential; IMF—intramuscular fat; RTN—Napole yield; TY—technological yield of cured meat in terminal processing; WHC—water-holding capacity.

Note: Subscript numbers denote time of measurement in hours postmortem.

with TG and an epistatic additive CSN1S1 × additive TG association, while predicted meat tenderness (by visible and near-infrared spectroscopy methods) showed a dominance association with TG, heterozygous TG meat being significantly more tender than meat from either homozygote. This highlights the complex associations that can occur between SNPs and MQ traits.

Epistasis can help explain the differences in the size of effects on MQ associated with different genotypes in different breeds. Genotype frequencies differ between breeds and populations in most of the studies in which these major genes have been found to segregate, which will contribute to the amount of epistasis observed (Barendse et al. 2007). Epistatic effects of a variety of genes may be influencing the effects of any of the major genes for MQ that can be identified, which makes their effects difficult to predict across breeds or populations. Once a major gene or QTL has been discovered and its effects have been validated, important practical questions arise as to how to use the gene to enhance rates of genetic progress. The theory for this was developed some time ago (Lande and Thompson 1990), but the answer also requires an economic appraisal.

10.4 USE OF SNPS AS CANDIDATE GENES

One of the challenges in meat science is to elucidate the genetic basis of phenotypic variation in MQ. Many measurable phenotypic traits reflect the cumulative influence of the inheritance of combinations of multiple genes or QTL. Some traits, such as stress susceptibility in pigs, are mainly affected by a single major gene. Other traits, such as drip loss or tenderness, are complex and multifactorial in character and are affected by multiple genes across the genome (Andersson and Georges 2004; Heidt et al. 2013) and also by processing conditions. Variation within the genome can involve changes of single nucleotides (SNPs), variation of repetitive sequences, for example, at mini- and microsatellites, and even variation in the numbers of regulatory sequences and genes (copy number variation or CNV) (Groenen et al. 2011).

Genomic microarrays (SNP chips) with varying densities have become commercially available in recent years for pigs, sheep, and cattle (e.g., Porcine SNP60, Ovine SNP50, Bovine SNP50 BeadChips, Illumina), making it possible to simultaneously identify thousands of SNP polymorphisms for individual animals. These genomic microarrays are a valuable supplement to investigations concerning transcriptomics, realized using expression microarrays and real-time PCR techniques (Kaminski et al. 2010, Soma et al. 2011, Lee et al. 2012). The preliminary results of pig genotyping, conducted using the SNiPORK microarray, which contains SNPs in candidate genes potentially associated with pork yield and quality, has allowed consistent investigations of single genes with specific influences on MQ traits. For example, Olenski et al. (2010) confirmed—through the use of these SNPs—the high value of boars of a newly selected Duroc breed for the production of commercial slaughter pigs with high-quality meat and exceptionally low drip loss. Additionally, Kaminski et al. (2010), identified a group of determining genes, the polymorphisms of which were linked with extreme differentiation of drip loss (CYP21, SFRS1) and GP (DECR1, PPARGC1, MC4R) of the *longissimus* muscle in porkers.

Similar work in other species is also underway. A recent example in Australian sheep involved studies to investigate the associations between MQ traits and 192 SNPs from a specially designed SNP panel (Meat Quality Research panel) (Knight et al. 2012, 2014). Association analysis found numerous SNPs associated with carcass and MQ traits, including shear force (SNPs in the CAST and CAPN2 genes) and omega-3 polyunsaturated fatty acid content of lamb (Knight et al. 2012). One of the SNPs in the CAST gene associated with shear force at 5 days postmortem was independently validated in another sheep population (Knight et al. 2014). Using data from the same sheep flocks, a region on ovine chromosome 6 was found to show associations with large effects for several carcass and MQ traits, including IMF (Daetwyler et al. 2012). This study also produced genomic breeding values, using the 50 K ovine SNP-chip, for a range of MQ traits, which were found to increase prediction accuracy over traditional breeding values based on pedigree information. As reflected in Section 10.2, when summarizing the major genes/QTL for MQ in the different species, there are fewer examples in sheep and cattle of the use of SNPs to identify individuals with the potential for improved MQ, compared to pigs. More research has been focused in this area in recent years in all three species, with the introduction of the new technologies described above. However, it is likely that the use of single genes/QTL will continue, alongside the use of emerging SNPs as candidate genes, until SNP technology has been further developed and validated or sequencing becomes a feasible approach.

10.5 USE OF MICROARRAYS IN RESEARCH ON THE DETERMINANTS OF MQ

Gene expression studies in farm animals can be useful to define the biological mechanisms involved in the response to experimental treatments or conditions, to analyze biological changes during development and to detect genes affecting production traits (Wimmers et al. 2010a; Fontanesi et al. 2011). Microarrays have been successfully used to find genes that are differentially expressed in muscle tissue due to developmental stage/age, breed, or phenotype in livestock species. Application-specific cDNA macro- and microarrays have enabled the expression of several hundreds or thousands of genes to be monitored (Wimmers et al. 2010b). Quantitative expression studies, using microarray technology, can indicate regulatory variation in genes for complex traits. QTL mapping and microarray analyses together make it possible to identify regulatory networks underlying the quantitative trait of interest and localize genomic variation (Heidt et al. 2013). Coupling genomic technologies for expression profiling with genome-wide genetic mapping using SNP markers can identify specific chromosomal regions that contain functional candidate genes (Heidt et al. 2013).

QTL analysis of expression levels of genes identifies genomic regions that are likely to contain at least one causal gene with a regulatory effect on the expression level, termed expression QTL (eQTL). The use of eQTL analyses has been demonstrated as a promising tool for narrowing the gap between detected phenotype-related QTL regions and confirmed causative variations in pigs (Steibel et al. 2011; Heidt et al. 2013).

Recently, research has focused on the analysis of the skeletal muscle transcriptome, comparing the gene expression profile in muscles of different quality, using expression microarrays and cDNA libraries (Davoli et al. 2002; Yao et al. 2002; Bai et al. 2003; Lobjois et al. 2008). These tools provide information on networks of expressed genes in muscle tissue and increase the knowledge of the biological pathways controlling MQ traits (Gorni et al. 2011). Studies examining the transcriptome of the porcine *longissimus* muscle, for example, have identified gene expression changes associated with MQ traits such as tenderness (Lobjois et al. 2008) and drip loss (Ponsuksili et al. 2008). Bai et al. (2003) described a porcine skeletal muscle microarray analysis to compare red (*psoas*) and white (*longissimus*) muscles. Da Costa et al. (2004) used the same microarray to profile molecular changes in response to dietary restriction in the skeletal muscle of young growing animals. te Pas et al. (2005) compared (using muscle-specific microarray) the muscle expression profiles of different porcine developmental stages.

Differences in gene expression can be used in reverse genetic studies to generate well-defined hypotheses regarding downstream effects on molecular, cellular, and functional networks, and finally at the phenotype level (Ciobanu et al. 2011). Further investigation of the functions of the novel differentially expressed sequences identified in these studies will add fundamental information to allow a better understanding of muscle metabolism and control of variation in MQ (Gorni et al. 2011). It should be stressed that the majority of identified gene expression changes are quite small (1.5- or 2-fold expression changes associated with extremes of a trait) and confirmation of the significance of expression fold changes by qPCR has been shown to be dependent on the choice of reference genes for normalization (Lobjois et al. 2008).

According to Fontanesi et al. (2011), the information content of porcine skeletal muscle RNA can be maintained intact for up to 24 h postmortem. Therefore, microarray data obtained from specimens collected in the processing plant, over a relatively long period, have the potential to identify preslaughter treatments that may affect skeletal muscle gene expression patterns, with no potential bias from RNA degradation. This opens new possibilities to develop gene expression biomarkers for meat production and meat product authentication and traceability using information derived from transcriptome analysis (Fontanesi et al. 2011).

10.6 ROLE OF GENOMICS, PROTEOMICS, AND METABOLOMICS IN FINDING NEW GENES AFFECTING MQ

Whereas genetics is defined as the study of heredity, genomics uses information from DNA or RNA to investigate genes, their functions, and interrelationships, in order to identify their combined influence on complex traits. Genomics does not study the more downstream products derived from the genome. This is investigated by proteomics, comprising larger-scale studies of proteins, particularly their structure and functions, with the proteome being the entire set of proteins produced by an organism. Metabolomics is the scientific study of metabolism and the metabolites involved in this process that provides life-sustaining chemical transformations within the cells of living organisms. It is clear that only a synergistic approach of all these disciplines will lead to a comprehensive understanding of complex biological organisms or systems. Correlative analysis between genes, their expression regulation, their products, and extensive metabolic profiling will ultimately provide the means to fully investigate these complex biological networks.

Recent trends in farm animal research, especially on carcass and MQ, involve the application of high-throughput technologies involving alternative "omics" disciplines, such as the study of miRNAs, protein posttranslational modification—PTM (PTMomics), mass spectrometry and NMR-based metabolomics, quantitative proteomics, and lipidomics (D'Alessandro and Zolla 2013). Explanation of the functional complexity of transcriptomes is extremely challenging, but the results of recent research show that much more of the genome is transcribed than previously thought and that the great majority of the genes are transcribed in a bidirectional manner (Katayama et al. 2005; Li et al. 2012). For a better understanding of the physiological complexity of the transcriptome of livestock species, expression and/or functional gene analysis studies need to be undertaken to gain knowledge on the effects of molecular, cellular, and functional networks, and finally how these affect traits of interest at the phenotypic level (Rothschild 2003; Ciobanu et al. 2011).

GWAS have identified a number of genetic polymorphisms associated with MQ traits in livestock species. However, besides the growing knowledge of genome sequences, the genetic mechanisms underlying MQ during the conversion of muscle to meat are not fully understood. The identification of biomarkers and the further development of rapid tests would be helpful for the control and improvement of MQ. Transcriptomics is of interest to study the development of complex phenotypic traits determined by genetic × environment interactions, such as MQ. The analysis of gene expression level appears highly relevant to improve knowledge about the biological mechanisms underlying the quality of meat and to identify biomarkers, that is, genes whose expression level is associated with MQ traits (te Pas et al. 2011; Damon et al. 2013). The investigations of Damon et al. (2013), using transcriptomics and further internal validation by quantitative RT-PCR, revealed 26 genes whose expression was correlated to at least one technological or sensory pork quality trait. However, it is necessary to refine identified genes as biomarkers for MQ assessment in various pork chains and then foresee the development of monitoring, decision-making, and management tools to allow these to be used in commercial pork industries. The same principles apply to other species.

Genes do not change during the lifetime of the animal, but their expression to mRNA and proteins is very dynamic and is regulated by a large number of factors, including environmental influences. The genome contains the information on which alleles are present, whereas the proteome contains information on which genes are actually being expressed (Hollung et al. 2007, 2009). Proteomics seems to be an effective tool to reflect the important mechanisms influencing the development of satisfactory MQ. In contrast to traditional methods that study only one or a few proteins at a time, proteomics can enable the study of several hundred proteins simultaneously. It has the potential to significantly enhance the understanding of molecular mechanisms underlying MQ (Hollung et al. 2007, 2009) and could provide further tools for the dissection of the molecular basis for phenotypic variation (Andersson and Georges 2004). However, many of these techniques are in their infancy. For example, the predictive ability of high-throughput proteomic profiles to determine MQ level is still largely unknown, as is the biological background of these proteins in relation to MQ (te Pas et al. 2013). Some examples are emerging in the literature. For example, Trott et al. (2013) identified SNPs in the porcine prolactin receptor gene that reduce protein expression and four haplotypes that suppress signaling and may differentially impact the phenotypic effects of prolactin *in vivo*.

Metabolomics, which is the rapidly evolving field of measuring all endogenous metabolites in a cell or body fluid, can contribute to this research area (Gieger et al. 2008). Biochemical measurements of particular intermediate phenotypes on a continuous scale can be expected to provide more detail on potentially affected metabolic pathways. The investigation of genetically determined metabotypes in their biochemical context (combination of genotyping and metabolic characterization) might help to better understand the gene-influencing mechanisms and gene–environment interactions.

The premise of genomic selection (GS) is that, with the availability of dense marker maps (the genome could be considered as many thousands of small segments, the phenotypic effects of which could be estimated), trait breeding values could potentially be predicted solely based on "marker" genotype and pedigree. Further advancement in marker-assisted BLUP, genomics and eventually proteomics will enable the development of more complex and accurate genetic improvement programs, leading to further development of differentiated meat value chains focused on ever-changing consumer needs (Sosnicki and Newman 2010).

10.7 GENERAL CONCLUSIONS

Evidence to date suggests that genetic improvement of MQ characteristics will be very difficult with traditional selection methods. Although MQ characteristics are generally moderate in heritability (see, e.g., Chapter 9), phenotypes are difficult and expensive to measure or predict in live animals or carcasses. An understanding of the genetic variants underlying the nutritional and sensory properties of meat will enable improvement in MQ through molecular genetic or GS. The advantages of these methods, over traditional selection, are greatest where traditional methods are difficult to implement. This makes MQ characteristics ideal candidate traits for efficient application of GS.

Most of the progress in MQ traits to date has been associated with the use of major gene effects, many of which are reviewed here. However, the Animal QTL database (AnimalQTLdb; http://www.animalgenome.org/cgi-bin/QTLdb/index) is constantly being updated and should be consulted for the most recent information on reported QTL. This knowledge provides opportunities to enhance selection responses, and some commercial genetic tests are already available, as documented in some of the examples given in this chapter. Their usefulness is associated with their classification into genetic tests based on causative mutations (direct markers), tests based on population-wide LD with the QTL (LD markers) or based on linked markers, which are in population-wide equilibrium with the QTL (LE markers) (Figure 10.1). Three relatively recent breakthroughs have resulted in the increasing use of DNA information in breeding programs: (i) GS methodology, as a form of marker-assisted selection on a genome-wide scale; (ii) the discovery of a vast number of SNPs; and (iii) cost-effective methods to genotype, with the outlook that in the near future genome sequence data may replace SNP genotypes as markers. GS is already widely implemented in dairy cattle, beef cattle, and pigs, and is beginning for sheep. For example, genomic breeding values are now commercially available in Australia for some sheep MQ traits, which are being used to refine selection decisions alongside traditional breeding values (see Chapter 9). Although the benefits from application of GS may be tremendous, especially with increasing market demands for quality, the cost of collecting the necessary experimental data to validate these methods is often prohibitive. GS in some species and countries is still a very long way from commercial implementation, due to small population sizes, lack of accuracy of estimated breeding values, and other structural difficulties in the industries. Therefore, the candidate gene approach is currently still useful to extend the panel of associated SNPs and obtain better estimates of SNP effects.

The use of identified QTL and genes for MQ is not restricted to GS and GWAS, but could have applications in different fields, such as traceability systems to combat food misdescription and adulteration, value-based marketing systems and/or for classification of carcasses into quality classes, SNP chip design or refinement, marker-assisted introgression, or possibly for future genetic modifications. This area of research is developing at an extremely fast pace and is providing information (knowledge on QTL and their effects) and tools (genetic tests, statistical methods to incorporate into genetic evaluations) that will be exceptionally useful for incorporation into livestock breeding programs, even before the implementation of GS. However, in all cases, the phenotypic effects (direct and indirect) of any QTL should be quantified in the target population and the genetic test should be accurate enough to facilitate efficient use.

REFERENCES

Abdulkhaliq, A.M., Meyer, H.H., Busboom, J.R., and Thompson, J.M. 2007. Growth, carcass and cooked meat characteristics of lambs sired by Dorset rams heterozygous for the Callipyge gene and Suffolk and Texel rams. *Small Rumin. Res.* 71:92–97.

Abdulkhaliq, A.M., Meyer, H.H., Thompson, J.M., Holmes, Z.A., Forsberg, N.E., and Davis, S.L. 2002. Callipyge gene effects on lamb growth, carcass traits, muscle weights and meat characteristics. *Small Rum. Res.* 45:89–93.

Abe, H., Morimatsu, M., Nikami, H., Miyashige, T., and Saito, M. 1997. Molecular cloning and mRNA expression of the bovine insulin-responsive glucose transporter (GLUT4). *J. Anim. Sci.* 75:182–188.

Albrecht, E., Teuscher, F., Ender, K., and Wegner, J. 2006. Growth- and breed-related changes of marbling characteristics in cattle. *J. Anim. Sci.* 84:1067–1075.

Allais, S., Journaux, L., Leveziel, H. et al. 2011. Effects of polymorphisms in the calpastatin and μ-calpain genes on meat tenderness in 3 French beef breeds. *J. Anim. Sci.* 89:1–11.

Allais, S., Leveziel, H., Payet-Duprat, N. et al. 2010. The two mutations, Q204X and nt821, of the myostatin gene affect carcass and meat quality in young heterozygous bulls of French beef breeds. *J. Anim. Sci.* 88:446–454.

Andersson, L. 2003. Identification and characterization of AMPK gamma 3 mutations in the pig. *Biochem. Soc. Trans.* 31:232–235.

Andersson, L., and Georges, M. 2004. Domestic-animal genomics: Deciphering the genetics of complex traits. *Nat. Rev. Genet.* 5:202–212.

Ansay, M., and Hanset, R. 1979. Anatomical, physiological and biochemical differences between conventional and double-muscled cattle in the Belgian Blue and White breed. *Livest. Prod. Sci.* 6:5–13.

Antosik, K., Krzecio, E., Sieczkowska, H., Zybert, A., Kocwin-Podsiadla, M., Kuryl, J., and Lyczynski, A. 2006. Relation between the polymorphism in H-FABP gene and chemical composition and quality of meat from Pietrain crossbreds heterozygous as regards RYR1 gene. *Anim. Sci. Pap. Rep.* 24:19–27.

Arnyasi, M., Grindflek, E., Javor, A., and Lien, S. 2006. Investigation of two candidate genes for meat quality traits in a quantitative trait locus region on SSC6: The porcine short heterodimer partner and heart fatty acid binding protein genes. *J. Anim. Breed. Genet.* 123:198–203.

Arthur, P.F. 1995. Double muscling in cattle: A review. *Aust. J. Agric. Res.* 46:1493–1515.

Arthur, P.F., and Hearnshaw, H. 1995. Effect of double muscling on meat yield and tenderness. In *Proceedings of the Australian Meat Industry Research Conference, "Meat '95"*, September 10–12, 1995, Abstract No. 74, Gold Coast, Queensland, Australia.

Aviles, C., Polvillo, O., Pena, F., Juarez, M., Martinez, A.L., and Molina, A. 2013. Associations between DGAT1, FABP4, LEP, RORC, and SCD1 gene polymorphisms and fat deposition in Spanish commercial beef. *J. Anim. Sci.* 91:4571–4577.

Bai, Q.F., McGillivray, C., Da Costa, N., Dornan, S., Evans, G., Stear, M.J., and Chang, K.C. 2003. Development of a porcine skeletal muscle cDNA microarray: Analysis of differential transcript expression in phenotypically distinct muscles. *BMC Genomics* 4:8.

Banks, R.G., Shands, C., Stafford, J.P., and Kenney, P. 1995. *Central Progeny Testing Results: 1991–1994 Matings.* Meat Research Corporation, Sydney, Australia.

Barendse, W., Bunch, R.J., and Harrison, B.E. 2010. The effect of variation at the retinoic acid receptor-related orphan receptor C gene on intramuscular fat percent and marbling score in Australian cattle. *J. Anim. Sci.* 88:47–51.

Barendse, W., Harrison, B.E., Hawken, R.J., Ferguson, D.M., Thompson, J.M., Thomas, M.B., and Bunch, R.J. 2007. Epistasis between calpain 1 and its inhibitor calpastatin within breeds of cattle. *Genetics* 176:2601–2610.

Bejerholm, C., and Barton-Gade, P. 1986. *Effect of Intramuscular Fat Level on Eating Quality of Pig Meat.* Danish Meat Research Institute, Roskilde, Denmark, Manuscript No. 720E.

Bellinge, R.H.S., Liberles, D.A., Iaschi, S.P., O'Brien, P.A., and Tay, G.K. 2005. Myostatin and its implications on animal breeding: A review. *Anim. Genet.* 36:1–6.

Bennett, G.L., Shackelford, S.D., Wheeler, T.L., King, D.A., Casas, E., and Smith, T.P.L. 2013. Selection for genetic markers in beef cattle reveals complex associations of thyroglobulin and casein1-S1 with carcass and meat traits. *J. Anim. Sci.* 91:565–571.

Bertram, H.C., Petersen, J.S., and Andersen, H.J. 2000. Relationship between RN^- genotype and drip loss in meat from Danish pigs. *Meat Sci.* 56:49–55.

Broad, T.E., Glass, B.C., Greer, G.J., Robertson, T.M., Bain, W.E., Lord, E.A., and McEwan, J.C. 2000. Search for a locus near to myostatin that increases muscling in Texel sheep in New Zealand. *Proc. NZ Soc. Anim. Prod.* 60:110–112.

Carr, C.C., Morgan, J.B., Berg, E.P., Carter, S.D., and Ray, F.K. 2006. Growth performance, carcass composition, quality, and enhancement treatment of fresh pork identified through deoxyribonucleic acid marker-assisted selection for the Rendement Napole gene. *J. Anim. Sci.* 84:910–917.

Casas, E., White, S.N., Wheeler, T.L., Shackelford, S.D., Koohmaraie, M., Riley, D.G., Chase, C.C., Johnson, D.D., and Smith, T.P.L. 2006. Effects of calpastatin and mu-calpain markers in beef cattle on tenderness traits. *J. Anim. Sci.* 84:520–525.

Ciobanu, D.C., Bastiaansen, J.W.M., Lonergan, S.M., Thomsen, H., Dekkers, J.C.M., Plastow, G.S., and Rothschild, M.F. 2004. New alleles in calpastatin gene are associated with meat quality traits in pigs. *J. Anim. Sci.* 82:2829–2839.

Ciobanu, D.C., Day, A.E., Nagy, A., Wales, R., Rothschild, M.F., and Plastow, G.S. 2001. Genetic variation in two conserved local Romanian pig breeds using type 1 DNA markers. *Genet. Sel. Evol.* 33:417–432.

Ciobanu, D.C., Lonergan, S.M., and Huff-Lonergan, E.J. 2011. Genetics of meat quality and carcass traits. In Rothschild, M.F. and Ruvinsky, A.J. (eds.), *The Genetics of the Pig.* CAB International, Wallingford, Oxfordshire, pp. 355–389.

Closter, A.M., Guldbrandtsen, B., Henryon, M., Nielsen, B., and Berg, P. 2011. Consequences of elimination of the Rendement Napole allele from Danish Hampshire. *J. Anim. Breed. Genet.* 128:192–200.

Cockett, N.E., Jackson, S.P., Shay, T.L., Farnir, F., Berghmans, S., Snowder, G.D., Nielsen, D.M., and Georges, M. 1996. Polar overdominance at the ovine callipyge locus. *Science* 273:236–238.

Cockett, N.E., Jackson, S.P., Snowder, G.D., Shay, T.L., Berghmans, S., Beever, J.E., Carpenter, C., and Georges, M. 1999. The callipyge phenomenon: Evidence for unusual genetic inheritance. *J. Anim. Sci.* 77:221–227.

Costello, S., O'Doherty, E., Troy, D.J., Ernst, C.W., Kim, K.S., Stapleton, P., Sweeney, T., and Mullen, A.M. 2007. Association of polymorphisms in the calpain I, calpain II and growth hormone genes with tenderness in bovine M. longissimus dorsi. *Meat Sci.* 75:551–557.

D'Alessandro, A., and Zolla, L. 2013. Meat science: From proteomics to integrated omics towards system biology. *J. Proteomics* 78:558–577.

Da Costa, N., McGillivray, C., Bai, Q., Wood, J.D., Evans, G., and Chang, K.C. 2004. Restriction of dietary energy and protein induces molecular changes in young porcine skeletal muscles. *J. Nutr.* 134:2191–2199.

Daetwyler, H., Swan, A., van der Werf, J., and Hayes, B. 2012. Accuracy of pedigree and genomic predictions of carcass and novel meat quality traits in multi-breed sheep data assessed by cross-validation. *Genet. Sel. Evol.* 44:33.

Damon, M., Denieul, K., Vincent, A., Bonhomme, N., Wyszynska-Koko, J., and Lebret, B. 2013. Associations between muscle gene expression pattern and technological and sensory meat traits highlight new biomarkers for pork quality assessment. *Meat Sci.* 95:744–754.

Davoli, R., Fontanesi, L., Zambonelli, P., Bigi, D., Gellin, J., Yerle, M., Milc, J., Braglia, S., Cenci, V., Cagnazzo, M., and Russo, V. 2002. Isolation of porcine expressed sequence tags for the construction of a first genomic transcript map of the skeletal muscle in pig. *Anim. Genet.* 33:168.

Dekkers, J.C.M. 2004. Commercial application of marker- and gene-assisted selection in livestock: Strategies and lessons. *J. Anim. Sci.* 82(13 Suppl.):E313–E328.

Dekkers, J.C.M., and Hospital, F. 2002. The use of molecular genetics in the improvement of agricultural populations. *Nat. Rev. Genet.* 3:22–32.

Destefanis, G., Brugiapaglia, A., Barge, M.T., and Dal Molin, E. 2008. Relationship between beef consumer tenderness perception and Warner–Bratzler shear force. *Meat Sci.* 78:153–156.

Dodds, K.G. 2007. Use of gene markers in the New Zealand sheep industry. *Proc. Assoc. Advmt. Anim. Breed. Genet.* 17:418–425.

Dodds, K.G., McEwan, J.C., and Davis, G.H. 2007. Integration of molecular and quantitative information in sheep and goat industry breeding programmes. *Small Rumin. Res.* 70:32–41.

Duckett, S.K., Klein, T.A., Dodson, M.V., and Snowder, G.D. 1998a. Tenderness of normal and callipyge lamb aged fresh or after freezing. *Meat Sci.* 49:19–26.

Duckett, S.K., Klein, T.A., Leckie, R.K., Thorngate, J.H., Busboom, J.R., and Snowder, G.D. 1998b. Effect of freezing on calpastatin activity and tenderness of callipyge lamb. *J. Anim. Sci.* 76:1869–1874.

Enfalt, A.C., Lundstrom, K., Karlsson, A., and Hansson, I. 1997. Estimated frequency of the RN$^-$ allele in Swedish Hampshire pigs and comparison of glycolytic potential, carcass composition, and technological meat quality among Swedish Hampshire, Landrace, and Yorkshire pigs. *J. Anim. Sci.* 75:2924–2935.

Fernandez, X., Naveau, J., Talmant, A., and Monin, G. 1990. Distribution du potentiel glycolytique dans une population porcine et relation avec le rendement "napole". *Journées de la Recherche Porcine en France* 22:97–100.

Fernandez, X., Tornberg, E., Naveau, J., Talmant, A., and Monin, G. 1992. Bimodal distribution of the muscle glycolytic potential in French and Swedish populations of hampshire crossbred pigs. *J. Sci. Food Agric.* 59:307–311.

Field, R.A., McCormick, R.J., Brown, D.R., Hinds, F.C., and Snowder, G.D. 1996. Collagen crosslinks in longissimus muscle from lambs expressing the callipyge gene. *J. Anim. Sci.* 74:2943–2947.

Fiems, L.O. 2012. Double muscling in cattle: Genes, husbandry, carcasses and meat. *Animal* 2:472–506.

Fontanesi, L., Davoli, R., Costa, L.N., Scotti, E., and Russo, V. 2003. Study of candidate genes for glycolytic potential of porcine skeletal muscle: Identification and analysis of mutations, linkage and physical mapping and association with meat quality traits in pigs. *Cytogenet. Genome Res.* 102:145–151.

Fontanesi, L., Davoli, R., Nanni Costa, L., Beretti, F., Scotti, E., Tazzoli, M., Tassone, F., Colombo, M., Buttazzoni, L., and Russo, V. 2008. Investigation of candidate genes for glycolytic potential of porcine skeletal muscle: Association with meat quality and production traits in Italian Large White pigs. *Meat Sci.* 80:780–787.

Fontanesi, L., Galimberti, G., Calo, D.G., Colombo, M., Astolfi, A., Formica, S., and Russo, V. 2011. Microarray gene expression analysis of porcine skeletal muscle sampled at several post mortem time points. *Meat Sci.* 88:604–609.

Freking, B.A., Keele, J.W., Beattie, C.W., Kappes, S.M., Smith, T.P., Sonstegard, T.S., Nielsen, M.K., and Leymaster, K.A. 1998. Evaluation of the ovine callipyge locus: I. Relative chromosomal position and gene action. *J. Anim. Sci.* 76:2062–2071.

Freking, B.A., Keele, J.W., Shackelford, S.D., Wheeler, T.L., Koohmaraie, M., Nielsen, M.K., and Leymaster, K.A. 1999. Evaluation of the ovine callipyge locus: III. Genotypic effects on meat quality traits. *J. Anim. Sci.* 77:2336–2344.

Fujii, J., Otsu, K., Zorzato, F., Deleon, S., Khanna, V.K., Weiler, J.E., Obrien, P.J., and MacLennan, D.H. 1991. Identification of a mutation in porcine ryanodine receptor associated with malignant hyperthermia. *Science* 253:448–451.

Fukumoto, H., Imura, H., and Seino, Y. 1989. Glucose transporter—A novel gene family. *Nihon Rinsho* 47:2619–2629.

Galve, A., Burgos, C., Varona, L., Carrodeguas, J.A., Canovas, A., and Lopez-Buesa, P. 2013. Allelic frequencies of PRKAG3 in several pig breeds and its technological consequences on a Duroc × Landrace-Large White cross. *J. Anim. Breed. Genet.* 130:382–393.

Geesink, G.H., and Koohmaraie, M. 1999a. Effect of calpastatin on degredation of myofibrillar proteins by u-calpain under postmortem conditions. *J. Anim. Sci.* 77:2685–2692.

Geesink, G.H., and Koohmaraie, M. 1999b. Postmortem proteolysis and calpain/calpastatin activity in callipyge and normal lamb biceps femoris during extended postmortem storage. *J. Anim. Sci.* 77:1490–1501.

Georges, M. 2011. The long and winding road from correlation to causation. *Nat. Genet.* 43:180–181.

Gerbens, F., Rettenberger, G., Lenstra, J.A., Veerkamp, J.H., and Tepas, M.F.W. 1997. Characterization, chromosomal localization, and genetic variation of the porcine heart fatty acid-binding protein gene. *Mamm. Genome* 8:328–332.

Gerbens, F., van Erp, A.J., Harders, F.L., Verburg, F.J., Meuwissen, T.H., Veerkamp, J.H., and te Pas, M.F. 1999. Effect of genetic variants of the heart fatty acid-binding protein gene on intramuscular fat and performance traits in pigs. *J. Anim. Sci.* 77:846–852.

Gieger, C., Geistlinger, L., Altmaier, E. et al. 2008. Genetics meets metabolomics: A genome-wide association study of metabolite profiles in human serum. *Plos Genet.* 4(11):e1000282.

Gorni, C., Garino, C., Iacuaniello, S., Castiglioni, B., Stella, A., Restelli, G.L., Pagnacco, G., and Mariani, P. 2011. Transcriptome analysis to identify differential gene expression affecting meat quality in heavy Italian pigs. *Anim. Genet.* 42:161–171.

Grindflek, E., Holzbauer, R., Plastow, G., and Rothschild, M.F. 2002. Mapping and investigation of the porcine major insulin sensitive glucose transport (SLC2A4/GLUT4) gene as a candidate gene for meat quality and carcass traits. *J. Anim. Breed. Genet.* 119:47–55.

Grisart, B., Coppieters, W., Farnir, F. et al. 2002. Positional candidate cloning of a QTL in dairy cattle: Identification of a missense mutation in the bovine DGAT1 gene with major effect on milk yield and composition. *Genome Res.* 12:222–231.

Grobet, L., Martin, L.J., Poncelet, D. et al. 1997. A deletion in the bovine myostatin gene causes the double-muscled phenotype in cattle. *Nat. Genet.* 17:71–74.

Grobet, L., Poncelet, D., Royo, L.J., Brouwers, B., Pirottin, D., Michaux, C., Menissier, F., Zanotti, M., Dunner, S., and Georges, M. 1998. Molecular definition of an allelic series of mutations disrupting the myostatin function and causing double-muscling in cattle. *Mamm. Genome* 9:210–213.

Groenen, M.A.M., Schook, L.B., and Archibald, A.L. 2011. Pig genomics. In Rothschild, M.F. and Ruvinsky, A.J. (eds.), *The Genetics of the Pig.* CAB International, Wallingford, Oxfordshire, pp. 179–199.

Hadjipavlou, G., Matika, O., Clop, A., and Bishop, S.C. 2008. Two single nucleotide polymorphisms in the myostatin (GDF8) gene have significant association with muscle depth of commercial Charollais sheep. *Anim. Genet.* 39:346–353.

Hanset, R. 1991. The major gene of muscular hypertrophy in the Belgian Blue cattle breed. In Owen, J.B. and Axford, R.F.E. (eds.), *Breeding for Disease Resistance in Farm Animals.* CABI, Oxford, pp. 467–478.

Hayes, B.J., and Goddard, M.E. 2001. The distribution of the effects of genes affecting quantitative traits in livestock. *Genet. Sel Evol.* 33:209–229.

Heidt, H., Cinar, M.U., Uddin, M.J. et al. 2013. A genetical genomics approach reveals new candidates and confirms known candidate genes for drip loss in a porcine resource population. *Mamm. Genome* 24:416–426.

Hill, W.G. 2014. Applications of population genetics to animal breeding, from Wright, Fisher and Lush to genomic prediction. *Genetics* 196:1–16.

Hill, W.G., and Knott, S. 1990. Identification of genes with large effects. In *Advances in Statistical Methods for Genetic Improvement of Livestock*; In *Proceedings of the International Symposium*; Armidale, 16–20 February; Gianola, D. and Hammond, K. (eds.), Springer-Verlag, Berlin, Heidelberg, New York, London, Paris, Advanced Series in Agricultural Sciences, Vol. 18, pp. 477–494.

Hocquette, J.F., Botreau, R.I., Picard, B., Jacquet, A., Pethick, D.W., and Scollan, N.D. 2012. Opportunities for predicting and manipulating beef quality. *Meat Sci.* 92:197–209.

Hollung, K., Veiseth, E., Jia, X.H., and Faergestad, E.M. 2009. Proteomics. In Nollet, L.M.L. and Toldrá, F. (eds.), *Handbook of Muscle Foods Analysis.* CRC Press, Taylor & Francis Group, Boca Raton, FL.

Hollung, K., Veiseth, E., Jia, X.H., Faergestad, E.M., and Hildrum, K.I. 2007. Application of proteomics to understand the molecular mechanisms behind meat quality. *Meat Sci.* 77:97–104.

Hope, M., Haynes, F., Oddy, H., Koohmaraie, M., Al-Owaimer, A., and Geesink, G. 2013. The effects of the myostatin g+6723G > A mutation on carcass and meat quality of lamb. *Meat Sci.* 95:118–122.

Hopkins, D.L., and Fogarty, N.M. 1998. Diverse lamb genotypes—2. Meat pH, colour and tenderness. *Meat Sci.* 49:477–488.

Hopkins, D.L., Fogarty, N.M., and Menzies, D.J. 1997. Meat and carcass quality traits of lambs from terminal sires. In *Proceedings of the 43rd International Congress of Meat Science and Technology.* July 27–August 1, 1997, Auckland, New Zealand, pp. 298–299.

Hopkins, D.L., Hegarty, R.S., Walker, P.J., and Pethick, D.W. 2006. Relationship between animal age, intramuscular fat, cooking loss, pH, shear force and eating quality of aged meat from sheep. *Aust. J. Exp. Agric.* 46:879–884.

Hopkins, D.L., Stanley, D.F., Martin, L.C., Ponnampalam, E.N., and van de Ven, R. 2007a. Sire and growth path effects on sheep meat production 1. Growth and carcass characteristics. *Aust. J. Exp. Agric.* 47:1208–1218.

Hopkins, D.L., Stanley, D.F., Martin, L.C., Toohey, E.S., and Gilmour, A.R. 2007b. Genotype and age effects on sheep meat production 3. Meat quality. *Aust. J. Exp. Agric.* 47:1155–1164.

Hopkins, D.L., Stanley, D.F., Toohey, E.S., Gardner, G.E., Pethick, D.W., and van de Ven, R. 2007c. Sire and growth path effects on sheep meat production 2. Meat and eating quality. *Aust. J. Exp. Agric.* 47:1219–1228.

Hopkins, D.L., Toohey, E.S., Kerr, M.J., and van de Ven, R. 2011. Comparison of two instruments (G2 tenderometer and a Lloyd texture analyser) for measuring the shear force of cooked meat. *Anim. Prod. Sci.* 51:71–76.

Houde, A., Pommier, S.A., and Roy, R. 1993. Detection of the ryanodine receptor mutation associated with malignant hyperthermia in purebred swine populations. *J. Anim. Sci.* 71:1414–1418.

Hu, Z.L., Park, C.A., Wu, X.L., and Reecy, J.M. 2013. Animal QTLdb: An improved database tool for livestock animal QTL/association data dissemination in the post-genome era. *Nucl. Acids Res.* 41(Database issue):D871–D879.

Hu, Z.L., and Reecy, J.M. 2007. Animal QTLdb: Beyond a repository. A public platform for QTL comparisons and integration with diverse types of structural genomic information. *Mamm. Genome* 18:1–4.

Hullberg, A., Johansson, L., and Lundstrom, K. 2005. Sensory perception of cured-smoked pork loin from carriers and noncarriers of the RN(-)allele and its relationship with technological meat quality. *J. Muscle Foods* 16:54–76.

Imamura, K., Noguchi, T., and Tanaka, T. 1986. Regulation of isozyme patterns of pyruvate kinase in normal and neoplastic tissues. In Staal, G.E.J. and Van Veelen, C.W.M. (eds.), *Markers of Human Neuroectodermal Tumors*. CRC Press, Boca Raton, FL, pp. 191–222.

Jiusheng, W., Yuehuan, L., and Ningying, X. 2009. Histological characteristics of musculus longissimus dorsi and their correlation with restriction fragment length polymorphism of the myogenin gene in Jinghua × Pietrain F2 crossbred pigs. *Meat Sci.* 81:108–115.

Johnson, P.L., McEwan, J.C., Dodds, K.G., Purchas, R.W., and Blair, H.T. 2005a. A directed search in the region of GDF8 for quantitative trait loci affecting carcass traits in Texel sheep. *J. Anim. Sci.* 83:1988–2000.

Johnson, P.L., Purchas, R.W., McEwan, J.C., and Blair, H.T. 2005b. Carcass composition and meat quality differences between pasture-reared ewe and ram lambs. *Meat Sci.* 71:383–391.

Jopson, N.B., Nicoll, G.B., Stevenson-Barry, J.M., Duncan, S., Greer, G.J., Bain, W.E., Gerard, E.M., Glass, B.C., Broad, T.E., and McEwan, J.C. 2001. Mode of inheritance and effects on meat quality of the rib-eye muscling (REM) QTL in sheep. *Proc. Assoc. Advmt. Anim. Breed. Genet.* 14:111–114.

Josell, A., Enfalt, A.C., von Seth, G., Lindahl, G., Hedebro-Velander, I., Andersson, L., and Lundstrom, K. 2003. The influence of RN genotype, including the new V199I allele, on the eating quality of pork loin. *Meat Sci.* 65:1341–1351.

Kaminski, S., Kocwin-Podsiadla, M., Sieczkowska, H., Help, H., Zybert, A., Krzecio, E., Antosik, K., Brym, P., Wojcik, E., and Adamczyk, G. 2010. Screening 52 single nucleotide polymorphisms for extreme value of glycolytic potential and drip loss in pigs. *J. Anim. Breed. Genet.* 127:125–132.

Katayama, S., Tomaru, Y., Kasukawa, T. et al. 2005. Antisense transcription in the mammalian transcriptome. *Science* 309:1564–1566.

Kijas, J.W., McCulloch, R., Edwards, J.E.H., Oddy, V.H., Lee, S.H., and van der Werf, J. 2007. Evidence for multiple alleles effecting muscling and fatness at the Ovine GDF8 locus. *BMC Genet.* 8:80

Kim, J.M., Choi, B.D., Kim, B.C., Park, S.S., and Hong, K.C. 2009. Associations of the variation in the porcine myogenin gene with muscle fibre characteristics, lean meat production and meat quality traits. *J. Anim. Breed. Genet.* 126:134–141.

Knight, M.I., Daetwyler, H.D., Hayes, B.J., Hayden, M.J., Ball, A.J., Pethick, D.W., and McDonagh, M.B. 2012. Discovery and trait association of single nucleotide polymorphisms from gene regions of influence on meat tenderness and long-chain omega-3 fatty acid content in Australian lamb. *Anim. Prod. Sci.* 52:591–600.

Knight, M.I., Daetwyler, H.D., Hayes, B.J., Hayden, M.J., Ball, A.J., Pethick, D.W., and McDonagh, M.B. 2014. An independent validation association study of carcass quality, shear force, intramuscular fat percentage and omega-3 polyunsaturated fatty acid content with gene markers in Australian lamb. *Meat Sci.* 96:1025–1033.

Kocwin-Podsiadla, M., Krzecio, E., and Kuryl, J. 2004a. Wplyw form polimorficznych wybranych genow na mi snosc oraz wlasciwosci fizykochemiczne i funkcjonalne tkanki miesniowej. In Switonski, M. (ed), *Postepy genetyki molekularnej bydla i trzody chlewnej.* Akademia Rolnicza, Poznan, pp. 263–328.

Kocwin-Podsiadla, M., Krzecio, E., and Zybert, A. 2004b. Effect of RN- and calpastatin gene as related to meat quality of stress resistant fatteners. In *Proceedings of the 50th International Congress of Meat Science and Technology*, August 8–13, 2004, Helsinki, Finland, p. 33.

Kocwin-Podsiadla, M., Krzecio, E., and Zybert, A. 2006a. Effect of calpastatin (CAST) gene on meat quality of stress resistant fatteners. *Anim. Sci.* S1(Suppl.):40–41.

Kocwin-Podsiadla, M., Krzecio, E., and Zybert, A. 2006b. The effect of RN⁻ gene on meat quality of stress resistant fatteners. *Anim. Sci.* S1(Suppl.):180–181.

Kocwin-Podsiadla, M., Krzecio, E., Zybert, A., Antosik, K., Sieczkowska, H., and Kuryl, J. 2006c. The interactive effect of RN⁻ gene and calpastatin (CAST) gene for meat quality traits. *Anim. Sci.* S1(Suppl.):198–199.

Kocwin-Podsiadla, M., Krzecio, E., Zybert, A., Sieczkowska, H., and Antosik, K. 2006d. The influence of PRKAG3 gene on meat quality of stress resistant fatteners. *Anim. Sci.* S1(Suppl.):196–197.

Kocwin-Podsiadla, M., Krzecio, E., Zybert, A., Sieczkowska, H., and Antosik, K. 2006e. The influence of PRKAG3 gene on meat quality of stress resistant fatteners. In *Proceedings of the 52nd International Congress of Meat Science and Technology*, August 13–18, 2006, Dublin, Ireland, pp. 79–80.

Kocwin-Podsiadla, M., and Kuryl, J. 2003. The effect interaction between genotypes at loci CAST, RYR1 and RN on pig carcass quality and pork traits—A review. *Anim. Sci. Pap. Rep.* 21:61–75.

Kocwin-Podsiadla, M., Kuryl, J., Krzecio, E., Zybert, A., and Przybylski, W. 2003. The interaction between calpastatin and RYR1 genes for some pork quality traits. *Meat Sci.* 65:731–735.

Kocwin-Podsiadla, M., Kuryl, J., and Przybylski, W. 1993. Physiological and genetic background of the occurence of stress-induced defective pork: A review. *Prace i Materiały Zootechniczne* 44:5–32.

Kocwin-Podsiadla, M., Przybyliski, W., Krzecio, E., Zybert, A., and Kaczorek, S. 2000. Effect of the RN⁻ gene on the growth rate carcass and meat quality in crossbreeding of Large White sows with P-76 boars. In Wenk, C., Fernandez, J.A., Dupuis, M. (eds.), *Quality of Meat and Fat in Pigs as Affected by Genetics and Nutrition*, EAAP Publication No. 100. Wageningen Academic Publishers, Zurich, Switzerland, pp. 165–170.

Kocwin-Podsiadla, M., Przybylski, W., Fernandez, X., and Monin, G. 1998. The comparison between NN and Nn pigs for meat quality and glycolytic potential measured before and after slaughter. In *Proceedings of the 44th International Congress of Meat Science and Technology*, August 30–September 4, 1998, Barcelona, Spain, pp. 278–279.

Kocwin-Podsiadla, M., Przybylski, W., Kuryl, J., Talmant, A., and Monin, G. 1995. Muscle glycogen level and meat quality in pigs of different halothane genotypes. *Meat Sci.* 40:121–125.

Kocwin-Podsiadla, M., Zybert, A., Krzecio, E., Antosik, K., and Sieczkowska, H. 2009. Polymorphism of selected genes and their relationship to the some traits of farm animals. In Zwierzchowski, L. and Switonski, M. (eds.), *The Genomics of Cattle and Pig.* Poznań University of Life Sciences Press, Poznań, Poland.

Koohmaraie, M. 1996. Biochemical factors regulating the toughening and tenderization processes of meat. *Meat Sci.* 43:S193–S201.

Koohmaraie, M., Shackelford, S.D., Wheeler, T.L., Lonergan, S.M., and Doumit, M.E. 1995. A muscle hypertrophy condition in lamb (callipyge): Characterization of effects on muscle growth and meat quality traits. *J. Anim. Sci.* 73:3596–3607.

Krzecio, E., Kocwin-Podsiadla, M., Kuryl, J., Zybert, A., Sieczkowska, H., and Antosik, K. 2007a. The effect of genotypes at loci CAST/MspI (calpastatin) and MYOG (myogenin) and their interaction on selected productive traits of porkers free of gene RYR1(T) I. Muscling and morphological composition of carcass. *Anim. Sci. Pap. Rep.* 25:5–16.

Krzecio, E., Kocwin-Podsiadla, M., Kuryl, J., Zybert, A., Sieczkowska, H., and Antosik, K. 2007b. The effect of genotypes at loci CAST/MspI (calpastatin) and MYOG (myogenin) and their interaction on selected productive traits of porkers free of gene RyR1(T). II. Meat quality. *Anim. Sci. Pap. Rep.* 25:17–24.

Krzecio, E., Kocwin-Podsiadla, M., Kuryl, J., Zybert, A., Sieczkowska, H., and Antosik, K. 2008. The effect of interaction between genotype CAST/RsaI (calpastatin) and MYOG/MspI (myogenin) on carcass and meat quality in pigs free of RYR1(T) allele. *Meat Sci.* 80:1106–1115.

Krzecio, E., Kuryl, J., Kocwin-Podsiadla, M., and Monin, G. 2004. The influence of CAST/RsaI and RYR1 genotypes and their interactions on selected meat quality parameters in three groups of four-breed fatteners with different meat content of carcass. *Anim. Sci. Pap. Rep.* 22:469–478.

Krzecio, E., Kuryl, J., Kocwin-Podsiadla, M., and Monin, G. 2005a. Association of calpastatin (CAST/MspI) polymorphism with meat quality parameters of fatteners and its interaction with RYR1 genotypes. *J. Anim. Breed. Genet.* 122:251–258.

Krzecio, E., Kuryl, J., Kocwin-Podsiadla, M., and Monin, G. 2005b. The preliminary study on influence of calpastatin and RYR1 genes polymorphism and their interactions on selected meat quality parameters of four-crossbreeds fatteners. *J. Anim. Breed. Genet.* 122:251–258.

Krzecio, E., Miszczuk, B., Kocwin-Podsiadla, M., Sieczkowska, H., Zybert, A., Antosik, K., and Lyczynski, A. 2006. Slaughter value of carcasses and technological quality of meat from porkers differing in drip loss from the longissimus lumborum muscle. *Anim. Sci. Pap. Rep.* 24:159–169.

Lambe, N.R., Haresign, W., Macfarlane, J.M., Richardson, R.I., Matika, O., and Bunger, L. 2010a. The effect of conditioning period on loin muscle tenderness in crossbred lambs with or without the Texel Muscling QTL (TM-QTL). *Meat Sci.* 85:715–720.

Lambe, N.R., Macfarlane, J.M., Richardson, R.I., Matika, O., Haresign, W., and Bunger, L. 2010b. The effect of a Texel muscling QTL (TM-QTL) on meat quality traits in crossbred lambs. *Meat Sci.* 85:684–690.

Lambe, N.R., Richardson, R.I., Macfarlane, J.M., Nevison, I., Haresign, W., Matika, O., and Bunger, L. 2011. Genotypic effects of the Texel Muscling QTL (TM-QTL) on meat quality in purebred Texel lambs. *Meat Sci.* 89:125–132.

Lande, R., and Thompson, R. 1990. Efficiency of marker-assisted selection in the improvement of quantitative traits. *Genetics* 124:743–756.

Laville, E., Bouix, J., Sayd, T., Bibe, B., Elsen, J.M., Larzul, C., Eychenne, F., Marcq, F., and Georges, M. 2004. Effects of a quantitative trait locus for muscle hypertrophy from Belgian Texel sheep on carcass conformation and muscularity. *J. Anim. Sci.* 82:3128–3137.

Laville, E., Sayd, T., Terlouw, C., Blinet, S., Pinguet, J., Fillaut, M., Glenisson, J., and Cherel, P. 2009. Differences in pig muscle proteome according to HAL genotype: Implications for meat quality defects. *J. Agric. Food Chem.* 57:4913–4923.

Le Rouzic, A., Alvarez-Castro, J.M., and Carlborg, O. 2008. Dissection of the genetic architecture of body weight in chicken reveals the impact of epistasis on domestication traits. *Genetics* 179:1591–1599.

Le Roy, P., Moreno, C., Elsen, J.M., Caritez, J.C., Billon, Y., Lagant, H., Talmant, A., Vernin, P., Amigues, Y., Sellier, P., and Monin, G. 2000. Interactive effects of HAL and RN major genes on carcass quality traits in pigs: Preliminary results. In Wenk, C., Fernandez, J.A., and Dupuis, M. (eds.), *Quality of Meat and Fat in Pigs as Affected by Genetics and Nutrition*, EAAP Publication No. 100. Wageningen Academic Publishers, Zurich, Switzerland, pp. 139–142.

Le Roy, P., Naveau, J., Elsen, J.M., and Sellier, P. 1990. Evidence for a new major gene influencing meat quality in pigs. *Genet. Res.* 55:33–40.

Lee, K.T., Lee, Y.M., Alam, M., Choi, B.H., Park, M.R., Kim, K.S., Kim, T.H., and Kim, J.J. 2012. A whole genome association study on meat quality traits using high density SNP chips in a cross between Korean native pig and landrace. *Asian Austral. J. Anim.* 25:1529–1539.

Li, X., Ekerljung, M., Lundstrom, K., and Lunden, A. 2013. Association of polymorphisms at DGAT1, leptin, SCD1, CAPN1 and CAST genes with color, marbling and water holding capacity in meat from beef cattle populations in Sweden. *Meat Sci.* 94:153–158.

Li, X.J., Yang, H., Li, G.X., Zhang, G.H., Cheng, J., Guan, H., and Yang, G.S. 2012. Transcriptome profile analysis of porcine adipose tissue by high-throughput sequencing. *Anim. Genet.* 43:144–152.

Lindholm-Perry, A.K., Rohrer, G.A., Holl, J.W., Shackelford, S.D., Wheeler, T.L., Koohmaraie, M., and Nonneman, D. 2009. Relationships among calpastatin single nucleotide polymorphisms, calpastatin expression and tenderness in pork longissimus(1). *Anim. Genet.* 40:713–721.

Lobjois, V., Liaubet, L., Sancristobal, M., Glenisson, J., Feve, K., Rallieres, J., Le Roy, P., Milan, D., Cherel, P., and Hatey, F. 2008. A muscle transcriptome analysis identifies positional candidate genes for a complex trait in pig. *Anim. Genet.* 39:147–162.

Lundstrom, K., Andersson, A., and Hansson, I. 1996. Effect of the RN gene on technological and sensory meat quality in crossbred pigs with Hampshire as terminal sire. *Meat Sci.* 42:145–153.

Lundstrom, K., Andersson, A., Maerz, S., and Hansson, I. 1994. Effect of the RN-gene on meat quality and lean meat content in crossbred pigs with Hampshire as terminal sire. In *Proceedings of the 40th International Congress of Meat Science and Technology*, August 28–September 2, 1994, S-IVA.07, The Hague, Netherlands.

Lundstrom, K., Malmfors, B., Stern, S., Rydhmer, L., Eliassonselling, L., Mortensen, A.B., and Mortensen, H.P. 1994. Skatole levels in pigs selected for high lean tissue-growth rate on different dietary-protein levels. *Liv. Prod. Sci.* 38:125–132.

Macfarlane, J.M., Lambe, N.R., Bishop, S.C., Matika, O., Rius Vilarrasa, E., McLean, K.A., Haresign, W., Wolf, B.T., McLaren, R.J., and Bünger, L. 2009. Effects of the Texel muscling quantitative trait locus on carcass traits in crossbred lambs. *Animal* 3:189–199.

Macfarlane, J.M., Lambe, N.R., Matika, O., Johnson, P.L., Wolf, B.T., Haresign, W., Bishop, S.C., and Bunger, L. 2014. Effect and mode of action of the Texel Muscling QTL (TM-QTL) on carcass traits in purebred Texel lambs. *Animal* 8:1053–1061.

Macfarlane, J.M., Lambe, N.R., Matika, O., McLean, K.A., Masri, A.Y., Johnson, P.L., Wolf, B.T., Haresign, W., Bishop, S.C., and Bünger, L. 2010. Texel loin muscling QTL (TM-QTL) located on ovine chromosome 18 appears to exhibit imprinting and polar overdominance. In *Proceedings of the 9th World Congress on Genetics Applied to Livestock Production*, August 1–6, 2010, Abstract No. 199, Leipzig, Germany.

Maedus, W.J., Maclinnis, R., Dugan, M.E.R., and Aalhus, J.L. 2002. PCR-RFLP method to identify the RN⁻ gene in retailed pork chops. *Can. J. Anim. Sci.* 82:449–451.

Masri, A.Y., Lambe, N.R., Macfarlane, J.M., Brotherstone, S., Haresign, W., and Bunger, L. 2011a. Evaluating the effects of a single copy of a mutation in the myostatin gene (c.*1232 G>A) on carcass traits in crossbred lambs. *Meat Sci.* 87:412–418.

Masri, A.Y., Lambe, N.R., Macfarlane, J.M., Brotherstone, S., Haresign, W., Rius-Vilarrasa, E., and Bunger, L. 2010. The effects of a loin muscling quantitative trait locus (LoinMAX™) on carcass and VIA-based traits in crossbred lambs. *Animal* 4:407–416.

Masri, A.Y., Macfarlane, J.M., Lambe, N.R., Haresign, W., Brotherstone, S., and Bunger, L. 2011b. Evaluating the effects of the c.*1232G > A mutation and TM-QTL in Texel × Welsh Mountain lambs using ultrasound and video image analyses. *Small Rumin. Res.* 99:99–109.

Matika, O., Sechi, S., Pong-Wong, R., Houston, R.D., Clop, A., Woolliams, J.A., and Bishop, S.C. 2011. Characterisation of OAR1 and OAR18 QTL associated with muscle depth in British commercial terminal sire sheep. *Anim. Genet.* 42:172–180.

McPherron, A.C., and Lee, S.J. 1997. Double muscling in cattle due to mutations in the myostatin gene. *Proc. Natl. Acad. Sci. U. S. A.* 94:12457–12461.

Mercade, A., Estelle, J., Perez-Enciso, M., Varona, L., Silio, L., Noguera, J.L., Sanchez, A., and Folch, J.M. 2006. Characterization of the porcine acyl-CoA synthetase long-chain 4 gene and its association with growth and meat quality traits. *Anim. Genet.* 37:219–224.

Michal, J.J., Zhang, Z.W., Gaskins, C.T., and Jiang, Z. 2006. The bovine fatty acid binding protein 4 gene is significantly associated with marbling and subcutaneous fat depth in Wagyu × Limousin F2 crosses. *Anim. Genet.* 37:400–402.

Milan, D., Jeon, J.T., Looft, C. et al. 2000. A mutation in PRKAG3 associated with excess glycogen content in pig skeletal muscle. *Science* 288:1248–1251.

Miller, M.F., Carr, M.A., Ramsey, C.B., Crockett, K.L., and Hoover, L.C. 2001. Consumer thresholds for establishing the value of beef tenderness. *J. Anim. Sci.* 79:3062–3068.

Moeller, S.J., Baas, T.J., Leeds, T.D., Emnett, R.S., and Irvin, K.M. 2003. Rendement Napole gene effects and a comparison of glycolytic potential and DNA genotyping for classification of Rendement Napole status in Hampshire-sired pigs. *J. Anim. Sci.* 81:402–410.

Monin, G., Brard, C., Vernin, P., and Naveau, J. 1992. Effects of the RN- gene on some traits of muscle and liver in pigs. In *Proceedings of the 38th International Congress of Meat Science and Technology*, August 23–28, 1992, Clermont-Ferrand, France, pp. 317–394.

Monin, G., and Sellier, P. 1985. Pork of low technological quality with a normal rate of muscle pH fall in the immediate post-mortem period—The case of the Hampshire breed. *Meat Sci.* 13:49–63.

Morris, C.A., Cullen, N.G., Hickey, S.M., Dobbie, P.M., Veenvliet, B.A., Manley, T.R., Pitchford, W.S., Kruk, Z.A., Bottema, C.D.K., and Wilson, T. 2006. Genotypic effects of calpain 1 and calpastatin on the tenderness of cooked M. longissimus dorsi steaks from Jersey × Limousin, Angus and Hereford-cross cattle. *Anim. Genet.* 37:411–414.

Naveau, J., 1986. Contribution a l'étude du déterminisme génétique de la qualité de viande porcine. Héritabilité du Rendement Technologique Napole. *Jour. Rech. Por. En France* 18:265–276.

Nechtelberger, D., Pires, V., Solkner, J., Stur, I., Brem, G., Mueller, M., and Mueller, S. 2001. Intramuscular fat content and genetic variants at fatty acid-binding protein loci in Austrian pigs. *J. Anim. Sci.* 79:2798–2804.

Nicoll, G.B., Burkin, H.R., Broad, T.E., Jopson, N.B., Greer, G.J., Bain, W.E., Wright, C.S., Dodds, K.G., Fennessy, P.F., and McEwan, J.C. 1998. Genetic linkage in microsatellite markers to the Carwell locus for rib-eye muscling in sheep. In *Proceedings of the 6th World Congress on Genetics Applied to Livestock Production*, January 11–16, 1998, Armidale, Australia, pp. 529–532.

Noguchi, T., Inoue, H., and Tanaka, T. 1986. The M1-type and M2-type isozymes of rat pyruvate-kinase are produced from the same gene by alternative RNA splicing. *J. Biol. Chem.* 261:3807–3812.

Olenski, K., Sieczkowska, H., Kocwin-Podsiadla, M., Help, H., and Kaminski, S. 2010. A subset of candidate polymorphisms identified by 52 SNPs mini-array in two Duroc subpopulations revealed significant differences in SNP allele distributions. *Ann. Anim. Sci.* 10:17–25.

Otto, G., Roehe, R., Looft, H., Thoelking, L., Knap, P.W., Rothschild, M.F., Plastow, G.S., and Kalm, E. 2007. Associations of DNA markers with meat quality traits in pigs with emphasis on drip loss. *Meat Sci.* 75:185–195.

Page, B.T., Casas, E., Heaton, M.P. et al. 2002. Evaluation of single-nucleotide polymorphisms in CAPN1 for association with meat tenderness in cattle. *J. Anim. Sci.* 80:3077–3085.

Pannier, L., Mullen, A.M., Hamill, R.M., Stapleton, P.C., and Sweeney, T. 2010. Association analysis of single nucleotide polymorphisms in DGAT1, TG and FABP4 genes and intramuscular fat in crossbred *Bos taurus* cattle. *Meat Sci.* 85:515–518.

Perez-Enciso, M., Clop, A., Folch, J.M., Sanchez, A., Oliver, M.A., Ovilo, C., Barragan, C., Varona, L., and Noguera, J.L. 2002. Exploring alternative models for sex-linked quantitative trait loci in outbred populations: Application to an Iberian × Landrace pig intercross. *Genetics* 161:1625–1632.

Perez-Enciso, M., Mercade, A., Bidanel, J.P., Geldermann, H., Cepica, S., Bartenschlager, H., Varona, L., Milan, D., and Folch, J.M. 2005. Large-scale, multibreed, multitrait analyses of quantitative trait loci experiments: The case of porcine X chromosome. *J. Anim. Sci.* 83:2289–2296.

Platter, W.J., Tatum, J.D., Belk, K.E., Chapman, P.L., Scanga, J.A., and Smith, G.C. 2003. Relationships of consumer sensory ratings, marbling score, and shear force value to consumer acceptance of beef strip loin steaks. *J. Anim. Sci.* 81:2741–2750.

Pommier, S.A., Pomar, C., and Godbout, D. 1998. Effect of the halothane genotype and stress on animal performance, carcass composition and meat quality of crossbred pigs. *Can. J. Anim. Sci.* 78:257–264.

Ponsuksili, S., Jonas, E., Murani, E., Phatsara, C., Srikanchai, T., Walz, C., Schwerin, M., Schellander, K., and Wimmers, K. 2008. Trait correlated expression combined with expression QTL analysis reveals biological pathways and candidate genes affecting water holding capacity of muscle. *BMC Genomics* 9:367.

Pösö, A.R., and Puolanne, E. 2005. Carbohydrate metabolism in meat animals. *Meat Sci.* 70:423–434.

Przybylski, W., Kocwin-Podsiadla, M., Kaczorek, S., Krzecio, E., and Kuryl, J. 1998. A comparative analysis between NN and Nn pigs for the carcass traits and also technological meat quality. *Polish J. Food Nutr. Sci.* 7/8:257–262.

Przybylski, W., Kocwin-Podsiadla, M., Kaczorek, S., Krzecio, E., and Monin, G. 2000. Interactive effects of HAL and RN genes in crossbreeding. In *Proceedings of the 46th International Congress of Meat Science and Technology*, August 27–September 1, 2000, Buenos Aires, Argentina, pp. 88–89.

Przybylski, W., Kocwin-Podsiadla, M., Kaczorek, S., Krzecio, E., Naveau, J., and Monin, G. 1996. Effect of the RN^- gene on meat quality traits in pigs originated from crossing of the Large White breed, Polish Landrace 23 with P-76 line and Hampshire boars. (In Polish) *Zesz. Nauk. Prz. Hod.* 26:143–149.

Qiao, R.M., Ma, J.W., Guo, Y.M., Duan, Y.Y., Zhou, L.H., and Huang, L.S. 2011. Validation of a paternally imprinted QTL affecting pH24h distinct from PRKAG3 on SSC15. *Anim. Genet.* 42:316–320.

Reardon, W., Mullen, A.M., Sweeney, T., and Hamill, R.M. 2010. Association of polymorphisms in candidate genes with colour, water-holding capacity, and composition traits in bovine M. longissimus and M. semimembranosus. *Meat Sci.* 86:270–275.

Rothschild, M.F. 2003. Advances in pig genomics and functional gene discovery. *Comp. Funct. Genomics* 4:266–270.

Rothschild, M.F., Hu, Z.L., and Jiang, Z.H. 2007. Advances in QTL mapping in pigs. *Int. J. Biol. Sci.* 3:192–197.

Rusc, A., Sieczkowska, H., Krzecio, E., Antosik, K., Zybert, A., Kocwin-Podsiadla, M., and Kaminski, S. 2011. The association between acyl-CoA synthetase (ACSL4) polymorphism and intramuscular fat content in (Landrace × Yorkshire) × Duroc pigs. *Meat Sci.* 89:440–443.

Schenkel, F.S., Miller, S.P., Jiang, Z., Mandell, I.B., Ye, X., Li, H., and Wilton, J.W. 2006. Association of a single nucleotide polymorphism in the calpastatin gene with carcass and meat quality traits of beef cattle. *J. Anim. Sci.* 84:291–299.

Sellier, P. 1998. Genetics of meat and carcass traits. In Rothschild, M.F. and Ruvinsky, A.J. (eds.), *The Genetics of the Pig.* CAB International, Wallingford, Oxfordshire, pp. 463–510.

Shackelford, S.D., Wheeler, T.L., and Koohmaraie, M. 1995. The effects of in-utero exposure of lambs to a beta-adrenergic agonist of prenatal and postnatal muscle growth, carcass cutability, and meat tenderness. *J. Anim. Sci.* 73:2986–2993.

Shackelford, S.D., Wheeler, T.L., and Koohmaraie, M. 1997. Effect of the callipyge phenotype and cooking method on tenderness of several major lamb muscles. *J. Anim. Sci.* 75:2100–2105.

Shackelford, S.D., Wheeler, T.L., and Koohmaraie, M. 1998. Can the genetic antagonisms of the Callipyge lamb be overcome? *Proc. Reciprocal Meat Conf.* 51:125–132.

Sherman, E.L., Nkrumah, J.D., Murdoch, B.M., Li, C., Wang, Z., Fu, A., and Moore, S.S. 2008. Polymorphisms and haplotypes in the bovine neuropeptide Y, growth hormone receptor, ghrelin, insulin-like growth factor 2, and uncoupling proteins 2 and 3 genes and their associations with measures of growth, performance, feed efficiency, and carcass merit in beef cattle. *J. Anim. Sci.* 86:1–16.

Shorthose, W.R., Powell, V.H., and Harris, P.V. 1986. Influence of electrical stimulation, cooling rates and aging on the shear force values of chilled lamb. *J. Food Sci.* 51:889–892.

Sieczkowska, H., Antosik, K., Zybert, A., Krzecio, E., Kocwin-Podsiadla, M., Kuryl, J., and Lyczynski, A. 2006a. The influence of H-FABP gene polymorphism on quality and technological value of meat from stress-resistant porkers obtained on the basis of Danish pigs and sharing Duroc blood. *Anim. Sci. Pap. Rep.* 24:259–265.

Sieczkowska, H., Kocwin-Podsiadla, M., Zybert, A., Krzecio, E., Antosik, K., Kaminski, S., and Wojcik, E. 2010a. The association between polymorphism of PKM2 gene and glycolytic potential and pork meat quality. *Meat Sci.* 84:180–185.

Sieczkowska, H., Krzecio, E., Antosik, K., Zybert, A., Kocwin-Podsiadla, M., and Kuryl, J. 2006b. The influence of H-FABP gene polymorphism on meatiness and carcass composition traits of stress-resistant fatteners produced with or without Duroc boars' share. *Anim. Sci. Pap. Rep.* 24:241–250.

Sieczkowska, H., Zybert, A., Antosik, K., Kocwin-Podsiadla, M., Kaminski, S., Wojcik, E., and Kmiec, M. 2007. The effect of interaction between PKM2 and GLUT4 genotypes on pork quality. In *Proceedings of the 53rd International Congress of Meat Science and Technology*, August 5–10, 2007, Beijing, China, pp. 279–280.

Sieczkowska, H., Zybert, A., Krzecio, E., Antosik, K., Kocwin-Podsiadla, M., Pierzchala, M., and Urbanski, P. 2010b. The expression of genes PKM2 and CAST in the muscle tissue of pigs differentiated by glycolytic potential and drip loss, with reference to the genetic group. *Meat Sci.* 84:137–142.

Smith, J.A., Lewis, A.M., Wiener, P., and Williams, J.L. 2000. Genetic variation in the bovine myostatin gene in UK beef cattle:allele frequencies and haplotype analysis in the South Devon. *Anim. Genet.* 31:306–309.

Soma, Y., Uemoto, Y., Sato, S., Shibata, T., Kadowaki, H., Kobayashi, E., and Suzuki, K. 2011. Genome-wide mapping and identification of new quantitative trait loci affecting meat production, meat quality, and carcass traits within a Duroc purebred population. *J. Anim. Sci.* 89:601–608.

Sosnicki, A.A., and Newman, S. 2010. The support of meat value chains by genetic technologies. *Meat Sci.* 86:129–137.

Soumillion, A., Erkens, J.F., Lenstra, J.A., Rettenberger, G., and te Pas, M.W. 1997. Genetic variation in the porcine myogenin gene locus. *Mamm. Genome* 8:564–568.

Steibel, J.P., Bates, R.O., Rosa, G.J. et al. 2011. Genome-wide linkage analysis of global gene expression in loin muscle tissue identifies candidate genes in pigs. *PLoS ONE* 6:e16766.

Taniguchi, M., Utsugi, T., Oyama, K., Mannen, H., Kobayashi, M., Tanabe, Y., Ogino, A., and Tsuji, S. 2004. Genotype of stearoyl-CoA desaturase is associated with fatty acid composition in Japanese Black cattle. *Mamm. Genome* 15:142–148.

Taylor, R.G., and Koohmaraie, M. 1998. Effects of postmortem storage on the ultrastructure of the endomysium and myofibrils in normal and callipyge longissimus. *J. Anim. Sci.* 76:2811–2817.

te Pas, M.F., de Wit, A.A., Priem, J., Cagnazzo, M., Davoli, R., Russo, V., and Pool, M.H. 2005. Transcriptome expression profiles in prenatal pigs in relation to myogenesis. *J. Muscle Res. Cell Motil.* 26:157–165.

te Pas, M.F., Hoekman, A.J., and Smits, M.A. 2011. Biomarkers as management tools for industries in the pork production chain. *J. Chain Netw. Sci.* 11:155–166.

te Pas, M.F., Kruijt, L., Pierzchala, M. et al. 2013. Identification of proteomic biomarkers in M. Longissimus dorsi as potential predictors of pork quality. *Meat Sci.* 95:679–687.

te Pas, M.F., Soumillion, A., Harders, F.L., Verburg, F.J., van den Bosch, T.J., Galesloot, P., and Meuwissen, T.H. 1999. Influences of myogenin genotypes on birth weight, growth rate, carcass weight, backfat thickness, and lean weight of pigs. *J. Anim. Sci.* 77:2352–2356.

te Pas, M.F.W., and Visscher, A.H. 1994. Genetic-regulation of meat production by embryonic muscle formation—A review. *J. Anim. Breed. Genet.* 111:404–412.

Thaller, G., Kuhn, C., Winter, A., Ewald, G., Bellmann, O., Wegner, J., Zuhlke, H., and Fries, R. 2003. DGAT1, a new positional and functional candidate gene for intramuscular fat deposition in cattle. *Anim. Genet.* 34:354–357.

Trott, J.F., Freking, B.A., and Hovey, R.C. 2013. Variation in the coding and 3′ untranslated regions of the porcine prolactin receptor short form modifies protein expression and function. *Anim. Genet.* 45:74–86.

Urban, T., Mikolasova¡, R., Kuciel, J., Ernst, M., and Ingr, I. 2002. A study of associations of the H-FABP genotypes with fat and meat production of pigs. *J. Appl. Genet.* 43:505–509.

Uytterhaegen, L., Claeys, E., Demeyer, D., Lippens, M., Fiems, L.O., Boucque, C.Y., Vandevoorde, G., and Bastiaens, A. 1994. Effects of double-muscling on carcass quality, beef tenderness and myofibrillar protein-degradation in Belgian Blue White bulls. *Meat Sci.* 38:255–267.

Van den Maagdenberg, K., Stinckens, A., Claeys, E., Buys, N., and De Smet, S. 2008. Effect of the insulin-like growth factor-II and RYR1 genotype in pigs on carcass and meat quality traits. *Meat Sci.* 80:293–303.

Van Eenennaam, A.L., Li, J., Thallman, R.M., Quaas, R.L., Dikeman, M.E., Gill, C.A., Franke, D.E., and Thomas, A.G. 2007. Validation of commercial DNA tests for quantitative beef quality traits. *J. Anim. Sci.* 85:891–900.

Verner, J., Humpolicek, P., and Knoll, A. 2007. Impact of MYOD family genes on pork traits in Large White and Landrace pigs. *J. Anim. Breed. Genet.* 124:81–85.

Voytik, S.L., Przyborski, M., Badylak, S.F., and Konieczny, S.F. 1993. Differential expression of muscle regulatory factor genes in normal and denervated adult rat hindlimb muscles. *Dev. Dyn.* 198:214–224.

Walling, G.A., Visscher, P.M., Wilson, A.D., McTeir, B.L., Simm, G., and Bishop, S.C. 2004. Mapping of quantitative trait loci for growth and carcass traits in commercial sheep populations. *J. Anim. Sci.* 82:2234–2245.

Wendt, M., Bickhardt, K., Herzog, A., Fischer, A., Martens, H., and Richter, T. 2000. [Porcine stress syndrome and PSE meat:clinical symptoms, pathogenesis, etiology and animal rights aspects]. *Berl. Munch. Tierarztl. Wochenschr.* 113:173–190.

Williams, J.L. 2008. Genetic control of meat quality traits. In Toldrá, F. (ed), *Meat Biotechnol.* Springer, New York, pp. 21–60.

Wimmers, K., Murani, E., and Ponsuksili, S. 2010a. Functional genomics and genetical genomics approaches towards elucidating networks of genes affecting meat performance in pigs. *Brief. Funct. Genomics* 9:251–258.

Wimmers, K., Murani, E., and Ponsuksili, S. 2010b. Pre- and postnatal differential gene expression with relevance for meat and carcass traits in pigs—A review. *Anim. Sci. Pap. Rep.* 28:115–122.

Winter, A., Kramer, W., Werner, F.A., Kollers, S., Kata, S., Durstewitz, G., Buitkamp, J., Womack, J.E., Thaller, G., and Fries, R. 2002. Association of a lysine-232/alanine polymorphism in a bovine gene encoding acyl-CoA:diacylglycerol acyltransferase (DGAT1) with variation at a quantitative trait locus for milk fat content. *Proc. Natl. Acad. Sci. U. S. A.* 99:9300–9305.

Wright, W.E., Sassoon, D.A., and Lin, V.K. 1989. Myogenin, a factor regulating myogenesis, has a domain homologous to MyoD. *Cell.* 56:607–617.

Wu, X.X., Yang, Z.P., Shi, X.K., Li, J.Y., Ji, D.J., Mao, Y.J., Chang, L.L., and Gao, H.J. 2012. Association of SCD1 and DGAT1 SNPs with the intramuscular fat traits in Chinese Simmental cattle and their distribution in eight Chinese cattle breeds. *Mol. Biol. Rep.* 39:1065–1071.

Yao, J.B., Coussens, P.M., Saama, P., Suchyta, S., and Ernst, C.W. 2002. Generation of expressed sequence tags from a normalized porcine skeletal muscle cDNA library. *Anim. Biotechnol.* 13:211–222.

Yuan, Z., Li, J., Li, J., Gao, X., Gao, H., and Xu, S. 2013. Effects of DGAT1 gene on meat and carcass fatness quality in Chinese commercial cattle. *Mol. Biol. Rep.* 40:1947–1954.

Zhu, L., and Li, X.W. 2005. The genetic effects of MyoG gene. *Yi Chuan.* 27:710–714.

Zybert, A., Antosik, K., and Sieczkowska, H. 2009. The accordance of RN^- phenotype with polymorphism of PRKAG3 gene of (Landrace × Yorkshire) × Hampshire fatteners. In *Proceedings of the 55th International Congress of Meat Science and Technology*, August 16–21, 2009, Copenhagen, Denmark, pp. 211–213.

Zybert, A., Sieczkowska, H., and Krzecio, E. 2007. The effect of interaction between PKM2 and CAST/HinfI genotypes on pork quality. In *Proceedings of the 53rd International Congress of Meat Science and Technology*, August 5–10, 2007, Beijing, China, pp. 281–282.

11 Beef Quality

Jean-François Hocquette, Dominique Bauchart, Didier Micol, Rod Polkinghorne, and Brigitte Picard

CONTENTS

11.1 INTRODUCTION

Despite industry efforts to control the eating and nutritional quality of beef, there remains a high level of variability in these quality traits, which is one reason for consumer dissatisfaction (Alfnes et al. 2008). In addition, there is a severe competition between beef and white meats, the price of the latter being low compared to beef. Consumers also have increasing concerns about safety and health factors, the latter being largely due to statements from the medical profession that beef may contain too much saturated (SFA) and *trans* monounsaturated (MUFA) fatty acids, which can be major risks for the development of various diseases in humans. All these factors and also media campaigns (illicit trading and use of hormones, the "mad cow" crisis, emphasis on environmental degradation by livestock, etc.) and economic and societal changes (industrialization, intensification of agriculture, and urbanization), have induced a decrease in beef consumption in most developed countries. This was, for instance, the case in Australia in the early 1990s. During this period, consumers were recording their dissatisfaction with Australian beef products by decreasing consumption (Polkinghorne et al. 2008). Consumers' future beef purchase intentions are key factors that impact on beef consumption levels. These levels are heavily influenced by consumers' previous experiences of beef eating quality (Banovic et al. 2009) that are dependent on a combination of their evaluations of taste, tenderness, and juiciness. Expected quality in terms of safety and health also impact on consumption levels. Using focus groups with consumers in Germany, Spain, France, and United Kingdom, it was found that consumers generally welcome the idea of a beef eating-quality guarantee, but that willingness to pay is conditional upon the system managing to deliver effectively upon its promises (Verbeke et al. 2010).

In reality, a great deal of research has been conducted to identify and better understand the role of the different factors that affect the sensory and nutritional traits of beef. However, despite this great amount of knowledge, there is still no reliable online tool to predict beef eating quality and deliver consistent quality beef to consumers, at least in Europe. In addition, consumers are increasingly aware of the relationships between diet, health, and well-being, resulting in choices of foods that are healthier and more nutritious (Verbeke et al. 2010). For meat, much attention has focused on intramuscular fat level that affects tenderness, flavor, and nutritional value. Attention has been focused also on fatty acid composition, along with the biological value of the protein, trace elements, and vitamins, which are key factors contributing to nutritional value. The WHO (2003) provided recommendations for total fat, SFA, n-6 PUFA, n-3 PUFA, and *trans* fatty acids in the diet. These recommendations focus on reducing the intake of SFA (considered to be associated with increased cholesterol) and on increasing the intake of omega-3 PUFA. The longer-chain n-3 PUFA, eicosapentaenoic acid (EPA, 20:5n-3), and docosahexaenoic acid (DHA; 22:6n-3) have beneficial effects in reducing the risk of cardiovascular disease, cancer, and type-2 diabetes, and have critical roles for proper brain function, visual development in the fetus, and for maintenance of neural and visual tissues (Simopoulos 1991). However, strategies do exist to regulate fatty acid composition of beef (to increase its PUFA and conjugated linoleic acid [CLA] contents) through animal type (age, gender, genotype, etc.) and its diet (level, nature of the basal diet, supplements in fat of plant origin). These will be outlined in this chapter.

It is recognized that science and innovation should play a major role in helping the industry respond to consumer concerns and expectations. However, science-based innovation still remains weak in the red meat sector and will need to be addressed for the sector to remain competitive (Troy and Kerry 2010). This chapter reviews knowledge in the field of beef quality, including the important relationships between the characteristics of different beef production systems and components of meat quality such as tenderness, flavor, overall palatability, and nutritional value. In the first part of this review, the different quality traits will be described. Then, the major factors affecting beef quality will be reviewed, namely, the effects of animal factors and breeding strategies on sensory quality traits and the effects of dietary factors on the nutritional quality of beef.

11.2 MAIN QUALITY CRITERIA OF BEEF

A major factor that needs consideration is the increasing number of quality attributes that must be considered, namely, intrinsic quality attributes, which includes, for instance, safety, health, sensory traits (e.g., tenderness, flavor, juiciness, overall liking), convenience, and so on, but also extrinsic quality traits, which are associated with the product, namely, (i) production system characteristics (from the animal to the processing stages, including, for example, animal welfare and carbon footprint), and (ii) marketing variables (including price, brand name, distribution, origin, packaging, labeling, and traceability). A new challenge in research now is combining all these criteria to provide accurate and understandable information to consumers (for review, see Hocquette et al. 2012b). As stated earlier, this chapter will focus only on the sensory and nutritional quality of beef.

11.2.1 Sensory Quality of Beef

Sensory properties of food are the characteristics that consumers can directly perceive through their senses (Figure 11.1). They are classified into three groups: (i) qualitative, which is characteristic of what it is perceived (salty, taste, off-flavor, etc.), (ii) quantitative, which represents the intensity of the sensation (little to intense), and (iii) characteristics that induce the hedonic pleasure of consumers. For meat from ruminants, the main sensory characteristics are color, tenderness, juiciness, and flavor liking (reviewed by Geay et al. 2001).

The first characteristic perceived by the consumer is color. This is often the only one trait he has to consider at the time of purchase, especially in the current circuits of distribution. The red color of meat, especially of beef, is conferred by a pigment, myoglobin. After prolonged exposure to air, the color is unstable because the pigment is oxidized to metmyoglobin, leading to a brownish color unpleasant to the eye

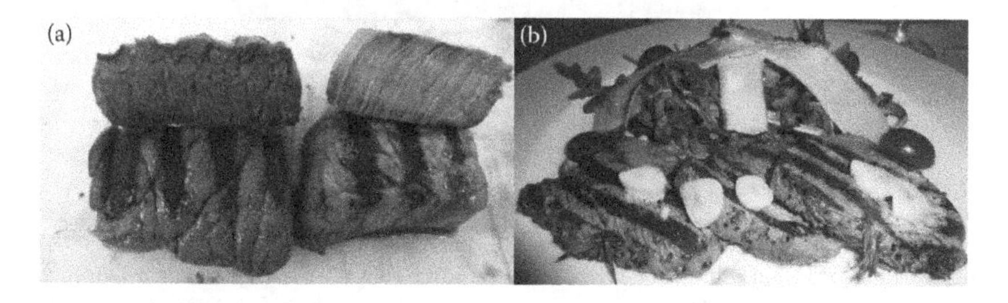

FIGURE 11.1 Beef cuts for consumer tests (a) and a typical meat dish in a restaurant (b). (Photo credit: J.F. Hocquette.)

of the buyer. Tenderness can be defined as the ease with which meat can be sliced or chewed. This is the most crucial sensory quality trait for beef consumers. This is also the quality criterion with the most variable multifactorial origin, and therefore the most difficult to control or predict. Juiciness represents the perception of meat as being more or less dry during consumption. A difference has been made between the initial juiciness, which is perceived at first bite, and the sustained juiciness during chewing. Flavor liking of meat is the combined result of two senses, taste and smell, in an olfactory–gustatory perception.

11.2.2 Fatty Acid Composition of Beef

In the current state of knowledge, the nutritional quality of food is directly related to its biochemical composition. The content of various favorable or unfavorable compounds for human health must be the highest or the lowest, respectively, to improve nutritional quality. Here, we will describe mainly the composition of the lipid fraction of beef, which is the subject of much research and debate. The WHO (2003) recommended that total fat and SFA should contribute <15%–30%, and <10%, respectively of total energy intake in the human diet. In addition, n-6 PUFA, n-3 PUFA, and *trans* fatty acids should contribute <5%–8%, <1%–2%, and <1% of total energy intake, respectively. More precisely, the n-6/n-3 PUFA ratio should be lower than 4 according to the WHO recommendations. In fact, on average, intramuscular fat in beef muscle consists proportionally on average of 45%–50% as SFA, 35%–45% MUFA, and up to 6% PUFA, respectively (Figure 11.2, Durand et al. 2005). The polyunsaturated:saturated fatty acid ratio (P:S) for beef is generally too low at around 0.1 except for very lean animals (<1% intramuscular fat) where P:S ratios are much higher ~0.5–0.7 and hence favorable for human health (Scollan et al. 2006).

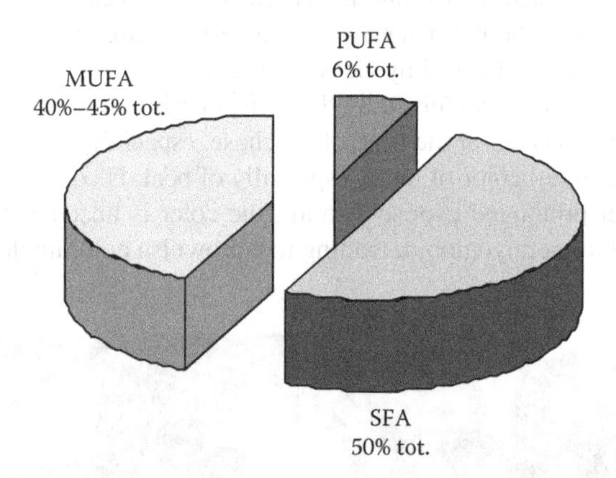

FIGURE 11.2 Mean composition of beef in fatty acids. SFA: saturated fatty acids (of which 3% C14:0, 25% C16:0, and 15% C18:0). MUFA: monounsaturated fatty acids. PUFA: polyunsaturated fatty acids (of which 5% C18:2 n-6 and 1% C18:3 n-3). (Adapted from Durand, D. et al. 2005. *55th Annual Meeting of the European Association for Animal Production*, Bled (Slovenia). *Indicators of Milk and Beef Quality*; EAAP Publication 112. Wageningen Academic Publishers, Wageningen, the Netherlands, pp. 135–150.)

The n-6:n-3 ratio for beef is beneficially low (usually <3), reflecting the significant amounts of desirable n-3 PUFA, such as α-linolenic acid (18:3n-3), but also EPA, docosapentaenoic acid (DPA; 22:5n-3), and DHA. Beef and other ruminant products are important dietary sources of CLA, mainly the *cis*-9,*trans*-11 isomer, which is supposed to have health-promoting beneficial properties (Salter 2013). Beef lipids also contain *trans* fatty acids of which the most dominant is *trans*-11 18:1 (vaccenic acid). In contrast to industrial *trans* fatty acids, *trans* fatty acids produced by ruminants have a potential protective effect against the development of coronary heart diseases (Salter 2013). The overall balance between the benefits and weaknesses of the nutritional value of beef intramuscular fat is consequently low to moderate (too much SFA, but presence of n-3 PUFA and fatty acid isomers beneficial for human health).

11.2.3 Health Benefits of Beef

As reviewed by Geay et al. (2001), meat represents, above all, an important source of proteins (17%–22% fresh tissue) and is rich in essential amino acids (55.2 g for 16 g N) and in lysine (9.1 g for 16 g N). Meat from ruminant animals, especially from bovines, is also an important source of heminic iron (about 2–5 mg · 100 mg^{-1} fresh tissue according to the type of muscle, which is, respectively, 3–4 times higher than that in meat from pork and chicken as reviewed by Geay et al. (2001). This heminic iron found in beef is 5–6 times more absorbed by humans than the nonheminic iron from plants, which is a major positive nutritional benefit of beef (and also sheep meat; see Chapter 13). Zinc is also abundant in bovine meat (3–11 mg · 100 g^{-1} according to cuts). Another positive nutritional benefit of beef is the fact it is an important source of vitamins of the B group: B1, B2, B6, B12, and niacin, especially vitamins B6 (0.3–0.4 mg · 100 mg^{-1}) and B12 (1.5–2.5 mg · 100 g^{-1}) virtually absent in plants but synthesized by microorganisms of the digestive tract of ruminants (for a review, see Geay et al. 2001).

To date, most of the communication to consumers has concerned negative nutritional properties of beef described above specifically related to its fatty acid composition. We strongly recommend that the positive health benefits of beef are promoted, a fact often not emphasized in public media.

11.3 EFFECTS OF ANIMAL CHARACTERISTICS ON SENSORY QUALITY

Beef production is characterized by a great diversity of breeds, some of them being early maturing with a high degree of fatness, others being late maturing with a higher muscle yield and less fatness. The quality of beef also differs according to the age, sex, and physiological stage of animals (Table 11.1). Beef production systems differ across countries according to all these factors, which means that beef from each part of the world has its own characteristics. For instance, beef contains more intramuscular fat in the United States (8%–11% for the best grades) and even more in Japan (Figure 11.3) than in France (less than 6% on average). In Japan, steers from the Japanese Black cattle breed (which is a breed selected for extreme

TABLE 11.1
Changes in Muscle Characteristics According to Management Factors and Expected Impacts on Sensory Qualities of Beef

	Intramuscular	Fibers			Collagen		Expected Impacts on Sensory Quality Score	
	Fat Level	Size	Type	Metabolism	Content	Solubility	Favorable	Unfavorable
Age	++	++	+ I	+ Oxidative	=	–	Color flavor	Tenderness
Muscular hypertrophy	–	++	+ IIX	+ Glycolytic	– or =	= or +	Tenderness	Color flavor
Gender (female/male)	++	–	+ I	+ Oxidative	–	+	Color tenderness	
Feeding level	++	+	+ IIX	+ Glycolytic	=	++	Tenderness flavor	
Pasture-based feeding system	– or =	=	– IIX	+ Oxidative	=	= or +	Color flavor	

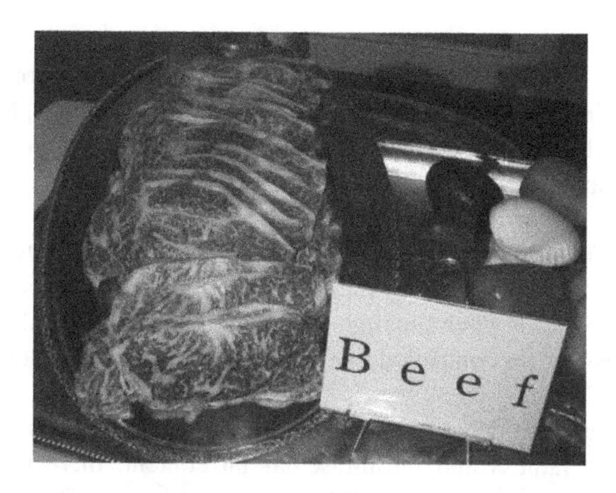

FIGURE 11.3 Marbled beef from Japan. (Photo credit: J.F. Hocquette.)

marbling fat development) are finished during a long period with a high-energy diet leading to very fat meat due to a combination of genetic and nutritional factors. The development of adipose tissues in some specific muscles appears to disorganize the muscle structure and contributes to tenderization of highly marbled beef from Japanese Black cattle during the late fattening period (Nishimura et al. 1999).

Similar systems called "feedlots" are common in North America. In this case, mainly British breeds (also early maturing) such as Aberdeen Angus and crosses are used. In continental European countries, especially France, a great diversity of breeds (more than 40 in France) exists with dairy breeds, beef breeds, and dual-purpose or hardy breeds. Generally, European beef breeds have been selected for their muscle growth potential and low fatness. In addition, production systems include not only steers but also heifers, young bulls, and cull cows fed with different types of diets according to local dietary resources, leading to a great diversity of beef products. A similar situation exists in China with a great diversity of regional environments and many local breeds that have been generally not extensively selected in contrast to France. Consumer habits also differ between countries. For instance, whereas beef from steers is consumed in many countries, French consumers consume beef mainly from females (heifers and cull cows). Differences in sensory assessment of beef are also derived from differences in cooking, but variations in experience and perception of consumers also contribute (Dransfield et al. 1984).

11.3.1 Effect of Breed Type

Between breed types, the resulting muscle color is mainly dependent on animal maturity; these differences are particularly expressed during the growing phase of the animal prior to reaching adult maturity. The Anglo-Saxon and dairy breeds on average mature earlier and faster than the continental beef breeds. Thus, the differences between breeds are lessened when they are compared at the same physiological stage of development or maturity (expressed as percentage of adult weight) (Renerre 1982a). In young French bull production conditions, significant differences in meat color are observed between bulls from early-maturing breeds (dairy type) and late-maturing breeds (beef breeds). As an example, a significant proportion of young Blonde d'Aquitaine cattle have a very pale meat unlike Normande, Montbeliard, and Holstein breeds. Finally, these differences are related to individual animal potential for color development or development rate, which results in relatively high color variability in the late-maturing breeds.

Comparisons of meat tenderness between breeds are difficult to interpret. Of all available studies, it appears that for meat quality, the differences between breeds are generally not significant given the high variability between animals within breeds if aging and processing are controlled to minimize confounding factors (Renand 1988; Dransfield et al. 2003).

The weak relationship between breed type and tenderness is predictable since even at the same fatness and intramuscular fat content, there is little difference between the sensory properties of meat from different breeds (Dransfield et al. 2003). However, from comparisons of meat sensory quality between breeds, it appears that, in all cases, the discriminatory factor is tenderness. Furthermore, overall liking from sensory panels strongly relates to tenderness. The appreciation of tenderness is generally positively associated with intramuscular fat content of beef (Renand 1988; Dransfield et al. 2003). As an illustration, Figure 11.4 presents the assessments of sensory qualities from 11 European breeds (EU-program GEMQUAL; Olleta et al. 2006). Pirenaica cattle produced meat that was very tender and juicy and with a desirable flavor, unlike Simmental animals that produced tougher meat with less flavor and juiciness. Overall in this study, despite small differences between breeds, beef breeds appear to have higher sensory qualities than dairy breeds. This figure also emphasizes the relationship between the assessments between the sensory qualities; the more tender the meat, the faster the water is released by chewing and juiciness is appreciated (Cross 1988).

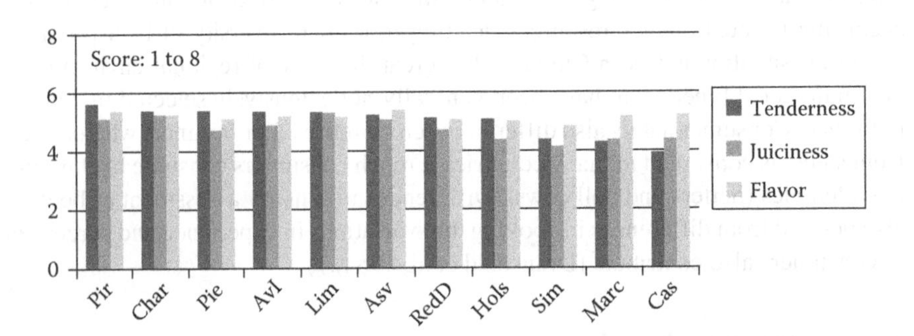

FIGURE 11.4 Evaluation of sensory qualities of the *Longissimus thoracis* muscle in 11 European breeds. Breeds in descending order of tenderness: Pir: Pirenaica; Char: Charolais; Pie: Piemontese; Avl: Avileña-Negra Ibérica; Lim: Limousin; Asv: Asturiana de los Valles; RedD: Danisch Red cattle; Hols: Holstein; Sim: Simmental; Marc: Marchigiana; Cas: Casina. (From Olleta, J.L. et al. 2006. *Mediterranean Livestock Production: Uncertainties and Opportunities.* Second Seminar of the Scientific-Professional Network on Mediterranean Livestock Farming (RME), 2006/05/18–20, Zaragoza, (Spain). Options Méditerranéennes: Série A. Séminaires Méditerranéens; 78. CIHEAM-IAMZ/CITA, pp. 297–300.)

Across animal types, ages, or genotypes, flavor liking of beef meat develops as animals grow older, coinciding with increased intramuscular fat content. A minimum amount of intramuscular fat is needed for flavor to be expressed. The relationship between intramuscular fat level and flavor liking is curvilinear and variation in intramuscular fat level explains variations of flavor liking in animals of different types (Dransfield et al. 2003): an increase of 1%–3% of fat index induces a higher lipid flavor of 0.7 point on a scale of 1–10. With data from only one muscle type and without any adjustment for fixed effects, about 16% of the variability in flavor liking can be explained by differences in intramuscular fat level from 0.3 to up to 15% (Thompson 2004). However, in another study with mainly young bulls, and adjusted for sex, breed, and age effects, no more than 3% of the variability in flavor liking was explained by differences in intramuscular fat level, using a dataset with little variability and low absolute values (on average, 1.5% of intramuscular fat level) (Hocquette et al. 2011b).

The lipid content also plays an important role on the juiciness of beef. Meat with more fat is always less dry than lean meat when chewing in the mouth.

11.3.2 Effect of Age

It is established that beef color and muscle myoglobin content increase with age. Color intensity increases significantly between 9 and 13 months of age, and especially between 13 and 17 months of age. Beyond this age, the color development is significantly lower (Renerre 1982b). In the production of young bulls from late-maturing beef breeds, meat is somewhat too light for the French market (10 ppm of iron) (Bastien 2003). Beef females in the same age group (9–24 months) have a more rapid increase in pigment content than males, which corresponds to their higher precocity (Renerre 1982b).

Tenderness of beef changes to a small extent in young animals and tends to decrease when the animals reach adulthood (reviewed by Oury et al. 2007a). This may be explained at least in part by an increase in collagen content and collagen crosslinks with animal age (Harper et al. 1999). Under conditions of French production, between the ages of 12 and 24 months in entire or castrated males, most recently listed tests do not show an increase in the toughness of meat assessed by sensory analysis or shear force (Oury et al. 2007b). However, between 15 and 19 months, an increase of toughness (+20%) in heavy entire males has been observed (Dransfield et al. 2003). Tenderness was not altered by increasing age in females between 12 and 35 months of age in line with no differences in total and soluble collagen contents in muscles (reviewed by Micol et al. 2010). In cull cows, results of sensory analysis showed no differences between 4 and 7 or 9 years of age on the scores of tenderness obtained from beef breeds (Jurie et al. 2006a).

Normal beef flavor depends on the animal and the dominant type of muscle fiber in the cut (reviewed by Geay et al. 2001). In general, oxidative and redder muscle fibers, associated with higher levels of intramuscular fat (Jurie et al. 2007), lead to a more fully developed flavor. The age of animals is an important factor for flavor: as the animals get older, it produces a more intense flavor. A minimum amount of fat in the muscle is needed to express flavor. Most authors agree that there is a curvilinear relationship between flavor score and intramuscular fat level although the level at which this plateaus depends on the study, that is, 2%–5% in mainly young bulls from France that are lean animals (Hocquette et al. 2011b), but 14% in cattle from Australia with fatter bovines with a higher range of intramuscular fat (Thompson 2004).

In a study conducted in France in four beef breeds (Aubrac, Charolais, Limousin, and Salers) (for review, see Picard et al. 2007), the total lipid content of three muscles, the *Longissimus thoracis* (LT), *Semitendinosus* (ST), and *Triceps brachii* (TB), was compared according to age and sex, in bulls (at 15, 19, 24 months of age) and in cull cows (at 4–5 years, 6–7 years, and 8–9 years of age) (Bauchart et al. 2002a,b). In growing Charolais bulls, muscles generally tended to accumulate fat (as the triglyceride form) with increasing age for the three muscles studied. In contrast, in the cull cows, intramuscular fat level increased only transiently for the ST and TB muscles and even tended to decrease in the case of LT muscle for older cows. In conclusion, in the case of young bulls, for the three muscles combined, intramuscular fat levels did not significantly differ between breeds, whereas they increased with age. In cull cows, these trends were opposite since intramuscular fat level varied according to the breed.

11.3.3 Effect of Gender

For beef cattle of the same age and type, animal gender (entire male, steer, or female) leads to differences in sensory qualities. Tenderness of beef from females and steers is always higher than that of entire animals. These differences are still important and significant for different cuts (Touraille 1982). However, the difference between castrated males and females is less clear, although in favor of females that produce more tender beef. In the study of Touraille (1982), castrated cattle had the highest juiciness compared to females and especially to entire males. For flavor, the ranking was the same as for juiciness (steers, females, and then young bulls). These rankings are consistent with differences in fatness between genders.

When females are slaughtered at the same age after or without gestation, nonpregnant heifers generally have a higher carcass weight. The studies conclude that gestation and parturition do not affect beef tenderness (Dumont et al. 1987). However, when females that have calved were slaughtered at an age and a weight significantly higher (which corresponded to adult cows), beef is less tender due to the older age (Patterson et al. 2002).

11.3.4 Recent Research to Better Understand the Determinism of Tenderness

Tenderness is a complex trait under multifactorial determinants and as such is difficult to control. Some biochemical factors such as muscle fibers, collagen, and lipid content are well known to be involved in tenderness (Oddy et al. 2001; Koohmaraie et al. 2002; Chriki et al. 2012). Moreover, several quantitative trait loci (QTLs) for tenderness have been detected (for a review, see Hocquette et al. 2007; see also Chapter 10). However, the control of tenderness variability remains a major challenge for the beef sector. For several years, various genomics programs (Cassar-Malek et al. 2008) have been conducted internationally in order to reveal genomic markers of tenderness (a biomarker is an indicator of a specific biological trait; here tenderness). Comparative transcriptomic (Bernard et al. 2007) and proteomic (Picard et al. 2012) analysis, conducted on bovine muscles with low or high tenderness scores estimated by sensory analysis and/or mechanical measurements, brought up a list of potential biological markers that may be used as phenotypic biomarkers to predict the "tenderness potential" of an animal or a carcass (for review, see Hocquette et al. 2007; Picard et al. 2012). A bioinformatic analysis showed that this list of biomarkers are involved mainly in four major biological functions: metabolic and contractile properties, muscle structure, calcium metabolism, and cellular stress (family of heat shock proteins: Hsps) (Guillemin et al. 2011; Picard et al. 2012).

In order to analyze these biomarkers on a large number of samples simultaneously, new molecular tools were developed. A DNA chip with specific genes involved in muscle biology or beef quality was developed (Hocquette et al. 2012a). At the protein level, a dot-blot tool was developed for the measure of relative abundance of proteins (Guillemin et al. 2009). This immunological tool allows measurement of the relative abundance of one protein for several muscle samples simultaneously. These tools are currently being used in several experiments with different breeds and muscles in order to establish prediction equations for beef tenderness. These biomarkers may be used as phenotypic markers to predict the "tenderness potential" of an animal or a carcass.

At the scientific level, these approaches allowed scientists to gain new insights about the biological functions involved in meat tenderness (Ouali et al. 2013). Moreover, it provided the development of a high-throughput screening method to analyze tenderness biomarkers. The final objective is to develop a tool to be used for a "paddock" application. This technology is already used in medical science as a diagnostic tool. It will be used on muscle samples taken by biopsies on live animals or on carcasses after slaughter.

11.4 EFFECTS OF LIVESTOCK PRACTICES ON SENSORY QUALITY OF BEEF

The management of a given animal (mainly food supply and modulation of its growth curve) may also affect meat sensory qualities (Table 11.1), but more strongly affect its nutritional qualities due to modulation of its intramuscular fat content and fatty acid profile. These effects can be explained by the nature and quantity of nutrients that can alter the characteristics and composition of the muscle.

11.4.1 Effect of Feeding Level on Sensory Qualities

Feed intake appears to modify the overall pigment content in the meat of ruminants as a reduction in intake results in an increase in the proportion of red oxidative fibers (Picard et al. 1995). Thus, the color of some muscles of restricted males is darker than animals receiving a high feeding level for the same slaughter weight.

Some authors have shown that the reduction of feeding level before slaughter decreases sensory meat quality, particularly tenderness. This reduction is associated with a reduction in the proportion of glycolytic fibers, which induces a rapid rate of aging (Picard et al. 1995). Also, the proportion of connective tissue is increased and the solubility of collagen is decreased (Fischell et al. 1985). In addition, it is known that a reduction in feeding level is accompanied by a reduction in carcass and muscle adiposity, which can interact with the assessment of the tenderness in connection with the *postmortem* cold contraction and aging of meat. The decrease in intramuscular lipid content can also reduce tenderness although it only explained from 3% to 10% of the differences in tenderness between samples in the study by Dransfield et al. (2003).

Because changes in feeding level induce differences in body weight gain of animals, the effect of feeding level on meat tenderness can be analyzed through the relationships with live weight gain. A study with 100 Charolais heifers concluded that increased carcass weight from 360 to 380 kg for the same slaughter age of 33 months and for the same carcass fatness level was accompanied by a 15% increase in the tenderness score by a sensory analysis panel. Thus, it appears that the increase in carcass weight and body weight gain, for the same age at slaughter and the same fatness will be in favor of beef tenderness (Oury 2007a). While increasing the feeding level will result in an increase in intramuscular fat content, some authors have observed no significant differences in the juiciness of the meat (Miller et al. 1987).

11.4.2 Effect of Compensatory Growth on Sensory Qualities

A high feeding phase after a period of feed restriction in ruminants leads to a growth rate higher than that of unrestricted animals (Hoch et al. 2003). This is called compensatory growth, which may result in an improvement of meat tenderness (reviewed by Geay et al. 2001). This improvement is partly due to an increase in synthesis of new collagen with a higher solubility or with different characteristics (McCormick 1994). In addition, Harper et al. (1999) showed that weight loss with grain realimentation favors tenderness in regard to the connective tissue component. The increase in tenderness also could be associated with an increase in the proportion of glycolytic muscle fibers, which induce a higher speed of aging compared to slow fibers (Picard et al. 1995). Finally, this increased tenderness could be associated with an increase of intramuscular fat content according to the higher physiological age of animals at the time of high feeding and finishing.

11.4.3 Effect of the Nature and Composition of the Diet

Variations in diet composition induce changes in the digestive processes that regulate the nature and proportions of nutrients absorbed by the ruminant. They are therefore likely to affect the sensory qualities of beef. In old cattle, a stronger pigmentation and a more intense red color were observed with grazed grass diets compared to diets rich in concentrate feed fed indoors (Priolo et al. 2001). This greater coloration may be related to a higher age in animals managed on pasture, but also to the more intense physical activity related to locomotion, leading to a more oxidative metabolism (Jurie et al. 2006b) and a higher proportion of oxidative muscle fibers (Vestergaard et al. 2000).

Many experiments have evaluated the effect of animal feeding during finishing on the tenderness of meat with similar or different feeding levels. The results are thus often the combined effects of the type of food and of the feeding level. A review by Oury summarizes the results obtained on meat tenderness between different foods: dry forage, corn silage or grass, and concentrate feed (Oury et al. 2007b). At the same feeding level, no clear differences in meat tenderness were observed according to the nature of food (dry or wet forage, mixed diets with cereals, etc.). Further work is thus needed to better discriminate the strict effects of the nature and composition of the diet on beef tenderness (Geay et al. 2001).

In Europe, some consumers consider that meat from animals raised on grass (Figure 11.5) have a specific flavor and taste better (Keane and Allen 1998) unlike in the United States where beef from animals finished on grass is less appreciated for its flavor than that of animals finished on concentrate diets (Melton 1990). Beef cattle raised and finished on grass are frequently discriminated by their "pastoral" stronger flavor, which persists even when animals are raised on grass and finished on concentrate diets.

Lipids contribute to the flavor of meat, in particular by the nature of their fatty acids, which in turn determines the nature of the oxidation products produced by cooking. For example, it is noted that SFA that are very resistant to oxidation at low temperature, unlike PUFA, are decomposed at high temperature. For example, when ruminants are fed with cereal-based diets, a high proportion of PUFA C18:2 n-6, which has escaped from ruminal hydrogenation, is deposited in tissues at the expense of SFA and in phospholipids at the expense of C18:3n-3. Thus, Larick and Turner (1989) showed that the flavor of the meat of finished steers on grass was

FIGURE 11.5 Limousin cattle at pasture. (Photo credit: J.F. Hocquette.)

changed when cereals were fed. A "sweetish" flavor in the case of animals from pasture declined in flavor of "beef flavor" when North Americans tested the meat. These authors also observed that products from lipid degradation such as aldehydes and ketones were present in higher levels in the volatiles of meat from steers finished on grass than in those fed with cereals. Terpenes are less abundant in beef from the latter and their concentration in beef is associated with changes in flavor (Coulon and Priolo 2002). For example, the content of the terpenoid neophytadiene is positively correlated with the "pastoral" beef flavor (Larick and Turner 1989).

From many reports, the nature of the diet did not affect juiciness of the meat between different types of food such as grazed grass, silage, or grain-based diet (for a review, see Geay et al. 2001). However, Listrat et al. (1999) reported that juiciness was slightly lower when growing Salers bulls were fed with hay compared to grass silage.

11.5 EFFECTS OF ANIMAL NUTRITION ON NUTRITIONAL QUALITY OF BEEF

Generally, in all species, dietary factors strongly influence the nutritional qualities of meat due to modulation of its intramuscular fat content and its fatty acid profile. It is generally acknowledged that other factors such as genetic factors have a smaller influence than dietary factors on the fatty acid composition of beef (De Smet et al. 2004; see also Chapter 4). These huge nutritional effects can be explained by the fact that dietary fatty acids absorbed by animals are found deposited in their tissues. However, in the specific case of herbivores, the potential to alter the fatty acid composition of muscle by nutrition is limited to a large extent by ruminal biohydrogenation of dietary lipids. Indeed, there is a great potential to markedly increase the concentration of n-3 PUFA in beef muscle when 18:3n-3 (as linseed oil) is infused directly into the small intestine, thereby by-passing the rumen (Durand et al. 2005). Similarly, Fortin et al. (2010) reported that abomasal infusion of fish oil (40 g/kg dry matter intake) increased the concentration of EPA in muscle phospholipids from 4.4 g/100 g in the control animals to 13.9 g/100 g in the infused animals. The corresponding data for DHA were 0.69 g/100 g and 3.9 g/100 g. The ongoing challenge is to achieve these high levels of enrichment by normal dietary means (which means without any infusion).

11.5.1 Compared Effects of Grass Feeding with Concentrate-Based Diets

Variations in the composition of the basal diet on beef fatty acids have been extensively studied in Europe and in America with the practical goal to determine the most favorable conditions for muscle growth in association with an improved nutritional value of deposited fatty acids. In 30–32-month-old Charolais steers raised on pasture (rye-grass rich in linolenic acid, 18:3n-3) or given a maize silage-based diet (rich in linoleic acid, 18:2n-6), total muscle lipids were not quantitatively modified at around 80 mg/g of dry tissue for *Rectus abdominus* muscle (RA, flank steak) and 60 mg/g for *Semitendinosus* muscle (ST, round steak) (Bauchart et al. 2001, Table 11.2).

On the other hand, fatty acid composition of muscle lipids (Table 11.2) in grass-fed steers is characterized by an 18:3n-3 level, which is 3 and 5 times higher in RA and ST muscles, respectively, when compared to that in maize silage steers. Similar variations

TABLE 11.2
Variations in Fatty Acid Composition of Total Lipids of the *Rectus abdominis* and *Semitendinosus* Muscles from 30–32 Month-Old Charolais Steers Given a Maize Silage or a Fresh Grass-Based Diet

	Rectus abdominis Muscle		*Semitendinosus* Muscle	
Fatty Acids (% Total FA)	Maize Silage	Fresh Grass	Maize Silage	Fresh Grass
16:0	24.0	24.0	22.8	21.3
18:0	12.4	12.4	11.6	11.8
18:1n-9	37.8	35.0	33.6	31.3
18:2n-6	4.4	3.8	6.4[a]	5.3[b]
18:3n-3	0.4[a]	1.4[b]	0.5[a]	2.3[b]
20:5n-3	0.2	0.6	0.5[a]	1.9[b]
22:5n-3	0.5	0.8	0.8[a]	1.8[b]
Σ n-6 PUFA	5.9	4.9	10.3	8.1
Σ n-3 PUFA	0.9[a]	2.7[b]	1.9[a]	6.1[b]
n-6/n-3 PUFA	7.0[a]	1.9[b]	5.6[a]	1.4[b]

Source: Adapted from Bauchart, D. et al. 2001. *Renc. Rech. Ruminants* 8:108.
[a,b] $P < 0.05$ within muscle.

were observed for long-chain n-3 PUFA, especially 20:5n-3 (EPA) and 22:5n-3 (DPA). The impact of dietary treatments on muscle individual n-3 PUFA is more pronounced for lipids of ST muscle (dominated by phospholipids) than for lipids of RA dominated by triglycerides (Table 11.2; Bauchart et al. 2001). Consequently, feeding steers with fresh grass led to a marked decrease in the n-6:n-3 PUFA ratio for both muscles (1.4–1.9 vs. 7.0–5.6), particularly beneficial to the health value of beef lipids for human consumers, the recommended value being lower than 4 (WHO 2003).

Other experiments conducted with other bovine breeds have confirmed these observations. Thus, a 2.5 times increase in 18:3n-3 proportion in total fatty acids of the *Longissimus thoracis* (LT, rib steak) muscle has been reported in Holstein × Simmental bulls given rye grass in the finishing period as compared to those fed a concentrate-based diet (Nürnberg et al. 2005). Additionally, grass feeding decreased the n-6:n-3 PUFA ratio by 3.4–4.2 to reach a value around 2, which has been confirmed in grass-fed bovines in two distinct systems of grass production (Razminowicz et al. 2006).

The beneficial effect of grass feeding on the preferential deposition of n-3 PUFA in ruminant muscles increases with the time on pasture. Thus, a beneficial effect was reported in bulls given grass from 0 to 158 days, due to the length of feeding on the nutritional value of beef FA, (i) by favoring n-3 PUFA deposition to the detriment of n-6 PUFA leading to a very low value of the n-6:n-3 PUFA ratio (1.3), (ii) by decreasing the level of SFA (rich in 16:0, known to be proatherogenous in humans; Moreno and Mitjavila 2003), (iii) by favoring deposition of 9*cis*,11*trans* 18:2 (rumenic acid) and its metabolic precursor 18:1 Δ11*trans* (vaccenic acid) (Noci et al. 2005), both being beneficial for human health by their hypocholesterolemic effect associated with the reduction of the denser low-density lipoproteins (LDL), particularly atherogenous in animal models for humans (Bauchart et al. 2007).

Concerning the isomers of CLA, known to have generally beneficial health properties for humans, Dannenberger et al. (2005) and Noci et al. (2005) found that grass feeding favored deposition of conjugated *trans,trans* 18:2, especially 11*trans*,13*trans* and 9*trans*,11*trans* forms, in addition to the dominant forms of CLA in beef FA (90%–95% of total CLA), which corresponds to *cis,trans* and *trans,cis* forms (+40 and +30% in Holstein and Simmental bulls, respectively). The 10*trans*,12*cis* form, the health value of which is still debated, represents a very minor form in beef CLA (<0.3% of total CLA), whereas it amounts to 50% of total CLA in synthetic commercial CLA mixtures produced by industry. Its level in beef among CLA isoforms was shown to be 4 times decreased in bovines given grass compared to a concentrate-based diet (Dannenberger et al. 2005).

The main form of CLA in beef FA is 9*cis*,11*trans* 18:2 (rumenic acid) generally considered to be very beneficial for human health. It represents around 80% of total CLA in bulls given a concentrate-based diet, but only 65%–76% in Holstein bulls fed on grass pasture, due to the higher deposition of 11*trans*,13*cis* 18:2 (Dannenberger et al. 2005). Additionally, these authors have shown a net effect of the animal breed on muscle CLA deposition, which was 2 times higher in Holstein than Simmental bulls for the two studied feed conditions.

The health value of total beef CLA for humans is, at the present time, not completely established. Additionally to its hypocholesterolemic effect (Bauchart et al. 2007), beef CLA would exert a beneficial protecting effect against cell carcinogenesis. Thus, with *in vitro* human tumoral cell cultures, De la Torre et al. (2006) have clearly shown a net apoptotic power of all CLA classes extracted from beef lipids, especially the 9*trans*,11*trans* 18:2 form, toward different human cancerous lines from breast, skin, ovaries, and colon.

Concerning the different *trans* isomers of 18:1 present in bovine LT, Dannenberger et al. (2004) have shown the preponderance of the Δ11 form (41.2%) associated to the Δ10 form (14.1%, known to be detrimental for human health by its hypercholesterolemic effect, Bauchart et al. 2007) and the Δ12 (15.2%) and Δ13/14 (12.4%) forms in beef FA of bulls given a concentrate-based diet. With grass feeding, the level of total *trans* 18:1 in beef FA was 25% higher, with a modified profile of total CLA characterized by a higher deposition of Δ 11*trans* (49%) and Δ 13/14*trans* (17.6%) to the detriment of Δ10 (3.8%) and Δ12*trans* (10.2%) forms (Dannenberger et al. 2004). In a similar study with Normande cull cows given a concentrate-based diet, Bispo-Villar et al. (2009) confirmed the domination of Δ11*trans* (36.1%), but associated with the Δ10*trans* isoform with a proportion (33.7%) 2 times higher than in bulls given the same concentrate-based diet. These data clearly show the importance of feeding × animal type (young bulls vs. older cull cow) interactions of animals on the profile of beef *trans* isomers of 18:1 and the consequences on the health value of beef fatty acids (bull > cull cow).

The nature and proportions of forages added to concentrate-based diets also modifies the nutritional value of beef fatty acids. In finishing young bulls, the effects of a straw + concentrate-based diet fed for 97 days on FA composition of the RA muscle were compared to feeding a corn silage + concentrate-based diet (60/40) (Bauchart et al. 2005). The results showed that a straw + concentrate-based diet improved the nutritional properties of beef lipids (i) by reducing the level of SFA (–13%), especially the proatherogenous 16:0 (–14%) to the benefit of n-6 (+52%) and n-3 (+57%) PUFA, and (ii) by increasing the deposition of *9cis,11trans* 18:2 CLA (+57%) and its metabolic precursor, the 18:1 Δ11*trans* (+48%). However, the n-6:n-3 PUFA ratio was not significantly modified by feeding and was 4.2, compatible with the recommendation for humans (<5.0).

11.5.2 Effects of Dietary Lipid Supplements

11.5.2.1 Oleaginous Seeds

Recommendations for the lipid supplementation of diets for bovine animals in the finishing period has been initially proposed in France as in other parts of Europe and in North America since lipids constitute a concentrate form having a high energetic density with a moderate cost due to their extensive production by the food industry. This dietary strategy allows the animals to be fattened, leading to the production of carcasses satisfying market requirements. This is especially true for cull cows for which conformation and fattening characteristics were frequently altered by a succession of lactations, especially in high-producing lactating cows. Sudden awareness by the beef industry of the additional interest in lipid supplements to improve the nutritional value of beef fatty acids has reinforced this feeding practice and encouraged research of lipid sources providing unsaturated fatty acids beneficial for human health, but frequently limited in alimentation (e.g., n-3 PUFA).

Different dietary sources of lipids have been tested in bovine animals (Clinquart et al. 1995), their incorporation in the basal diet being carried out as free purified oils (the use of animal fats such as beef tallow or lard is not used in France), calcium soaps, or micronized lipids (palmitostearin, saturated fraction of palm oil), but mainly as oleaginous seeds (linseed, rapeseed, sunflower seed, etc.). The use of additional lipids tends to reduce the intensity of ruminal fermentations associated with higher production of propionic acid, but these dietary supplements are generally well tolerated by ruminants if the administrated amount does not exceed 5%–6% of ingested dry matter. Nutrition-controlled experiments have been conducted (e.g., Scollan et al. 2005) utilizing linseed due to its high level of alpha linolenic acid (18:3n-3), an essential PUFA very beneficial for human health (favoring brain development and its cognitive functions, body growth, and fertility), but frequently quantitatively limited in human feeding. Bauchart et al. (2003) conducted an experiment with Charolais × Salers steers fed for 70 days on a straw + concentrate-based diet (45/55) supplemented with lipids (4% diet DM) from extruded linseed. Compared to the straw/concentrate-based diet, dietary lipid supplements induced an increase in the level of 18:3n-3 leading to a decrease of the n-6:n-3 ratio in beef neutral lipids (–18%) and especially of beef polar lipids (–38%) associated to a concomitant of Δ11*trans* 18:1 (+42%) and of *9cis,11trans* CLA (+50%) in neutral lipids. A net stimulating effect of oleaginous seeds rich in PUFA on the deposition of CLA in beef has been regularly reported in the literature (Scollan et al. 2005), the more marked effects being observed with n-6 PUFA sources (sunflower and soybean) than with n-3 PUFA sources (fresh grass, linseed).

The impact of lipid supplements from linseed on beef fatty acids varies with the type of basal diet. Thus, the effects of two types of basal diets composed of straw + concentrate (70/30) or of maize silage + concentrate (60/40) supplemented with extruded linseed have been compared in Charolais bulls (Bauchart et al. 2005). The results showed a positive effect of the linseed supplement more marked on CLA and 18:1 Δ11*trans* with the straw + concentrate-based diet, whereas the two basal diets supplemented with linseed led to similar enrichment in 18:3n-3 fatty acids. Similar studies have been carried out on Normande 5-year-old or more cull cows during the finishing period in order to verify the precise impact of linseed supplement on beef fatty acid composition in relation to the age of animals and their level of body fat stores. Cull cows were given straw + concentrate (70/30) supplemented with lipids (40 g/kg diet DM) from extruded linseed or from a mixture of extruded rapeseed oil (66%) and linseed (33%). Fatty acid analysis of the LT muscle showed a lower 18:3n-3 enrichment in cull cows, (0.6 vs. 0.4% total fatty acids, Habeanu et al. 2014) than in bulls (3.6 vs. 1.1%, Bauchart et al. 2005), which are leaner animals.

Similarly, lower effects of linseed supplementation on total *trans* 18:1 and CLA levels were reported in cull cows compared to bulls. This would indicate a lower uptake of these FA by muscle cells and by inter- and intramuscular adipose cells, which would depend on the age of animals and/or the lipid content of their tissues. Furthermore, concerning the effects of lipid supplements on the profile in *trans* isomers of 18:1, addition of linseed in a concentrate (70%) and straw (30%)-based diet led to modifications in Normande cull cows more favorable for human health when compared to that with the control diet, especially in case of supplementation with a lipid mixture composed of extruded rapeseed (66%) and linseed (33%) (Habeanu et al. 2014). Indeed, with the diet only supplemented with linseed, deposition of Δ9*trans* and Δ10*trans* were 1.7 and 2.4 times lower than with the control diet, which was beneficial for the health value of beef lipids since these *trans* 18:1 isoforms are known to have proatherogenous properties in the rabbit, an animal model for humans (Bauchart et al. 2007). On the other hand, incorporation of the mixture of rapeseed and linseed to the same diet led to negative effects on the health value of beef lipids since it tended to decrease the level of the beneficial Δ11*trans* in relation to the undesirable isoform Δ10*trans* (Habeanu et al. 2014). Such results show the complexity of the effects of dietary conditions on the formation of *trans* MUFA by rumen bacteria, and their subsequent deposition in the muscle tissues, mainly as parts of neutral lipids (Habeanu et al. 2014).

Dietary lipid supplements also affected the metabolism of other *trans* 18:1 isoforms such as Δ12 to Δ16*trans* of which the deposition in muscles was stimulated by dietary linseed supplementation and not by a mixture of linseed + rapeseed (Habeanu et al. 2014). Further investigations will be needed to establish the precise health value of Δ12 to Δ16*trans* 18:1 isomers, which can represent up to 40% of the total *trans* 18:1 in beef.

11.5.2.2 Combined Supply of Linseed Rich in n-3 PUFA and Vitamin E

Incorporation in diets of a large amount of vitamin E (2500 units/animal) to ensure a better protection of PUFA toward peroxidation led to modifications of beef fatty acid composition (Bauchart et al. 2005). Thus, a decrease of PUFA deposition in bovine tissues has been noted (18:2n-6 [–20%] and 18:3n-3 [–6 to –13%]) in finishing young bulls, given straw + concentrate- or a corn silage + concentrate-based diet (60/40) supplemented with extruded linseed and vitamin E (Bauchart et al. 2005). Such an effect could be explained by a stimulation of bacterial biohydrogenation reactions toward dietary PUFA in the rumen by dietary vitamin E, leading to a higher production of 18:1 Δ11*trans* subsequently deposited in beef tissues (+17% to +36%) together with CLA (mainly 9*cis*,11*trans* 18:2). A similar effect was noted with the same ratio both supplemented with rapeseed and linseed (Bauchart et al. 2005).

11.5.2.3 Supply of Protected Lipids

In order to more efficiently increase PUFA deposition in bovine tissues, different treatments of dietary supplements to limit bacterial biohydrogenation of fatty acids have been proposed by the food industry such as (i) chemical or thermic treatments applied to oleaginous seeds, (ii) emulsification or encapsulation of oils by proteins, and (iii) formation of calcium soaps (Gulati et al. 2005). Feeding cattle protected oils constituted by a 50/50 mixture of n-3 and n-6 PUFA led to a decrease in the value of the n-6:n-3 ratio of beef fatty acids from 3.6 to 1.9 (Scollan et al. 2004). Use of protected fish oil rich in long-chain n-3 PUFA favored deposition of very beneficial PUFA for human health, especially EPA and DHA, associated with a beneficial decrease of the n-6:n-3 ratio of beef fatty acids (Richardson et al. 2004).

The real potential of n-3 PUFA uptake and deposition by muscle tissues has been shown experimentally by a direct and continuous infusion in the duodenum for 70 days of free linseed oil in a similar amount (400 g/day) to that of extruded linseed incorporated in the basal diet (Bauchart et al. 2003, Figure 11.6). The results revealed a very large rush of 18:3n-3 in the blood compartment as well as in muscle tissues. Thus, the 18:3n-3 level in beef lipids was 16 times higher with the duodenal oil infusion than in the control treatment without any oil addition (8.0% vs. 0.5% total fatty acids). However, such duodenal oil infusion led to a strong fish taste of

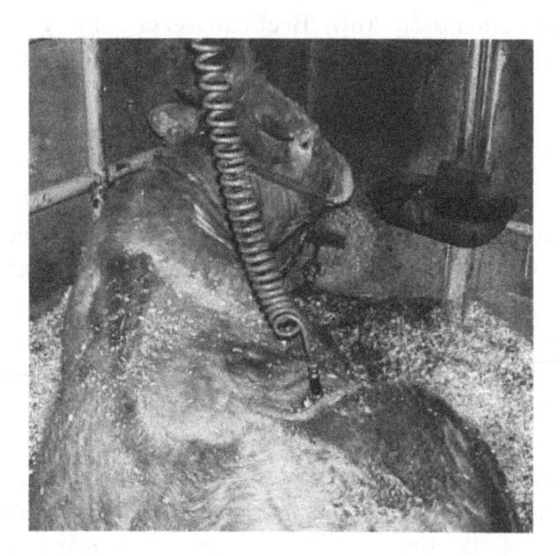

FIGURE 11.6 Direct and continuous infusion in the duodenum of free linseed oil in cows. (Photo credit: J.F. Hocquette.)

beef, totally undesirable for human consumers, which emphasized the necessity of a good control of the level of 18:3n-3 in beef. Moreover, it makes beef lipids very sensitive to peroxidation (Durand et al. 2005) leading to the formation of peroxidized derivates such as alkenals known to be toxic for animals as well for human consumers. Such alterations of beef lipids were shown to be efficiently prevented by incorporation in diets of antioxidant mixtures combining vitamin E with plant extracts rich in polyphenols (Gobert et al. 2010).

11.6 MEAT LABELS

11.6.1 Carcass Grading and Branding Systems

Grading is defined as the placing of different values on meat for pricing purposes, depending on the market and requirements of traders and such grades are often associated with brands. There are different grading systems in the world depending on the country (Table 11.3), but almost all of these systems are carcass based. The European system to describe carcasses (the EUROP grid) is mainly based on yield estimation to pay producers. While the EUROP system may adequately describe carcass muscling characteristics, it does not predict eating quality (Bonny et al. 2013). Most grading systems describe carcasses with various traits such as carcass weight, age, or maturity of the animal, sex, fatness, fat color, carcass conformation, and sometimes marbling and lean color and finally saleable meat yield usually predicted by measurements of fatness and/or muscling. USDA quality grades are used to predict the palatability of meat from a beef carcass, using carcass physiological maturity and marbling (USDA 1996). In North American and Asian countries, emphasis has been put on maturity and marbling. Most of the current grading and classification schemes still use these variables and in some ways, they are indicators of finish or fatness rather than indicators of real beef palatability at the consumer level (reviewed by Polkinghorne and Thompson 2010). Beef carcass grading systems are described in more detail in Chapter 15.

11.6.2 Examples of Meat Labeling Systems

Consumer grading schemes were also developed with the aim to define or predict consumer satisfaction with cooked and ready to eat meat. Different strategies were developed in various countries such as origin or quality grading schemes in Europe. In other situations, carcass grading schemes have been replaced by cuts-based grading schemes as, for instance, in Australia (Table 11.3). Only a few examples will be given in the next sections and more details are given in Chapter 15.

In the United States, Prime, Choice, and Select are the three most important grades to know as a consumer. USDA Prime is the top grade of beef available in the market. When beef is graded Prime, it is guaranteed to be tasty, tender, and extra juicy. Prime beef has a buttery flavor that makes it a cut above any other grade. Prime has the most amount of marbling, so it is easy to cook this steak to perfection. USDA Choice beef is the second-highest grade and is typically lower in cost and quality, but still provides a juicy, tender, flavorful meat product. USDA Select beef is the lowest

TABLE 11.3
Principal Components of Selected Grading Schemes in Some Countries around the World

Country	**Europe**	**South Africa**	**Canada**	**Japan**	**South Korea**	**USA**	**Australia**
Scheme	**EUROP**	**South Africa**	**Canada**	**JMGA**	**Korea**	**USDA**	**MSA**
Grading unit	**Carcass**	**Carcass**	**Carcass**	**Carcass**	**Carcass**	**Carcass**	**Cut**
Preslaughter factors							HGP implants and *Bos Indicus*
Slaughter floor	Carcass weight and sex	Carcass weight and sex	Carcass weight and sex	Carcass weight and sex	Carcass weight and sex	Carcass weight and sex	Carcass weight and sex
Slaughter floor	Conformation, fat cover	Dentition, rib fat	Conformation				Electric stimulation, hang
Chiller			Marbling score, meat color	Marbling score, meat color	Marbling score, meat color	Marbling score, meat color	Marbling score, meat color
Chiller			Fat color and fat thickness	Fat color and fat thickness	Fat color and fat thickness	Ossification score	Ossification score
Chiller				Eye muscle area	Eye muscle area	Eye muscle area	Fat thickness
Chiller			Texture	Meat brightness	Texture	Meat texture	Hump height
Chiller				Fat luster, texture and firmness	Firmness and lean maturity	Kidney fat and perirenal fat	Ultimate pH
Chiller				Rib thickness		Rib fat	
Post chiller							Ageing time
Post chiller							Cooking method

Source: Adapted from Polkinghorne, R.J., and Thompson, J.M. 2010. *Meat Sci*. 86:227–235.

grade of beef that you will find at a grocery store or restaurant. Select beef is much leaner than Prime or Choice and does not have the same flavor or texture. Select has little marbling, which makes it less juicy and tender, but if you cook it correctly, you will still have an enjoyable piece of meat. Select beef will be much cheaper in price than USDA Prime or USDA Choice grades (Firestine 2009). In addition, the U.S. beef industry has developed more than 100 beef brands, but probably not for all cuts, some using palatability assurance critical control point plans, total quality management approaches, USDA certification, and so on, including fundamental factors (such as *Bos indicus* content, hormone growth promotant [HGP] use, and days aging), or combinations of different systems to further differentiate fresh beef products (Smith et al. 2008). USDA grade will probably not be affected, but this type of approach clearly does affect consumer satisfaction.

In France, the animal food product market is much segmented due to the proliferation of quality labels. First, official labels identify a superior quality (Label Rouge), an environmental quality (organic farming), or quality linked to origin (European labels). Second, nonofficial labels include certified products differentiated from standard products by some specific characteristics or products highlighting a specific feature (such as meat produced from grass-fed animals).

For official labels, professionals voluntarily undertake to set up and monitor a quality-focused approach individually (organic farming) or collectively (protected designation of origin, protected geographical indication Label Rouge). Independent and competent bodies carry out regular checks, and the public authorities supervise the system (Hocquette et al. 2013). The foundation of the French scheme of labels identifying quality and origin of products is a joint commitment of the state and industry professionals (farmers, processors) to guarantee the quality of food for consumers in order to meet their expectations and offer them more choice with more information to guide them. The development of such a scheme is based on the assumption that consumers attach great importance to taste, to the pleasure of eating, to traditional gastronomy, and to the sustainable development of agriculture.

Among the quality labels, the French "Label Rouge" certifies that the raw or processed product possesses a specific set of characteristics guaranteeing a higher quality level than that of a similar standard product (INAO 2010). Two aspects play an important role in the Label Rouge: palatability and quality associated with the image of the products. In practice, this implies that farmers must follow specific rules to breed meat-producing animals so that a significant part of the benefits are for them as primary producers. This label is famous mostly in the poultry industry because it is relatively easy to make a difference between Label Rouge and standard products in terms of palatability. In the case of beef, less than 2% of beef is sold with the Label Rouge label. More generally, 85%–93% of the volume of French production has no official quality label (Hocquette et al. 2013).

In the United Kingdom, there are also many beef schemes related to areas of geographical origin, brands, and breeds (e.g., specialist Hereford or Aberdeen Angus beef and beef products). In Wales, Celtic Pride Beef was established in 2003 to provide a specialized and differentiated premium beef product (http://www.celticpride.co.uk/home/gtwp_section_leader.htm). No such product was available at that time from Wales. The key issue was to establish a strong brand name linked with a beef production and processing protocol, which would consistently deliver a high eating quality experience for consumers. The production protocol includes many preslaughter and postslaughter factors, but a major feature is extended maturation of the prime cuts.

More generally, in Europe, there are schemes known as PDO (protected designation of origin), PGI (protected geographical indication), and TSG (traditional specialty guaranteed), which promote and protect names of quality agricultural products and foodstuffs (Figure 11.7). These EU schemes encourage diverse agricultural production, protect product names from misuse and imitation, and help consumers by giving them information concerning the specificity of the products. The concept of designation of origin and geographical indication is based on a combination (i) of the characteristics of the natural environment where production takes place, (ii) of biological factors such as animal breeds, and (iii) of human factors, mainly the know-how of producers. This combination is considered as the key that determines the quality of the final product. More precisely, PDO covers agricultural products and foodstuffs that are produced, processed, and prepared in a given geographical area using recognized know-how. PGI covers agricultural products and foodstuffs closely linked to the geographical area. At least one of the stages of production, processing or preparation must take place in the area. TSG highlights the traditional character, either in the composition or means of production. Besides this system, the "organic farming" label certifies that the product derives from a mode of production and processing that is protective of natural balances and animal welfare as defined in a highly stringent set of specifications backed by systematic controls (Figure 11.7).

In each country, there are other systems related to certification of product conformity according to predefined requirements as well as many commercial labels. These quality labels are not official. Products are differentiated from a standard product by some specific characteristics. This approach can be individually led (in contrast with official quality marks). In France, Product Conformity Certification (PCC) provides confidence that specific products will meet their contractual requirements. Inspection is performed by an independent third-party organization, which

	Protected designation of origin (PDO)	Protected geographical indication (PGI)	Traditional speciality guaranteed (STG)	Organic farming	Label rouge
EU					
France					

FIGURE 11.7 European and French logos for official quality labels.

is under the responsibility of the Ministry of Agriculture. Generally, this approach is quick and simple compared to official labels. Consequently, PCC products are often less expensive. PCC products have boomed during food-health crises and the role of supermarket chains in the development of PCCs has been important. PCC is a success mainly for animal products. Besides this system, plenty of other labels such as "on-farm processed" or "mountain produce," highlights a specific feature (Hocquette et al. 2013).

It is well known that muscle type and cooking method have a great impact on palatability (Thompson 2002; Sullivan and Calkins 2011; Modzelewska-Kapitula et al. 2012). These factors are thus worth being included in quality labels. Although *postmortem* handling of the carcasses may be much more effective in controlling beef tenderness than *premortem* factors (Juarez et al. 2012), all ante- and *postmortem* factors have to be combined together for a better prediction of beef tenderness. Such an integrative strategy was built up in Australia from 1996, with the development of the Meat Standards Australia (MSA) grading scheme to predict beef quality for consumers. One reason for the MSA success is standardization of the consumer evaluation protocols (Watson et al. 2008a). Another reason is the accumulation of large amounts of data over time, that have been subjected to vigorous statistical analyses in order to identify and quantify the main factors and their interaction affecting beef palatability (Watson et al. 2008b). Various other countries or regions of the world have tested or are testing the MSA system: Korea (Thompson et al. 2008), the United States (Smith et al. 2008), France (Hocquette et al. 2011a; Legrand et al. 2013), Japan (Polkinghorne et al. 2011), South Africa (Thompson et al. 2010), New Zealand, Northern Ireland (Farmer et al. 2009), the Irish Republic, and Poland. Overall results indicate that the methodology behind this approach may be universal, but probably requires some minor adjustments to be applied in each country. In France, scientists and professionals recognized many qualities of MSA, which was judged comprehensive, consistent, and scientifically supported. However, the adaptability of the MSA system to France would be difficult due to the complexity of the French beef industry and market (beef from different animal types: young bulls, steers, heifers, cows; beef from the dairy herd or from the beef herd with a great number of breeds) and due to the existence of preexisting quality marks such as the Label Rouge (Hocquette et al. 2011a). In fact, the prediction of the final ratings for palatability (3*, 4*, or 5*) by the French consumers using the MSA system was correct for more than 70% of the samples which is, at least, similar to the Australian experience. Similar results were obtained comparing "rare" and "medium" beef (Legrand et al. 2013). A further research challenge is to combine the MSA approach with biochemical and genomic, which data in modeling approaches in order to improve the accuracy of the prediction of beef sensory quality (Hocquette et al. 2014).

11.7 TAKE-HOME MESSAGES

Beef meat is a very attractive food, especially due to its sensory qualities: color, tenderness, juiciness, and flavor. These qualities are the result of complex biological mechanisms involved in muscle biochemistry during the animals' life and after slaughtering during aging. Muscle characteristics and hence sensory qualities depend on genotype, animal type, age, and sex. Genotype/breed is a major factor that affects muscle characteristics and hence beef quality: breeds are more or less early or late maturing, thereby affecting muscle growth and carcass fatness. The most important effects are observed in the case of *Bos indicus* (for which beef quality is decreased) and double-muscled cattle (which produce lean but tender meat). Some biomarkers are associated with tenderness differences and correspond with large sensory differences within Brahman cattle in particular.

Animal management also significantly influences subsequent muscle eating quality by varying weight and fatness in relation to age and physiological maturity at slaughter. Increasing age seems to be favorable for juiciness and flavor (due to more intramuscular fat), but unfavorable to tenderness due to connective tissue characteristics despite an attenuation of this effect by more intramuscular fat. Sex plays an important role: females provide tastier, more tender, and redder beef than steers or bulls. Compared to beef from young bulls, meat from steers contains more intramuscular fat, which is favorable for flavor and tenderness. Livestock practices also play a role mainly by level of food intake, and also by the nature of diets. Dietary restriction is detrimental to the tenderness and flavor of beef. A high level of feed intake before slaughter promotes and increases tenderness and intramuscular fat content, thus improving the flavor of meat. The favorable effect of increasing feed intake is more sensitive when the animal undergoes compensatory growth after a period of undernutrition.

The nutritional value of beef depends on the content and composition in long-chain fatty acids of neutral (triglycerides) and polar lipids (phospholipids) in muscles. These characteristics differ depending on the location and metabolic activity of muscles and, for a given muscle, they vary widely according to animal factors (genotype, age, sex) and especially according to animal nutrition. Thus, the nature and proportion of nutritional elements in the composition of basal diets, as well as in lipid supplements (such as oil seeds) given to animals have large effects on the characteristics of fatty acids, which are deposited within tissues. These factors affect especially the ratio of SFA (C16:0 and 18:0), the contents of *cis* and *trans* MUFA, and the distribution of PUFA between n-6 and n-3 families. Improving the health value of beef fatty acids must not be thought of only through essential fatty acids for human nutrition (especially n-3 fatty acids often deficient in the human diet) directly from animal diets. It must also take into account the production of fatty acids (essential fatty acids and *trans* fatty acids) from microbial biohydrogenation processes. Some of these fatty acids of microbial origin are beneficial for human health such as CLAs, especially rumenic acid (9*cis*,11*trans* 18:2) and its precursor vaccenic acid (18:1 Δ11*trans*). However, others are frankly harmful by their proatherogenic properties (18:1 Δ9*trans* and Δ10*trans*). Ongoing studies on interactions between the basal diet and lipid supplements for cattle breeds that differ by their fatness should define the best nutritional conditions for the production of beef of enhanced nutritional quality for the consumer.

Meat labels vary between countries, but, in any case, they have been set up mainly to describe positive attributes of carcasses, which come from animal × livestock practice combinations. Therefore, such systems were not initially designed to satisfy consumers at the level of consumption. Consequently, there is still great potential to improve beef quality at the consumer level especially by modeling approaches, which integrate the effects of all factors known to affect beef quality, especially palatability. Of course, this type of research should be conducted with the ultimate goal to provide added value to all players along the entire supply chain from producers to consumers.

REFERENCES

Alfnes, F., Rickertsen, K., and Ueland, Ø. 2008. Experimental evidence of risk aversion in consumer markets: The case of beef. *Appl. Econ.* 40:3039–3049.

Banovic, M., Grunert, K.G., Barreira, M.M., and Fontes, M.A. 2009. Beef quality perception at the point of purchase: A study from Portugal. *Food Qual. Prefer.* 20:335–342.

Bastien, D. 2003. Inventory of tests conducted on the production of young cattle. Compte rendu Institut de l'Elevage n 2033225.

Bauchart, D., Durand, D., and Gruffat, D. 2003. Effects of linseed-supplemented diets on specific fatty acids in total lipids and in neutral and polar components of lipids of *Rectus abdominis* and *Longissimus thoracis* muscles and of intermuscular adipose tissue of finishing crossbred Charolais × Salers steers, In *Proceedings of 6th Scientific Meeting of the European Program "HealthyBeef" (5th PCRD)*, June 2003, Dublin, Ireland, pp. 10.

Bauchart, D., Durand, D., Gruffat-Mouty, D., Dozias, D., Ortigues-Marty, I., and Micol, D. 2001. Concentration and fatty acid composition of lipids in muscles and liver of fattening steers fed a fresh grass based diet. *Renc. Rech. Ruminants* 8:108.

Bauchart, D., Durand, D., Martin, J.F., Jailler, Rd., Geay, Y., and Picard, B. 2002a. Effects of breed and age on lipids in muscles *Longissimus thoracis, Triceps brachii* and *Semitendinosus* of bulls. *Renc. Rech. Ruminants* 9:268.

Bauchart, D., Durand, D., Martin, J.F., Jailler, Rd., Picard, B., and Geay, Y. 2002b. Effects of age and production type on lipids in *Longissimus thoracis, Triceps brachii* and *Semitendinosus* muscles of purebred Charolais animals. *9èmes Journées des Sciences du Muscle et Technologie de la Viande*, October 15–16, 2002. Clermont-Ferrand, France, *Viandes Prod. Carnés* 9 (Hors série), pp. 127–128.

Bauchart, D., Gladine, C., Gruffat, D., Leloutre, L., and Durand, D. 2005. Effects of diets supplemented with oil seeds and vitamin E on specific fatty acids of *rectus abdominis* muscle in charolais fattening bulls. *55th Annual Meeting of the European Association for Animal Production*, Bled (Slovenia). In Hocquette, J.F. and Gigli, S. (eds.), *Indicators of Milk and Beef Quality*; EAAP Publication 112, pp. 431–436. Wageningen Academic Publishers, Wageningen, the Netherlands.

Bauchart, D., Roy, A., Lorenz, S., Ferlay, A., Gruffat, D., Chardigny, J.M., Sébédio, J.L., Chilliard, Y., and Durand, D. 2007. Dietary supply of butter rich in *trans* 18:1 isomers or in 9*cis*,11*trans* conjugated linoleic acid affects plasma lipoproteins in hypercholesterolemic rabbits. *Lipids* 42:123–133.

Bernard, C., Cassar-Malek, I., Le Cunff, M., Dubroeucq, H., Renand, G., and Hocquette, J.F. 2007. New indicators of beef sensory quality revealed by expression of specific genes. *J. Agric. Food Chem.* 55:5229–5237.

Bispo-Villar, E., Thomas, A., Lyan, B., Gruffat, D., Durand, D., and Bauchart, D. 2009. Lipid supplements rich in n-3 polyunsaturated fatty acids deeply modify *trans* 18:1 isomers in the *Longissimus thoracis* muscle of finishing bovine. In *The Proceedings of the 11th International Symposium on Ruminant Physiology*, September 6–9, 2009, Clermont-Ferrand, France, pp. 464–466.

Bonny, S.P.F., Legrand, I., Polkinghorne, R.J., Gardner, G.E., Pethick, D.W., and Hocquette, J.F. 2013. The EUROP carcase grading system does not predict the eating quality of beef. *Abstracts of the 64th Annual Meeting of the European Association for Animal Production*, Session 12, Theatre 8, Nantes, France, p. 96.

Cassar-Malek, I., Picard, B., Bernard, C., and Hocquette, J.F. 2008. Application of gene expression studies in livestock production systems: A European perspective. *Aust. J. Exp. Agric.* 48:701–710.

Chriki, S., Gardner, G.E., Jurie, C., Picard, B., Micol, D., Brun, J.P., Journaux, L., and Hocquette, J.F. 2012. Cluster analysis application in search of muscle biochemical determinants for beef tenderness. *BMC Biochem.* 13:29.

Clinquart, A., Micol, D., Brundseaux, C., Dufrasne, I., and Istasse, L. 1995. Use of fat in fattening diets for cattle. *INRA Prod. Anim.* 8:29–42.

Coulon, J.B., and Priolo, A. 2002. Sensory properties of meat and dairy products are affected by the forages consumed by the animals. *INRA Prod. Anim.* 15:333–342.

Cross, H.R. 1988. Factors affecting sensory properties of meat. In Cross, H.R. (ed.), *Meat Science, Milk Science and Technology*. Elsevier Science Publisher BV, Amsterdam, pp. 158–161.

Dannenberger, D., Nürnberg, K., Nürnberg, G., Scollan, N., Steinhart, H., and Ender, K. 2005. Effect of pasture vs concentrate diet on CLA isomer distribution in different tissue lipids of beef cattle. *Lipids* 40:589–598.

Dannenberger, D., Nürnberg, G., Scollan, N., Schabbel, W., Steinhart, H., Ender, K., and Nürnberg, K. 2004. Effect of diet on the deposition of n-3 fatty acids, conjugated linoleic and C18:1 *trans* fatty acid isomers in muscle lipids of German Holstein bulls. *J. Agric. Food Chem.* 52:6607–6615.

De La Torre, A., Debiton, E., Durand, D., Juanéda, P., Durand, D., Chardigny, J.M., Barthomeuf, C., Bauchart, D., and Gruffat, D. 2006. Beef conjugated linoleic acid isomers reduce human cancer cell growth even when associated to other beef fatty acids. *Brit. J. Nutr.* 95:346–352.

De Smet, S., Raes, K., and Demeyer, D. 2004. Meat fatty acid composition as affected by genetic factors. *Anim. Res.* 53:81–88.

Dransfield, E., Martin, J.F., Bauchart, D., Abouelkaram, S., Lepetit, J., Culioli, J., Jurie, C., and Picard, B. 2003. Meat quality and composition of three muscles from French cull cows and young bulls. *Anim. Sci.* 76:387–399.

Dransfield, E., Nute, G.R., and Roberts, T.A. 1984. Beef quality assessed at European research centres. *Meat Sci.* 10:1–20.

Dumont, R., Teissier, J.H., Bonnemaire, J., and Roux, M. 1987. Early calving heifers versus maiden heifers for beef production from dairy herd. II. Physicochemical and sensorial characteristics of meat. *Livest. Prod. Sci.* 16:21–35.

Durand, D., Scislowski, V., Chilliard, Y., Gruffat, D., and Bauchart, D. 2005. High fat rations and lipid peroxidation in ruminants; consequences on animal health and quality of products. *55th Annual Meeting of the European Association for Animal Production*, Bled (Slovenia). In Hocquette, J.F. and Gigli, S. (eds.), *Indicators of Milk and Beef Quality*; EAAP Publication 112. Wageningen Academic Publishers, Wageningen, the Netherlands, pp. 135–150.

Farmer, L.J., Devlin, D.J., Gault, N.F.S., Gee, A., Gordon, A.W., Moss, B.W., Polkinghorne, R., Thompson, J., Tolland, E.L.C., and Tollerton, I.J. 2009. Prediction of eating quality using the Meat Standards Australia system for Northern Ireland. *Proceedings of the 55th International Congress of Meat Science and Technology*, August 2009, Copenhagen, p. PE7.34.

Firestine, M. http://animalsciencenews.blogspot.fr/2009/07/prime-choice-select-beef-what-does-it.html.

Fischell, V.K., Abelery, E.D., Judge, M.D., and Perry, T.W. 1985. Palatibility and muscles properties of beef influenced by preslaughter growth rate. *J. Anim. Sci.* 61:151–157.

Fortin, M., Julien, P., Couture, Y., Dubreuil, P., Chouinard, P.Y., Latulippe, C., Davis, T.A., and Thivierge, M.C. 2010. Regulation of glucose and protein metabolism in growing steers by long-chain n-3 fatty acids in muscle membrane phospholipids is dose-dependent. *Animal* 1:89–101.

Geay, Y., Bauchart, D., Hocquette, J.F., and Culioli, J. 2001. Effect of nutritional factors on biochemical structural and metabolic characteristics of muscles in ruminants, consequences on dietetic value and sensorial qualities of meat. *Reprod. Nutr. Dev.* 41:1–26. Erratum, 41:377.

Gobert, M., Gruffat, D., Hăbeanu, M., Parafita, E., Bauchart, D., and Durand, D. 2010. Plant extracts combined with vitamin E in PUFA-rich diets of cull cows protect beef against lipid oxidation. *Meat Sci.* 85:676–683.

Guillemin, N., Bonnet, M., Jurie, C., and Picard, B. 2011. Functional analysis of beef tenderness. *J. Proteomics* 75:352–365.

Guillemin, N., Meunier, B., Jurie, C., Cassar-Malek, I., Hocquette, J.F., Levéziel, H., and Picard, B. 2009. Validation of a dot-blot quantitative technique for large-scale analysis of beef tenderness biomarkers. *J. Physiol. Pharmacol.* 60:91–97.

Gulati, S.K., Garq, M.R., and Scott, T.W. 2005. Rumen protected protein and fat produced from oil seeds and/or meals by formaldehyde treatment; their role in ruminant production and product quality. *Aust. J. Exp. Agric.* 45:1189–1203.

Habeanu, M., Thomas, A., Bispo Villar, E., Gobert, M., Gruffat, D., Durand, D., and Bauchart, D. 2014. Extruded linseed and rapeseed both influenced fatty acid composition of total lipids and their polar and neutral fractions in *longissimus thoracis* and *semitendinosus* muscles of finishing Normand cows. *Meat Sci.* 96:99–107.

Harper, G.S., Allingham, P.G., and Le Feuvre, R.P. 1999. Changes in connective tissue of *M. semitendinosus* as a response to different growth paths in steers. *Meat Sci.* 53:107–114.

Hoch, T., Begon, C., Cassr-Malek, I., Picard, B., and Savary-Auzeloux, I. 2003. Mechanisms and consequences of compensatory growth in ruminants. *INRA Prod. Anim.* 16:49–59.

Hocquette, J.-F., Bernard-Capel, C., Vidal, V., Jesson, B., Levéziel, H., and Cassar-Malek, I. 2012a. The GENOTEND chip: A new tool to analyse gene expression in muscles of beef cattle for beef quality prediction. *BMC Vet. Res.* 8:135.

Hocquette, J.F., Botreau, R., Picard, B., Jacquet, A., Pethick, D.W., and Scollan, N.D. 2012b. Opportunities for predicting and manipulating beef quality. *Meat Sci.* 92:197–209.

Hocquette, J.F., Jacquet, A., Giraud, G., Legrand, I., Sans, P., Mainsant, P., and Verbeke, W. 2013. Quality of food products and consumer attitudes in France. In Klopcic, M., Kuipers, A., and Hocquette, J.F. (eds.), *Consumer Attitudes to Food Quality Products*, EAAP Publication 133, Wageningen Academic Publishers, Wageningen, the Netherlands, pp. 67–82.

Hocquette, J.F., Legrand, I., Jurie, C., Pethick, D.W., and Micol, D. 2011a. Perception in France of the Australian system for the prediction of beef quality (MSA) with perspectives for the European beef sector. *Anim. Prod. Sci.* 51:30–36.

Hocquette, J.F., Lehnert, S., Barendse, W., Cassar-Malek, I., and Picard, B. 2007. Recent advances in cattle functional genomics and their application to beef quality. *Animal* 1:159–173.

Hocquette, J.F., Meurice, P., Brun, J.P., Jurie, C., Denoyelle, C., Bauchart, D., Renand, G., Nute, G.R., and Picard, B. 2011b. The challenge and limitations of combining data: A case study examining the relationship between intramuscular fat content and flavor intensity based on the BIF-BEEF database. *Anim. Prod. Sci.* 51:975–981.

Hocquette, J.F., Van Wezemael, L., Chriki, S. et al. 2014. Modelling of beef sensory quality for a better prediction of palatability. *Meat Sci.* 97:316–322.

INAO Leaflet. 2010. Ministère de l'Agriculture et Institut National de l'Origine et de la Qualité. The official quality and origin signs. http://www.inao.gouv.fr/public/home.php?pageFromIndex=textesPages/Supports_de_presentation412.php~mnu=412.

Juarez, M., Basarab, J.A., Baron, V.S., Valera, M., Larsen, I.L., and Aalhus, J.L. 2012. Quantifying the relative contribution of *ante-* and *post-mortem* factors to the variability in beef texture. *Animal* 6:1878–1887.

Jurie, C., Cassar-Malek, I., Bonnet, M., Leroux, C., Bauchart, D., Boulesteix, P., Pethick, D.W., and Hocquette, J.F. 2007. Adipocyte fatty acid-binding protein and mitochondrial enzyme activities in muscles as relevant indicators of marbling in cattle. *J. Anim. Sci.* 85:2660–2669.

Jurie, C., Martin, J.-F., Listrat, A., Jailler, R., Culioli, J., and Picard, B. 2006a. Carcass and muscle characteristics of beef cull cows between 4 and 9 years of age. *Anim. Sci.* 82:415–421.

Jurie, C., Ortigues-Marty, I., Picard, B., Micol, D., and Hocquette, J.F. 2006b. The separate effect of the nature of diet and grazing mobility on metabolic potential of muscles from Charolais steers. *Livest. Prod. Sci.* 104:182–192.

Keane, M.G., and Allen, P. 1998. Effects of production system intensity on performance, carcass composition and meat quality of beef cattle. *Livest. Prod. Sci.* 56:203–214.

Koohmaraie, M., Kent, M.P., Shackelford, S.D., Veiseth, E., and Wheeler, T.L. 2002. Meat tenderness and muscle growth: Is there any relationship? *Meat Sci.* 62:345–352.

Larick, D.K., and Turner, B.E. 1989. Influence of finishing diet on phospholipid composition and fatty acid profile of individual phospholipids in lean muscles of beef cattle. *J. Anim. Sci.* 67:2282–2293.

Legrand, I., Hocquette, J.-F., Polkinghorne, R.J., and Pethick, D.W. 2013. Prediction of beef eating quality in France using the Meat Standards Australia system. *Animal* 7:524–529.

Listrat, A., Rakadjiyski, N., Jurie, C., Picard, B., Touraille, C., and Geay, Y. 1999. Effect of the type of diet on muscle characteristics and meat palatability of growing Salers bulls. *Meat Sci.* 53:115–124.

McCormick, R.J. 1994. The flexibility of the collagen compartment of muscle. *Meat Sci.* 36:79–91.

Melton, S.L. 1990. Effects of feeds on flavor of red meat: A review. *J. Anim. Sci.* 68:4421–4435.

Micol, D., Jurie, C., and Hocquette, J.F. 2010. Sensory qualities of beef. Impacts of livestock factors. In *Muscle and Meat from Ruminants.* Bauchart, D. and Picard, B. (coords.). Quae, Collections Synthèses, France, Versailles, Chapitre 13, pp. 165–174.

Miller, R.K., Cross, H.R., Crouse, J.D., and Tatum, J.D. 1987. The influence of diet and time on feed on carcass traits and quality. *Meat Sci.* 19:303–313.

Modzelewska-Kapituła, M., Dąbrowska, E., Jankowska, B., Kwiatkowska, A., and Cierach, M. 2012. The effect of muscle, cooking method and final internal temperature on quality parameters of beef roast. *Meat Sci.* 91:195–202.

Moreno, J., and Mitjavila, M.T. 2003. The degree of unsaturation of dietary fatty acids and the development of atherosclerosis (review). *J. Nutr. Biochem.* 14:182–195.

Nishimura, T., Hattori, A., Takahashi, K. 1999. Structural changes in intramuscular connective tissue during the fattening of Japanese Black cattle: Effect of marbling on beef tenderization. *J. Anim. Sci.* 77:93–104.

Noci, F., Monahan, F.J., French, P., and Moloney, A.P. 2005. The fatty acid composition of muscle fat and subcutaneous adipose tissue of pasture-fed heifers: Influence of the duration grazing. *J. Anim. Sci.* 83:1167–1178.

Nürnberg, K., Dannenberger, D., Nürnberg, G., Ender, K., Voigt, J., Scollan, N., Wood, J.D., Nute, G., and Richardson, I. 2005. Effect of a grass-based and a concentrate feeding systems on meat quality characteristics and fatty acid composition of *longissimus* muscle in different cattle breeds. *Livest. Prod. Sci.* 94:137–147.

Oddy, V.H., Harper, G.S., Greenwood, P.L., and McDonagh, M.B. 2001. Nutritional and developmental effects on the intrinsic properties of muscles as they relate to the eating quality of beef. *Aust. J. Exp. Agric.* 41:921–942.

Olleta, J.L., Sañudo, C., Monsón, F. et al. 2006. Sensory evaluation of several European cattle breeds. In Olaizola, A., Boutonnet, J.P., and Bernues, A. (eds.), *Mediterranean Livestock Production: Uncertainties and Opportunities.* Second Seminar of the

Scientific-Professional Network on Mediterranean Livestock Farming (RME), 2006/05/18–20, Zaragoza, (Spain). Options Méditerranéennes: Série A. Séminaires Méditerranéens; 78. CIHEAM-IAMZ/CITA, pp. 297–300.

Ouali, A., Gagaoua, M., Boudida, Y., Becila, S., Boudjellal, A., Herrera-Mendez, C.H., and Sentandreu, M.A. 2013. Biomarkers of meat tenderness: Present knowledge and perspectives in regards to our current understanding of the mechanisms involved. *Meat Sci.* 95:854–870.

Oury, M.P., Agabriel, J., Agabriel, C., Micol, D., Picard, B., Blanquet, J., Laboure, H., Roux, M., and Dumont, R. 2007a. Relationship between rearing practices and eating quality traits of the muscle *rectus abdominis* of Charolais heifers. *Livest. Sci.* 111:242–254.

Oury, M.P., Picard, B., Istasse, L., Micol, D., and Dumont, R. 2007b. Effect of rearing management practices on tenderness of bovine meat. *INRA Prod. Anim.* 20:309–326.

Patterson, D.C., Moore, C.A., Moss, B.W., and Kilpartrick, D.J. 2002. Parity associated changes in slaughter weight and carcass characteristics of Charolais crossbred cows kept on a lowland grass/grass silage feeding and management system. *Anim. Sci.* 75:221–235.

Picard, B., Jurie, C., Bauchart, D., Dransfield, E., Ouali, A., Martin, J.F., Jailler, R., Lepetit, J., and Culioli, J. 2007. Muscle and meat characteristics from the main beef breeds of the Massif Central. *Sci. Alim.* 27:168–180.

Picard, B., Lefèvre, F., and Lebret, B. 2012. Meat and fish flesh quality improvement with proteomic applications. *Anim. Frontiers* 2:18–25.

Picard, B., Robelin, J., and Geay, Y. 1995. Influence of castration and postnatal energy restriction on the contractile and metabolic characteristics of bovine muscle. *Ann. Zootech.* 44:347–357.

Polkinghorne, R., Nishimura, T., Neath, K.E., and Watson, R. 2011. Japanese consumer categorisation of beef into quality grades, based on Meat Standards Australia methodology. *Anim. Sci. J.* 82:325–333.

Polkinghorne, R., Thompson, J.M., Watson, R., Gee, A., and Porter, M. 2008. Evolution of the Meat Standards System (MSA) beef grading system. *Aust. J. Exp. Agric.* 48:1351–1359.

Polkinghorne, R.J., and Thompson, J.M. 2010. Meat standards and grading: A world view. *Meat Sci.* 86:227–235.

Priolo, A., Micol, D., and Agabriel, J. 2001. Effects of grass feeding systems on ruminant meat color and flavor. A review. *Anim. Res.* 50:185–200.

Razminowicz, R.H., Kreuzer, M., and Scheeder, M.R.L. 2006. Quality of retail beef from two grass-based production systems in comparison with conventional beef. *Meat Sci.* 73:351–361.

Renand, G. 1988. Genetic variability of muscle growth and consequences on meat quality of cattle. *INRA Prod. Anim.* 1:115–121.

Renerre, M. 1982a. La couleur de la viande—Les principaux types de production de viande. *Bull. Tech. CRZV Theix* 48:42–46.

Renerre, M. 1982b. Effects of age and slaughter weight on the color of beef (Friesian and Charolais breeds). *Sci. Aliments* 2:17–30.

Richardson, R.I., Hallett, K., Robinson, A.M., Nute, G.R., Enser, M., Wood, J.D., and Scollan, N.D. 2004. Effect of free and ruminally-protected fish oils on fatty acid composition, sensory and oxidative characteristics of beef loin muscle. In *Proceedings of the 50th International Congress of Meat Science and Technology*, Helsinki, Finland.

Salter, A.M. 2013. Dietary fatty acids and cardiovascular disease. *Animal* 7:163–171.

Scollan, N., Enser, M., Gulati, S., Hallett, K.G., Nute, G.R., and Wood, J.D. 2004. The effects of ruminally protected dietary lipids on the fatty acid composition and quality of beef muscle. In *Proceedings of the 50th International Congress of Meat Science and Technology*, Helsinki, Finland.

Scollan, N., Hocquette, J.F., Nuernberg, K., Dannenberger, D., Richardson, I., and Moloney, A., 2006. Innovations in beef production systems that enhance the nutritional and health value of beef lipids and their relationship with meat quality. *Meat Sci.* 74:17–33.

Scollan, N., Richardson, I., De Smet, S., Moloney, A.P., Doreau, M., Bauchart, D., and Nürnberg, K. 2005. Enhancing the content of beneficial fatty acids in beef and consequences for meat quality. *55th Annual Meeting of the European Association for Animal Production*, Bled, Slovenia. In Hocquette, J.F. and Gigli, S. (eds.), *Indicators of Milk and Beef Quality*; EAAP Publication 112, Wageningen Academic Publishers, Wageningen, the Netherlands, pp. 151–162.

Simopoulos, A.P. 1991. Omega-3 fatty acids in health and disease and in growth and development. *Am. J. Clin. Nutr.* 54:438–463.

Smith, G.C., Tatum, J.D., and Belk, K.E. 2008. International perspective: Characterisation of United States Department of Agriculture and Meat Standards Australia systems for assessing beef quality. *Aust. J. Exp. Agric.* 48:1465–1480.

Sullivan, G.A., and Calkins, C.R. 2011. Ranking beef muscles from Warner–Bratzler shear force and trained sensory panel ratings from published literature. *J. Food Quality* 34:195–203.

Thompson, J., Polkinghorne, R., Gee, A., Motiang, D., Strydom, P., Mashau, M., Ng'ambi, J., deKock, R., and Burrow, H. 2010. Beef palatability in the Republic of South Africa: Implications for niche-marketing strategies. ACIAR Technical Report 72, Australian Centre for International Agricultural Research, Canberra, Australia.

Thompson, J.M. 2002. Managing meat tenderness. *Meat Sci.* 60:365–369.

Thompson, J.M. 2004. The effects of marbling on flavour and juiciness scores of cooked beef, after adjusting to a constant tenderness. *Aust. J. Exp. Agric.* 44:645–652.

Thompson, J.M., Polkinghorne, R., Hwang, I.H., Gee, A.M., Cho, S.H., Park, B.Y., and Lee, J.M. 2008. Beef quality grades as determined by Korean and Australian consumers. *Aust. J. Exp. Agric.* 48:1380–1386.

Touraille, C. 1982. Influence du sexe et de l'âge à l'abattage sur les qualités organoleptiques des viandes de bovins limousins abattus entre 16 et 33 mois. *Bull. Tech. CRZV Theix* 48:83–89.

Troy, D.J., and Kerry, J.P. 2010. Consumer perception and the role of science in the meat industry. *Meat Sci.* 86:214–226.

Verbeke, W., VanWezemael, L., de Barcellos, M.D., Kügler, J.O., Hocquette, J.-F., Ueland, Ø., and Grunert, K.G. 2010. European beef consumers' interest in a beef eating-quality guarantee: Insights from a qualitative study in four EU countries. *Appetite* 54:289–296.

Vestergaard, M., Therkildsen, M., Henckel, P., Jensen, L.R., Andersen, H.R., and Sejrsen, K. 2000. Influence of feeding intensity, grazing and finishing feeding on meat and eating quality of young bulls and the relationship between muscle fibre characteristics, fibre, fragmentation and meat tenderness. *Meat Sci.* 54:187–196.

Watson, R., Gee, A., Polkinghorne, R., and Porter, M. 2008a. Consumer assessment of eating quality—Development of protocols for Meat Standards Australia (MSA) testing. *Aust. J. Exp. Agric.* 48:1360–1367.

Watson, R., Polkinghorne, R., and Thompson, J.M. 2008b. Development of the Meat Standards Australia (MSA) prediction model for beef palatability. *Aust. J. Exp. Agric.* 48:1368–1379.

World Health Organisation. 2003. Diet, nutrition and the prevention of chronic diseases. Report of the Joint WHO/FAO Expert Consultation. WHO Technical Report.

12 Pork Quality

Andrzej Sośnicki

CONTENTS

12.1 INTRODUCTION

This chapter is not intended to be all inclusive, it delineates pig breeding and genetics, farm and processing plan environments, and overall industry and global pork consumption aspects of pork production, its value, and quality. Both the scientific aspects of pork quality; that is, genetic and environmental factors impacting carcass and meat quality; and their practical ramifications; that is, food industry consolidation leading to the creation of vertically integrated and/or coordinated food production systems; are discussed. These systems have been focused on development of differentiated "value chains," where the main goal is to achieve sustainable competitiveness through focusing resources on efficiently producing goods that offer superior consumer-recognized quality and value. A closely aligned value chain often contains vertically and horizontally linked entities such as genetics and genetic improvement program(s), farmer(s), processor(s), distributor(s), and retailer(s). In this chapter it is postulated that the underlying foundation of the success of pork value chain accomplishments has been through substantial development of swine genetic technologies. These have enabled sustainable and affordable production of protein-based pork consumer products of desirable quantity and quality. It is plausible to assume that further advancement in genomic selection will enable implementation of more complex genetic improvement programs. This may lead to further development of differentiated pork value chains focused on the ever increasing needs for global pork consumption and changing consumer desires.

12.2 IMPACT OF PORK PRODUCTION PRACTICES AND GLOBAL MEAT TRADE ON SUPPLY OF ANIMAL PROTEIN WITH THE EMPHASIS ON PORK

The global market for meat, including pork, is directly related to the demand for added-value foods. More disposable income translates into higher meat consumption, whereas a reduction in income normally results in consumers switching to cheaper meats, such as poultry, and then gradually toward smaller amounts of meat. Also, purchasing meat is not only an economic or even a dietary decision, it is also a cultural one and the highest-status meat consumption varies from country to country. In China it is pork, as it is across most of East Asia. Beef is by far the highest status meat in the United States and in Latin America. In many parts of Europe little pork is consumed fresh because it is normally converted into a variety of products such as bacon, ham, or sausage.

Coupled with this expansion, the role of meat has emerged as a primary commodity in the typical diet, where developed countries may fill 70% of their protein consumption with animal-based products, sometimes reaching over 300 g of meat per person daily. Thus, developing countries have nearly doubled their meat intake per capita as a result of growing incomes, urbanization, and shifts in food preferences. Simultaneously, world meat exports have skyrocketed (and are projected to continue climbing) to meet demand (FAO 2011). Those export demands are driven by the regional and/or cultural preferences for quality of cuts/pork products, not always necessary by price.

Meat-animals, including pigs, are raised in a cycle that is longer than most agricultural crops. For instance, there is a basic production cycle that exists in all pig farms or enterprises regardless of their size and function. At each point on the production cycle, events happen in a logical, animal-biology-driven sequence (Figure 12.1). Genetic product (breed/line) decisions, along with the management of the production cycle, are the foundation of any live-production system. Possible trade-offs among sow lifetime prolificacy, average daily gain, and feed conversion efficiency (cost of weight gain), carcass yield, backfat thickness, loin muscle depth, boneless loin and ham yield, and pork quality attributes such as color, water holding capacity, tenderness, and taste are being continuously evaluated. This cyclical nature of meat/pork production creates an inbuilt inflexibility to supply, which cannot change nearly as fast as demand. The high input requirements also add economic inflexibility. Recent trends in higher feed costs, lower consumer incomes, and other high farm costs such as fuel, have greatly increased the production cost of meat, and along with extreme weather events such as droughts have further restricted input supplies (Morrison-Paul 2001; Deloitte Consumer Business 2007; European Pork Chains 2009).

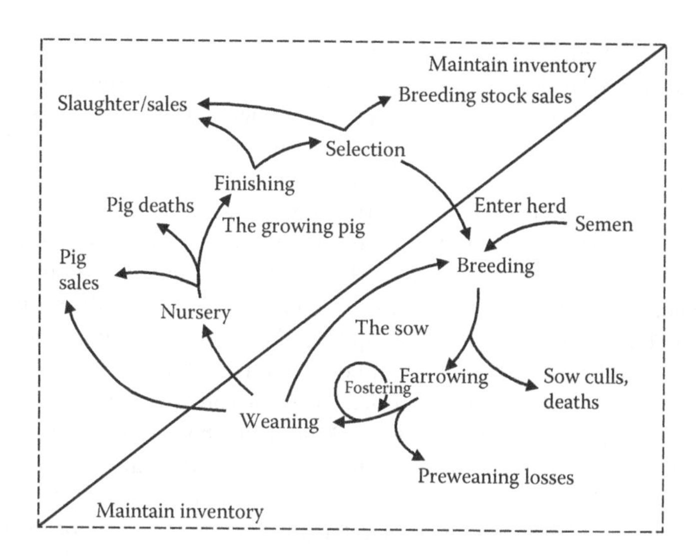

FIGURE 12.1 The pork production cycle: biological and financial events; revenue: sales number weight and value; variable costs: feed amount and value, breeding stock and semen, veterinary and medicine, supplies, utilities, labor and benefits, transport; fixed costs: depreciation, finance, service, taxes.

Pork production, as many other meat production chains, is primarily a commodity-driven industry. Pig carcass quantity characteristics, obtained at the least cost of production, have been the major driver for economic value due to the relative ease of measuring carcass (meat content) weight and lean percentage/yield postslaughter. However, in the last decade the meat/pork industry has undergone consolidation leading to bigger and more complex vertically integrated and/or coordinated pork production systems. Much of the consolidation has been driven by the need for strategic positioning in an increasingly globalized agribusiness world. As supply chain customers become larger and trade liberalization continues to open new market opportunities for exporting countries, agribusiness corporations are positioning themselves to become truly global agribusinesses (Martinez and Zering 2004; Reiner 2006; Deloitte Consumer Business 2007).

Simultaneously to the industry consolidation, enhanced coordination has been taking place in all segments of the pork industry, including those between (1) input suppliers, such as pig genetics and animal feed providers, and producers; (2) pig production segments of farrowing, nursery, and finishing; (3) pig producers and meat processing companies; and (4) meat processors and retail and food service markets. These events have led to the development of many differentiated "value chains" from the traditional commodity business-oriented "supply chains." The most obvious differences between supply chain and value chain management are: (1) supply chain management focuses mostly on increasing the efficiency of current operations, such as minimizing transportation or production costs in isolation of other factors; its core focus is on reducing costs while retaining the systems and processes already in place; and (2) value chain management principles are based on management decisions, which consider whether an operation creates value from a consumers' perspective; its core focus is on developing the systems (resources, infrastructure, processes, and relationships) necessary to satisfy or exceed consumers' expectations; cost reduction is an outcome of this approach, as is superior quality and competitiveness. Thus, the main goal of value chain activities is to achieve sustainable competitiveness through focusing resources on effectively and efficiently producing goods that offer superior consumer-recognized value to that of competing products. A closely aligned value chain often contains vertically and horizontally linked players such as genetics and genetic improvement program(s), farmer(s), processor(s), distributor(s), and retailer(s); (The Boston Consulting Group Inc. 2001, 2006; Rabobank International 2006; Gooch and Felfel 2008; European Pork Chains 2009).

At the same time retail and food service businesses are also becoming more interested in consistently sized "case-ready" pork products, better tasting product varieties, and cuts of pork that are suited to the cooking, nutritional, and eating quality demands of today's consumer. Thus, the latest consumer demands have also led to pork product differentiation and a greater pressure on the value of meat quality parameters, especially tenderness, juiciness, and flavor of fresh and value-added pork (Deloitte Consumer Business 2007). The value-added strategies of the leading retail, food service, and meat industry entities will dominate the animal protein industry for years to come, with a renewed interest in high raw material quality produced at least cost (Morrison-Paul 2001). Given the shift away from a commodity pork market to a consumer-driven one, it has become much more important for the pork value chains to understand consumer preference, attitude, and acceptance of meat (pork) as driving factors. As the competition for shelf space increases, so will the motivation to produce more value-added products having high consumer appeal (regardless of the reason). This will further foster a close collaboration in product development and marketing between value chain players, especially between meat processors and the retail and food service industry (Martinez 1999; Martinez and Zering 2004).

Pork is the biggest meat sector globally, not least because this is the meat of choice in the world's biggest country, China. Pork production in China is half of all global production, double the size of the EU, despite the important pork industry in Germany and France, and five times the size of the United States (Table 12.1). International trade accounts for about 7% of global production, which has shown a small increase since 2008, but among major exporting countries only the United States presents a high export orientation (about 25% of total production in 2010–2012). Output in China is increasing, assisted by improvements in pork meat yields and type of products as the Chinese government encourages better management and industrialization of pork production. As affluence increases and the potential for further efficiency improvements declines, China may need to further increase their pork imports. By comparison, output in the established pork producer countries has been static in the last decade. Europe has been increasing its exports but by relatively little in absolute terms (Innova Analysis 2013; Table 12.2). The major global importer, Japan, has a policy of supporting local farmers but not at the expense of much higher prices for its mainly urban population, and levels of imports seem to have stabilized despite the stagnant Japanese economy in the last several years. Russia is in a different situation, where the agricultural economy went through a period of near-collapse while urban

TABLE 12.1
Pork Production in the World (Million Tons)

	2008	2009	2010	2011	2012	2013-est.
			Global Trade			
Production	97.9	100.5	102.9	102.0	104.4	104.7
Imports	6.3	5.5	5.9	6.6	6.8	6.8
Exports	6.2	5.7	6.1	7.0	7.2	7.3
			Biggest Producers			
China	46.2	48.9	51.1	49.5	51.4	52.0
EU 27	22.6	22.4	22.6	22.9	22.8	22.6
USA	10.6	10.4	10.2	10.3	10.6	10.4
Brazil	3.0	3.1	3.2	3.2	3.3	3.3
			Biggest Consumers			
China	46.7	48.8	51.1	50.0	51.9	52.6
EU 27	21.0	21.0	20.8	20.7	20.5	20.3
USA	8.8	9.0	8.7	8.3	8.5	8.4
Russia	2.8	2.7	2.8	3.0	3.0	3.1
Brazil	2.4	2.4	2.6	2.6	2.7	2.7
Japan	2.5	2.5	2.5	2.5	2.5	2.5
			Biggest Importers			
Japan	1.3	1.1	1.2	1.3	1.3	1.3
Russia	1.1	0.9	0.9	1.0	1.0	1.0
			Biggest Exporters			
USA	2.1	1.9	1.9	2.4	2.4	2.5
EU 27	1.8	1.4	1.8	2.2	2.3	2.4

Source: Adapted from United States Department of Agriculture. 2012. *USDA Agricultural Projections to 2021* (OCE-2012-1). Office of the Chief Economist, World Agricultural Outlook Board, Long-term Projections Report, OCE-2012-1; February 2012. http://www.ntis.gov.

incomes rose, drawing in imports. There are major economic initiatives to modernize farming and food processing and to improve pig genetics, but to date Russia is not near self-sufficiency in pork, which is the main meat consumed in both fresh and processed form. The relative scale of imports by two major markets shows the dominance of intra-EU trade. Both Germany and France import substantial volumes of meat but non-EU imports even in Germany are much lower than EU imports, with the exception of sheep meat where EU supplies have been falling fast (FAO 2011).

Providing food to the world's growing human population with limited resources will require rapid application of scientific advances in agricultural production. Food safety, along with new meat/pork product development and processing technologies requiring an adequate meat quality (i.e., high protein functionality), will continue to be paramount. Animal production methods will continuously have to be morally and ethically

TABLE 12.2
Value of Output in EU—2005 Index

	Pigs				
	2005	2009	2010	2011	2012
EU 27	100.0	103.7	101.9	111.3	N/A
Poland	100.0	119.0	102.9	121.6	146.1
United Kingdom	100.0	140.2	135.9	138.7	143.8
Hungary	100.0	115.8	108.2	118.8	141.0
Romania	100.0	113.1	109.5	115.9	131.1
Italy	100.0	107.4	106.3	120.9	130.7
Finland	100.0	109.5	106.7	113.8	127.2
Denmark	100.0	99.6	105.2	112.4	126.2
Netherlands	100.0	102.0	99.8	111.9	126.1
Austria	100.0	99.3	98.6	108.6	124.4
France	100.0	97.3	96.3	108.7	117.9
Spain	100.0	97.2	101.4	108.7	117.2
Germany	100.0	99.4	98.3	106.4	N/A

Source: Adapted from Innova Analysis: Briefing Series, 2013. The Future of Meat: Coping With Shifting Market Dynamics and Supply Pressures. Feb 2013.

Note: N/A, data not available.

acceptable to the consuming public. Consumer attitudes toward food, especially meat, will determine the acceptance of novel food items in the future, and to some degree, the implementation of new meat processing technologies. To be accepted, new technologies must often be put into the context of the familiar (Sosnicki et al. 2003; Reiner 2006; Floros et al. 2010; Ros-Freixedes et al. 2010; Sosnicki and Newman 2010).

12.3 IMPACT OF PIG BREEDING AND GENETIC DETERMINANTS ON PORK QUALITY AND CARCASS VALUE

12.3.1 Innovations in Pig Breeding Programs Enabling Improvement in Biological Efficiency of Pork Production and Carcass Value

In the face of the rapidly changing global market place, the pig (Sus scrofa) has changed enormously to meet new demands and the science of pig breeding has developed a long way from Robert Bakewell's initial breeding program of the eighteenth century when he advocated breeding "the best males to the best females." Prior to that time most animal breeding programs were relatively random affairs. However, Bakewell's example was followed by many and, therefore, today's pig breeds were established as a result of selection for uniformity of phenotypic characteristics such as coat color patterns, but accompanied by some crude method of identifying the "best." Over the next 250 years there was a gradual improvement in techniques to measure the "best," leading to today's sophisticated, science-based industry estimating the genetic components of an animal's live performance, carcass and muscle/meat quality, and producing improvements against defined objectives in every subsequent generation.

In the second half of the twentieth century, pig breeders used performance testing of candidates to the local live slaughter weights to measure growth rate, backfat, loin depth, and sometimes feed intake to enable selection within contemporary groups. These traits are easy to measure and have a medium to high genetic component or heritability; meaning that a significantly large part of the superiority measured in the selected parents will be passed to their offspring in the next generation. This resulted in considerable gains in production efficiencies and carcass and meat quality value realized by all segments of the pork supply chain including consumers worldwide. The achievement of today's breeding and selection programs were dramatically accelerated in the 1990s when a widespread implementation of Best Linear Unbiased Prediction (BLUP) and Multi-trait Index objectives took place (Harris and Newman 1994; Henderson 1975). This allowed the ranking of the candidate's ancestors and relatives. The major benefit of BLUP was to enable improvement in difficult, low heritable traits such as the reproductive ability of sows. The addition of many new traits into the breeding programs also meant that the rate of change in carcass leanness also slowed down. However, pig breeders developed a mechanism to rank a living candidate on postmortem traits using data measured on the carcasses of the candidate's relatives; that is, typically litter-mates of the animals kept for rebreeding; who were surplus to breeding requirements. The adverse correlated responses in meat quality could be prevented by including meat quality traits in the line selection index. Simultaneously, the rates of realized genetic change at the commercial level were improved as pig line improvement programs expanded performance testing to the crossbreeding of Genetic Nucleus (GN) tested pure line pigs. The performance of crossbred pigs grown in commercial environments and harvested in commercial processing plants enhanced the traditional genetic improvement of pork quality based on within-line selection by including such meat quality traits as ultimate pHu and meat color (typically measured in the loin or ham muscles 24 h postmortem) in the selection index (de Vries et al. 2000a,b; Sosnicki et al. 2003; Binder et al. 2004; Plastow et al. 2005; Barbut et al. 2008; Sosnicki and Newman 2010).

The most significant technology to contribute to pig breeding since the implementation of BLUP has been the contribution of genomics to breed improvement. This has rapidly developed from early Marker Assisted Selection (MAS) applications as a prelude to Genomic Selection (GS), Genome Wide Association Studies (GWAS), and the development of the Genomic Relationship Matrix (GRM) that can distinguish among animals with the same parents at birth by their genetic code. GS simultaneously evaluates genotypes from any number of SNP markers (the current porcine SNP chip is 60,000 markers) in order to cover the entire genome in a sufficiently dense manner so all the genes are in linkage disequilibrium (nonrandom assortment) with some of the markers. The next step extended the application to improving all the objective traits simultaneously in all lines using a GRM and Single Step Genetic Evaluation (SSGE; Aguilar et al. 2010). Next Generation Sequencing (NGS) is also becoming a practical research tool, aiding identification of specific genes controlling quantitative traits. With ever more powerful computers and laboratory techniques, animal scientists are learning more about the genetic architecture and biology underlying key performance traits while biotechnology tools are already available for modifying genomes with an increased efficiency and precision than was possible in the past.

As a result of this history of selection, the global pork industry today is predominantly supplied by international breeding companies whose genotypes can grow quickly, efficiently, and produce lean, high quality carcasses at over 135 kg liveweight, which results in economically affordable pork in the global market place. Dam-lines are developed with a primary focus on reproductive ability and two or three dam-lines are combined into the female used as the dam of progeny destined for harvest. Each dam-line is chosen for their suitability for the local environment, but also for the pork market that their progeny will supply, simply because the dam contributes 50% of the genetic material and resulting performance and carcass value to its progeny. The purebred dam-lines are identified and combined in a crossbreeding program designed to maximize heterosis and their reproductive ability. The dam-lines have to be selected so that pork quality is adequate for the market place both in pork quantity and quality. Most modern dam-lines are not selected to be extremely lean because this results in lower fertility either as a lower number of pigs in a litter or more often failure to cycle and conceive a litter. The emphasis of muscle/meat quality and quantity (and in a few cases fat quantity and quality) is, therefore, usually focused on the development of the terminal sire-line, which is also improved for growth rate and efficiency of converting feed into body or lean muscle mass. The terminal sire-lines of today are more extreme in their carcass characteristics so that their average carcass value is well above the average of the market place. Therefore, when the terminal sire-line is bred to the crossbred dam-line females the resulting progeny (i.e., slaughter generation) will meet or be above the average market requirements (Sosnicki and Newman 2010).

12.3.2 Genetic Parameters of Pork Quality Traits

The first heritability (h^2) for pig meat quality were reported in the early 1960s (Duniec 1961), and since that time numerous h^2 estimates have been published for all important meat quality attributes, including eating quality. Simultaneously, a genetic antagonism has been postulated between growth and carcass performance and pork quality, and thus, many authors concluded that selection on low backfat thickness, high lean content, and high growth rate, without consideration of any meat quality traits, may cause a decline in meat eating quality (Hovenier 1993; Sellier 2010). Inclusion in breeding programs of such pork quality traits as ultimate pH, meat color, and intramuscular fat (IMF) content has enabled to alleviate this biological antagonism (Hovenier 1993; Sellier 2010). Extensive reviews of the heritabilities of pork quality traits and their genetic and phenotypic correlations have been provided by many authors (for review, see Sellier and Monin 1994; Morel 1995; Knapp et al. 1997; Fernandez et al. 2003; Suzuki et al. 2005; Rotshchild and Ruvinsky 2010; Sellier 2010). A summary of genetic correlations between eating quality, backfat thickness and IMF, and heritability of meat quality attributes is provided in Table 12.3.

TABLE 12.3
Relationship between Meat Processing and Eating Quality, Backfat Thickness and IMF, and Heritability Values of Meat Quality Attributes

	Color	IMF	WHC	Tenderness	pHu	C18:2
h^2	0.30	0.50	0.20	0.30	0.20	0.55
Range	0.10/0.60	0.25/0.85	0.05/0.65	0.20/0.40	0.10/0.40	0.40/0.70
r_gADG	–0.15	0.40	0.00	–	–0.15	–0.20
Range	–0.50/0.15	0.15/0.60	–0.80/0.50	–	–0.40/0.50	–0.40/0.0
r_gBF	0.20	0.20	0.20	0.20	0.15	–0.70
Range	–0.20/0.60	–0.20/0.60	–0.20/0.80	–0.20/0.40	–0.40/0.50	–0.50/–0.90
r_gLean	–0.25	–0.20	–0.35	–0.20	–0.10	0.60
Range	–0.60/0.20	–0.50/0.40	–0.80/0.25	–0.60/0.20	–0.70/0.35	0.40/0.80
r_gIMF	–0.20	–	–0.05	0.25	0.0	–0.30
Range	–0.10/0.35	–		0.20/0.30	–0.20/0.40	

Note: Heritability values (h^2) for meat color (color), IMF, water holding capacity (WHC), tenderness (Tenderness), pH ultimate (pHu) and linoleic acid (C18:2) and their genetic correlation (r_g) with average daily gain (ADG), backfat thickness (BF), lean percentage (Lean), and IMF.

12.3.3 Pig Breed Effects on Carcass and Pork Quality

It is well known that a combination of postmortem muscle pH and temperature decline explains a major part of pork eating and processing quality, and those decline patterns differ between pig breeds, lines, and crosses (Gusse 1993). It must be noted, however, that although breed effects are still important in many pig genetic improvement programs, the within-line selection can and very often does change the set of attributes that have previously been attributed to the particular breed. Therefore, there are several inconsistent results published in the scientific literature regarding the actual breed effects of pork quality (McGloughlin et al. 1988; Barton-Gade 1989; Wood et al. 1989, 1996, 2004; Cameron et al. 1990; Warkup et al. 1990; Cameron and Ensner 1991; Ellis et al. 1995; Warris et al. 1996; Lonergan et al. 2001; Hviid et al. 2002; Chang et al. 2003; Suzuki et al. 2003; Straadt et al. 2013; Kim et al. 2014). Reports from Canada, Denmark, France, and New Zealand indicated that pigs produced from "white" hybrid mothers and Duroc sires produced meat with tenderness advantages over pigs sired by Large White (Barton-Gade 1989; Brewer et al. 2001, 2002; Piao et al. 2004; Suzuki et al. 2003; Fortin et al. 2005). In British and Irish studies no significant tenderness advantage for Duroc cross vs. Large White cross pigs was found (Cameron et al. 1990; Cameron and Enser 1991). Some authors in Great Britain found that Duroc meat was less tender, but more juicy than pure Landrace pork, whereas in America and Denmark Duroc meat was found to be more tender than Yorkshire pork, and Pietrain or Hampshire × Duroc. Others found no difference in tenderness, but better flavor and overall acceptability for 100% Duroc pigs, or that the *Longissimus* muscle from the Large White had higher tenderness (but poorer flavor attributes) than a commercially available Duroc line or Tamworth, with Berkshire meat quality being intermediate. Duroc genetic-background meat is still considered as a superior in eating quality in the Asian (Japan, Korea, China) and in American niche markets, and this superiority is often attributed to a higher IMF content.

The effect of IMF level on pork eating quality may be confusing, however, because some published results pointed out that there was no difference between pigs sired by Duroc boars and Landrace or Large White boars for tenderness, juiciness, and overall meat acceptability (Font-i-Furnols et al. 2012). Several publications also reported that meat from Duroc purebred or crossbred pigs was paler and had a slightly lower ultimate pH than meat from Large White purebreds or crosses. Difference in ultimate pH between Landrace and Large White pigs are considered to be generally low, and dependent of the HAL-1843 gene status (currently most of the modern populations of Landrace and Large White pigs are free of this genetic mutation). The Belgian Landrace breed was described as having a relative fast rate of pH fall whereas the Pietrain breeds were characterized by a fast rate of pH decline that, at least partially, was a consequence of a high HAL-1843 gene frequency, with a rather low ultimate pH when compared to other breeds (McGloughlin et al. 1988; Hviid et al. 2002; Sellier 2010). It was also noted that Pietrain and Belgian Landrace pigs provide meat with inferior quality (particularly tenderness) when compared to Large White or French Landrace breeds. The Hampshire breed was characterized by a slow pH decline early postmortem, similar to that described in Large White breeds, but it was also characterized by markedly lower ultimate pH due to the presence of the RN(–) gene (these rigor mortis conditions typically result in paler meat color, higher cooking losses but significantly better meat tenderness). Elimination of the RN(–) mutation from several Hampshire-based lines of pigs made meat quality quite acceptable in several global markets.

Several comparisons of meat eating quality between European and American breeds with Chinese purebred or crossbred pigs revealed that the latter produces pork products that are more tender, juicier, and tastier (Ellis et al. 1995; Channon et al. 2004; Piao et al. 2004; Suzuki et al. 2004). The amount of visible fat was judged as excessive in meat from Chinese crossbreeds, which counterbalanced the perceived better meat eating quality.

12.3.4 Pig Sex Effects on Carcass and Pork Quality

There is resistance in most parts of the world to the use of boars for meat production, and castration is employed at 2–5 days of age to minimize risks of boar taint and reduce aggression and enhance the quality of pork products. The main issue with the eating quality of boar meat is the risk of odors/flavors, which some consumers find unacceptable ("boar taint"). In fact, other quality attributes (tenderness, juiciness, and sometimes flavor) are often found to be higher in boars than gilts (Jeremiah and Weiss 1984, 1999a,b; Martel et al. 1988; Brewer et al. 2001, 2002; D'Souza and Mullan 2002; Channon et al. 2004; Latorre et al. 2004; Piao et al. 2004; Fortin et al. 2005; Lonergan et al. 2014). The so-called boar taint odor/flavor is caused mainly by a higher incidence of two compounds in meat from boars—skatole and androstenone. Androstenone reduces the breakdown of skatole from the liver in entire male pigs and skatole therefore accumulates in fat tissue. Androstenone is stored in the lipid component of tissues (Gower 1972) and is exacerbated upon cooking of fresh pork as well as processed pork products, including bacon. An EU funded research project that examined the perception of boar taint across seven European countries showed that, of male pigs slaughtered in Britain, 22.3% had androstenone levels above 1 ppm (the notional threshold for detection) and 8.9% had skatole levels above 0.25 ppm (the notional threshold for detection) (Jensen 1998; Walstra et al. 1999; Matthews et al. 2000). Looking at consumers across the seven countries in the same project, the percentage who disliked the flavor and odor only differed between boars and gilts by 3% and 4%, respectively. Skatole proved to be the more important of the two compounds in affecting acceptability (Matthews et al. 2000). Research also showed that meat from entire males, gilts, and castrates was equivalent in tenderness and juiciness. Trained taste panellists did detect taints in some entire male carcases, but typically the consumer panel scores for flavor and odor were not different. This is inline with much of the literature where consumers did not detect differences between the sexes (Ros-Freixedes et al. 2010).

Although untrained consumers have determined the sensory thresholds for boar taint of 1.0 μg/g for androstenone and 0.2 μg/g for skatole (Bonneau 1982, 1998; Desmoulin and Bonneau 1982; Claus et al. 1994), there is no clear EU agreement on these values for androstenone and skatole (Merks et al. 2009). A range of threshold levels in fat of 0.2–0.25 μg/g for skatole and 0.5–1.0 μg/g for androstenone were used in different studies (Lundström et al. 2009). Lunde et al. (2010) found that consumers who were sensitive to boar taint gave a lower liking score for pork with androstenone levels of 0.3 μg/g during frying whilst for skatole, consumers differentiated skatole samples for flavor at 0.15 μg/g—lower than the Norwegian threshold level of 0.21 μg/g. The large range in threshold values for androstenone may be influenced by intrinsic differences in an individual's ability to detect androstenone depending on their origin, age, sex, or sensitivity to androstenone, differences in methodology used to conduct sensory evaluations as well as analytical methods used in different laboratories to determine androstenone. Leong et al. (2011) determined that the threshold level for skatole for Singaporean consumers was 0.03 μg/g, whilst threshold levels for androstenone were not determined.

Consistent methods for sensory evaluation of boar taint are needed to ensure consistency and comparability between studies. As the majority of studies investigating the influence of boar taint on consumer acceptability of pork have been conducted on precooked samples, consumers have not being exposed to volatiles released during cooking. It has also been suggested that higher internal temperatures could reduce the sex effect on flavor when pork from entire males and females is compared, as volatile compounds may have a smaller effect on abnormal flavor at higher temperatures (Wood et al. 1995).

Finishing entire males at a lower carcass weight has not been found to be an effective method to reduce the risk of boar taint in male pigs (D'Souza et al. 2011), with weak correlations reported for both androstenone and skatole compared with hot carcass weight (ranging from 60–80 kg). This suggests that managing boar taint risks by using carcass weight strategies for entire males has its limitations. Considerable variability between genetic lines on sexual maturity (Wilson et al. 1977) and the onset of puberty (Prunier et al. 1987) may also exist. Furthermore, as it is more economically feasible to market pigs at a heavier live weight, this places additional restrictions on the use of such strategies to manage boar taint risks.

An alternative approach to the production of entire males (leaving aside the unlikely solution of re-introducing castration) is immunization against the gonadotropin releasing hormone. This is often termed "immunocastration" and a commercial product is marketed under the trade name "Improvac." This has been found to be extremely effective at reducing both androstenone and skatole levels in pork from otherwise entire male pigs (Jaros et al. 2005; Einarsson 2006). The technology requires two injections administered during the finisher phase, with the second injection administered from 4–5 weeks (Dunshea et al. 2001) to 2 weeks (Lealiifano et al. 2011) prior to slaughter. Management practices may need to be amended to allow this to be achieved in group housed pigs near their slaughter weight. The timing of the injections is thought to be critical and ensuring correct timing could also be difficult to manage. The generally negative consumer reaction to the animal injections and the notion of "immunocastration" has made the implementation of this technology very limited in the global pork industry thus far. Immunocastration can also have animal welfare benefits on-farm, due to reduced aggressive and riding behaviors following the administration of the second vaccine, compared with entire males, resulting in reduced incidence and severity of carcass lesions (Lealiifano et al. 2011).

12.3.5 The Effects of Pig Rearing Methods on Carcass and Pork Quality

There is a lack of consistent knowledge of the effects of pig rearing conditions, especially outdoors, on pork eating quality. In many cases, the reported studies were subjected to confounding of factors such as pig genotype (Gentry et al. 2002a,b, 2004; Guy and Edwards 2002; Olsson et al. 2003; Lambooij et al. 2004; Millet et al. 2005). An evaluation of four finishing systems: indoor slatted buildings, indoor deep bedded buildings, outdoor on "dirt," and outdoor on alfalfa pasture vs. indoor rearing showed that the outdoor finished pigs had lower juiciness and also slightly lower off-flavor scores of the loin (*Longissimus lumborum*) muscle. Also, outdoor rearing resulted in darker meat and lower shear force; that is, more tender meat; as compared to the confinement reared pigs (the outdoor pigs were on alfalfa pasture). There were also reports observing higher drip loss, but improved juiciness in pigs bedded on sawdust and given access to an outdoor area compared to conventional (slatted) housed pigs (Guys and Edwards 2002). Similarly, there are no consistent reports on the effects of organic production systems on pork quality. Hansen et al. (2006) found no effect of three organic systems compared to a conventional system. Millet et al. (2005) found no effect of Belgian organic housing or an organic diet on carcass or meat quality traits (sensory traits were not assessed). Some authors have reported negative effects of organic rearing on pork eating quality (Olsson et al. 2003; Olsson and Pickova 2005).

Another element of management of pigs on farm relates to reducing the risk of high skatole levels (see the section on the sex effect on pork quality). Skatole is produced in the hind gut of pigs from the amino acid, tryptophan, by fermentation. This means that pig feces is high in skatole and it can be ingested or reabsorbed through the skin. Skatole levels can be reduced by dietary changes to decrease its formation in the hind gut by the following management practices: (a) avoid use of brewers yeast; consider a withdrawal period of 7 days prior to slaughter; (b) avoid overfeeding proteins; this can be achieved by the use of phase feeding; (c) provide clean lying areas; pigs lying in manure/urine can absorb skatole through the skin; (d) maintain adequate ventilation rates; inadequate ventilation will result in pigs choosing to lie in the moister areas of the pen, thus absorption of skatole for those pigs can be substantial; (e) design housing to minimize the use of excrement area by the pigs; (f) provide correctly positioned water drinkers; (g) use specific feed ingredients to reduce skatole if economically justifiable; that is, coconut cake (at 10% of diet), sugar beet pulp (at 20% of the diet or more), wheat bran (at 20% of the diet or more), raw potato starch (at 10% of diet), lupins (at 10% of diet), protein sources that are readily digestible in the small intestine (such as casein). In addition, the beneficial effect of chicory (feeding crude chicory roots, dried chicory, and extracted inulin), fed for as little as 1 week prior to slaughter has been reported (Jensen and Hansen 2006), no effect of feeding chicory on reducing androstenone levels of entire males have been shown (Hansen et al. 2008).

12.3.6 The Effects of Pig Age and Weight on Carcass and Pork Quality

There has been a considerable implementation of increased slaughter weights of pigs in many countries, primarily because modern genetics has bred for efficient muscle deposition at heavy weights; that is, 125 kg or heavier; and because of the economic benefits of spreading fixed costs of pig slaughter and carcass/meat processing over a greater weight of saleable meat. Research has shown that the effect of slaughter weight on pork quality is very small. This is explained by (a) within-line genetic improvement programs implemented by some breeding companies that include ultimate pH as an economical trait in the selection index (Sosnicki et al. 2003, Sosnicki and Newman 2010); (b) the processing industry implementing high animal welfare/preslaughter handling standards reducing ante-mortem stress; and (c) the processing industry implementing carcass-blast chilling practices; that is, a fast reduction of internal muscle temperature to below 35°C prevents, to a large extent, aberrations in pork quality. For instance, Canadian research comparing liveweights of 90 and 110 kg at slaughter (Jeremiah and Weiss 1984) found the pork from heavier pigs was more juicy, had a less desirable flavor, but there was no difference in meat tenderness. Also, no difference in shear force was found in a Spanish study where the experimental pigs were reared to the slaughter weights of 116, 124, and 133 kg; although increased weights resulted in darker, redder meat with higher myoglobin contents (Latorre et al. 2004). Others have also found no effect of slaughter weight and it was even reported that heavier pigs can have better eating quality (Purchas et al. 1990, Ramaswami et al. 1993; Ellis and Bertol 2001; Piao et al. 2004). Slower chilling rate of heavier vs. lighter carcasses must always be considered when one studies the effects of pig/carcass weight on pork quality.

12.4 IMPACT OF PIG FEEDING PROGRAMS ON CARCASS AND PORK QUALITY

Approximately 75% of the cost of production in modern swine production is the cost of feed. Thus, it is very important that growing-finishing pigs are fed to optimize growth rate, feed efficiency, carcass composition, and pork quality. Most nutrient requirements have been well defined (NRC 2012). However, a good understanding of the biology of the pig is necessary to ensure that some nutrients (particularly amino acids) are provided at the correct proportions to complement efficient growth and desirable carcass composition (PIC 2010). As pigs grow their requirement for energy and amino acids increase; however, the requirement for amino acids as a percentage of the diet decreases as the pig gets older because of the increased feed intake. In an ideal production situation the nutrient concentrations of the diet would be changed weekly to precisely meet the pig's requirement, thus fully optimizing growth potential. However, this is not practical because one typically feeds barns with 500–3000 pigs that could have body weight variations of 5–15 kg, for instance, and the feed processing and delivery on such a frequent basis is not feasible.

In Figure 12.2 the growth, feed intake, back at, and loin depth curves of pigs fed either low (3179 kcal ME/kg) or high (3430 kcal ME/kg) energy diets that were formulated to contain the same proportion of standardized ileal digestible (SID) lysine to metabolizable energy (i.e., g SID Lys/Mcal ME) is depicted. Feeding high energy diets results in faster growth rates using less feed which results in a better feed conversion than in pigs fed low energy diets (PIC 2010). However, pigs will eat to their energy requirements, thus energy efficiency (Mcal ME/kg gain) will be very similar between

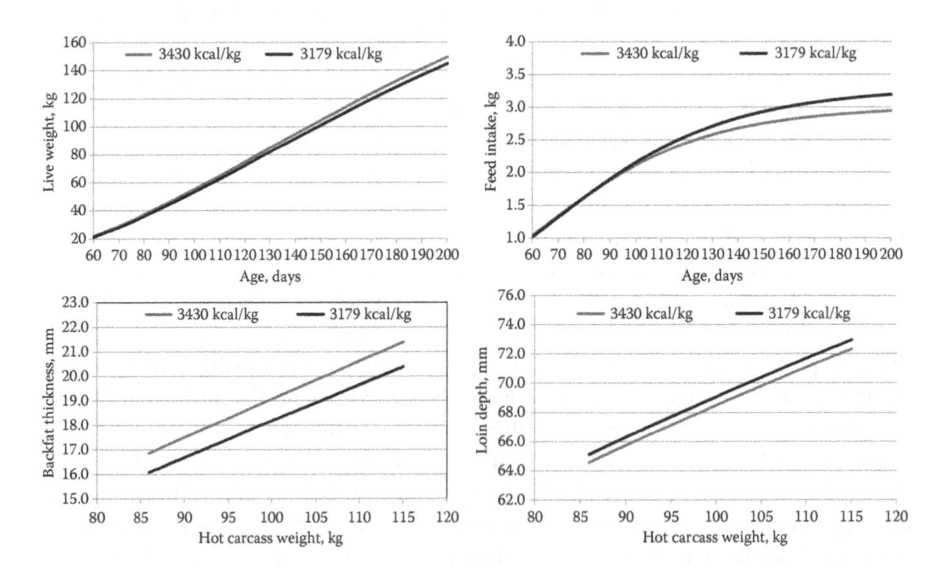

FIGURE 12.2 Effect of dietary metabolizable energy levels on growth and carcass composition of pigs. (Adapted from PIC. 2010. Growth Curves for PIC337RG sired pigs fed high and low energy diets. Executive Summary 051—Tech. Memo 344. PIC North America, Hendersonville, TN. www.genusplc.com.)

pigs fed low and high energy diets unless the dietary energy level is so low that pigs fed low energy diets will not physically consume their daily energy requirement (PIC 2010). Feeding high energy diets results in increased fat deposition and lower muscle deposition than pigs fed low energy diets (based on backfat and loin depth measures). Measures of pork quality (i.e., pH-ultimate, color, eating quality) were not affected by the energy level of the diet (PIC 2010). Thus, energy level of the diet fed to pigs clearly has an effect on growth and carcass composition without affecting pork quality (an exception is the IMF content that is typically positively correlated with higher energy diets promoting fast growth and more backfat and internal muscle fat deposition).

Feeding diets deficient in lysine/protein has been shown to consistently improve IMF levels in the loin muscle (*M. longissimus*) of the pigs (Boyd et al. 1997; Witte et al. 2000; Apple et al. 2004). Also, Warkup et al. (1990) found that pigs fed *ad libitum* produced meat that was more tender and juicy than that of pigs that were restricted fed (82% of *ad libitum*). However, this was at the expense of feed efficiency and carcass leanness and could cost as much as 1.34 € per 0.1 increase in subjective marbling score.

Many factors contribute to fat composition, and thus to its quality, and those include diet, pig/carcass leanness/fatness, age/body weight, gender, breed/genotype, fat location, and pig growth rate (Villegas et al. 1973; Matthews et al. 2014). Swine diets are one of the most important factors in ensuring or changing the fatty acid profile because the diet can be altered more readily than most of the nondietary influences on fat quality. Most carcass fat in pigs is generated from dietary fats and carbohydrates (Mayes 1996). Dietary fats are readily converted to body/carcass fat and carcass fat formed in this manner acquires the general characteristics of the dietary fat (soft dietary fat = soft carcass fat). Dietary carbohydrates are converted to carcass fat through the process of *de novo* fatty acid synthesis, forming predominantly saturated fatty acids (Mayes 1996), which yields a firmer carcass fat. Dietary fat additions dramatically inhibit and/or totally stop *de novo* fat synthesis and therefore, as the percentage of fat is increased in the diet *de novo* fatty acid synthesis is further inhibited resulting in a less saturated fat (softer); and as the fatty acid profile of dietary fat becomes less saturated (softer), carcass fat is also less saturated (softer; Mayes 1996).

Recent research has evaluated the effects of feed form (meal/mash diet vs. pelleted diet) on fat quality (Matthews et al. 2014). This research indicated that pelleted diets have a detrimental effect on fat quality (i.e., fat containing a higher proportion of unsaturated fatty acids). This is likely due to the pelleting process, which improves energy digestibility of the pigs, therefore, dietary fatty acids, especially mono and poly-unsaturated fatty acids, are more readily available to undergo metabolic changes.

12.5 PORK QUALITY STANDARDS AND THEIR IMPORTANCE IN GLOBAL PORK TRADE

12.5.1 Carcass Grading and Classification Systems

Carcass leanness is the quality characteristic most commonly used to classify pork carcasses. The increasing consumer demand for lean meat and the growing rejection of a large fat consumption has driven the industry to encourage the production and processing of lean pork. Today these pork carcass classification methods are aimed to guarantee a fair payment system between the producers and the processor, and to regulate the market permitting price comparisons of carcasses from different markets at the same class or category. Those carcass class categories differ from region to region, country to country, or even within market segments, and therefore, no detailed information can be provided in this chapter.

It is generally agreed that carcass lean content is defined as the ratio of the weight of the edible muscle tissue compared to the hot carcass weight. Therefore, very clear definitions of the dissection method to obtain the actual lean content of a carcass and the definition of the carcass presentation are necessary to compare carcass lean percentages between different companies, countries, or even classification programs within a company. The carcass lean percentage is higher when the dissection methods include all the muscle, compared with methods when only the main commercial cuts are dissected (Branscheid and Dobrowolski 1999). Also, the carcass lean percentage is higher when the weight of the head is not included in carcass weight, compared with results including the head in carcass weight. In the EU a common reference dissection method and carcass presentation in pig carcass classification is used to compare results from different countries (Walstra and Merkus 1995). In the United States market carcass lean content is often referred to as fat free lean yield and in the Canadian market as saleable meat content.

In the 1990s, to reduce labor costs and operator influence on the results, fully automatic measuring equipment was developed. The most commonly used is the AutoFOM (Carometec A/S Herlev, Denmark), an ultrasound system capable of measuring up to 3200 fat and muscle depths per carcass. It is used in pork abattoirs to predict carcass lean percentage and also can predict ham, loin, shoulder, and belly carcass cutting yields (i.e., in Germany, carcass payment is based on the weight of primal cuts [ham, loin, shoulder, belly] and on belly leanness). Other automatic equipment includes the Vision Carcass System (VCS e + V Technology GmbH, Oranienburg, Germany), which uses video imaging from three specialized cameras to capture two-dimensional (2D) and three-dimensional (3D) images of the carcasses. From the images, up to 126 different fat and muscle depths, areas, lengths, and body angles are taken to assess the conformation and to be used as carcass lean predictors. Ultimately, no single best system has emerged and most countries adhere to one type of device. The EU have allowed for such variance in measurement technique by instituting accuracy requirements and verification systems to ensure commonality (Branscheid and Dobrowolski 1999).

12.5.2 Pork Quality Standards and Global Trade Requirements

Pork quality is not a unique trait, but is usually considered to be a mixture of traits, influenced by both environmental and biological factors, that cause it to be either desirable or undesirable to the end user. Pork quality may also be described simply as the perception of the product by an individual. Given that standards for pork quality are typically highly dependent on local markets and consumers, each region of the world and each market segment have their own unique qualifiers as to what good quality pork is. For instance, the cooked, ready-to-eat meat products market relies primarily on the technological meat quality attributes (water binding capacity, texture), whereas the fresh meat market relies primarily on the visual (color) and sensory (tenderness, juiciness, taste) traits. Food safety experts look at the hygienic and toxicological factors whereas dieticians look at the nutritional components (Nakai et al. 1975; Kauffman et al. 1986, 1990; Gusse 1993; Hofmann 1994; Honikel 1993, 1998; NPPC 1999). Ultimately, and regardless of market segment, there are three primary "classes" for pork quality. Those are: reddish-pink, firm, non-exudative (RFN), dark, firm, dry (DFD), and pale, soft, exudative (PSE). Other combinations exist such as pale, firm, non-exudative (PFN) or reddish-pink, soft, exudative (RSE), but are of less significance due to the low proportion of meat falling into these categories. Chapter 4 has detailed description of the biochemistry of these types.

There are numerous methods to measure meat quality, both objectively and subjectively. Most are accepted globally although certain methods may have more value in certain regions or markets. The easiest method is simple subjective evaluation, commonly used for traits such as color (both for lean and fat tissue), marbling (or IMF), and firmness/texture. Color may be the most important of these as it is often the first factor in purchase intent at the retail level. The most widely used subjective color standards are the Japanese Color Standards (Nakai et al. 1975) and the National Pork Board (USA) Color Standards (National Pork Producers Council—NPCC 1999). The NPPC also has standards for marbling, whereas there is no adequately defined or widely accepted system for firmness. Most subjective assessments of firmness range from soft to firm in either a 3 or 5 point scale and encompass a mixture of bending, poking, and squeezing of the muscle during the evaluation. Color can be objectively measured using a variety of instruments that measure light reflectance from the surface. Typically reported as a 3D color space such as the CIE $L^*a^*b^*$ (Minolta Co., Ltd. 1994), values allow for calculations such as chromaticity and hue angle that may be of relevance to how meat color is perceived by the consumer; that is, consumers typically do not distinguish color variations within one Minolta L^* unit (NPPC 1999).

Water holding capacity is the physical ability of the muscle to either retain or absorb additional water. This is a very important trait not only in fresh meat, but also for the processed meat industry. For a fresh product, the ability of the muscle to retain its inherent moisture is essential to ensuring juicy and tender pork. Meat with low water holding capacity also exhibiting high drip loss will be unappealing at the retail level due to the high exudate observed in the package and will be less acceptable upon eating due to the lack of moisture in the meat. It will be dryer and tougher than pork with a low drip loss. Numerous methods are described for the measurement of water holding capacity including the filter paper test, suspension methods, and the drip-tube method (Kauffman et al. 1986; Honikel 1998).

When unable to directly measure color, firmness, or water holding capacity due to constraints such as the inability to excise a portion of meat from an intact carcass, one can utilize perhaps the best overall predictor of meat quality—the pH–ultimate, typically measured either 24 or 48 h postmortem, due to its practicality rather than scientific importance; that is, although pork pH at 48 h postmortem is slightly lower that at 24 h postmortem due to meat aging processes, these pH difference are of little practical significance in defining pork quality (Barbut et al. 2008). Typically, the pH is measured via an invasive handheld probe. Therefore, it can be inserted into the desired muscle without having to make extra cuts that may damage or reduce the carcass value or remove the cut/muscle from the carcass. The pH of muscle is strongly correlated with its color, water holding capacity, firmness, tenderness, and so on. (Huff-Lonergan et. al. 2002) (Table 12.4, PIC 2008). Many companies manufacture and sell pH meters and probes capable of use in meat. The two main types of probes are the Ion-Sensitive, Field-Effect Transistor (ISFET) and glass. ISFET probes are considered more durable as there are no glass components to break, yet are often less reliable due to the small surface area of the transistor and the need for frequent cleaning to remove fat and protein build-up on the surface. Glass electrodes have proven to be more reliable in real-world environments and offer longer working lives than their ISFET counterparts.

As previously described, muscle/meat pH at 24–48 h postmortem is the best overall indicator of pork quality traits with the exception of marbling. Research using common factor analysis has shown pH to be the best predictor of the eating quality of fresh pork loin, followed by meat color and marbling score (Table 12.5; PIC 2008). Loin muscle pH measured at 24 h postmortem and with values greater than 5.7 will generally have the most desirable color and firmness for quality markets and will exhibit improved eating quality over loin with lower pH (i.e., below 5.6). Meat with pH below 5.5 is generally considered of unacceptable quality and it is associated with PSE pork conditions. When ultimate pH exceeds 6.1, a decrease in shelf life may occur as the closer the pH is to the neutral point, the more conducive it is for bacterial and microbial growth (Gusse 1993). Moreover, the meat may exhibit signs of DFD characteristics. Therefore, the most desirable range and the target for most swine breeding organizations and processing companies alike is an ultimate pH of 5.7–6.0.

Summarizing, while the measurement techniques and desired values of the various pork quality traits may vary between geographical regions and the tastes of local consumers, all segments of the industry seek reddish-pink pork that does not drip and has a desirable flavor. Therefore, all sectors of the pork chain must work together to implement the proper genetics and management practices to produce high quality lean pork at low cost. These products must then be delivered through the supply chain to satisfy consumers according to the best quality control and Hazard Analysis Critical Control Point (HACCP) practices. As a result, the consumer will reap the benefits of an endless supply of affordable, consistent, high quality pork products.

12.6 CONCLUSIONS

Pork quality is not a unique trait, but it is usually considered to be a mixture of traits, influenced by both environmental and biological factors, that cause it to be either desirable or undesirable to the end user, given that, standards for pork quality are typically highly dependent on local markets and consumers. Pork quality issues have been vastly reduced by industry, especially by the commercial value chains, by practical applications of science and technology aiming at fulfilling the needs of targeted market segments. Today, the global production-to-consumption

TABLE 12.4
Phenotypic Correlations between Longissium Longissimus Meat Quality Traits[a]

	pH22	pH14	MinL	Mina	Minb	Color	Marble	Firm	Drip	Purge	Shear	Moist	IMF	Juicy	Tender	BF
pH14	0.70															
MinL	–0.29	–0.24														
Mina	–0.43	–0.21	–0.19													
Minb	–0.43	–0.25	0.64	0.44												
Color	0.47	0.36	–0.63	–0.06	–0.55											
Marble	0.26	0.21	–0.01	–0.13	–0.09	0.18										
Firm	0.60	0.46	–0.30	–0.35	–0.41	0.61	0.29									
Drip	–0.44	–0.30	0.07	0.33	0.18	–0.20	–0.03	–0.35								
Purge	–0.40	–0.35	0.19	0.27	0.30	–0.22	–0.07	–0.27	0.21							
Shear	–0.15	0.09	–0.04	0.38	0.13	–0.11	–0.10	–0.17	0.19	0.08						
Moist	0.22	0.22	–0.13	–0.13	–0.18	0.12	–0.32	0.05	–0.01	–0.26	0.06					
IMF	–0.01	–0.01	0.16	0.05	0.15	–0.07	0.50	0.04	–0.03	0.05	–0.09	–0.70				
Juicy	0.11	0.11	–0.12	0.01	–0.02	0.12	0.09	0.03	–0.14	–0.11	–0.05	0.01	0.08			
Tender	0.17	0.19	–0.08	–0.09	–0.09	0.08	0.07	0.08	–0.17	–0.14	–0.19	0.01	0.12	0.60		
BF	–0.03	0.02	0.02	0.14	0.12	–0.04	0.28	–0.02	–0.06	0.09	0.05	–0.31	0.36	0.08	0.15	
Days	–0.03	0.06	0.04	–0.07	0.03	0.00	0.02	0.04	0.18	0.08	0.01	–0.04	–0.03	–0.18	–0.13	–0.28

[a] Trait explanation: pH14 = pork pH measured 14 h postmortem; MinL = Minolta lightness; Mina = Minolta redness; Minb = Minolta yellowness; Color = subjective 1–6 color score where 1 = extremely pale and 6 = extremely dark; Marble = subjective marbling, 1–10 score where 1 = devoid of fat and 10 = extreme fat deposition; Firm = subjective firmness 1–3 score where 1 = very soft and 3 = very firm; Drip = drip loss measured on a cube of meat suspended in a plastic bag for 24 h in a 4°C cooler; the difference between the initial and end weight delineates drip loss; Purge = purge loss measured on a 2–3 cm-thick cross section of the loin ("pork chop") placed on a retail trail and kept in a retail-display cooler at 4°C for 5 days; the difference between the initial and the end weight delineates purge loss; Shear = Warner–Bratzler shear force, kg; Moist = meat moisture or water content; IMF = intra muscular fat content measured chemically; Juicy = subjective 1–5 juiciness score where 1 = very dry and 5 = very juicy; Tender = subjective 1–5 tenderness score where 1 = very tough and 5 = very tender; BF = backfat thickness measured with FOM; Days = life-time average daily gain, grams per day.

TABLE 12.5
Common Factorial Analysis of pH Ultimate vs. Pork Quality Traits[a]

Trait	Factor 1 (pH)	Factor 2 (Muscle Color)	Factor 3 (Marbling)
pH22	0.88	−0.25	−0.01
pH14	0.73	−0.15	−0.04
MinL	−0.08	0.83	0.09
Minb	−0.24	0.69	0.08
Color	0.38	−0.72	0.02
Marble	0.30	−0.06	0.54
Firm	0.59	−0.40	0.13
Moist	0.23	−0.07	−0.76
IMF	0.03	0.13	0.91

[a] Trait explanation: pH22 = pH measured 22 h postmortem; pH 14 = pH measured 14 h postmortem; MinL = Minolta lightness; Mina = Minolta redness; Minb = Minolta yellowness; Color = subjective 1–6 color score where 1 = extremely pale and 6 = extremely dark; Marble = subjective marbling, 1–10 score where 1 = devoid of fat and 10 = extreme fat deposition; Firm = subjective firmness 1–3 score where 1 = very soft and 3 = very firm; Moist = meat moisture or water content; IMF = intra muscular fat content measured chemically.

animal-originated food system is complex, and our food is safe, tasty, nutritious, abundant, diverse, convenient, and less costly than ever before. Contemporary food science and technology contributed greatly to the success of this modern food system. That has been accomplished by integrating biology, chemistry, physics, engineering, materials science, microbiology, nutrition, toxicology, biotechnology, genetics and genomics, computer science, and many other disciplines. Future aspects of controlling pork quality must focus on balancing desirable quality traits with the economics of pig production, processing, and distribution of pork products to ultimate consumers worldwide.

ACKNOWLEDGMENT

The scientific contribution of Andrew Coates, Alejandro Diestre, Brandon Fields, and Neal Matthews is greatly appreciated.

REFERENCES

Aguilar, I., Misztal, I., Johnson, D., Legarra, A., Tsuruta, S., and Lawlor, T. 2010. A unified approach to utilize phenotypic, full pedigree, and genomic information for genetic evaluation of Holstein final score. *J. Dairy Sci.* 93:743–752.

Apple, J.K., Maxwell, C.V., Brown, D.C., Friesen, K.G., Musser, R.E., Johnson, Z.B., and Armstrong, T.A. 2004. Effects of dietary lysine and energy density on performance and carcass characteristics of finishing pigs fed ractopamine. *J. Anim. Sci.* 82:3277–3287.

Barbut, S., Sosnicki, A.A., Lonergan, S., Knapp, T., Ciobanu, D.C., Gatcliffe, L.J., Huff-Lonergan, E., and Wilson, E.W. 2008. Progress in reducing pale, soft, exudative (PSE) problem in pork and poultry meat. *Meat Sci.* 79: 46–63.

Barton-Gade, P.A. 1989. The effect of breed on meat quality characteristics in pigs. *35th International Congress of Meat Science and Technology*, Danish Meat Research Institute, Roskilde, Denmark.

Binder, B., Ellis, M., Brewer, M.S., Campion, D., Wilson, E.R., and McKeith, F.K. 2004. Effect of ultimate pH on the quality characteristics of pork. *J. Muscle Foods* 15:139–154.

Boyd, R.D., Johnston, M.E., Scheller, K., Sosnicki, A.A., and Wilson, E.R. 1997. Relationship between dietary fatty acid profile and body composition in growing pigs. *PICUSA T&D Technical Memo 153*. Pig Improvement Company, Franklin, KY.

Branscheid, W., and Dobrowolski, A.1999. Evaluation of market value: comparison between different techniques applied on pork carcasses. *Arch. Anim. Breed.* 43:131–137.

Brewer, M.S., Jensen, J., Sosnicki, A.A., Fields, B., Wilson, E., and McKeith, F.K. 2002. The effect of pig genetics on palatability, color and physical characteristics of fresh pork chops. *Meat Sci.* 61:249–256.

Brewer, M.S., Zhu, L.G., and McKeith, F.K. 2001. Marbling effects on quality characteristics of pork loin chops: Consumer purchase intent, visual and sensory characteristics. *Meat Sci.* 59:153–163.

Bonneau, M. 1982. Compounds responsible for boar taint, with special emphasis on androstenone: A review. *Livest. Prod. Sci.* 9(6):687–705.

Bonneau, M. 1998. Use of entire males for pig meat in the European Union. *Meat Sci.* 49(Suppl. 1):S257–S272.

Cameron, N.D., and Enser, M.B. 1991. Fatty acid composition of lipid in Longissimus Dorsi muscle of Duroc and British Landrace pigs and its relationship with eating quality. *Meat Sci.* 29:295–307.

Cameron, N.D., Warriss, P.D., Porter, S.J., and Enser, M.B. 1990. Comparison of Duroc and British Landrace pigs for meat and eating quality. *Meat Sci.* 27:227–247.

Chang, K.C., da Costa, N., Blackley, R., Southwood, O., Evans, G., Plastow, G., Wood, J.D., and Richardson, R.I. 2003. Relationships of myosin heavy chain fibre types to meat quality traits in traditional and modern pigs. *Meat Sci.* 64:93–103.

Channon, H.A., Kerr, M.G., and Walker, P.J. 2004. Effect of Duroc content, sex and ageing period on meat and eating quality attributes of pork loin. *Meat Sci.* 66:881–888.

Claus, R., Weiler, U., and Herzog, A. 1994. Physiological aspects of androstenone and skatole formation in the boar—A review with experimental data. *Meat Sci.* 38:289–305.

Deloitte Consumer Business. 2007. Profitable growth and value creation in the meat industry. www.deloitte.com.

D'Souza, D.N., Dunshea, F.R., Hewitt, R.J.E. et al. 2011. High boar taint risk in entire male carcases. In van Barneveld, R.J. (ed.), *Manipulating Pig Production XIII*.Australasian Pig Science Association, Werribee, Australia, Adelaide, South Australia.

D'Souza, D.N., and Mullan, B.P. 2002. The effect of genotpye, sex and management strategy on the eating quality of pork. *Meat Sci.* 60:95–101.

Desmoulin, B., and Bonneau, M. 1982. Consumer testing of pork and processed meat from boars: The influence of fat androstenone level. *Livest. Prod. Sci.* 9:707–715.

Dunshea, F.R., Colantoni, C., Howard, K. et al. 2001. Vaccination of boars with a GnRH vaccine (Improvac) eliminates boar taint and increases growth performance. *J. Anim. Sci.* 79:2524–2535.

de Vries, A.G., Faucitano, L., Sosnicki, A., and Plastow, G.S. 2000a. The use of gene technology for optimal development of pork meat quality. *Food Chem.* 69:397–405.

de Vries, A.G., Faucitano, L., Sosnicki, A., and Plastow, G.S. 2000b. Influence of genetics on pork quality. In Wenk, C., Fernandez, J.A., Dupuis, M. (eds.), *Quality of Meat and Fat in Pigs as Affected by Genetics and Nutrition*. Wageningen Press, Wageningen, the Netherlands.

Duniec, H. 1961. Heritability of chemical fat content in the loin muscle of baconers. *Anim. Prod.* 3:195–198.

Einarsson, S. 2006. Vaccination against GnRH: Pros and cons. *Acta Vet. Scan.* 48(Suppl 1):S10.

Ellis, M., and Bertol, T.M. 2001. Effects of slaughter weight on pork and fat quality. *Second International Virtual Conference on Pork Quality.* University of Illinois, Urbana-Champaign, USA.

Ellis, M., Lympany, C., Haley, C.S., Brown, I., and Warkup, C.C. 1995. The eating quality of pork from Meishan and Large White pigs and their reciprocal crosses. *Anim. Sci.* 60(1):125–131.

Floros, J.D., Newsome, R., Fisher, W. et al. 2010. *Comprehensive Reviews in Food Science and Food Safety* 9(5):572–599.

FAO. 2011. OECD-FAO *Agricultural Outloo*k 2011–2020, OECD Publishing and FAO. http://dx.doi.org/10.1787/agr_outlook-2011.

Fernandez, A., de Pedro, E., Nenez, N., Silio, L., Garcia-Casco, J., and Rodriquez, C. 2003. Genetic parameters for meat and fat quality and carcass composition traits in Iberian pigs. *Meat Sci.* 64:405–410.

Font-i-Furnols, M., Tous, N., Esteve-Garcia, E., and Gispert, M. 2012. Do all the consumers accept marbling in the same way? The relationship between eating and visual acceptability of pork with different intramuscular fat content. *Meat Sci.* 91:448–453.

Fortin, A., Robertson, W.M., and Tong, A.K.W. 2005. The eating quality of Canadian pork and its relationship with intramuscular fat. *Meat Sci.* 69:297–305.

Gentry, J.G., McGlone, J.J., Blanton Jr. J.R., and Miller, M.F. 2002a. Alternative housing systems for pigs: Influences on growth, composition, and pork quality. *J. Anim. Sci.* 80:1781–1790.

Gentry, J.G., McGlone, J.J., Miller, M.F., and Blanton Jr. J.R. 2002b. Diverse birth and rearing environment effects on pig growth and meat quality. *J. Anim. Sci.* 80:1707–1715.

Gentry, J.G., McGlone, J.J., Miller, M.F., and Blanton, Jr. J.R. 2004. Environmental effects on pig performance, meat quality and muscle characteristics. *J. Anim. Sci.* 82:209–217.

Guy, J.H., and Edwards, S.A. 2002. Consequences for meat quality of producing pork under organic standards. *Pig News and Information* 23(3):75N–80N.

Gooch, M., and Felfel, A. 2008. Characterizing the ideal model of value chain management and barriers to its implementation. *George Morris Center.* 225–150. Research Lane, Guelph, Ontario, Canada. www.georgemorris.com.

Gower, D.B. 1972. 16-Unsaturated C 19 steroids. A review of their chemistry, biochemistry and possible physiological role. *J. Steroid Biochem.* 3(1):45–103.

Gusse, M.D. 1993. *A Comparison and Prediction of Carcass and Cut Moisture Loss for PSE, Normal and DFD Pork.* MS thesis. Univ. Illinois, Urbana-Champaign.

Hansen, L.L., Claudi-Magnussen, C., Jensen, S.K., and Andersen, H.J. 2006. Effect of organic pig production systems on performance and meat quality. *Meat Sci.* 74:605–615.

Hansen, L.L., Stolzenbach, S., Jensen, J.A., Henckel, P., Hansen-Møller, J., Syriopoulos, K., and Byrne, D.V. 2008. Effect of feeding fermentable fibre-rich feedstuffs on meat quality with emphasis on chemical and sensory boar taint in entire male and female pigs. *Meat Sci.* 80:1165–1173.

Harris, D.L., and Newman, S. 1994. Breeding for profit: Synergism between genetic improvement and livestock production (a review). *J. Anim. Sci.* 72:2178–2200.

Henderson, C.R. 1975. Best linear unbiased estimation and prediction under a selection model. *Biometrics* 31(2):423–448.

Hofmann, K. 1994. What is quality? *Meat Focus Int.* 2:73–82.

Honikel, K.O. 1993. Quality of fresh pork—Review. In Puolanne, E., Demeyer, D.I., Ruusunen, M., Ellis, S. (eds.), *Pork Quality: Genetic and Metabolic Factors.* CAB International, Oxfordshire, UK.

Honikel, K.O. 1998. Reference methods for the assessment of physical characteristics of meat. *Meat Sci.* 49:447–457.

Hovenier, R. 1993. *Breeding for Meat Quality in Pigs.* PhD thesis, Department of Animal Breeding, Wageningen, Agricultural University, The Netherlands.

Huff-Lonergan, E., Baas, T.J., Malek, M., Dekkers, J.C.M., Prusa, K., and Rothschild, M.F. 2002. Correlations among selected pork quality traits. *J. Anim. Sci.* 80:617–627.

Hviid, M., Barton-Gade, P., Oksama, M., and Aaslyng, M.D. 2002. Effect of using Pietrain, Duroc or HD as sire line on eating quality in pork loin. *7th World Congress on Genetics Applied to Livestock Production. Montpellier*, France, INRA, Session 11, pp. 0–4.

Innova Analysis: Briefing Series, 2013. The Future of Meat: Coping With Shifting Market Dynamics and Supply Pressures. Feb 2013.

Jaros, P., Burgi, E., Stark, K.D.C., Claus, R., Hennessy, D., and Thun, R. 2005. Effect of active immunization against GnRH on androstenone concentration, growth performance and carcass quality in intact male pigs. *Livest. Prod. Sci.* 92:31–38.

Jensen, M.T., and Hansen, L.L. 2006. Feeding with chicory roots reduces the amount of odorous compounds in colon and rectal contents of pigs. *Anim. Sci.* 82:369–376.

Jensen, W.K. (ed.), 1998. Skatole and boar taint: Results from an integrated national research project investigating causes of boar taint in Danish pigs. Danish Meat Research Institute, Roskilde.

Jeremiah, L.E., Gibson, J.P., Gibson, L.L., Ball, R.O., Aker, C., and Fortin, A. 1999a. The influence of breed, gender and PSS (halothane) genotype on meat quality, cooking loss and palatability of pork. *Food Res. Int.* 32(1):59–71.

Jeremiah, L.E., Sather, A.P., and Squires, E.J. 1999b. Gender and diet influences on pork palatability and consumer acceptance. I. Flavor and texture profiles and consumer acceptance. *J. Muscle Foods* 10:305–316.

Jeremiah, L.E., and Weiss, G.M. 1984. The effects of slaughter and sex on the cooking losses from and palatability attributes of pork loin chops. *Canadian J. Anim.* Sci. 64:39–43.

Kauffman, R.G., Eikelenboom, G., Van Der Wal, P.G., Merkus, G., and Zaar, M. 1986. The use of filter paper to estimate drip loss in porcine musculature. *Meat Sci.* 18:191–200.

Kauffman, R.G., Sybesma, W., and Eikelenboom, G. 1990. In search of quality. *J. Inst. Can. Sci. Technol. Aliment.* 23(4/5):160–164.

Kim, G.-D., Ryu, Y.-C., Jo, C., Lee, J.-G., Yang, H.S., Jeong, J.-Y., and Joo, S.-T. 2014. The characteristics of myosin heavy chain-based fiber types in porcine longissimus dorsi muscle. *Meat Sci.* 96:712–718.

Knapp, P., William, A., and Solkner, J. 1997. Genetic parameters for lean meat content and meat quality traits in different pig breeds. *Livest. Prod. Sci.* 52:69–73.

Lambooij, E., Hulsegge, B., Klont, R.E., Winkelman-Goedhart, H.A., Reimert, H.G.M., and Kranen, R.W. 2004. Effects of housing conditions of slaughter pigs on some post mortem muscle metabolites and pork quality characteristics. *Meat Sci.* 66:855–862.

Latorre, M.A., Lazaro, R., Valencia, D.G., Medel, P., and Mateos, G.G. 2004. The effects of gender and slaughter weight on the growth performance, carcass traits, and meat quality characteristics of heavy pigs. *J. Anim. Sci.* 82:526–533.

Lonergan, S.M., Huff-Lonergan, E., Dekkers, J.C.M., and Rothschild, M.F. 2014. Relationship between gilt behavior and meat quality using principal component analysis. *Meat Sci.* 96:264–269.

Lonergan, S.M., Huff-Lonergan, E., Rowe, L.J., Kuhlers, D.L., and Jungst, S.B. 2001. Selection for growth efficiency on Duroc pigs influences pork quality. *J. Anim. Sci.* 79:2075–2085.

Lealiifano, A.K., Pluske, J.R., Nicholls, R.R., Dunshea, F.R., Campbell, R.G., Hennessy, D.P., Miller, D.W., Hansen, C.F., and Mullan, B.P. 2011. Reducing the length of time between harvest and the secondary gonadotropin-releasing factor immunization improves growth performance and clears boar taint compounds in male finishing pigs. *J. Anim. Sci.* 89:2782–2792.

Leong, J., Morel, P.C.H., Purchas, R.W., and Wilkinson, B.H.P. 2011. Effects of dietary components including garlic on concentrations of skatole and indole in subcutaneous fat of female pigs. *Meat Sci.* 88:45–50.

Lunde, K., Skuterud, E., Hersleth, M., and Egelandsdal, B. 2010. Norwegian consumers' acceptability of boar tainted meat with different levels of androstenone or skatole as related to their androstenone sensitivity. *Meat Sci.* 86:706–711.

Lundström, K., Matthews, K.R., and Haugen, J.-E. 2009. Pig meat quality from entire males. *Animal* 3:1497–1507.

Martel, J., Minvielle, F., and Poste, L.M. 1988. Effects of crossbreeding and sex on carcass composition, cooking properties and sensory characteristics of pork. *J. Anim. Sci.* 66:41–46.

Martinez, S.W. 1999. Vertical coordination in the pork and broiler industries: Implications for pork and chicken products. *Food and Rural Economics Division. Economic Research Service, U.S. Department of Agriculture. Agricultural Economic.* Report No. 777.

Martinez, S.M., and Zering, K. 2004. Pork quality and the role in market organizations. *United States Department of Agriculture.* AER. 835:10.

Matthews, K.R., Homer, D.B., Punter, P. et al. 2000. An international study on the importance of androstenone and skatole for boar taint: III. Consumer survey in seven European countries. *Meat Sci.* 54:271–283.

Matthews, N., Greiner, L., Neill, C.R., Fields, B., Jungst, S., Johnson, R.C., and Sosnicki, A. 2014. Effect of feed form (mash vs. pellets) and ractopamine on pork fat quality. *J. Anim. Sci.* 92(Suppl. 2):148 Abstr.

Mayes, P.A. 1996. Biosynthesis of fatty acids. In Murray, R.K., Granner, D.K., Mayes, P.A., and Rodwell, V.W. (eds.), *Harper's Biochemistry*. 24th edition. SouthMimms, England. Universities Federation for Animal Welfare. pp. 216–223.

Merks, J.W.M., Hanenberg, E., Bloemhof, S. and Knol, E.F. 2009. Genetic opportunities for pork production without castration. *Anim. Welfare* 18(4):539–544.

Minolta Co., Ltd. 1994. *Precise Color Communication.*

Morel, P. 1995. Meat quality series: Relationship between eating quality, backfat thickness and intramuscular fat. *Newsbrief (Australia),* April: 2.

Morrison-Paul, C.J. 2001. Cost economics and market power. The case of the U.S. meat packing industry. *Rev. Econ. Stat.* 83:531.

McGloughlin, P., Allen, P., Tarrant, P.V., Joseph, R.L., Lynch, P.B., and Hanrahan, T.J. 1988. Growth and carcase quality of crossbred pigs sired by Duroc, Landrace and Large White boars. *Livest. Prod. Sci.* 18:275–288.

Millet, S., Raes, K., Van den Broeck, W., De Smet, S., and Janssens, G.P.J. 2005. Performance and meat quality of organically versus conventionally fed and housed pigs from weaning till slaughtering. *Meat Sci.* 69:335–341.

Nakai, H., Saito, F., Ikeda, T., Ando, S., and Komatsu, A. 1975. Standard models of pork color. *Bull. Natl. Inst. Anim. Industr.* 29:69–74.

NPPC. 1999. *Composition & Quality Assessment Procedures.*

NRC. 2012. *Nutrient Requirements of Swine*. 11th Rev. edition. National Academy Press, Washington D.C.

Olsson, V., Andersson, K., Hansson, I., and Lundstrom, K. 2003. Differences in meat quality between organically and conventionally produced pigs. *Meat Sci.* 64:287–297.

Olsson, V., and Pickova, J. 2005. The influence of production systems on meat quality, with emphasis on pork. *Ambio* 34(4/5):338–343.

Piao, J.R., Tian, J.Z., Kim, B.G., Choi, Y.I., Kim, Y.Y., and Han, I.K. 2004. Effects of sex and market weight on performance, carcass characteristics and pork quality of market hogs. Asian-Austrlasian *J. Anim. Sci.* 17:1452–1458.

Plastow, G.S., Carrion, D., Gil, M. et al. 2005. Quality pork genes and meat production. *Meat Sci.* 70:409–421.

Purchas, R.W., Smith, W.C., and Pearson, G. 1990. A comparison of the Duroc, Hampshire, Landrace and Large White as terminal sire breeds of crossbred pigs slaughtered at 85 kg liveweight. 2. Meat quality. *NZ J. Agric. Res.* 33:97–104.

PIC. 2008. Common factors that determine pork quality. *Cutting Edge*, Second Quarter 2008. www.genusplc.com.

PIC. 2010. Growth Curves for PIC337RG sired pigs fed high and low energy diets. Executive Summary 051—Tech. Memo 344. PIC North America, Hendersonville, TN. www.genusplc.com.

Prunier, A., Bonneau, M., and Etienne, M. 1987. Effects of age and live weight on the sexual development of gilts and boars fed two planes of nutrition. *Reprod. Nutr. Dev.* 27:689–700.

Rabobank International. 2006. *Pork in the Third Millennium: Outlook for the Future.* www.rabobank.com/en/home/index.html

Ramaswami, A.M., Jayaprasad, I.A., Shanmugam, A.M., and Abraham, R.J.J. 1993. Influence of slaughter weight on eating quality of pork. *Cheiron* 22(4):126–130.

Reiner, J.J. 2006. Vertical integration in the pork industry. *Am. J. Agric. Econ.* 88(1):234–248.

Ros-Freixedes, R., Sadler, L.J., Onteru, S.K., Smith, R.M., Young, J.M., Johnson, A.K., and Simmons, J. 2010. Why agriculture needs technology to help meet a growing demand for safe, nutritious and affordable food? *Elanco Anim. Health.*

Rotshchild, M.F., and Ruvinsky, A. 2010. *The Genetics of the Pig*. CABI, Oxfordshire, UK.

Sellier, P. 2010. Genetics of meat and carcass traits. In Rotshchild, M.F., Ruvinsky, A. (eds.), *The Genetics of the Pig*. CABI, Oxfordshire, UK.

Sellier, P., and Monin, G. 1994. Genetics of pig meat quality: A review. *Meat Focus Int.* 5(2):287–219.

Sosnicki, A.A., and Newman, S. 2010. The support of meat value chains by genetic technologies. *Meat Sci.* 86(1):129–137.

Sosnicki, A.A., Pommier, S., Klont, R., Newman, S., and Plastow, G.S. 2003. Best-cost production of high quality pork: Bridging the gap between pig genetics, muscle biology/meat science and consumer trends. In *Proceedings of Manitoba Pork Seminar*, Winnipeg, Manitoba, July 28–30, 2003, Vol. 17, pp. 1–11.

Straadt, I.K., Aaslyng, M.D., and Bertram H.C. 2013. Sensory and consumer evaluation of pork loins from crossbreeds between Danish Landrace, Yorkshire, Duroc, Iberian and Mangalitza. *Meat Sci.* 95:27–35.

Suzuki, K., Irie, M., Kadowaki, H., Shibata, T., Kumagai, M., and Nishida, A. 2005. Genetic parameter estimates of meat quality traits in Duroc pigs selected for average daily gain, longissimus muscle area, backfat thickness, and intramuscular fat content. *J. Anim. Sci.* 83:2058–2065.

Suzuki, K., Nakagawa, M., Katoh, K., Kadowaki, H., Shibata, T., Uchida, H., Obara, Y., and Nishida, A. 2004. Genetic correlation between serum insulin-like growth factor-1 concentration and performance and meat quality traits in Duroc pigs. *J. Anim. Sci.* 82(4):994–999.

Suzuki, K., Shibata, T., Kadowaki, H., Abe, H., and Toyoshima, T. 2003. Meat quality comparison of Berkshire, Duroc and crossbred pigs sired by Berkshire and Duroc. *Meat Sci.* 64:35–42.

The Boston Consulting Group, Inc. 2001. Procurement: An Untapped Opportunity for Improving Profits.

The Boston Consulting Group, Inc. 2006. The Battle for Europe's Grocery Shoppers.

Trienekens, J., Petersen, B., Wognum, N., and Brinkmann, D. (eds.), *Diversity and Quality Challenges in Consumer-Oriented Production and Distribution.* Wageningen Academic Publishers.

United States Department of Agriculture. 2012. *USDA Agricultural Projections to 2021* (OCE-2012-1). Office of the Chief Economist, World Agricultural Outlook Board, Long-term Projections Report, OCE-2012-1; February 2012. www.ntis.gov.

Villegas, F.J., Hedrick, H.B., Veum, T.L., McFate, K.L., and Bailey, M.E. 1973. Effect of diet and breed on fatty acid composition of porcine adipose tissue. *J. Anim. Sci.* 36:663–668.

Walstra, P., Claudi-Magnussen, C., Chevillon, P., von Seth, G., Diestre, A., Matthews, K.R., Homer, D.B., and Bonneau, M. 1999. Skatole and androstenone levels in entire male pigs: Seasonal effects and differences between six European countries. *Livest. Prod. Sci.*62:15–28.

Walstra, P., and Merkus, G.S.M. 1995. *Procedure for the Assessment of Lean eat Percentage as a Consequence of the New EU Reference Dissection Method in Pig Carcass classification. DLO Research Institute of Animal Science and Health* (IDDLO). Zeist, The Netherlands.

Warkup, C.C., Dilworth, A.W., Kempster, A.J., and Wood, J.D. 1990. The effect of sire type, company source, feeding regime and sex on eating quality of pig meat. *Anim. Prod.* 52:559.

Warriss, P.D., Kestin, S.C., Brown, S.N., and Nute, G.R. 1996. The quality of pork from traditional pig breeds. *Meat Focus Int.* May/June 1996, 179–182.

Wilson, E.R., Johnson, R.K., and Wetterman, R.P. 1977. Reproductive and testicular characteristics of purebred and crossbred boars. *J. Anim. Sci.* 44:939–947.

Witte, D.P., Ellis, M., McKeith, F.K., and Wilson, E.R. 2000. Effect of dietary lysine Level and environmental temperature during the finishing phase on intramuscular fat content of pork. *J. Anim. Sci.* 78:1272–1276.

Wood, J.D., Brown, S.N., Nute, G.R., Whittington, F.M., Perry, A.M., Johnson, S.P., and Enser, M. 1996. Effects of breed, feed level and conditioning time on the tenderness of pork. *Meat Sci.* 44(1/2):105–112.

Wood, J.D., Enser, M., Whittington, F.M., Moncrieff, C.B., and Kempster, A.J. 1989. Backfat composition in pigs: Differences between fat thickness groups and sexes. *Livest. Prod. Sci.* 22:351–362.

Wood, J.D., Nute, G.R., Fursey, G.A.J., and Cuthbertson, A. 1995. The effect of cooking conditions on the eating quality of pork. *Meat Sci.* 40:127–135.

Wood, J.D., Nute, G.R., Richardson, R.I., Whittington, F. M., Southwood, O., Plastow, G., Mansbridge, T., da Costa, N., and Chang, K.C. 2004. Effects of breed, diet and muscle on fat deposition and eating quality in pigs. *Meat Sci.* 67:651–667.

13 Sheep Quality

Effect of Breed, Genetic Type, Gender, and Age on Meat Quality

David Hopkins

CONTENTS

13.1 INTRODUCTION

Meat quality includes many factors including palatability, water-holding capacity, color, and nutritional value (Hopkins and Geesink 2009) and it can be affected by changing the genetics and the production and processing environments. The relative importance of meat quality traits varies according to the user of the product and the type of product. For example, tenderness is more important for beef (Thompson 2002) than sheep meat (Hopkins et al. 2005b). As improvements are made in individual traits their relative importance changes (Thompson 2004), which impacts on their emphasis in breeding programs. Market research indicates that meat quality traits are becoming more important to consumers (Pethick et al. 2006; Bermingham et al. 2008) and this will increase the focus on methods to improve them.

Genetic change can occur through cross-breeding and selection for quantitative traits directly (using phenotypic records and pedigree), or using marker-assisted selection and genomic selection. The void of information on genetic variation for meat quality traits in sheep was highlighted by Safari et al. (2005), who reported only two estimates of heritability for pH and meat color, both for Merino rams. In recent years, more information has become available for both genetic variation (Mortimer et al. 2010, 2014) and major gene effects on meat quality traits and for the use of molecular markers (see Chapter 10). Marker-assisted selection has the potential to significantly increase the rate of gain from selection for meat traits (Meuwissen and Goddard 1996). In Australia, large data sets have been generated from the Cooperative Research Centre for Sheep Industry Innovation (Sheep CRC) Information Nucleus (Fogarty et al. 2007; van der Werf et al. 2010) and the Sheep Genomics (Oddy et al. 2007) programs. They are providing estimates of genetic parameters for a large range of traits including meat quality as well as developing molecular markers and evaluating whole genome selection using single nucleotide polymorphism technology. As highlighted by Fogarty (2009) the development of strategies to combine quantitative and molecular information into effective breeding programs is required and this is beginning to occur.

Other production factors that can impact on meat quality and which will be considered in this chapter are sex (gender) and animal age. Nutritional aspects will be considered elsewhere (Chapter 4) and processing factors have been documented previously (Hopkins 2010). The effect of gender on traits like tenderness is not clear with no effect reported in some studies (e.g., Kemp et al. 1981; Lee 1986), whereas others have shown meat from entire male lambs (Johnson et al. Blair 2005) or castrates (Hopkins et al. 2007a) to be tougher than that from ewe lambs. The impact of animal age on meat quality traits is of particular importance as it can affect marketing decisions, but clarifying the extent of "true" age effects is not straightforward as highlighted by Purchas (2007). This is because often older animals are also heavier and this can impact on cooling rates, thus pH decline and subsequently traits such as tenderness and color.

The purpose of this chapter is to discuss the impact of sheep genotypes and genetics (breeds, cross-breeds, and genetic parameters), gender, and animal age on meat quality traits of lamb meat. The review of Hopkins and Mortimer (2014) will form the foundation of the chapter.

13.2 EFFECT OF GENOTYPES ON MEAT QUALITY

13.2.1 IMPACT ON TENDERNESS

Tenderness can be evaluated by objectively measuring shear force (Hopkins et al. 2010) and using trained panelists (Safari et al. 2001) or consumers (Hopkins et al. 2005b). The differences between genotypes may vary with the method used, as each detects subtle differences in tenderness. Some studies have shown either no differences in objectively measured tenderness between breeds and crossbreds (Dransfield et al. 1979; Hopkins and Fogarty 1998; Hopkins et al. 2005b, 2007a) or inconsistent differences that were not explained by variation in other traits that influence tenderness, such as pH, sarcomere length, carcass weight, or fat levels (Purchas et al. 2002). Different strategies have been used to minimize the impact of processing on tenderness, including conditioning (holding at temperatures above chilling for a period of time) after slaughter and aging (Dransfield et al. 1979), electrical stimulation and aging (Hopkins et al. 2005b), and aging for 7 days (Hopkins and Fogarty 1998). Such approaches are needed to estimate genetic variation because of the potential confounding due to processing factors.

No sire breed effects on taste panel-assessed tenderness were reported by Dransfield et al. (1979) or Safari et al. (2001) in comparisons of Merino lambs and other breeds, including Texel × Merino or Poll Dorset × Merino (PDM). Hopkins et al. (2005c) reported minimal differences in consumer-assessed tenderness between genotypes, except that the Merinos had lower sensory scores than Border Leicester × Merino (BLM) lambs for two different muscles, which may have reflected a slower rate of pH decline in the Merino lambs. More recent work by Pannier et al. (2014a) showed by contrast that male Terminal (meat breeds) sired lambs had lower tenderness scores (~5 points on a 0–100 scale) for the loin and topside compared to the male Maternal and Merino sired lambs which had similar scores. This effect could reflect the fact that the Terminal sires used by Pannier et al. (2014a) had estimated breeding values that indicated that these sires were on average leaner than their breed average. It is known that this can lead to a decline in tenderness (Hopkins et al. 2007b), but it should also be stressed that the manifestation of effects in progeny will be influenced by processing conditions as demonstrated by Hopkins et al. (2007b).

Rambouillet lambs produced tougher leg steaks than Karakul and crossbred (Suffolk or Hampshire × Rambouillet) lambs (Edwards et al. 1982), although the reason cannot be confirmed as other traits, such as pH, were not reported. This was also the case in a comprehensive study of lighter weight lambs, in which Merino lambs were rated more tender by trained panelists than Rasa Aragonesa and Churra breeds as slaughter weight increased (Martínez-Cerezo et al. 2005). Merino lambs had the most tender *M. longissimus* in the work of Young et al. (1993b), which was attributed to significantly higher pH, although this was not found by Hopkins and Fogarty (1998).

There were no effects of sire breed on sensory tenderness of lamb from three sire breeds (Charollais, Suffolk, and Texel) sampled over three years (Ellis et al. 1997). Similarly, Esenbuga et al. (2001) found no difference in shear force or sensory-assessed tenderness between four fat-tailed types (Awassi, Red Karaman, Tushin, and Awassi × Tushin) when slaughtered at similar weights. Likewise, Hoffman et al. (2003) reported tougher meat (shear force of the *M. semimembranosus*; SM) for only one of the six genotypes they studied, with the effect associated with the dam breed (Dohne Merino), although it did not affect sensory traits. In hill breeds, Carson et al. (2001) reported no difference in shear force of loin meat from six genotypes, although the low absolute shear force values indicate that the meat had been aged for an extended period, which may have reduced any differences between the genotypes.

Inconsistent effects were reported for taste panel tenderness of roasted hind leg lamb meat from three Greek dairy breeds (Arsenos et al. 2002), although slaughter days were confounded with breed and few animals were evaluated. In a larger study, Navajas et al. (2008) reported a reduction in taste panel tenderness for both the loin and the SM from pure Texel compared to Scottish Blackface lambs. The authors suggested that it was due to the lower intramuscular fat (IMF) levels in the Texel, although it was not analyzed and surprisingly there was no difference between breeds for muscularity, with the latter trait derived from computer tomography measures of the hindleg.

In a large study across six countries, lamb meat from the Icelandic breed was the most tender, whether determined by objective or subjective means, whereas the Bergamasca breed was the toughest (Berge et al. 2003; Sañudo et al. 2003). However, the data suggested that some of the effect was due to differences in sarcomere length (Berge et al. 2003), final pH, and lambs raised under different production systems and slaughtered in different countries over a wide range in carcass weights (5.4–30.5 kg). In another study, genotype was confounded with feeding system and age (Fisher et al. 2000). These reports are not informative for understanding any genetic differences in tenderness between genotypes and overall no large genotype effects on tenderness are apparent. There is a need for more controlled studies, where sources of variation are controlled so the true influence of genotype is quantified. In this regard the recent work reported by Mousel et al. (2014) is a good example of a study where sufficient sire representation was applied and the meat was aged before shear testing. As such no sire breed effect on shear force was reported.

13.2.2 Impact on Eating Quality

Young et al. (1993b) reported no differences in the juiciness, flavor, and overall acceptability of loin meat from six genotypes when tested by trained panelists. A comparison of roasted legs from Romney, Border Leicester × Romney, Perendale, Corriedale, and Merino animals by Kirton et al. (1974) found that those from Merinos rated the lowest for overall preference, although they had very light carcass weights with minimal fat cover. Safari et al. (2001) reported no difference in overall acceptability, tenderness, or juiciness for roasted loin meat from first cross (BLM), Merino, or second cross lambs. In another study of the hindleg (*M. biceps femoris)*, Merino lambs had lower juiciness, flavor liking, and overall liking scores than BLM and second cross lambs, but were similar to PDM lambs (Hopkins et al. 2005b) when assessed by consumers.

Dransfield et al. (1979), Edwards et al. (1982), Crouse et al. (1981), Crouse (1983), Ellis et al. (1997), and Esenbuga et al. (2001) reported no significant differences between genotypes in eating quality. Hoffman et al. (2003) did find that the initial juiciness of the SM was lower from Suffolk × Merino than other genotypes, but it was of no practical significance. In other work from South Africa, Webb et al. (1994) reported that roasted loin meat from South African Mutton Merino (SAMM) lambs had better flavor and overall acceptability than from Dorper lambs. This was attributed to the significantly higher fat levels in the Dorper, but it is noteworthy that the subcutaneous fat of the Dorpers also had higher levels of unsaturated fatty acids which the authors suggested could have depressed the flavor scores of the meat. The results may also have reflected the composition of the taste panel which has been shown to impact on scores even when they are trained (Sañudo et al. 1998). The work of Cloete et al. (2012) reported that on first bite loin meat from Dohne Merino sheep was more tender than that from Merino, SAMM, and Dormer sheep (20 months of age). This was consistent with the ranking for shear force for these genotypes, but the small numbers of animals sampled does limit interpretation. A decrease in overall liking for meat from Texel over Scottish Blackface lambs (Navajas et al. 2008) seems to have been due to differences in tenderness, with inconsistent differences for juiciness across the various cuts. The impact of IMF on eating quality was clearly shown in the work of Komprda et al. (2012) and Jandesk et al. (2014). In the work of Komprda et al. (2012) lambs sired by Zwartbles rams exhibited much lower eating quality (tenderness and juiciness) in the *quadriceps femoris* muscle (round or knuckle) than Suffolk or Oxford Down sired lambs and the level of IMF was also significantly lower, 1.7% versus 3.1% and 2.8% respectively, but interestingly the Zwartble sired progeny had the lowest collagen concentration. The results of Jandasek et al. (2014) concurred with this where Charollais sired lambs had significantly higher levels of IMF in the *M. longissimus* than Oxford Down, Texel, Suffolk, and Merinolandschaf sired lambs, which was matched by a better eating quality (texture and juiciness) in meat from Charollais sired lambs, although there was some overlap in sensory scores between sire breeds. Overall, there does not appear to be consistent genotype effects *per se* on eating quality traits, but it is clear that lower levels of IMF will reduce eating quality.

13.2.3 Impact on IMF

For lamb, it was suggested that a target of 5% IMF is required to help ensure a "good every day" (grade 3 out of a 5 level system) score despite the literature suggesting that IMF is only one component of eating quality (Hopkins et al. 2006). In more recent work by Pannier et al. (2014a) the importance of IMF for eating quality was further supported. For all sensory traits, increasing levels of loin IMF were associated with increasing sensory scores within both loins and topsides. Across the 4.5% IMF range (2.5%–7.0%), the sensory scores for the loins increased by 10.7, 10.0, 9.1, and 5.9 units for juiciness, overall liking, flavor, and tenderness respectively on a 0–100 scale. For the topside, the response was 6.7 and 6.6 for juiciness and overall liking respectively. These authors also reported that the 5% target may be unnecessarily high and suggested that a level of 3.9% is sufficient to ensure that the "good every day" grade is achieved. Either way, IMF has important effects on organoleptic traits.

There were no differences in loin IMF between Merino and various crossbred types with a mean level of 4.8% (Hopkins et al. 2005b). However, Díaz et al. (2005) showed variation in IMF between breeds for loins, but the comparison was confounded by differences in production systems. There were some differences in IMF between five genotypes (Merino, BLM, and first and second cross types with sires having different breeding values for growth and muscling) in a rigorous study in which the animals were run together and slaughtered at four ages from 4 to 22 months (McPhee et al. 2008). The BLM had the highest IMF across both ewes and wethers and the highest increases in IMF between 14 and 22 months. A dam effect of the BLM contrasted with Merino dams was further supported by recent data (Pannier et al. 2014b).

The claim for Merino sired lambs to have 0.42% more IMF than Terminal and Maternal sired lambs (Pethick et al. 2010) was contrary to Hopkins et al. (2007b) who found that Merino lambs actually had the lowest levels of IMF. Merinos in the Pethick et al. (2010) study were older than the comparative types and the more recent data of Pannier et al. (2014b) which also showed Merino sired progeny as having higher IMF levels, suggests again that the comparison is confounded by differences in age. In a study which examined several sire breeds with a good number of sires represented per breed, Charollais sired lambs had significantly higher levels of IMF in the *M. longissimus* (Jandasek et al. 2014) as already outlined. Lambe et al. (2008) reported Texel lambs selected for muscling and reduced fatness had very low IMF levels (1.60%) in loins compared to Scottish Blackface (2.3%), although the difference was unlikely to be significant. Indeed the low values for IMF suggest the meat was suboptimal for good eating quality. This report of Lambe et al. (2008) illustrates that in some specific cases extreme genotypes can be detrimental to traits such as eating quality and IMF. In light of these results, the sheep industry needs either screening systems to prevent the production of such genotypes, breeding programs for the stud sector that account for these animals or production systems that can elevate IMF levels as appropriate.

13.2.4 Impact on pH and Color

High mean pH values for meat affect keeping quality and can adversely affect flavor and aroma (Young et al. 1993b). Shelf life is reduced when pH exceeds 5.8 (Egan and Shay 1988) and as pH increases meat becomes darker (Fogarty et al. 2000), affecting consumer purchase decisions.

Higher muscle ultimate pH for Merino and BLM lambs compared to Terminal sired second cross lambs has been reported (Hopkins and Fogarty 1998; Gardner et al. 1999; Fogarty et al. 2000; Hopkins et al. 2007a). Under high stress commercial slaughter conditions, the Merino loses a greater amount of muscle glycogen than other types (Gardner et al. 1999), but under "low stress" slaughter the meat from Merinos can have a similar ultimate pH as that from other genotypes (Hopkins et al. 2005b). The *M. semitendinosus* muscle has more type 2X fast glycolytic fibers than the *M. longissimus* (Greenwood et al. 2007) and as the content of these fibers increases so do the final pH levels (Gardner et al. 2006). This effect was clearly evident in the Merino lambs studied by Hopkins et al. (2007a), which had a higher ultimate pH indicative of a faster rate of glycogen depletion during the preslaughter period. However, it does not appear that these Merinos were more agitated than the other genotypes as measured by an isolation test (Warner et al. 2006). Related to this recent work studying the behavior (flight speed and agitation scores) of more than 11,000 lambs at weaning found no increased flight speed or agitation in Merino lambs (Dodd et al. 2014). Hopkins et al. (2005a) reported significant differences in muscle pH among Merino wether bloodlines for both the *M. longissimus* and the *M. semitendinosus* with higher muscle pH for the Merino superfine wool bloodlines. In contrast, Fogarty et al. (2003) found that the broad wool strain of Merino rams had a higher loin pH, than both the fine and medium wool strains.

The higher muscle pH for Merinos compared to other genotypes was consistent across three different muscles for lambs sourced and slaughtered in New South Wales, but the effect was not found in lambs sourced and slaughtered in Victoria

(Hopkins et al. 2005b). The Victorian animals had a much greater weight gain than those in New South Wales which may have led to higher levels of muscle glycogen and, together with a lower stress slaughter environment, may have resulted in low pH meat from the Merinos. The results of Gardner et al. (1999) support this because Merinos slaughtered at a research abattoir did not exhibit high pH values. The higher ultimate pH reported by Hopkins et al. (2005b) cannot be ascribed to an increasing proportion of Merino genes as the pH levels for the BLM and Merino animals were similar for the loin at least, which is also supported by other reports (Hopkins and Fogarty 1998; Hopkins et al. 2007a).

However, Hopkins et al. (2007a) reported a higher muscle pH in the *M. semitendinosus* of Merino lambs compared with all other crosses including BLM lambs. The samples were aged for 5 days and measured after homogenization to ensure an accurate measure of ultimate pH, overcoming electrode placement issues which can arise *in situ*. A New Zealand experiment with various breeds of Terminal sires joined to Merino ewes, reported that loins from Merino lambs had a significantly higher pH than BLM and PDM lambs (Young et al. 1993b). Loin pH values were also greater for BLM compared to PD sired lambs (Fogarty et al. 2000; Hopkins et al. 2007a). Higher loin pH values have also been reported in lambs from Merino dams compared to those from Dohne Merino or SAMM dams (Hoffman et al. 2003), but a more recent report by Cloete et al. (2012) showed an inconsistent effect with pure bred Merinos and Dohne Merinos having significantly higher pH levels than SAMMs.

There is little evidence of differences in muscle pH between British breed crossbred lambs (Dransfield et al. 1979; Carson et al. 2001), although purebred Scottish Blackface lambs had higher loin pH levels than crossbred lambs in the latter study with no explanation for the result. Other studies involving purebreds and crossbreds have not found any differences in muscle pH levels with Romney and East Friesian breeds (Purchas et al. 2002), Elliottdale carpet wool breed (Hopkins et al. 1992) or crossbred lambs from Columbia, USMARC-Composite, Suffolk, or Texels sires (Mousel et al. 2014). Unfortunately, in some studies where genotype has been found to impact on traits such as pH and meat color there is inadequate detail provided (e.g., Cloete et al. 2012; Kuchtík et al. 2012) about sire breed representation to conclude whether the differences are a genuine breed effect or due to individual sires.

While meat from Merinos is more susceptible to high pH than meat from other types, there is little evidence of genetic type impacting on objectively measured fresh color (Dransfield et al. 1979; Fogarty et al. 2000, 2003; Hopkins et al. 2005b, 2007a). Even when Merino lambs produce meat with a higher pH than other types, they do not produce darker fresh meat, which may reflect the low phenotypic correlation between pH and L^* values (Menzies and Hopkins 1996). Loin muscle from Merinos browns quicker and to a greater extent through the formation of metmyoglobin than muscle from the other types (Warner et al. 2007) and this has been further confirmed by recent work (Jacob et al. 2014). In this latter case, data on color stability were obtained for 2700 lamb loins and three factors were found to explain the majority of the variation in the R630/R580 ratio (as a proxy for brownness development). These were breed type, pH at 24 h postslaughter and linoleic acid (LA) concentration, such that Merino breed type, high pH, and high LA reduced color stability (increased brownness). These results could explain the industry perception that meat from Merinos has poor color. Loin meat from Romney lambs was darker than that from crossbred Texel lambs (Purchas et al. 2002), which was independent of pH, with time of slaughter having a greater effect. The loin meat from these Romney lambs was below the acceptable threshold lightness value of 34 (*L**) established by consumer evaluation (Khliji et al. 2010). There is variation within the Merino breed for muscle pH, but the physiological basis of higher pH levels in the Merino remains still to be fully established and, given the importance of this breed to sheep production in a number of countries, is an area worthy of further research.

13.2.5 Impact on Iron, Zinc, and Omega-3 Fatty Acids

The importance of red meat in human diets in achieving recommended levels of iron (Fe), zinc (Zn), and omega-3 fatty acids has been proposed as a future key marketing tool (Pethick et al. 2006). Iron is a component of several important proteins such as hemoglobin and myoglobin, zinc is a component of various enzymes that helps maintain the structural integrity of proteins (see Chapter 1), regulates gene expression and omega-3 fatty acids, particularly eicosapentaenoic (EPA) and docosahexaenoic (DHA), and has cardiovascular and antiinflammatory benefits (NHMRC 2006).

There is little information on the impact of genotype on these traits and the muscle type needs to be considered as levels vary especially for Fe (Lin et al. 1989) and omega-3 fatty acids (Ponnampalam et al. 2010). Pearce et al. (2009) reported BLM had the highest Fe level in the *M. semimembranosus,* but not in two other muscles, with smaller differences for Zn. There was an interaction between genotype and muscle, with Merino and BLM *M. semimembranosus* and *M. semitendinosus* having 5% higher Zn levels than PD progeny (Pearce et al. 2009). In a study where lambs were given excess levels of dietary Fe and Zn, Field et al. (1985) found that Suffolk-sired lambs had a lower level of Zn in their muscle than Southdown-sired lambs, but there was no difference in Fe levels. In a larger study based on 2000 lambs of three main genotypes by 94 sires, Pannier et al. (2010) reported no differences between genotypes for Fe levels, but elevated levels of Zn in BLM compared to Merino and Terminal sired cross lambs. While this was indicative of more oxidative muscle fiber type (Gardner et al. 2007), the absolute levels in this study were approximately half that previously reported from retail sampling of lamb in Australia (Williams 2007) and elsewhere (Lin et al. 1989). More recently in an expansion of the Pannier et al. (2010) study it was confirmed that Zn levels in 5600 lambs from 270 sires were lower than reported from retail monitoring (Pannier et al. 2014c) and maternal genotypes (e.g., BLM) did have higher Zn levels than Merino and Terminal sired cross lambs. It is noteworthy that in the larger study of Pannier et al. (2014c) they did report that Terminal sired lambs had about 3% and 6% less Fe compared to maternal and Merino-sired lambs respectively, an outcome in contrast with the earlier report of Pannier et al. (2010). This illustrates the point made by Purchas (2007) about large experiments where because of sheer numbers you can find significant results, but it must be asked, are the differences of practical importance? Hoffman et al. (2003) found much higher Zn levels across six different types (with no type effects) than those reported by either Pearce et al. (2009) or Pannier et al. (2010, 2014c). However, Hoffman et al. (2003) did report genotype differences for Fe, with lower absolute levels than those reported by others (Pearce et al. 2009; Pannier et al. 2010). The study by Hoffman et al. (2003) was only based on 42 lambs, which was considerably less than those of Pearce et al. (2009) or Pannier et al. (2010). Overall, it appears that genotype differences for minerals such as Fe and Zn are not great and it is considered that other factors such as animal age are likely to be more important.

Ponnampalam et al. (2009) reported a decrease in the ratio of polyunsaturated to saturated fats as the level of Merino genes decreased, even after adjustment for IMF%, although there was no genotype effect for the levels of individual fatty acids. Similarly, Hoffman et al. (2003) found no genotype effect for the level of EPA, but Suffolk (S) cross SAMM *M. semimembranosus* had lower DHA and docosapentaenoic acid (DPA) than from S × M, which is of interest given that EPA is synthesized into DHA. In contrast, the S × SAMM had the second highest level of DPA and the S × M the lowest (Hoffman et al. 2003). The results strongly suggest that differences due to genotype are likely to occur in the wider sheep population and Ponnampalam et al. (2009) suggested that crossbred lambs may require a higher level of dietary intervention to attain the same level of omega-3 fatty acids as that found in Merinos. In a study of two Turkish breeds, Demirel et al. (2006) reported an interaction between breed type and the type of nutrition for the level of DHA in the *M. longissimus*, suggesting an effect on the incorporation of fatty acids into phospholipids. This could reflect a difference in fat levels (De Smet et al. 2004) specifically IMF%, but Demirel et al. (2006) did not adjust for the significant differences in IMF% between breeds. This same problem is evident in the data of Fisher et al. (2000), where the different fat levels of the Welsh Mountain and Soay were not considered in the comparison of the levels of the various fatty acids, where the Soay was much leaner and had a significantly higher level of EPA in the *M. semimembranosus*. By contrast, more recently Komprda et al. (2012) reported no effect of genotype on EPA and DHA levels when data were adjusted for differing IMF levels. Ponnampalam et al. (2014b) showed that genotype effects on the levels of EPA and DHA were meditated by the sire breed by dam breed combination, such that progeny from Poll Dorset sires mated to Merino dams had higher levels of these omega-3 fatty acids than if crossbred dams (e.g., BLM) were used. However, if White Suffolk sires were used the differences disappeared. In relation to other factors such as production site, these differences however were of minor importance.

Given the claims that a greater ratio of polyunsaturated to saturated fat and a lower ratio of omega-6 to omega-3 in meat would be desirable to lower the incidence of human metabolic diseases such as heart disease, inflammation, and mental health (Wood et al. 2003; Scollan et al. 2006), it is important to develop both genetic and nongenetic approaches to manipulate the level of the omega-3 fatty acids.

13.3 HERITABILITY AND GENETIC CORRELATIONS

Until recently there were few estimates of heritability for meat quality traits in sheep (Safari et al. 2005), but some extensive recent reports have dramatically expanded

TABLE 13.1
Estimates of Heritability (h^2), Standard Error (s.e.), and Phenotypic Variance (σ^2_P) for Shear Force, Eating Quality, and IMF

Trait	h^2	s.e.	σ^2_P	Number of Records	Breed Base	Reference
Shear Force						
(kg)—2 day aged	0.28	0.10		802[a]	Rambouillet, Columbia, and Corriedale crosses	Botkin et al. (1969)
Initial (kg)—10 day aged	0.39	0.16	3.84	349	Scottish Blackface	Karamichou et al. (2006b)
(N)	0.44			586[a,b]	Merino and meat crosses	Cloete et al. (2008)
Peak (N)—1 day aged	0.27	0.07	59.30	1637	Merino, BLM, Terminal × Merino, and Terminal × BLM	Mortimer et al. (2010)
Peak (N) —5 day aged	0.38	0.08	39.20	1759		
Peak (N) —5 day aged	0.27	0.04	51.24	5572	Merino, BLM, Terminal × Merino, and Terminal × BLM	Mortimer et al. (2014)
Eating Quality (0–100)						
Tenderness	0.15	0.13	99.7	349	Scottish Blackface—trained panels	Karamichou et al. (2006b)
Juiciness	0.21	0.12	35.1	349		
Flavor	0.11	0.11	23.8	349		
Overall liking	0.22	0.13	51.7	349		
IMF						
mg/100 g muscle	0.32	0.09		349	Scottish Blackface	Karamichou et al. (2006b)
%	0.39	0.05	0.63	3811	Merino, BLM, Terminal × Merino, and Terminal × BLM	Mortimer et al. (2010)
%	0.48	0.16	0.11	348	Terminal × Nor	Lorentzen and Vangen (2012)
%	0.48	0.05	0.68	5735	Merino, BLM, Terminal × Merino, and Terminal × BLM	Mortimer et al. (2014)

Note: Derived for the *M. longissimus* unless otherwise indicated.

[a] No information given to indicate if peak or initial yield.

[b] Aging period not given.

our knowledge in this area. A summary of these heritabilities is given in Tables 13.1 through 13.3. Shear force appears to be moderately to highly heritable (Table 13.1) with some higher estimates for aged meat indicated in the work of Mortimer et al. (2010), although with a greater numbers the estimate was decreased (Mortimer et al. 2014) and the precision of the estimate increased. The heritability estimates for IMF are also moderate to high (Karamichou et al. 2006b; Mortimer et al. 2010, 2014). Estimates of the genetic correlation between IMF and shear force of -0.54 ± 0.24 (Karamichou et al. 2006b), -0.69 ± 0.11 (Mortimer et al. 2010), and -0.62 ± 0.07 Mortimer et al. (2014) suggest that selection for increasing IMF will have a favorable effect on shear force.

The heritability estimates for eating quality traits were low (Karamichou et al. 2006b, 2007), albeit from a very small data set. Positive genetic correlations were found between IMF and the sensory traits juiciness and flavor (0.12 ± 0.06 and 0.20 ± 0.06 respectively) (Karamichou et al. 2006b). In this area there is an obvious need for more estimates if at any stage selection for eating quality is to be included in breeding programs.

The estimates of heritability for meat pH and measures of meat color were generally low to moderate (Table 13.2), with low phenotypic variation which suggests that selection response for these traits would be slow. There appears to be a favorable genetic correlation (−0.6) between pH and L^* color values in Merino hogget rams (Fogarty et al. 2003; Greeff et al. 2008) and −0.30 for Scandinavian crossbred lambs (Lorentzen and Vangen 2012), whereas others have reported correlations not significantly different from zero in meat sheep breeds (Ingham et al. 2007; Payne et al. 2009) and similarly across all breeds (Mortimer et al. 2014). This latter study did however show that the heritability of a measure of brownness/redness for meat on display was moderate (Mortimer et al. 2014). Given this trait has an impact on consumer acceptability (Khliji et al. 2010); this finding is of significance as it indicates that this trait could be incorporated into a breeding program to enable genetic change if a genomic breeding value was derived. Genetic correlations between behavior traits (flight speed and agitation assessed at weaning) and meat quality traits have shown that pH (at 24 h) in the *M. longissimus* is significantly related to agitation ($r_g = 0.27 \pm 0.14$) (Dodd et al. 2014), but flight speed was not related.

The preliminary reports on the minerals iron and zinc in meat suggested low to moderate heritabilities (Table 13.3), although there was one very high estimate for zinc from limited data (Bennett and Field 1985). More precise estimates are now available, thanks to the work of Mortimer et al. (2014) and these suggest that selection response would occur if animals were selected for these traits. The estimates of heritability for the omega-3 fatty acids EPA and DHA range from 0.16 to 0.29, with relatively high phenotypic variation, supporting the conclusions of Karamichou et al. (2006a) and Greeff et al. (2007) that it may be possible to improve omega-3 fatty acids through selection. It is probably more efficient however, to manipulate these fatty acids through nutrition as it explains a large proportion of the variation in these compounds (Ponnampalam et al. 2014b). Karamichou et al. (2007) reported negative residual correlations between both EPA and DHA with eating quality (overall liking), but more extensive studies are required to determine the significance of this finding.

TABLE 13.2
Estimates of Heritability (h^2), Standard Error (s.e.), and Phenotypic Variance (σ^2_P) for pH and Meat Color Traits

Trait	h^2	s.e.	σ^2_P	Number of Records	Breed Base	Reference
pH						
Measured 24 h postmortem	0.27	0.09		957	Merino	Fogarty et al. (2003)
Measured 24 h postmortem	0.21	0.14	0.01	349	Scottish Blackface	Karamichou et al. (2006b)
Measured 24 h postmortem	0.18	0.07	0.108	1330	Merino cross	Ingham et al. (2007)
Measured 24 h and 48 h postmortem	0.22	0.03		5700	Merino	Greeff et al. (2008)
Measured 48 h postmortem	0.09	0.09		672	Merino and meat crosses	Cloete et al. (2008)
Measured 24 h post-mortem	0.12			6565	Terminal cross	Payne et al. (2009)
Measured 19–24 h postmortem	0.10	0.03	0.006	3709	Merino, BLM, Terminal × Merino, and Terminal × BLM	Mortimer et al. (2010)
	0.09	0.04	0.025	3766	Measured on the *M. semitendinosus*	
Measured 48 h postmortem	0.20	0.12	0.005	349	Terminal × Nor	Lorentzen and Vangen (2012)
	0.30	0.15	0.002	348	Measured on the *M. semimembranosus*	
Measured 24 h postmortem	0.08	0.02	0.009	7805	Merino, BLM, Terminal × Merino, and Terminal × BLM	Mortimer et al. (2014)
Color[a] (Lightness, L^*)						
Calibrated with a white tile	0.14	0.07		1035	Merino	Fogarty et al. (2003)
Calibration details not given	0.15	0.12	3.43	349	Scottish Blackface	Karamichou et al. (2006b)
Calibration details not given	0.23	0.07	2.05	1913	Merino cross	Ingham et al. (2007)
Calibration details not given	0.45	0.19		580	Merino and meat crosses	Cloete et al. (2008)
Calibrated with a white tile	0.18	0.03		5107	Merino	Greeff et al. (2008)
Calibrated with a white tile	0.29			6565	Terminal cross	Payne et al. (2009)

(Continued)

TABLE 13.2 (*Continued*)
Estimates of Heritability (h^2), Standard Error (s.e.), and Phenotypic Variance (σ^2_P) for pH and Meat Color Traits

Trait	h^2	s.e.	σ^2_P	Number of Records	Breed Base	Reference
Calibrated with a white tile	0.21	0.04	3.21	3432	Merino, BLM, Terminal × Merino, and Terminal × BLM	Mortimer et al. (2010)
Calibrated with a white tile	0.53	0.18	2.29	285	Terminal × Nor	Lorentzen and Vangen (2012)
Calibrated with a white tile	0.18	0.03	3.99	7198	Merino, BLM, Terminal × Merino, and Terminal × BLM	Mortimer et al. (2014)
Redness (a*)[a]	0.02	0.06		1011	Merino	Fogarty et al. (2003)
	0.45	0.19	2.17	349	Scottish Blackface	Karamichou et al. (2006b)
	0.10	0.06	2.01	1331		Ingham et al. (2007)
	0.04	0.10		580	Merino and meat crosses	Cloete et al. (2008)
	0.10	0.03		5080	Merino	Greeff et al. (2008)
	0.19			6565	Terminal cross	Payne et al. (2009)
	0.06	0.03	1.43	3431	Merino, BLM, Terminal × Merino, and Terminal × BLM	Mortimer et al. (2010)
	0.17	0.14	1.35	284	Terminal × Nor	Lorentzen and Vangen (2012)
	0.08	0.03	1.83	7200	Merino, BLM, Terminal × Merino, and Terminal × BLM	Mortimer et al. (2014)
R630/R580[b] (indicator of brownness/redness)	0.40	0.10	0.28	1156	Merino, BLM, Terminal × Merino, and Terminal × BLM	Mortimer et al. (2010)
	0.27	0.04	0.37	4459	Merino, BLM, Terminal × Merino, and Terminal × BLM	Mortimer et al. (2014)

Note: Derived for the *M. longissimus* unless otherwise indicated.
[a] Measured under the same conditions as the pH, thus on fresh meat.
[b] Measured on meat during simulated retail display.

TABLE 13.3
Estimates of Heritability (h^2), Standard Error (s.e.), and Phenotypic Variance (σ^2_P) for Minerals (Fe and Zn) and Fatty Acids

Trait	h^2	s.e.	σ^2_P	Number of Records	Breed Base	Reference
Minerals						
Iron (mg/kg dried muscle tissue)	0.21	0.38		100	Suffolk or Southdown sires × Perendale, Romney, Coopworth, Merino-Perendale, and Merino-Romney ewes	Bennett and Field (1985)
Iron (mg/kg wet muscle tissue)	0.12	0.05	12.77	1915	Merino, BLM, Terminal × Merino, and Terminal × BLM	Mortimer et al. (2010)
Iron (mg/kg wet muscle tissue)	0.21	0.04	8.31	5716	Merino, BLM, Terminal × Merino, and Terminal × BLM	Mortimer et al. (2014)
Zinc (mg/kg dried muscle tissue)	0.92	0.48		100		Bennett and Field (1985)
Zinc (mg/kg wet muscle tissue)	0.21	0.06	12.83	1915	Merino, BLM, Terminal × Merino, and Terminal × BLM	Mortimer et al. (2010)
Zinc (mg/kg wet muscle tissue)	0.27	0.04	15.15	5716	Merino, BLM, Terminal × Merino, and Terminal × BLM	Mortimer et al. (2014)
Long Chain Fatty Acids						
Eicosapentaenoic acid (EPA, mg/100 g wet muscle tissue)	0.21	0.13		350	Scottish Blackface	Karamichou et al. (2006a)
EPA	0.18	0.07	0.129	1109	Merino	Greeff et al. (2007)
EPA	0.29	0.07	0.039	1919	Merino, BLM, Terminal × Merino, and Terminal × BLM	Mortimer et al. (2010)
EPA	0.17	0.03	0.066	5722	Merino, BLM, Terminal × Merino, and Terminal × BLM	Mortimer et al. (2014)
Docosahexaenoic acid (DHA, mg/100 g wet muscle tissue)	0.16	0.10		350	Scottish Blackface	Karamichou et al. (2006a)
DHA	0.19	0.08	0.032	1069	Merino	Greeff et al. (2007)
DHA	0.25	0.06	0.051	1915	Merino, BLM, Terminal × Merino, and Terminal × BLM	Mortimer et al. (2010)
DHA	0.22	0.03	0.088	5718	Merino, BLM, Terminal × Merino, and Terminal × BLM	Mortimer et al. (2014)

Note: Derived for the *M. longissimus* unless otherwise indicated.

13.4 GENDER EFFECTS

13.4.1 Meat Quality Traits

There are production advantages (faster growth, leaner carcasses) in retaining entire males or rendering them "crytorchid" (Wilson et al. 1970; Lee 1986; Hopkins et al. 1990), but the impact on meat quality traits is less clear. In the work of Corbett et al. (1973), no effect on pH or shear force was reported when comparing cryptorchid, wether or ewe crossbred or straight bred lambs, but these were lightweight animals. However, when older (20 months) entire rams were compared with ewes the former was reported to have higher pH in the *M. longissimus* (Cloete et al. 2012), although the mixing of these animals just prior to slaughter is the probable explanation for this effect. This also probably contributed to the tougher meat (i.e., higher shear force) of the rams and this could have been verified by including pH as a covariate in the analysis. In the report of Hopkins et al. (2001), cryptorchids had a higher pH in the *M. longissimus* than wether or ewe carcasses such that 19% had a pH above the critical 5.8, but this did not translate into an effect on color. By contrast Hopkins et al. (2007a) did report that *M. longissimus* from wether lambs was lighter colored than that from ewe lambs, but it is unlikely that consumers would have been able to detect the differences and further studies showed no such difference (Hopkins et al. 2007b).

An important trait is shear force and the early work of Corbett et al. (1973) found no gender effect on this trait when measured in the *M. longissimus* or *M. semimembranosus,* but it should be stressed that these lambs were relatively young. The interaction with age is important as illustrated by the work of Channon et al. (1993) in which they showed no difference in the shear force of either the *M. longissimus* or *M. semimembranosus* between cryptorchids and wethers at 8 months of age, but thereafter the levels were on average higher for the cryptorchids. In older animals, Cloete et al. (2012) also reported a 9% increase in shear force for *M. longissimus* from rams compared to ewes, with the absolute values indicative of very tough meat. The report of Johnson et al. (2005) also showed that *M. semimembranosus* from rams had significantly higher shear force values compared to ewes in lambs 8 months or younger. Small differences were reported by Hopkins et al. (2007a) across a number of genotypes with wethers producing significantly tougher *M. longissimus* than ewe lambs over an age range of 4–22 months.

Less attention has been paid toward IMF, but in a recent study Pannier at el. (2014b) found that ewe lambs had significantly higher IMF levels (0.10%) than male lambs, irrespective of correction for carcass weight, but it is suspected that this effect simply reflects the large number of lambs sampled in the study, given such a difference has not been reported in other studies (e.g., Hopkins et al. 2007a; Tejeda et al. 2008). There is one report however that shows wethers to have a higher amount of IMF in the *M. longissimus* than ewes over a wide age range (4–22 months; McPhee et al. 2008), but this was derived with respect to the proportional development of total carcass fat over the entire age range which is very different from other studies. Interestingly, Craigie et al. (2012) found females to have higher IMF levels compared to males, though entire rams were used instead of castrates, while Solomon et al. (1990) found higher IMF levels in wethers than ram lambs.

The sex effect on fatty acid concentrations is potentially of more importance, with the work of Solomon et al. (1990) showing lower levels of polyunsaturated fatty acids in the *M. longissimus* from wethers compared to rams, with cryptorchids intermediate. Although very small in magnitude, there was a gender effect found on the health claimable fatty acid content (EPA + DHA) in lamb *M. longissimus* (Ponnamplam et al. 2014), such that females showed greater levels than males. One explanation that was proposed was that as female lambs approach their reproductive stage, it is possible that they synthesize more long chain omega-3 fatty acid in the body for the production of series-3 eicosanoids which are associated with the ovulation process, conception, and pregnancy (Mattos et al. 2000). It is apparent that the level and types of hormones impact on the expression of fatty acids, but the magnitude of the effect in relation to other production factors is small (Chapter 4).

13.4.2 Eating Quality

Given that entire male lambs in particular are more likely to produce meat with a higher pH (e.g., Cloete et al. 2012), it is surprising that this often does not translate into a negative impact on eating quality. In fact there are a number of studies which have shown no practical difference between the meat of entire lambs, crytorchids and wethers (Kirton et al. 1982; Butler-Hogg et al. 1984; Lee 1986; Cloete et al. 2012). However, Hopkins et al. (2001) did report on a gender by nutrition interaction which showed the *M. longissimus* from cryptorchid lambs fed a pasture/oats/sunflower ration as being less acceptable than that of wethers fed the same ration. This appeared to be related to stronger flavor and aroma scores in the meat from cryptorchids, a finding that was also in part supported by the work of Corbett et al. (1973). Interestingly in a recent large study based on consumer assessment of both *M. longissimus* and *M. semimembranosus* it was found within Terminal sired lambs that females had better sensory scores than male lambs (Pannier et al. 2014a). The effect was not observed in Merino or Maternal sired lambs, and was small, and only evident within the *M. longissimus* where the scores were 1.8, 1.5, 0.9, and 0.9 units higher for tenderness, overall liking flavor, and odor, when compared to the wether lambs (on a 0–100 scale). Where production systems utilize wether and ewe lambs, there are no apparent grounds for discounting the meat from the former class of lambs, but if entire or cyptorchid production systems are adopted more attention needs to be given to the final product.

13.5 AGE EFFECTS

13.5.1 Meat Quality Traits

Hopkins et al. (2007a) reported a clear advantage in the ability of sucker lambs (4 months of age and still on their mothers) over older weaned lambs to withstand stress or reduce the preslaughter depletion of glycogen as seen in the *semitendinosus* muscle through pH measures. There is previous evidence that this is a general finding (Hopkins et al. 2005b). Given that fiber typing of this muscle using antibodies against myosin heavy chains did not indicate a significant change in the ratio of glycolytic to oxidative fiber types as the animals aged (Greenwood et al. 2007), this supports the theory that glycogen depletion was reduced. Another notable finding from the work of Hopkins et al. (2007a) was the change in color as animal age increased. Using a threshold value of 34 for lightness as established by consumer evaluation (Hopkins 1996; Khliji et al. 2010) the results indicated that as sheep approach 12–13 months of age the meat color on average became unacceptable to consumers. After this age the meat will, based on the results of Hopkins et al. (2007a), in general be too dark and red for acceptance at the retail counter. The redness of meat has been shown in other studies to increase with animal age (Dawson et al. 2002). The superior meat color from sucker lambs was again demonstrated in line with previous reports (Hopkins et al. 2005b) and is in part a reflection of lower myoglobin levels (Gardner et al. 2007) with a clear increase in myoglobin as animals age (Ledward and Shorthose 1971). This reflects the increase in muscle oxidative capacity as animals become older (Greenwood et al. 2007), which is seen in a higher concentration of iron (Pannier et al. 2010).

For some meat quality traits it is difficult to separate "age" effects from potentially confounding factors such as carcass weight and slaughter day (Purchas 2007). However, the data of Hopkins et al. (2007a) suggest that these factors were not operative for meat color, because once past 8 months of age there was, for example, no effect on pH and so this could not be implicated in the increasingly darker, redder muscle as the animals became older. If anything, a fatter, heavier carcass would be expected to produce lighter colored meat from a faster rate of pH decline. An increase in IMF as animals increase in age is expected and consistent with previous reports (Martínez-Cerezo et al. 2005; Pethick et al. 2005); although there is a large degree of overlap between animals of differing ages (Hopkins et al. 2005b).

With respect to traits like shear force, factors such as carcass weight and slaughter day can have a significant effect. Consequently, strategies like applying aging periods (e.g., Hopkins et al. 2007a) can be used to lessen the impact of sources of variation associated with processing so as to establish "true" differences due to increasing animal age. In the study reported by Jeremiah et al. (1971) which examined five muscles from the hindleg of sheep (ewes and wethers) ranging in age from 74 to 665 days for shear force and tenderness, a positive correlation was shown between animal age and decreasing tenderness. Across five muscles the correlation was −0.46 and between shear force of the *M. semimembranosus* and animal age it was 0.33.

Data provided by Furnival et al. (1977) for lambs slaughtered at 32 kg live weight, but varying in age from 98 to 363 days showed that those with an average age of 275 days (9 months) had significantly ($P < 0.05$) tougher *semimembranosus* muscles than those aged 226 and 209 days. There was little difference for those slaughtered between 145 and 226 days of age. After approximately 250 days of age the shear force values tended to be over 50 N, a level above which it is suggested individuals would find the meat "tough" (Hopkins et al. 2006). This response is consistent with the data of Hopkins et al. (2007a) which showed much higher shear force values for the *M. semimembranosus* from animals aged 14 and 20 months of age, versus those aged 8 and 14 months of age. This outcome is consistent with expectations

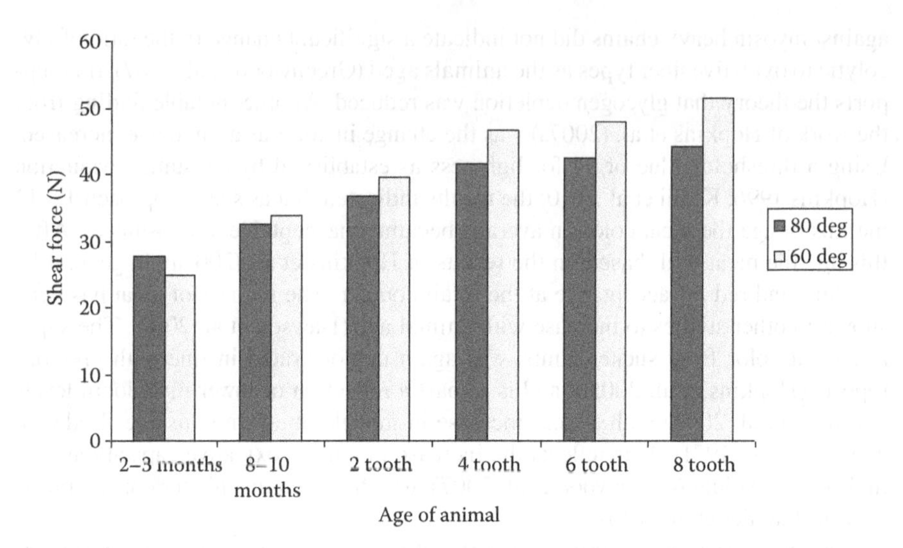

FIGURE 13.1 Shear force for muscle samples cooked at 60°C or 80°C for 90 min from sheep of various ages (weaners 2–3 months, 0 tooth 8–10 months, 2, 4, 6, and 8 tooth). There were no differences within age groups due to cooking temperature in animals younger than four tooth. (Adapted from Bouton, P.E. et al. 1978. *J. Food Sci.* 43:1038–1039.)

based on the data of Young et al. (1993a) which demonstrated a reduction in collagen solubility as animal age increased and a commensurate increase in shear force. The reduction in solubility is due to increased cross-linking between collagen molecules. Bouton et al. (1978) combined shear force data for four leg muscles and examined the effect of age of the animal. It was found that shear force did increase with increasing animal age, and there were smaller differences between ages if the meat was cooked at 80°C (Figure 13.1). It should be noted that in this work the carcasses were hung by the pelvis to avoid shortening and this may have reduced the effect of increasing cross-linking of collagen. There was no effect within age groups of the temperature used for cooking the meat, if the animals were younger than four tooth.

13.5.2 Eating Quality

Given the impact of animal age on shear force it is expected that this would also impact on eating quality and this is what the literature confirms. Kirton et al. (1983) reported that sheep over 2 years of age produced tougher meat compared to animals 1 year old as assessed by a test panel, but this was not consistent with the results from shear force measurements. In a study to examine the eating quality of the *M. semimembranosus* from sheep aged from 4 months to 5 years, Young and Braggins (1993) found that the decrease in eating quality as animals became older was due to an increase in collagen concentration, more than a decrease in collagen solubility and this was attributed to the highly insoluble nature of collagen in the muscle. More recent work has focused on the use of consumer panels.

For example, Thompson et al. (2005) reported on an experiment in which the eating quality of meat from 120 sheep was examined. These sheep comprised 80 young lambs (weaned second cross, Poll Dorset × (Border Leicester × Merino) ewe lambs, approximately 6 months old) and 40 old (Merino wethers, approximately 48 months old) sheep. Of all the variables examined in the work (animal age, muscle type, aging period, hanging method, and electrical stimulation) animal age accounted for the largest proportion of variance in sensory scores, with muscle being the next most important factor. A muscle × animal age interaction was most evident for tenderness score, whereby the decrease in tenderness score in lamb compared with mutton was greatest for the *M. biceps femoris* and *M. serratus ventralis*, compared with the *M. longissimus* muscle. Related work (Pethick et al. 2005) showed that eating quality scores for the *M. longissimus* and *M. biceps femoris* derived from Merinos ranging in age from 8 months to 68.5 months declined significantly when the animals were older than 32.5 months, but did not decline thereafter.

There has been a general contention that as age increases so does the intensity of flavor (Young et al. 1997). Again the evidence at times is apparently contradictory (Sink and Caporaso 1977). Butler-Hogg and Francombe (1985) found that meat from lamb at 16 weeks of age was less flavorsome than meat from lambs (hogget) at 43 weeks of age and in a subsequent study suggested this may have been due to higher concentrations of muscle lipid (Butler-Hogg and Buxton 1986). Kirton et al. (1974) showed that meat from 15 month old sheep was as acceptable as that from 5 month old sheep when preference included tenderness, flavor, and juiciness. The study of Pethick et al. (2005) shed light on this apparent contradiction as they found no effect of increasing animal age on the liking of flavor, but pointed out that all cuts were denuded of subcutaneous and intermuscular fat in their work and hence suggested that when this is applied the potential age affects are minimized. One of the most important factors impacting on the flavor of meat is the nutrition of the animal prior to slaughter and readers are referred to a recent review of this topic (Watkins et al. 2013).

13.6 CONCLUSIONS

Differences in experimental design have constrained some of the robustness of the conclusions which can be drawn from the studies reviewed. However there is good evidence that Merinos do have a propensity to produce meat with a higher pH and in some cases specific muscles have reduced color stability. Given the importance of this breed to sheep production in a number of countries, research to establish the physiological basis of the higher pH levels in the Merino is worthy of further investigation. There is evidence that sires selected for extremes in breeding values for muscling or fatness (e.g., highly muscled and lean) will produce progeny with lower eating quality. Thus, systems need to be established to manage these affects. Estimates of genetic parameters are now becoming available and indicate there is genetic variation for most meat quality traits. Availability of more accurate estimates of the genetic correlations between meat quality and production traits will allow the development of appropriate objectives and selection criteria for use in meat sheep breeding programs. All the evidence suggests that castration will provide greater surety for the production of high quality meat compared to the retention of testes and that younger animals will provide the highest quality meat, but there is potential to improve the eating quality of meat from older sheep by removal of subcutaneous and intermuscular fat.

REFERENCES

Arsenos, G., Banos, G., Fortomaris, P., Katsaounis, N., Stamataris, C., Tsaras, L., and Zygoyiannis, D. 2002. Eating quality of lamb meat: Effects of breed, sex, degree of maturity and nutritional management. *Meat Sci.* 60:379–387.

Bennett, G.L., and Field, R.A. 1985. A note on the influence of breed and sire differences on iron and zinc concentration of lamb muscle. *Anim. Prod.* 41:421–424.

Berge, P., Sañudo, C., Sanchez, A., Alfonso, A., Stamataris, C., Thorkelsson, G., Piasentier, E., and Fisher, A.V. 2003. Comparison of muscle composition and meat quality traits in diverse commercial lambs. *J. Muscle Foods* 14:281–300.

Bermingham, E.N., Coy, N.C., Anderson, R.C., Barnett, M.P.G., Knowles, S.O., and McNabb, W.C. 2008. Smart foods from the pastoral sector—Implications for meat and milk producers. *Aust. J. Exp. Agric.* 48:726–734.

Botkin, M.P., Field, R.A., Riley, M.L., Nolan, J.C., and Roehrkasse, G.P. 1969. Heritability of carcass traits in lambs. *J. Anim. Sci.* 29:251–255.

Bouton, P.E., Harris, P.V., Ratcliff, D., and Roberts, D.W. 1978. Shear force measurements on cooked meat from sheep of various ages. *J. Food Sci.* 43:1038–1039.

Butler-Hogg, B.W., and Francombe, M.A. 1985. Carcass and meat quality in spring and hogget lamb. *Anim. Prod.* 40:527.

Butler-Hogg, B.W., Francombe, M.A., and Dransfield, E. 1984. Carcass and meat quality of ram and ewe lambs. *Anim. Prod.* 39:107–113.

Butler-Hogg, B.W., and Buxton, P.J. 1986. Preliminary assessment of muscle quality in spring and hogget lamb. *Anim. Prod.* 42:461.

Carson, A.F., Moss, B.W., Dawson, L.E.R., and Kilpatrick, D.J. 2001. Effects of genotype and dietary forage to concentrate ratio during the finishing period on carcass characteristics and meat quality of lambs from hill sheep systems. *J. Agric. Sci.* 137:205–220.

Channon, H.A., Thatcher, L.P., and Copper, K.L. 1993. Effect of age on meat-quality attributes of second-cross cryptorchid and wether lambs. In *Proc. Aust. Meat Ind. Res. Conf.*, Session 3A, Gold Coast, Queensland, Australia, pp. 1–3.

Cloete, J.J.E., Hoffman, L.C., and Cloete, S.W.P. 2012. A comparison between slaughter traits and meat quality of various sheep breeds: Wool, dual-purpose and mutton. *Meat Sci.* 91:318–324.

Cloete, S.W.P., Cloete, J.J.E., and Hoffman, L.C. 2008. Heritability estimates for slaughter traits in South African terminal crossbred lambs. In *Proc. 54th Inter. Congr. Meat Sci. Tech.*, Session 4, Cape Town, South Africa, pp. 1–3. Available on line at: http://www.icomst.helsinki.fi/icomst2008/CD%20Papers/General%20speakers+posters-3p%20papers/Session4/4.3.Cloete.pdf.

Corbett, J.L., Furnival, E.P., Southcott, W.H., Park, R.J., and Shorthose, W.R. 1973. Induced cryptorchidism in lambs. Effects on growth rate, carcasses and meat characteristics. *Anim. Prod.* 16:157–163.

Craigie, C.R., Lambe, N.R., Richardson, R.I., Haresign, W., Maltin, C.A., Rehfeldt, C., Roehe, R., Morris, S.T., and Bunger, L. 2012. The effect of sex on some carcass and meat quality traits in Texel ewe and ram lambs. *Anim. Prod. Sci.* 52:601–607.

Crouse, J.D. 1983. The effects of breed, sex, slaughter weight, and age on lamb flavour. *Food Tech.* 37:264–268.

Crouse, J.D., Busboom, J.R., Field, R.A., and Ferrell, C.L. 1981. The effects of breed, diet, sex, location and slaughter weight on lamb growth, carcass composition and meat flavor. *J. Anim. Sci.* 53:376–386.

Dawson, L.E.R., Carson, A.F., and Moss, B.W. 2002. Effects of crossbred ewe genotype and ram genotype on lamb meat quality from the lowland sheep flock. *J. Agric. Sci.* 139:195–204.

Demirel, G., Ozpinar, H., Nazli, B., and Keser, O. 2006. Fatty acids of lamb meat from two breeds fed different forage: Concentrate ratio. *Meat Sci.* 72:229–235.

De Smet, S., Raes, K., and Demeyer, D. 2004. Meat fatty acid composition as affected by fatness and genetic factors: A review. *Anim. Res.* 53:81–98.

Díaz, M.T., Álvarez, I., De la Fuente, J., Sañudo, C., Campo, M.M., Oliver, M.A., Font i Furnols, M., Montossi, F., San Julián, R., Nute, G.R., and Cañeque, V. 2005. Fatty acid composition of meat from typical lamb production systems of Spain, United Kingdom, Germany and Uruguay. *Meat Sci.* 71:256–263.

Dodd, C.L., Hocking-Edwards, J.E., Hazel, S.J., and Pitchford, W.S. 2014. Flight speed and agitation in weaned lambs: Genetic and non-genetic effects and relationships with carcass quality. *Livest. Sci.* 160:12–20.

Dransfield, E., Nute, G.R., MacDougall, D.B., and Rhodes, D.N. 1979. Effect of sire breed on eating quality of cross-bred lambs. *J. Sci. Food and Agric.* 30:805–808.

Edwards, R.L., Crenwelge, D.D., Savell, J.W., Shelton, M., and Smith, G.C. 1982. Cutability and palatability of Rambouillent, Blackface-crossbred and Karakul lambs. *Int. Goat Sheep Res.* 2:77–80.

Egan, A.F., and Shay, B.J. 1988. Long-term storage of chilled fresh meats. In *Proc. 34th Int. Cong. Meat Sci. Tech.,* Brisbane, Australia, pp. 476–481.

Ellis, M., Webster, G.M., Merrell, B.G., and Brown, I. 1997. The influence of terminal sire breed on carcass composition and eating quality of crossbred lambs. *Anim. Sci.* 64:77–86.

Esenbuga, N., Yanar, M., and Dayioglu, H. 2001. Physical, chemical and organoleptic properties of ram lamb carcasses from four fat-tailed genotypes. *Small Rum. Res.* 39:99–105.

Field, R.A., Bennett, G.L., and Munday, R. 1985. Effect of excess zinc and iron on lamb carcass characteristics. *NZ J. Agric. Res.* 28:349–355.

Fisher, A.V., Enser, M., Richardson, R.I., Wood, J.D., Nute, G.R., Kurt, E., Sinclair, L.A., and Wilkinson, R.G. 2000. Fatty acid composition and eating quality of lamb types derived from four diverse breed × production systems. *Meat Sci.* 55:141–147.

Fogarty, N.M. 2009. Meat sheep breeding—Where we are at and future challenges. In *Proc. Assoc. Advmt. Anim. Breed. Genet.* 18:414–421.

Fogarty, N.M., Banks, R.G., van der Werf, J.H.J., Ball, A.J., and Gibson, J.P. 2007. The Information Nucleus—A new concept to enhance sheep industry genetic improvement. In *Proc. Assoc. Advmt. Anim. Breed. Genet.* 17:29–32.

Fogarty, N.M., Hopkins, D.L., and van de Ven, R. 2000. Lamb production from diverse genotypes. 2. Carcass characteristics. *Anim. Sci.* 70:147–156.

Fogarty, N.M., Safari, E., Taylor, P.J., and Murray, W. 2003. Genetic parameters for meat quality and carcass traits and their correlation with wool traits in Australian Merino sheep. *Aust. J. Agric. Res.* 54:715–722.

Furnival, E.P., Corbett, J.L., and Shorthose, W.R. 1977. Meat properties of lambs grown at 32 kg at various rates on phalaris or lucerne pastures and an apparent effect of preslaughter ambient temperature. *J. Agric. Sci.* 88:207–216.

Gardner, G.E., Hopkins D.L., Greenwood, P.L., Cake, M.A., Boyce, M.D., and Pethick, D.W. 2007. Sheep genotype, age, and muscle type affect the expression of metabolic enzyme markers. *Aust. J. Exp. Agric.* 47:1180–1189.

Gardner, G.E., Kennedy, L., Milton, J.W., and Pethick, D.W. 1999. Glycogen metabolism and ultimate pH of muscle in Merino, first-cross and second-cross wether lambs as affected by stress before slaughter. *Aust. J. Agric. Res.* 50:175–181.

Gardner, G.E., Pethick, D.W., Greenwood, P.L., and Hegarty, R.S. 2006. The effect of genotype and plane of nutrition on the rate of pH decline in lamb carcasses and the expression of metabolic enzymatic markers. *Aust. J. Agric. Res.* 57:661–670.

Greenwood, P.L., Harden, S., and Hopkins, D.L. 2007. Age and genotypic influences on myofibre characteristics in *longissimus* and *semitendinosus* muscles of sheep. *Aust. J. Exp. Agric.* 47:1137–1146.

Greeff, J.C., Harvey, M., Young, P., Kitessa, S., and Dowling, M. 2007. Heritability estimates of individual fatty acids in Merino meat. *Proc. Assoc. Advmt. Anim. Breed. Genet.* 17:203–206.

Greeff, J.C., Safari, E., Fogarty, N.M., Hopkins, D.L., Brien, F.D., Atkins, K.D., Mortimer, S.I., and van der Werf, J.H.J. 2008. Genetic parameters for carcass and meat quality traits and their relationships to liveweight and wool production in hogget Merino rams. *J. Anim. Breed. Genet.* 125:205–215.

Hoffman, L.C., Muller, M., Cloete, S.W.P., and Schmidt, D. 2003. Comparison of six crossbred lamb types; sensory, physical and nutritional meat quality characteristics. *Meat Sci.* 65:1265–1274.

Hopkins, D.L. 1996. An assessment of lamb meat colour. *Meat Focus Inter.* 5:400–401.

Hopkins, D.L. 2010. Processing of sheep and sheep meat. In Cottle, D.J. (ed.), *International Sheep and Wool Handbook.* Nottingham University Press, UK, pp. 691–710.

Hopkins, D.L., and Fogarty, N.M. 1998. Diverse lamb genotypes- 2. Meat pH, colour and tenderness. *Meat Sci.* 49:477–488.

Hopkins, D.L., and Geesink, G.H. 2009. Protein degradation post mortem and tenderisation. In Du, M. and McCormick, R. (eds.), *Applied Muscle Biology and Meat Science.* CRC Press, Taylor & Francis Group, USA, pp. 149–173.

Hopkins, D.L., and Mortimer, S.I. 2014. Sheep quality—Effect of breed, genetic type, gender and age on meat quality and a case study illustrating integration of knowledge. *Meat Sci.* 98:544–555.

Hopkins, D.L., Gilbert, K.D., Pirlot, K.L., and Roberts, A.H.K. 1992. Elliottdale and crossbred lambs: Growth rate, wool production, fat depth, saleable meat yield, carcass composition and muscle content of selected cuts. *Aust. J. Exp. Agric.* 32:429–434.

Hopkins, D.L., Gilbert, K.D., and Saunders, K.L. 1990. The performance of short scrotum and wether lambs born in winter or spring and run at pasture in Northern Tasmania. *Aust. J Exp. Agric.* 30:165–170.

Hopkins, D.L., Hall, D.G., Channon, H.A., and Holst, P.J. 2001. Meat quality of mixed gender lambs grazing pasture and supplemented with, roughage, oats or oats and sunflower meal. *Meat Sci.* 59:277–283.

Hopkins, D.L., Hatcher, S., Pethick, D.W., and Thornberry, K.J. 2005a. Carcass traits, meat quality and muscle enzyme activity in strains of Merino hoggets. *Aust. J. Exp. Agric.* 45:1225–1230.

Hopkins, D.L., Hegarty, R.S., Walker, P.J., and Pethick, D.W. 2006. Relationship between animal age, intramuscular fat, cooking loss, pH, shear force and eating quality of aged meat from sheep. *Aust. J. Exp. Agric.* 46:878–884.

Hopkins, D.L., Stanley, D.F., Martin, L.C., Toohey, E.S., and Gilmour, A.R. 2007a. Genotype and age effects on sheep meat production. 3. Meat quality. *Aust. J. Exp. Agric.* 47:1155–1164.

Hopkins, D.L., Stanley, D.F., Toohey, E.S., Gardner, G.E., Pethick, D.W., and van de Ven, R. 2007b. Sire and growth path effects on sheep meat production. 2. Meat and eating quality. *Aust. J. Exp. Agric.* 47:1219–1228.

Hopkins, D.L., Toohey, E.S., Warner, R.D., Kerr, M.J., and van de Ven, R. 2010. Measuring the shear force of lamb meat cooked from frozen samples: A comparison of 2 laboratories. *Anim. Prod. Sci.* 50:382–385.

Hopkins, D.L., Walker, P.J., Thompson, J.M., and Pethick, D.W. 2005b. Effect of sheep type on meat eating quality of sheep meat. *Aust. J. Exp. Agric.* 45:499–507.

Ingham, V.M., Fogarty, N.M., Gilmour, A.R., Afolayan, R.A., Cummins, L.J., Gaunt, G.M., Stafford, J., and Hocking Edwards, J.E. 2007. Genetic evaluation of crossbred lamb production 4. Genetic parameters for first-cross animal performance. *Aust. J. Agric. Res.* 58:839–846.

Jacob, R.H., D'Antuono, M.F., Gilmour, A.R., and Warner, R.D. 2014. Phenotypic characterisation of colour stability of lamb meat. *Meat Sci.* 96:1040–1048.

Jandasek, J., Milerski, M., and Lichovnikova, M. 2014. Effect of sire breed on physico-chemical and sensory characteristics of lamb meat. *Meat Sci.* 96:88–93.

Jeremiah, L.E., Smith, A.C., and Carpenter, Z.L. 1971. Palatability of individual muscles from ovine leg steaks as related to chronological age and marbling. *J. Food Sci.* 35:45–47.

Johnson, P.L., Purchas, R.W., McEwan, J.C., and Blair, H.T. 2005. Carcass composition and meat quality differences between pasture-reared ewe and ram lambs. *Meat Sci.* 71:383–391.

Karamichou, E., Richardson, R.I., Nute, G.R., Gibson, K.P., and Bishop, S.C. 2006a. Genetic analyses and quantitative trait loci detection, using a partial genome scan, for intramuscular fatty acid composition in Scottish Blackface sheep. *J. Anim. Sci.* 84:3228–3238.

Karamichou, E., Richardson, R.I., Nute, G.R., McLean, K.A., and Bishop, S.C. 2006b. Genetic analyses of carcass composition, as assessed by X-ray computer tomography, and meat quality traits in Scottish Blackface sheep. *Anim. Sci.* 82:151–162.

Karamichou, E., Richardson, R.I., Nute, G.R., Wood, J.D., and Bishop, S.C. 2007. Genetic analyses of sensory characteristics and relationships with fatty acid composition in the meat from Scottish Blackface lambs. *Animal* 1:1524–1531.

Kemp, J.D., Mahyuddin, M., Ely, D.G., Fox, J.M., and Moody, W.G. 1981. Effect of feeding systems, slaughter weight and sex on organoleptic properties and fatty acid composition of lamb. *J. Anim. Sci.* 51:321–330.

Khliji, S., van de Ven, R., Lamb, T.A., Lanza, M., and Hopkins, D.L. 2010. Relationship between consumer ranking of lamb colour and objective measures of colour. *Meat Sci.* 85:224–229.

Kirton, A.H., Clarke, J.N., and Hickey, S.M. 1982. A comparison of the composition and carcass quality of Kelly and Russian castrate, ram, wether and ewe lambs. *Proc. NZ Soc. Anim. Prod.* 42:117–118.

Kirton, A.H., Dalton, D.C., and Ackerley, I.R. 1974. Performance of sheep on New Zealand hill country. *NZ. J. Agric. Res.* 17:283–293.

Kirton, A.H., Winger, R.J., Dowie, J.L., and Duganzich, D.M. 1983. Palatability of meat from electrically stimulated carcasses of yearling and older entire male and female sheep. *J. Food Tech.* 18:639–649.

Komprda, T., Kuchtík, J., Jarošová, A., Dračková, E., Zemánek, L., and Filipčík, B. 2012. Meat quality characteristics of lambs of three organically raised breeds. *Meat Sci.* 91:499–505.

Kuchtík, J., Zapletal, D., and Šustová, K. 2012. Chemical and physical characteristics of lamb meat related to crossbreeding of Romanov ewes with Suffolk and Charollais sires. *Meat Sci.* 90:426–430.

Lambe, N.R., Navajas, E.A., Schofield, C.P., Fisher, A.V., Simm, G., Roehe, R., and Bünger, L. 2008. The use of various live animal measurements to predict carcass and meat quality in two divergent lamb breeds. *Meat Sci.* 80:1138–1149.

Ledward, D.A., and Shorthose, W.R. 1971. A note on the haem pigment concentration of lamb as influenced by age and sex. *Anim. Prod.* 13:193–195.

Lee, G.J. 1986. Growth and carcass characteristics of ram, cryptorchid and wether Border Leicester × Merino lambs: Effects of increasing carcass weight. *Aust. J. Exp. Agric.* 26:153–157.

Lin, K.C., Cross, H.R., Johnson, H.K., Breidenstein, B.C., Randecker, V., and Field, R.A. 1989. Mineral composition of lamb carcasses from the United States and New Zealand. *Meat Sci.* 24:47–59.

Lorentzen, T.K., and Vangen, O. 2012. Genetic and phenotypic analysis of meat quality traits in lamb and correlations to carcass composition. *Livest. Sci.* 143:201–209.

Martínez-Cerezo, S., Sañudo, C., Panea, B., Medel, I., Delfa, R., Sierra, I., Beltrán, J.A., Cepero, R., and Olleta, J.L. 2005. Breed, slaughter weight and ageing time effects on physico-chemical characteristics of lamb meat. *Meat Sci.* 69:325–333.

Mattos, R., Staples, C.R, and Thatcher, W.W. 2000. Effects of dietary fatty acids on reproduction in ruminants. *Rev. Rep.* 5:38–45.

Menzies, D.J., and Hopkins, D.L. 1996. Relationship between colour and pH in lamb loins. *Proc. Aust. Soc. Anim. Prod.* 21:353.

Meuwissen, T.H.E., and Goddard, M.E. 1996. The use of marker halotypes in animal breeding schemes. *Genet. Sel. Evol.* 28:161–176.

McPhee, M.J., Hopkins, D.L., and Pethick, D.W. 2008. Intramuscular fat levels in sheep muscle during growth. *Aust. J. Exp. Agric.* 48:904–909.

Mortimer, S.I., van der Werf, J.H.J., Jacob, R.H., Pethick, D.W., Pearce, K.L., Warner, R.D., Geesink, G.H. et al. 2010. Preliminary estimates of genetic parameters for carcass and meat quality traits in Australian sheep. *Anim. Prod. Sci.* 50:1135–1144.

Mortimer, S.I., van der Werf, J.H.J., Jacob, R.H., Pannier, L., Hopkins, D.L., Gardner, G., Pearce, K.L. et al. 2014. Genetic parameters for meat quality traits of Australian lamb meat. *Meat Sci.* 96:1016–1024.

Mousel, M.R., Notter, D.R., Leeds, T.D., Zerby, H.N., Moeller, S.J., Taylor, J.B., and Lewis, G.S. 2014. Evaluation of Columbia, USMARC-Composite, Suffolk, and Texel rams as terminal sires in an extensive rangeland production system: VIII. Quality measures of lamb longissimus dorsi. J. Anim. Sci. 92:2861–2868.

Navajas, E.A., Lambe, N.R., Fisher, A.V., Nute, G.R., Bünger, L., and Simm, G. 2008. Muscularity and eating quality of lambs: Effects of breed, sex and selection of sires using muscularity measurements by computed tomography. *Meat Sci.* 79:105–112.

National Health and Medical Research Council 2006. *Nutrient Reference Values for Australia and New Zealand Including Recommended Dietary Intakes.* Canberra: Commonwealth Department of Health and Ageing.

Oddy, H., Dalrymple, B., McEwan, J., Kijas, J., Hayes, B., van der Werf, J., Emery, D. et al. 2007. Sheep genomics and the International sheep genomics consortium. *Proc. Assoc. Advmt. Anim. Breed. Genet.* 17:411–417.

Pannier, L. Ponnampalam, E.N., Gardner, G.E., Butler, K.L., Hopkins, D.L., Ball, A.J., Jacob, R.H., Pearce, K.L., and Pethick, D.W. 2010. Prime Australian lamb supplies key nutrients for human health. *Anim. Prod. Sci.* 50:1115–1122.

Pannier, L., Gardner, G.E., Pearce, K.L., McDonagh, M., Ball, A.J., Jacob, R.H., and Pethick, D.W. 2014a. Associations of sire estimated breeding values and objective meat quality measurements with sensory scores in Australian lamb. *Meat Sci.* 96:1076–1087.

Pannier, L., Pethick, D.W., Geesink, G.H., Ball, A.J., Jacob, R.H., and Gardner, G.E. 2014b. Intramuscular fat in the *longissimus* muscle is reduced in lambs from sires selected for leanness. *Meat Sci.* 96:1068–1075.

Pannier, L., Pethick, D.W., Boyce, M.D., Ball, A.J., Jacob, R.H., and Gardner, G.E. 2014c. Associations of genetic and non-genetic factors with concentrations of iron and zinc in the *longissimus* muscle of lamb. *Meat Sci.* 96:1111–1119.

Payne, G.M., Campbell, A.W., Jopson, N.B., McEwan, J.C., Logan, C.M., and Muir, P.D. 2009. Genetic and phenotypic parameter estimates for growth, yield and meat quality traits in lamb. *Proc. NZ. Soc. Anim. Prod.* 69:210–214.

Pearce, K.L., Pannier, L., Williams, A., Gardner, G.E., Ball, A., and Pethick, D.W. 2009. Factors affecting the iron and zinc content of lamb. In Cronje, P. and Richards, N. (eds.), *Recent Advances in Animal Nutrition – Australia.* . Animal Science, University of New England, Vol 17, 37–42.

Pethick, D.W., Hopkins, D.L., D'Souza, D.N., Thompson, J.M., and Walker, P.J. 2005. Effect of animal age on the eating quality of sheep meat. *Aust. J Exp. Agric.* 45:491–498.

Pethick, D.W., Banks, R.G., Hales, J., and Ross, I.R. 2006. Australian prime lamb—A vision for 2020. *Int. J. Sheep & Wool Sci.* 54:66–73.

Pethick, D.W., Pannier, L., Gardner, G.E., Geesink, G.H., Ball, A.J., Hopkins, D.L., Jacob, R.H., Mortimer, S.I., and Pearce, K.L. 2010. Genetic and production factors that influence the content of intramuscular fat in the meat of prime lambs. In Matteo Crovetto, G. (ed.), *Energy and Protein Metabolism and Nutrition*. Wageningen Academic Publishers, The Netherlands, EAAP publication No.127: *Presented at the International Symposium on energy and protein metabolism and nutrition*, PARMA (Italy), September 6–10, 2010, pp. 673–674.

Ponnampalam, E.N., Hopkins, D.L., Butler, K.L., Dunshea, F.R., Sinclair, A.J., and Warner, R.D. 2009. Polyunsaturated fats in meat from Merino, first- and second-cross sheep slaughtered at yearling stage. *Meat Sci.* 83:314–319.

Ponnampalam, E.N., Warner, R.D., Kitessa, S., McDonagh, M.B., Pethick, D.W., Allen, D, and Hopkins, D.L. 2010. Influence of finishing systems and sampling site on fatty acid composition and retail shelf-life of lamb. *Anim. Prod. Sci.* 50:775–781.

Ponnampalam, E.N., Butler, K.L., Pearce, K.M., Mortimer, S.I., Pethick, D.W., Ball, A.J., and Hopkins, D.L. 2014. Sources of variation of health claimable long chain omega-3 fatty acids in meat from Australian lamb slaughtered at similar weights. *Meat Sci.* 96:1095–1103.

Purchas, R.W., Sobrinho, A.G.S., Garrick, D.J., and Lowe, K.I. 2002. Effects of age at slaughter and sire genotype on fatness, muscularity, and the quality of meat from ram lambs born to Romney ewes. *NZ J. Agric. Res.* 45:77–86.

Purchas, R.W. 2007. Opportunities and challenges in meat production from sheep. *Aust. J. Exp. Agric.* 47:1239–1243.

Safari, E., Fogarty, N.M., Ferrier, G.R., Hopkins, D.L., and Gilmour, A. 2001. Diverse lamb genotypes-3. Eating quality and the relationship between its objective measurement and sensory assessment. *Meat Sci.* 57:153–159.

Safari, E., Fogarty, N.M., and Gilmour, A.R. 2005. A review of genetic parameter estimates for wool, growth, meat and reproduction traits in sheep. *Livest. Prod. Sci.* 92:271–289.

Sañudo, C., Nute, G.R., Campo, M.M., Mafia, G., Baker, A., Sierra, I., Enser, M.E., and Wood, J.D. 1998. Assessment of commercial lamb meat quality by British and Spanish taste panels. *Meat Sci.* 48:91–100.

Sañudo, C., Alfonso, A.M., Sanchez, A., Berge, P., Dransfield, E., Zygoyiannis, D., Stamataris, C. 2003. Meat texture of lambs from different European production systems. *Aust. J. Agric. Res.* 54:551–560.

Scollan, N., Hocquette, J., Nuernberg, K., Dannenberger, D., Richardson, I., and Moloney, A. 2006. Innovations in beef production systems that enhance the nutritional and health value of beef lipids and their relationship with meat quality: A review. *Meat Sci.* 74:17–33.

Sink, J.D., and Caporaso, F. 1977. Lamb and mutton flavour; contributing factors and chemical aspects. *Meat Sci.* 1:119–127.

Solomon, M.B., Lynch, G.P., Ono, K., and Paroczay, E. 1990. Lipid composition of muscle and adipose tissue from crossbred ram, wether and cryptorchid lambs. *J. Anim. Sci.* 68:137–142.

Tejeda, J.F., Peña, R.E., and Andrés, A.I. 2008. Effect of live weight and sex on physico-chemical and sensorial characteristics of Merino lamb meat. *Meat Sci.* 80:1061–1067.

Thompson, J.M. 2002. Managing meat tenderness. *Meat Sci.* 62:295–308.

Thompson, J.M. 2004. The effects of marbling on flavour and juiciness scores of cooked beef, after adjusting to a constant tenderness. *Aust. J. Exp. Agric.* 44:645–652.

Thompson, J.M., Hopkins, D.L., D'Sousa, D., Walker, P.J., Baud, S.R., and Pethick, D.W. 2005. The impact of processing on sensory and objective measurements of sheep meat eating quality. *Aust. J Exp. Agric.* 45:561–573.

van der Werf, J.H.J. 2009. Potential benefits of genomic selection in sheep. *Proc. Assoc. Advmt. Anim. Breed Genet.* 18:38–41.

van der Werf, J.H.J., Kinghorn, B.P., and Banks, R.G. 2010. Design and role of an information nucleus in sheep breeding programs. *Anim. Prod. Sci.* 50:998–1003.

Warner, R.D., Dunshea, F.R., Ponnampalam, E., Ferguson, D., Gardner, G., Martin, K.M., Salvatore, L., Hopkins, D.L., and Pethick, D.W. 2006. Quality meat from Merinos. *Int. J. Sheep Wool Sci.* 54:48–53.

Warner, R.D., Ponnampalam, E.N., Kearney, G.A., Hopkins, D.L., and Jacob, R.H. 2007. Genotype and age at slaughter influence the retail shelf-life of the loin and knuckle from sheep carcasses. *Aust. J Exp. Agric.* 47:1190–1200.

Watkins, P.J., Frank, D., Singh, T.K., Young, O.A., and Warner, R.D. 2013. Sheepmeat flavor and the effect of different feeding systems: A review. *J. Agric. Food Chem.* 61:3561–3579.

Webb, E.C., Bosman, M.J.C., and Casey, N.H. 1994. Dietary influences on subcutaneous fatty acid profiles and sensory characteristics of Dorper and SA Mutton Merino wethers. *SA J. Food Sci. Nutr.* 6:45–50.

Williams, P. 2007. Nutritional composition of red meat. *Nutr. Diet.* 64:S113–S119.

Wilson, L.L., Ziegler, J.H., Rugh, MC., Watkins, J.L., Merritt, T.L., Simpson, M.J., and Kreuzberger, F.L. 1970. Comparison of live, slaughter and carcass characteristics of rams, induced cryptorchids and wethers. *J. Anim. Sci.* 31:455–458.

Wood, J.D., Richardson, R.I., Nute, G.R., Fisher, A.V., Campo, M.M., Kasapidou, E., Sheard, P.R., and Enser, M. 2003. Effects of fatty acids on meat quality: A review. *Meat Sci.* 66:21–32.

Young, O.A., Berdagué, J.L., Viallon, C., Rousset-Akrim, S., and Theriez, M. 1997. Fat-borne volatiles and sheepmeat odour. *Meat Sci.* 45:183–200.

Young, O.A., and Braggins, T.J. 1993. Tenderness of ovine semimembranosus: Is collagen concentration or solubility the critical factor? *Meat Sci.* 35:213–222.

Young, O.A., Hogg, B.W., Mortimer, B.J, and Waller, J.E. 1993a. Collagen in two muscles of sheep. *NZ J. Agric. Res.* 36:143–150.

Young, O.A., Reid, D.H, and Scales, G.H. 1993b. Effect of breed and ultimate pH on the odour and flavour of sheep meat. *NZ J. Agric. Res.* 36:363–370.

14 Transgenic Animal Technology and Meat Quality

Paul E. Mozdziak and James N. Petitte

CONTENTS

14.1 INTRODUCTION

Since the emergence of recombinant DNA technology about 40 years ago, genetic engineering has been applied to every class of organism from single cell prokaryotes to plants and animals (DePamphilis et al. 1988; Rexroad et al. 1990). Today, genetic engineering impacts almost every facet of biology and is an essential tool for biomedical research, pharmaceutical development, agriculture, and food production. Meat science, taken broadly to include preharvest animal production, postharvest processing, and value-added product development, can also take advantage of the advances in transgenic technology to provide a means of understanding the biology of muscle, the efficient production of muscle-meats, and new approaches toward improving meat quality for the consumer. At the same time, direct application to the production of genetically engineered animal food products has been slow and to-date none have been approved to enter the human food chain. However, the anticipated demand for animal protein will increase globally, particularly in developing nations, and transgenic technology will play a part in animal food production systems of the future. The purpose of this chapter is to provide an overview of animal transgenic technology in food production, and its potential applications to meat science. This will cover basic approaches toward making transgenic animals used in muscle-food production and include mammals, birds, and fish. In addition, it will also address current efforts to produce meat with better quality. Finally, consideration will be given to the regulatory demands of incorporating transgenic animal protein into human food systems.

14.2 TRANSGENIC ANIMAL TECHNOLOGY

14.2.1 Mammals

The first transgenic mammals to be developed in 1980 began with the laboratory mouse through the microinjection of DNA into a newly fertilized egg followed by embryo transfer into a recipient female (DePamphilis et al. 1988). Since that time, refinement of DNA microinjection has been applied to the production of transgenic cattle, sheep, pigs, goats, and rabbits to the point where the ability to harvest, manipulate, microinject, and transfer embryos has become a routine laboratory procedure for the production of transgenic animals (Pinkert 2002) (Figure 14.1). In addition to microinjection, nuclear transfer technology has also been used as a vehicle for the production of transgenic animals and has the advantage of allowing for specific targeted changes to the genome that cannot be accomplished with DNA microinjection alone. A third option is to use cloning, that is, a method to produce genetically identical individuals. Strictly speaking, cloning is not transgenesis; however, cloned food animals are often viewed through the same lens as transgenic animals (meat biotechnology).

14.2.2 Poultry

The techniques used to produce transgenic mammals could not be readily applied to the production of transgenic poultry due to the large, hard-shelled egg produced by birds. Early attempts at microinjection required specialized *ex ovo* culture systems and direct DNA injection yielded few transgenic birds. Until recently, transgenic poultry could only be produced using retroviral or lentiviral vectors. An advantage of lentiviral technology is that the lentiviral subclass of retroviral vectors will transduce both dividing and quiescent cells. The perceived risks associated with retroviral vectors has limited the production of transgenic poultry to biomedical applications such as the production of pharmaceutical proteins in eggs. However, recently the use of germ line stem cells has allowed the production of transgenic chickens without the use of viral vectors. This should allow the production of transgenic poultry with more agricultural applications (Petitte and Mozdziak 2002).

14.2.3 Fish

Aquaculture production is expected to increase about 33% over the next decade to meet the world's steady demand for fish as healthy and nutritious food. World

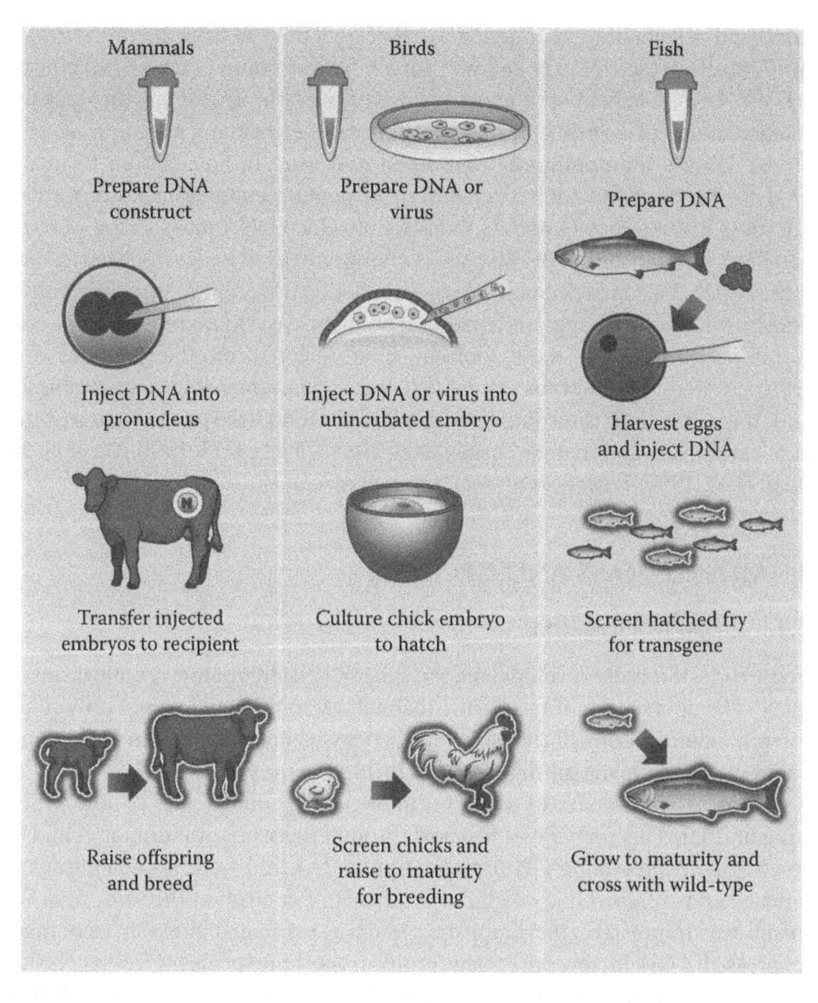

FIGURE 14.1 An outline of the general methods of generating transgenic mammals, birds, and fish (e.g., beef cattle, chickens, salmon) For mammals, the gene of choice is constructed in the laboratory and the plasmid DNA purified. A newly fertilized egg or zygote is harvested from a superovulated animal or an egg is fertilized *in vitro* and injected with the DNA construct at the pronuclear stage. The zygote is then implanted into a recipient surrogate animal, which is the standard procedure for transgenic ruminants and pigs. The offspring are evaluated for proper gene function and then bred to establish a line. For poultry, the process is more complex since the DNA is packaged in a virus that is injected into the unincubated chicken egg. Upon hatching the chicks are screened for the transgene, raised to sexual maturity, and then bred with wild-type birds. For fish, DNA expressing the gene of interest is prepared and eggs are harvested from gravid females and fertilized. The fertilized eggs are injected with the DNA expression construct and the hatched fry are screened for expression of the transgene. Transgenic fry are grown to maturity and bred with wild-type fish. In all three cases, establishment of a line of transgenic animals requires a significant number of generations. (From Pinkert, C.A. 2002. *Transgenic Animal Technology: A Laboratory Handbook*, 2nd edition. Academic Press, Amsterdam; Boston, MA.)

fisheries and aquaculture production is projected to rise from 66 million tons to about 172 million tons in 2021 and will supply 50% or more of the global consumption (FAO 2012). Coupled with concerns over depletion of wild caught stocks, efficient aquaculture of finfish will drive the growing industry. Catfish, trout, tilapia, carp, and salmon are commonly consumed and used in aquaculture to meet the demand for high-quality finfish protein. Transgenic technology also has a role to play in these diverse species and methods to introduce DNA into fish are well established. DNA microinjection of fish eggs is commonly used to develop transgenic fish (Pinkert 2002). This is facilitated by the fact that fertilization occurs externally and the transparency of fish eggs/embryos makes them easy to manipulate and screen. DNA can be injected into the cytoplasm, microinjected into the germinal vesicle, or electroporated into embryos. Many fish species are amenable to transgenesis. In general, microinjection is the preferred method for food fish species such as Atlantic salmon, carp, trout, and tilapia (Disney et al. 1988; Zhang et al. 1990; Du et al. 1992; Maclean et al. 1992; Powers et al. 1992; Cao et al. 2014).

14.3 MUSCLE MASS AND GROWTH

14.3.1 Growth Hormone

The regulation of muscle development, growth, and differentiation is multifaceted and complex. The application of transgenic technology to this process not only requires a thorough understanding of the biological processes involved, but an understanding of the process for a given animal species. In 1982, it was believed that simply overexpression of a gene, such as growth hormone (GH), could create a more-efficient, growing animal. This image was fostered through the work of Palmiter et al. (1982) in which mice expressing a GH transgene grew 2–4 fold faster than nontransgenic litter mates and obtained body weights twice that of controls at the same age. While this result was impressive, the simplistic view was taken and attempts were made to overexpress the GH in several agriculturally important species. Overexpression of GH in pigs did lead to some instances of improved growth (Pursel et al. 1990). This was also accompanied by changes in carcass composition. For example, total carcass fat, *Longissimus* muscle fat, backfat thickness, and loin eye area were reduced compared to the nontransgenic animals, while carcass protein and water content of *Longissimus* muscle were higher (Pursel et al. 1997). However, overexpression was deleterious to the general health of most of the transgenic pigs and included general lethargy and lameness, and anestrous (Pursel et al. 1990). Likewise, transgenic sheep expressing GH exhibited physiological problems (Rexroad et al. 1989, 1990, 1991; Rexroad 1991). The lesson learned from these studies indicated that any increased GH production in mammals must be tightly regulated to obtain the appropriate secretion of GH from pituitary or peripheral tissues to avoid deleterious effects on the overall physiology of the animal (Pursel et al. 2004).

Attempts were also made to express GH in chickens, again with the idea to promote more efficient growth. Bosselman et al. (1989) injected retroviral vectors into unincubated chicken embryos to insert GH into the genome. Analysis of the embryos showed increased circulating levels of GH at day 15 of incubation. However, no expression of GH in adult birds was reported, suggesting that a sustained production of GH during embryonic development was detrimental to embryo survival (Briskin et al. 1991).

In contrast to the mixed effects of overexpression of GH in mammals and birds, transgenic fish have been found to readily respond to increased production of GH. Overexpression of mammalian forms of GH in medaka, salmon, catfish, tilapia, silver sea bream, and Arctic charr all resulted in fish that grew larger than their non-transgenic counter parts (Du et al. 1992; Ozato et al. 1992; Krasnov et al. 1999; Lu et al. 2002; Caelers et al. 2005). Most importantly, the fish appear normal except for size and show no reproductive problems. Of the fish species overexpressing GH, Atlantic salmon has undergone the most rigorous analysis and efforts are underway to market the fish to consumers pending regulatory approval.

14.3.2 Myostatin

In 1997, growth/differentiating factor 8 (myostatin), was shown to be a negative regulator of muscle mass in mice (Bergin et al. 1997). When the gene for myostatin was disrupted, muscle size was increased 2–3 times over normal mice without any apparent reproductive effects and was due to both muscle hypertrophy and hyperplasia (Bergin et al. 1997). Inversely, overexpression of myostatin causes a lower muscle mass in mice (Reisz-Porszasz et al. 2003). A natural deletion in the myostatin gene in cattle was reported to be responsible for the double-muscled phenotype in cattle and several mutations have been associated with muscling in sheep (see Chapter 10; Clop et al. 2006; Boman et al. 2010; Hu et al. 2013). Hence, downregulation or disruption of myostatin expression is often viewed as a means to increase muscle mass in food animals. In fish, the inhibition of myostatin using RNA interference (RNAi), resulted in the increased size of zebrafish through muscle hypertrophy and hyperplasia, including a double muscle phenotype (Acosta et al. 2005; Lee et al. 2009). Likewise, a dominant-negative form of myostatin or RNAi also increased muscle fiber number in medaka (Sawatari et al. 2010; Terova et al. 2013). Interestingly, transgenic salmon overexpressing GH also showed decreased expression of myostatin in faster growing fish (Roberts et al. 2004). Overexpression of follistatin, a negative regulator of myostatin in trout also leads to increased muscling (Medeiros et al. 2009). Myostatin knockdown transgenic calves were generated with one animal showing some increased muscle mass (Tessanne et al. 2012).

Recently, double muscled cattle were produced through zinc-finger nuclease editing of the myostatin gene. A 50% reduction in myostatin expression resulted in the double-muscled phenotype appearing at one month of age (Luo et al. 2014). Likewise, transgenic sheep were generated using RNAi knockdown of myostatin. In this case, the transgenic sheep grew faster than control sheep with an increase in myofiber diameter (Hu et al. 2013). The results with fish, cattle, and sheep are the most promising to date for manipulating muscle mass via reduction in myostatin.

Future attempts to manipulate muscle mass in meat animals will have to consider the emerging picture associated with the signaling pathways involved in the control of skeletal muscle hypertrophy and atrophy (Figure 14.2). Generally, IGF-I signaling increases muscle protein synthesis and acts through Akt, which

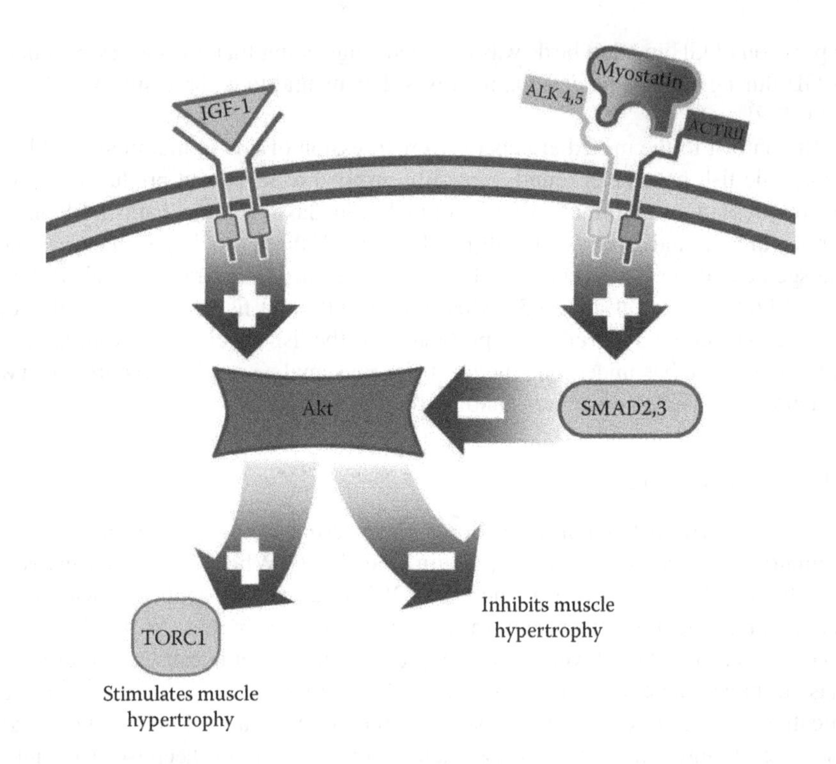

FIGURE 14.2 Insulin-like growth factor-1 (IGF-1) and myostatin control of muscle size. Signaling by IGF-1 positively regulates muscle mass through the downstream signaling of Akt and the TORC1 complex, a multiprotein complex that includes the protein raptor, which can be inhibited by rapamycin. IGF-1 stimulates protein synthesis in muscle. Myostatin and a few other members of the transforming growth factor-beta family inhibits muscle size through the phosphorylation of SMAD2,3 and inhibits Akt. Transgenic animals where the myostatin signaling pathway is downregulated one approach toward increasing muscle mass in mammals. (From Egerman, M.A., and Glass, D.J. 2014. *Crit. Rev. Biochem. Mol. Biol.* 49:59–68; Lee, S.J. 2004. *Annu. Rev. Cell. Dev. Biol.* 20:61–86.)

in-turn activates mTOR complex 1 and mediates muscle hypertrophy (Egerman and Glass 2014). The latter complex requires the protein raptor for function and is inhibited by rapamycin. Conversely, the myostatin pathway negatively regulates muscle size by stimulating SMAD2/3 which inhibits Akt (Egerman and Glass 2014). Hence, inhibition of TGF-beta signaling mediators such as the SMAD family can also result in muscle hypertrophy (Lee 2004; Sartori et al. 2013). Rather than simple overexpression or disruption of gene function, the most useful transgenic animals for meat production could result from minor/subtle changes in gene expression.

14.4 FAT CONTENT/COMPOSITION

In addition to an increase in overall muscle mass, manipulation of muscle fat composition is another goal associated with the potential for transgenic technology in meat science. Consumers have become more conscious of the amount and type of dietary fat consumed with a trend toward reducing saturated fats and increasing unsaturated fats (Solomon et al. 1994; Wang et al. 2012). Several approaches have emerged for the manipulation of muscle fat using transgenic technology. For example, early work on transgenic animals also resulted in changes in the lipid composition of the carcass. In pigs the overexpression of GH resulted in animals with less total carcass fat, less saturated fatty acids, and less monounsaturated fatty acids, and polyunsaturated fatty acids. The effect became more pronounced as the animals grew and suggested that the increase in lean meat was responsible for a dilution of total carcass fat (Solomon et al. 1994). When transgenic pigs were generated to express IGF-I specifically in muscle, the transgenic pigs had less fat and more lean tissue than control pigs without significant differences in body weight (Pursel et al. 2004). Therefore, attempts to change muscle mass can also have consequences on the fat content of meat.

14.4.1 FAT-1 Gene

In some cases it could be desirable to alter the composition of animal fats. Unlike fish, mammals cannot produce omega-3 fatty acids from omega-6 fatty acids and must obtain omega-3 fatty acids from the diet. In 2004, the n-3 fatty acid desaturase gene from *Caenorhabditis elegans* (fat-1) was expressed in mice and allowed the conversion of n-6 to n-3 fatty acids in the absence of dietary n-3 fatty acids (Kang et al. 2004). This observation in mice suggested that a similar approach in livestock animals could yield meat with an altered fatty acid composition. Subsequently, Lai et al. (2006) described the generation of transgenic pigs that expressed a humanized fat-1 gene that resulted in high levels of n-3 fatty acids in their tissues. Likewise, transgenic pigs expressing a nematode fat-1 gene that resulted in a lowered n-6/n-3 ratio in muscle and major organs was created (Pan et al. 2010). Recently, the fat-1 gene from another nematode species was overexpressed in pigs and resulted in an enrichment of n-3 polyunsaturated fatty acids confirming previous work (Zhou et al. 2014). Similar results of the overexpression of fat-1 have been demonstrated for cattle and sheep (Duan et al. 2012; Wang et al. 2012). However, gene silencing in fat-1 sheep illustrates the need to carefully choose functional promoters (Duan et al. 2012). Nevertheless, these reports show that the fatty acid composition of meat can be altered toward a more unsaturated fatty acid composition.

Other genes show promise for the manipulation of the fat content in meat. Pigs transgenic for the delta12 fatty acid desaturase from spinach resulted in 20% more linoleic acid present in adipose than that of wild-type pigs and show that the functional expression of a plant gene for a fatty acid desaturase is functional in mammals (Saeki et al. 2004). Other candidate mammalian genes that alter lipid composition include peroxisome proliferator-activated receptor gamma-2 (PPARγ2), interleukin-15 (Il-15), and 1,2-acyl CoA:diacylglyceroltransferase-1 (DGAT1). In mice, intramuscular fat content increases substantially with overexpression of DGAT1, mainly through an increase in triglycerides (Roorda et al. 2005). This observation immediately suggests the possibility of generating meat animals with a high intramuscular fat content to improve meat quality. Conversely, overexpression of Il-15, a "myokine" normally expressed in muscle inhibits adipose deposition in mice; thereby, influencing carcass composition (Quinn et al. 2009). PPARy2 regulates adipocyte differentiation and fat storage. In transgenic mice, overexpression of PPARy2 increased triacylglycerol content and promoted the accumulation of more polyunsaturated fatty acids (Huang et al. 2012). Efforts are now underway to generate transgenic sheep overexpressing PPARy (Qin et al. 2013). A similar effort in pigs is likely on the horizon.

14.5 EXTENDING SHELF LIFE

Most of the focus on transgenic animals and meat quality has focused on preharvest growth and muscle/carcass composition, however postharvest storage of meat could also benefit from transgenic technology. Oxidation of red meat manifests itself via the conversion of the red muscle pigment myoglobin to brown metmyoglobin (see Chapter 2). In addition, fatty acid oxidation occurs during meat storage over 4–7 days and leads to the development of rancid odors and flavors from the degradation of the polyunsaturated fatty acids in the tissue membranes. Meanwhile, health conscious consumers tend to prefer more polyunsaturated fats in meat products, thereby exacerbating the opportunity for fatty acid oxidation. One of the approaches to ameliorate product oxidation during storage is to add naturally occurring antioxidants in the diet of preslaughter animals, such as vitamin E (see Chapter 4). Increasing the intramuscular levels of vitamin E inhibits oxidation in beef, lamb, pork, and poultry meat (Galvin et al. 2000; Blatt et al. 2001; Rebolé et al. 2006; Boler et al. 2009; Rymer and Givens 2010; Cardenia et al. 2011; Juarez et al. 2012; Gonzalez-Calvo et al. 2015). However, vitamin E accumulation in meat varies and is probably dependent upon the efficiency of alpha-tocopherol transport proteins and other regulatory molecules (Blatt et al. 2001). Vitamin E supplementation in cattle increases beef color stability, and inhibits fatty acid oxidation. Therefore, it may be possible to engineer cattle to accumulate higher levels of alpha-tocopherol in muscle and fat to inhibit fatty acid oxidation and increase shelf-life.

14.6 REGULATION OF TRANSGENIC MEAT

No matter what the future prospects are for transgenic animal technology for meat production and storage, interactions with regulatory bodies remains a large hurdle and is in some ways more difficult than the science. In the United States, engineered foods (microbe, plant, or animal) are regulated by the USDA, the Environmental Protection Agency (EPA) and the Food and Drug Administration (FDA). While genetically engineered plants have penetrated the world markets and transgenic plants are common in human and animal food stuffs, it is taking much longer for general acceptance of genetically engineered animals for meat production. The FDA Center for Veterinary Medicine (FDA-CWM) specifically regulates transgenic animals and views such animals as containing a new drug. Transgenic salmon expressing GH produced by the company AquaBounty is the first transgenic meat product under consideration for FDA approval. The review process began in 1996 with rather limited regulatory guidance. In 2009, some 13 years later the FDA-CVM released guidance on the development and approval process for transgenic animals. Publication CVM GFI #187, “The Regulation of Genetically Engineered Animals Containing

Heritable rDNA Construct" provides recommendations to producers of transgenic animals to help navigate the review process. The FDA guidance is very similar to CAC/GL 68-2008 "Guideline for the Conduct of Food Safety Assessment of Foods Derived from Recombinant-DNA Animals" published by the Codex Alimentarius International Food Standards of the Food and Agriculture Organization and the World Health Organization of the United Nations. The review process for the FDA-CVM has been and is rather arduous and includes documenting molecular characterization of both the recombinant DNA aspect of the animal and its lineage; comprehensive data on the characteristics of the animal and its health; safety assessments for human consumption; demonstration of effectiveness of the transgenic; and assessment of any environmental impact. All these evaluation criteria were addressed in the case of transgenic salmon. Subsequently, the final review of the transgenic salmon and the required public comment ended in May 2013. Currently, the FDA has yet to issue a final ruling on marketing the fish for human consumption. However, should approval be granted, this will pave the way for the development and marketing of transgenic meat in the future.

14.7 CONCLUSIONS

Transgenic technology as applied to the production of muscle meats has been slow compared to the use of transgenics in plants for crop production. In addition to consumer resistance to transgenic products entering the human food chain, technological limitations and regulatory issues have contributed to the relatively slow adaptation of transgenic food animals. Early transgenic animals were created based upon limited knowledge of the physiology and biochemical mechanisms of growth and yielded mixed results. Today new insights to muscle development and growth have allowed better approaches toward increasing muscle mass and improved efficiency. So far, candidate products from transgenic mammals and poultry still await further development. However, transgenic fish stocks specifically developed for aquaculture have been developed which show significant promise for the increasing demand for fish. In this regard, transgenic salmon appear to be the first transgenic food animal awaiting regulatory approval in the United States. Once approved, transgenic fish should pave the way for the use of transgenic mammals and poultry in efforts to meet the growing global demand for animal protein.

ACKNOWLEDGMENTS

The authors would like to thank Jennifer Petitte for the illustrations.

REFERENCES

Acosta, J., Carpio, Y., Borroto, I., Gonzalez, O., and Estrada, M.P. 2005. Myostatin gene silenced by RNAi show a zebrafish giant phenotype. *J. Biotechnol.* 119:324–331.

Bergin, A., Kim, G., Price, D.L., Sisodia, S.S., Lee, M.K., and Rabin, B.A. 1997. Identification and characterization of a mouse homologue of the spinal muscular atrophy-determining gene, survival motor neuron. *Gene* 204:47–53.

Blatt, D.H., Leonard, S.W., and Traber, M.G. 2001. Vitamin E kinetics and the function of tocopherol regulatory proteins. *Nutrition* 17:799–805.

Boler, D.D., Gabriel, S.R., Yang, H., Balsbaugh, R., Mahan, D.C., Brewer, M.S., McKeith, F.K., and Killefer, J. 2009. Effect of different dietary levels of natural-source vitamin E in grow-finish pigs on pork quality and shelf life. *Meat Sci.* 83:723–730.

Boman, I.A., Klemetsdal, G., Nafstad, O., Blichfeldt, T., and Vage, D.I. 2010. Impact of two myostatin (MSTN) mutations on weight gain and lamb carcass classification in Norwegian White Sheep (*Ovis aries*). *Genet. Sel. Evol.* 42:4:1–7.

Bosselman, R.A., Hsu, R.Y., Boggs, T. et al. 1989. Replication-defective vectors of reticuloendotheliosis virus transduce exogenous genes into somatic stem cells of the unincubated chicken embryo. *J. Virol.* 63:2680–2689.

Briskin, M.J., Hsu, R.Y., Boggs, T., Schultz, J.A., Rishell, W., and Bosselman, R.A. 1991. Heritable retroviral transgenes are highly expressed in chickens. *Proc. Natl. Acad. Sci. USA*. 88:1736–1740.

Caelers, A., Maclean, N., Hwang, G., Eppler, E., and Reinecke, M. 2005. Expression of endogenous and exogenous growth hormone (GH) messenger (m) RNA in a GH-transgenic tilapia (*Oreochromis niloticus*). *Transgenic Res.* 14:95–104.

Cao, M., Chen, J., Peng, W. et al. 2014. Effects of growth hormone over-expression on reproduction in the common carp *Cyprinus carpio* L. *Gen. Comp. Endocrinol.* 195:47–57.

Cardenia, V., Rodriguez-Estrada, M.T., Cumella, F., Sardi, L., Della Casa, G., and Lercker, G. 2011. Oxidative stability of pork meat lipids as related to high-oleic sunflower oil and vitamin E diet supplementation and storage conditions. *Meat Sci.* 88:271–279.

Clop, A., Marcq, F.H., Takeda, P.D. et al. 2006. A mutation creating a potential illegitimate microRNA target site in the myostatin gene affects muscularity in sheep. *Nat. Genet.* 38:813–818.

DePamphilis, M.L., Herman, S.A., Martinez-Salas, E. et al. 1988. Microinjecting DNA into mouse ova to study DNA replication and gene expression and to produce transgenic animals. *Biotechniques* 6:662–680.

Disney, J.E., Johnson, K.R., Banks, D.K., and Thorgaard, G.H. 1988. Maintenance of foreign gene expression and independent chromosome fragments in adult transgenic rainbow trout and their offspring. *J. Exp. Zool.* 248:335–344.

Du, S.J., Gong, Z.Y., Fletcher, G.L. et al. 1992. Growth enhancement in transgenic Atlantic salmon by the use of an "all fish" chimeric growth hormone gene construct. *Biotechnology (N Y)* 10:176–181.

Duan, B., Cheng, L., Gao, Y. et al. 2012. Silencing of fat-1 transgene expression in sheep may result from hypermethylation of its driven cytomegalovirus (CMV) promoter. *Theriogenology* 78:793–802.

Egerman, M.A., and Glass, D.J. 2014. Signaling pathways controlling skeletal muscle mass. *Crit. Rev. Biochem. Mol. Biol.* 49:59–68.

FAO. 2012. FAO yearbook. *Fishery and Aquaculture Statistics*. Rome, FAO. 76 pp. Available at: http://www.fao.org/3/a-i3720e/index.html

Galvin, K., Lynch, A.M., Kerry, J.P., Morrissey, P.A., and Buckley, D.J. 2000. Effect of dietary vitamin E supplementation on cholesterol oxidation in vacuum packaged cooked beef steaks. *Meat Sci.* 55:7–11.

Gonzalez-Calvo, L., Ripoll, G., Molino, F., Calvo, J.H., and Joy, M. 2015. The relationship between muscle alpha-tocopherol concentration and meat oxidation in light lambs fed vitamin E supplements prior to slaughter. *J. Sci. Food. Agric.* 95(1):103–110.

Hu, S., Ni, W., Sai, W., Zi, H., Qiao, J., Wang, P., and Sheng, J. 2013. Knockdown of myostatin expression by RNAi enhances muscle growth in transgenic sheep. *PLOS ONE* 8(3):e58521.

Huang, J., Xiong, Y., Li, T., Zhang, L., Zhang, Z., Zuo, B., Xu, D., and Ren, Z. 2012. Ectopic overexpression of swine PPARγ2 upregulated adipocyte genes expression and triacylglycerol in skeletal muscle of mice. *Transgenic Res.* 21:1311–1318.

Juarez, M., Dugan, M.E., Aldai, N., Basarab, J.A., Baron, V.S., McAllister, T.A., and Aalhus, J.L. 2012. Beef quality attributes as affected by increasing the intramuscular levels of vitamin E and omega-3 fatty acids. *Meat Sci.* 90:764–769.

Kang, J.X., Wang, J., Wu, L., and Kang, Z.B. 2004. Transgenic mice: Fat-1 mice convert n-6 to n-3 fatty acids. *Nature* 427:504.

Krasnov, A., Agren, J.J., Pitaknen, T.I., and Molsa, H. 1999. Transfer of growth hormone (GH) transgenes into Arctic charr. (*Salvelinus alpinus L.*) II. Nutrient partitioning in rapidly growing fish. *Genet. Anal.* 15:99–105.

Lai, L., Kang, J.X., Li, R. et al. 2006. Generation of cloned transgenic pigs rich in omega-3 fatty acids. *Nat. Biotechnol.* 24:435–436.

Lee, C.Y., Hu, S.Y., Gong, H.Y., Chen, M.H., Lu, J.K., and Wu, J.L. 2009. Suppression of myostatin with vector-based RNA interference causes a double-muscle effect in transgenic zebrafish. *Biochem. Biophys. Res. Commun.* 387:766–771.

Lee, S.J. 2004. Regulation of muscle mass by myostatin. *Annu. Rev. Cell. Dev. Biol.* 20:61–86.

Lu, J.K., Fu, B.H., Wu, J.L., and Chen, T.T. 2002. Production of transgenic silver sea bream (*Sparus sarba*) by different gene transfer methods. *Mar. Biotechnol. (NY)* 4:328–337.

Luo, J., Song, Z., Yu, S., Cui, D., Wang, B., Ding, F., Li, S., Dai, Y., and Li, N. 2014. Efficient Generation of Myostatin (MSTN) Biallelic Mutations in Cattle Using Zinc Finger Nucleases. *PLOS ONE* 9:e95225.

Maclean, N., Iyengar, A., Rahman, A., Sulaiman, Z., and Penman, D. 1992. Transgene transmission and expression in rainbow trout and tilapia. *Mol. Mar. Biol. Biotechnol.* 1:355–365.

Medeiros, E.F., Phelps, M.P., Fuentes, F.D., and Bradley, T.M. 2009. Overexpression of follistatin in trout stimulates increased muscling. *Am. J. Physiol. Regul. Integr. Comp. Physiol.* 297:R235–R242.

Ozato, K., Wakamatsu, Y., and Inoue, K. 1992. Medaka as a model of transgenic fish. *Mol. Mar. Biol. Biotechnol.* 1:346–354.

Palmiter, R.D., Brinster, R.L., Hammer, R.E. et al. 1982. Dramatic growth of mice that develop from eggs microinjected with metallothionein-growth hormone fusion genes. *Nature* 300:611–615.

Pan, D., Zhang, L., Zhou, Y. et al. 2010. Efficient production of omega-3 fatty acid desaturase (sFat-1)-transgenic pigs by somatic cell nuclear transfer. *Sci. China Life Sci.* 53:517–523.

Petitte, J.N., and Mozdziak, P.E. 2002. Production of transgenic poultry. In: Pinkert, C.A. (ed.), *Transgenic Technology: A Laboratory Handbook*, 2nd edition. Elsevier Science, New York, pp. 279–306.

Pinkert, C.A. 2002. *Transgenic Animal Technology: A Laboratory Handbook*, 2nd edition. Academic Press, Amsterdam; Boston, MA.

Powers, D.A., Hereford, L., Cole, T., Chen, T.T., Lin, C.M., Kight, K., Creech, K., and Dunham, R. 1992. Electroporation: A method for transferring genes into the gametes of zebrafish (*Brachydanio rerio*), channel catfish (*Ictalurus punctatus*), and common carp (*Cyprinus carpio*). *Mol. Mar. Biol. Biotechnol.* 1:301–308.

Pursel, V.G., Hammer, R.E., Palmiter, R.D., Brinster, R.L., Bolt, D.J., Miller, K.F., and Pinkert, C.A. 1990. Expression and performance in transgenic pigs. *J. Reprod. Fertil. Suppl.* 40:235–245.

Pursel, V.G., Mitchell, A.D., Bee, G., Elsasser, T.H., McMurtry, J.P., Wall, R.J., Coleman, M.E., and Schwartz, R.J. 2004. Growth and tissue accretion rates of swine expressing an insulin-like growth factor I transgene. *Anim. Biotechnol.* 15:33–45.

Pursel, V.G., Wall, R.J., Solomon, M.B., Bolt, D.J., Murray, J.E., and Ward, K.A. 1997. Transfer of an ovine metallothionein-ovine growth hormone fusion gene into swine. *J. Anim. Sci.* 75:2208–2214.

Qin, Y., Chen, H., Zhang, Y. et al. 2013. Cloning of the Xuhuai goat PPARγ gene and the preparation of transgenic sheep. *Biochem. Genet.* 51:543–553.

Quinn, L.S., Anderson, B.G., Strait-Bodey, L., Stroud, A.M., and Argiles, J.M. 2009. Oversecretion of interleukin-15 from skeletal muscle reduces adiposity. *Am. J. Physiol. Endocrinol. Metab.* 296:E191–E202.

Rebolé, A., Rodríguez, M.L., Ortiz, L.T., Alzueta, C., Centeno, C., Viveros, A., Brenes, A., and Arija, I. 2006. Effect of dietary high-oleic acid sunflower seed, palm oil and vitamin E supplementation on broiler performance, fatty acid composition and oxidation susceptibility of meat. *Br. Poult. Sci.* 47:581–591.

Reisz-Porszasz, S., Bhasin, S., Artaza, J.N., Shen, R., Sinha-Hikim, I., Hogue, A., Fielder, T.J., and Gonzalez-Cadavid, N.F. 2003. Lower skeletal muscle mass in male transgenic mice with muscle-specific overexpression of myostatin. *Am. J. Physiol. Endocrinol. Metab.* 285:E876–E888.

Rexroad, C.E. Jr. 1991. Production of sheep transgenic for growth hormone genes. *Biotechnology.* 16:259–263.

Rexroad, C.E. Jr., Behringer, R.R., Palmiter, R.D., Brinster, R.L., Miller, K.F., Mayo, K., Bolt, D.J., and Elsasser, T.H. 1991. Transferrin- and albumin-directed expression of growth-related peptides in transgenic sheep. *J. Anim. Sci.* 69:2995–3004.

Rexroad, C.E. Jr., Hammer, R.E., Behringer, R.R., Palmiter, R.D., and Brinster, R.L. 1990. Insertion, expression and physiology of growth-regulating genes in ruminants. *J. Reprod. Fertil. Suppl.* 41:119–124.

Rexroad, C.E. Jr., Hammer, R.E., Bolt, D.J., Mayo, K.E., Frohman, L.A., Palmiter, R.D., and Brinster, R.L. 1989. Production of transgenic sheep with growth-regulating genes. *Mol. Reprod. Dev.* 1:164–169.

Roberts, S.B., McCauley, L.A., Devlin, R.H., and Goetz, F.W. 2004. Transgenic salmon overexpressing growth hormone exhibit decreased myostatin transcript and protein expression. *J. Exp. Biol.* 207:3741–3748.

Roorda, B.D., Hesselink, M.K., Schaart, G., Moonen-Kornips, E., Martínez-Martínez, P., Losen, M., De Baets, M.H., Mensink, R.P., and Schrauwen, P. 2005. DGAT1 overexpression in muscle by *in vivo* DNA electroporation increases intramyocellular lipid content. *J. Lipid Res.* 46:230–236.

Rymer, C., and Givens, D.I. 2010. Effects of vitamin E and fish oil inclusion in broiler diets on meat fatty acid composition and on the flavour of a composite sample of breast meat. *J. Sci. Food Agric.* 90:1628–1633.

Saeki, K., Matsumoto, K., Kinoshita, M., Suzuki, I., Tasaka, Y., Kano, K., Tagushi, Y., Mikami, K., Hirabayashi, M., and Kashiwazaki, N. 2004. Functional expression of a Delta12 fatty acid desaturase gene from spinach in transgenic pigs. *Proc. Natl. Acad. Sci. USA.* 101:6361–6366.

Sartori, R., Schirwis, E., Blaauw, B. et al. 2013. BMP signaling controls muscle mass. *Nat. Genet.* 45:1309–1318.

Sawatari, E., Seki, R., Adachi, T., Hashimoto, H., Uji, S., Wakamatsu, Y., Nakata, T., and Kinoshita, M. 2010. Overexpression of the dominant-negative form of myostatin results in doubling of muscle-fiber number in transgenic medaka (*Oryzias latipes*). *Comp. Biochem. Physiol. A Mol. Integr. Physiol.* 155:183–189.

Solomon, M.B., Pursel, V.G., Paroczay, E.W., and Bolt, D.J. 1994. Lipid composition of carcass tissue from transgenic pigs expressing a bovine growth hormone gene. *J. Anim. Sci.* 72:1242–1246.

Terova, G., Rimoldi, S., Bernardini, G., and Saroglia, M. 2013. Inhibition of myostatin gene expression in skeletal muscle of fish by *in vivo* electrically mediated dsRNA and shRNAi delivery. *Mol. Biotechnol.* 54:673–684.

Tessanne, K., Golding, M.C., Long, C.R., Peoples, M.D., Hannon, G., and Westhusin, M.E. 2012. Production of transgenic calves expressing an shRNA targeting myostatin. *Mol. Reprod. Dev.* 79:176–185.

Wang, W., Guo, X.M., Wang, J., and Lai, S.J. 2012. Product fat-1 transgenic simmental cattle endogenously synthesizing omega-3 polyunsaturated fatty acid using OSM. *J. Anim. Vet. Adv.* 11:1041–1045.

Zhang, P.J., Hayat, M., Joyce, C., Gonzalez-Villaseñor, L.I., Lin, C.M., Dunham, R.A., Chen, T.T., and Powers, D.A. 1990. Gene transfer, expression and inheritance of pRSV-rainbow trout-GH cDNA in the common carp, *Cyprinus carpio* (*Linnaeus*). *Mol. Reprod. Dev.* 25:3–13.

Zhou, Y., Lin, Y., Wu, X., Feng, C., Long, C., Xiong, F., Wang, N., Pan, D., and Chen, H. 2014. The high-level accumulation of n-3 polyunsaturated fatty acids in transgenic pigs harboring the n-3 fatty acid desaturase gene from *Caenorhabditis briggsae*. *Transgenic Res.* 23:89–97.

15 Production of High-Quality Meat

Yuan H. Brad Kim, Heather A. Channon, Darryl N. D'Souza, and David Hopkins

CONTENTS

15.1 INTRODUCTION

Providing consistently high-quality and wholesome meat products to consumers, the ultimate end-users of these products, is crucial to the continued success of the meat industry (Smith et al. 2008a). Numerous studies have reported that consumers are willing to pay premiums for meat products with guaranteed eating quality (palatability or often simply called "taste" by consumers) (Miller et al. 2001; Shackelford et al. 2001; Polkinghorne 2006; Polkinghorne and Thompson 2010). The production of high-quality meat is a complicated multistep process, which can be also called "a whole value chain" as each individual step from live animal production/management to postslaughter events can contribute to influence meat quality. In this regard, various quality assurance programs overseeing entire meat production systems such as live-animal factors, carcass-treatment factors, and carcass-trait constraints have been developed and implemented in the meat industry around the world (Smith et al. 2008b). Different carcass classification/quality grading systems have been implemented globally to segment carcasses into groups with similar eating quality attributes (Davis et al. 1979). However, these typical carcass quality grading systems do not account for variation in palatability at the consumer level, as this requires more complicated combinations of different cut types and various cooking methods (Cho et al. 2010). This limitation led the Australian red meat industry to develop a more comprehensive/accurate grading system (e.g., Meat Standards Australia; MSA) that can predict eating quality and palatability while being "user-friendly."

As some of the previous chapters cover the important topics in live animal production and its relevance to meat quality (e.g., Chapters 2, 4, and 5), this chapter will focus on discussing some of the critical postslaughter factors that influence eating quality attributes, and introduce carcass classification/grading systems and quality assurance/management programs currently applied in the meat industry around the world. Due to the physiological, biochemical, and meat quality similarities of beef and lamb meat, one combined discussion on the postslaughter factors will be made for meat from those two species, with another separate subsection for pork. The primary purpose of this chapter is to introduce brief fundamental concepts and some of essential postharvest factors rather than an intensive literature review of underlying biochemical/biophysical mechanisms of the pre/postrigor muscle (see Chapters 2 and 3), and to discuss pathways to ensure delivery of a consistent quality eating experience to the consumers based on the current industry practices.

15.2 PRODUCING BEEF AND LAMB WITH HIGH QUALITY

15.2.1 Critical Postharvest Practices

Substantial biochemical/biophysical changes occur during the conversion process of muscle into meat, and these changes are greatly influenced by various postharvest practices such as carcass chilling regimes (Savell et al. 2005), electrical inputs (Hwang et al. 2003), carcass hanging methods and/or stretching (Hostetler et al. 1970; Bouton et al. 1973), and postmortem aging conditions (Gruber et al. 2006). As these postharvest processing factors have direct impacts on meat quality attributes, the brief fundamental concepts and importance of postharvest factors are discussed as below.

15.2.1.1 Control of pH and Temperature Decline

The rate of pH and temperature decline of prerigor muscle greatly influence meat quality characteristics such as tenderness, color, and water-holding capacity (Hwang and Thompson 2001; Huff-Lonergan and Lonergan 2005; Savell et al. 2005; Thompson et al. 2006; Kim et al. 2014). During rigor development, the chilling regimes (temperature) affect the rate of glycolytic reactions and thus impact on the rate of pH decline of the muscle (Marsh et al. 1987; Pike et al. 1993). Furthermore, the rate of pH decline under different prerigor temperatures has a substantial influence on the degree of shortening (Marsh 1954; Locker and Hagyard 1963) and the aging potential (proteolytic enzyme activity) (Dransfield et al. 1992; Hwang and Thompson 2001; Kim et al. 2014), which subsequently impact on meat tenderness (Marsh et al. 1981). Smith et al. (1976) reported that lamb carcasses with increased quantities of subcutaneous fat had slower chilling rates, less shortening of sarcomeres, and enhanced proteolytic enzyme activities, resulting in improved tenderness compared to leaner lamb carcasses. The classical research work of Locker and Hagyard (1963) emphasized the importance of optimizing the rate of decline in pH prerigor relative to temperature for tenderness, where minimal shortening was observed when the carcass entered rigor at 15–20°C. In contrast, excessive shortening occurred when prerigor muscles were exposed to either lower or higher temperatures. The former case traditionally has been called "cold-shortening," where the rapid temperature decline triggers an increased leakage of calcium from the sarcoplasmic reticulum (Locker and Hagyard 1963; Bendall 1978), resulting in more severe shortening (about 50%) than the normal rigor shortening (Locker and Hagyard 1963). The incidence of cold-shortening of beef muscles has been remarkably diminished in the current meat industry due to the development of chilling systems, application of electrical stimulation, and/or increased carcass weight and fat thickness (e.g., heavy grain-fed cattle).

On the other hand, the incidence of the latter case where the prerigor muscle is exposed to a higher than normal temperature accompanied by accelerated pH decline appears to be high. This phenomenon has traditionally been called heat shortening or heat-induced toughening (Locker and Hagyard 1963), which normally occurs at carcass temperature >35°C at the point at which the loin pH reaches 6.0 (Thompson 2002). Warner et al. (2014) reported that the high rigor temperature incidence based on measurements in the *Longissimus lumborum* muscle was 74.6% across seven beef processing plants (ranging from 56% to 94%) in Australia. They reported high correlations between days in the feedlot, heavier carcass weight, longer duration of electric inputs at the hide puller, fatness of cattle, and a higher temperature at pH 6 (Warner et al. 2014). The combined incidence of rapid pH decline at a higher than normal prerigor temperature can result in extensive denaturation of myofibrillar and sarcoplasmic proteins, which in turn cause adverse impacts on meat quality attributes (Kim et al. 2014). A classic example of the quality defects associated with high temperature/rapid pH decline is pale, soft, and exudative (PSE) meat, which predominantly occurs in porcine muscle. However, PSE-like phenomena (particularly lighter in color and with weep on the surface) has also been observed in beef muscles which have undergone high temperature/rapid pH decline conditions postslaughter (Ledward 1985; Roeber et al. 2000; Sammel et al. 2002). Decreased tenderness of beef muscles due to heat shortening and/or reduced aging potential (early exhaustion of proteolytic enzyme activities) also has been reported (Lee and Ashmore 1985; Hertzman et al. 1993; Hwang and Thompson 2001; Kim et al. 2012).

In light of the critical pH and temperature relationships with meat quality, the pH/temperature window (or "abattoir window") was established by MSA. In this case the target is for the pH to be 6 (Temp@pH6) in the *Longissimus thoracis et lumborum* muscle between the 2nd and 5th lumbar vertebrae between 12°C and 35°C (Figure 15.1) to prevent the incidence of either cold-shortening or heat-toughening

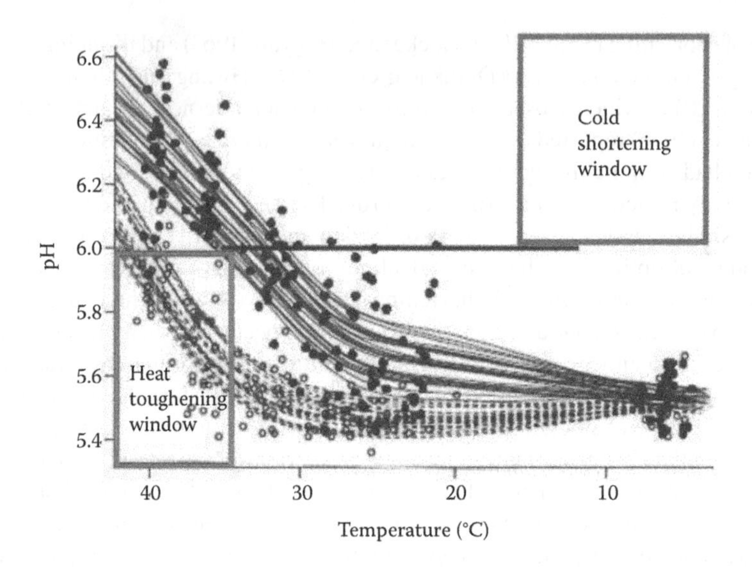

FIGURE 15.1 The rates of pH/temperature decline of beef carcass sides that were stimulated (open circles) or nonstimulated (black circles) with the different windows illustrating heat toughening (temperatures at pH 6 (Temp@pH6) of the *Longissimus muscle* >35°C), cold-shortening (Temp@pH6 < 12°C), and ideal window (Temp@pH6 < 35°C and >12°C). (Adapted with permission from Hopkins, D.L. 2014. *Beef Cattle Production and Trade.* CSIRO Publishing, Melbourne, pp. 17–46, with modification.)

(Polkinghorne et al. 2008b) in beef. However, the excessive use of electrical inputs (rigid probes and/or carcass stimulation) possibly coupled with heavier carcasses (with greater muscle mass and/or subcutaneous fat thickness) can result in high temperature/rapid pH decline conditions as discussed above. Therefore, finding an optimal pH/temperature decline condition for prerigor muscles through the appropriate usage of electrical inputs and chilling regimes is a crucial postharvest factor that governs meat quality and eating characteristics. In this regard, hot-boning (excised prerigor muscles within 45 min after slaughter often from carcasses being electrically stimulated) can be a good way to facilitate the optimization of pH/temperature decline rate of prerigor muscles to ensure the eating quality of fresh meat products. However, hot-boning is not practically feasible in those countries where grading of beef carcasses occurs after a few days postslaughter. Traditionally hot boning has been utilized on older beef carcasses that are of generally lower eating quality, although partial hot-boning of younger beef carcasses has produced encouraging results (Taylor et al. 2012).

15.2.1.2 Carcass Suspension and Stretch

In general, the length of the sarcomere (the individual unit of a myofibril that is responsible for muscle contraction) is positively related to tenderness, where the longer the sarcomere, the more tender the meat (Locker and Hagyard 1963; Herring et al. 1965). Altering carcass suspension procedures by the pelvis or aitchbone (the obturator foramen), also generally known as tenderstretching, can substantially improve meat tenderness by minimizing muscle shortening and increasing postrigor sarcomere length (Hostetler et al. 1970, 1973). Tenderstretching straightens and stretches the vertebral column, which prevents the *Longissimus* muscle from shortening as much as the traditional Achilles-tendon (leg) suspension. Stretching also impacts the round muscles due to the femur being pulled forward by its weight (Hostetler et al. 1970). Despite these positive effects on eating quality, pelvic suspension has not been adopted widely in the meat industry, due to the alteration of carcass conformation, requirement for more chiller space for carcasses and additional labor to further process cuts for the retail sector (Hopkins 2011). The severance of the vertebral column in five different locations and of the *ligamentum nuchae* of prerigor beef carcasses (also called "tendercut") is also known to improve meat tenderness by increasing sarcomere length (Smith et al. 1971; Wang et al. 1994). A similar stretching concept has been further developed for the development of technology for the hot-boned prerigor muscle called SmartStretch™ or SmartStretch™/SmartShape™ (SS). This SS machine stretches prerigor muscle by a compression force generated through pumped air within a vacuum chamber of the machine and tightly packages it immediately after in order to maintain the shape and stretching of the muscle (Pen et al. 2012). Increased muscle length and sarcomere length of the beef muscles have been reported (Pen et al. 2012) with improved meat tenderness and appearance (Toohey et al. 2012a,b,c).

15.2.1.3 Postmortem Aging

Substantial improvement in eating quality attributes (e.g., tenderness, juiciness, and flavor) occurs during postmortem aging through endogenous proteolytic enzymatic activities (Koohmaraie et al. 1984). However, different muscles have different aging patterns with respect to the extent of tenderization mainly due to different proteolytic enzyme activities and/or different amounts of connective tissue (Melody et al. 2004; Bratcher et al. 2005; Gruber et al. 2006). Gruber et al. (2006) also found that muscle-to-muscle tenderness differences were influenced by the USDA quality grade and aging times, where in general beef muscles from the USDA Choice grade (more marbled) tenderized more rapidly than USDA Select grade muscles (less marbled). This observation suggested that different postmortem aging protocols should be managed by the meat industry based on individual muscle and quality grade (Gruber et al. 2006).

15.2.2 Beef Grading and Quality Assurance Programs

The overall perceptions of the "taste" of beef products that consumers consider as high quality are associated the three primary eating quality attributes—tenderness, juiciness, and flavor (Smith et al. 2008a,b). Among those sensory attributes, tenderness has been identified as the primary factor affecting consumers' satisfaction and thus determining whether consumers will repeat purchase beef (Shackelford et al. 2001; Smith et al. 2008a,b). Studies have found that consumers are willing to pay a premium for guaranteed tender beef (Miller et al. 2001; Shackelford et al. 2001). Various beef carcass classification and grading systems (Table 15.1) and quality assurance programs (Table 15.2) to provide consumers with red meat of a guaranteed

TABLE 15.1
Summarized Principal Components of Beef Carcass/Cut Grading Schemes in Selected Countries around the World

		Grading Scheme		
Country	**Unit**	**Yield Grade**	**Quality Grade**	**Carcass/Quality Trait**
Canada	Carcass	3—Canada 1, Canada 2, and Canada 3	4—Prime, AAA, AA, and A	Carcass weight, sex, conformation, marbling, meat color, firm lean texture, fat color (exclude yellow fat from grading), fat thickness
Europe	Carcass	5—carcass conformation—E (extremely muscled), U, R, O, and P (very poorly muscled)	N/A	Carcass weight, sex, fat cover, conformation
Japan	Carcass	3—A (72% and above cut yield), B (69%–72%), and C (under 69%)	5—1 to 5	Carcass weight, sex, marbling, meat color/brightness, fat color, fat texture, fat firmness, luster
South Korea	Carcass	3—A (67.5 and above), B (62.7%–67.5%), and C (under 62.7%)	5—1++, 1+, 1, 2, and 3	Carcass weight, sex, marbling, meat color, fat color, firmness, texture, lean maturity, fat thickness
USA	Carcass	5—1 to 5	8—Prime, Choice, Select, Standard, Commercial, Utility, Cutter, and Canner	Carcass weight, sex, marbling, ossification score, meat color, meat texture, ribfat, kidney, pelvic, and heart fat
Australia				
AUS-MEAT	Carcass	N/A	Carcass classification	Carcass weight, dentition, P8 fat, sex, butt shape, marbling, meat color, fat color
Meat Standards Australia (MSA)	Cut	N/A	3—(5* premium, 4* better than every day, and 3* good every day)	Bos indicus%, hormone growth promotant implant, carcass weight, fat depth, sex, whether milk-fed veal, hanging method, marbling score, ossification score (based on the USDA system), meat color, hump height, ultimate pH (must be ≤5.7), aging time, cooking method.

Source: Adapted from Polkinghorne, R.J., and Thompson, J.M. 2010. *Meat Sci.* 86:227–235. With permission.

eating quality have been developed around the world, although almost all of these systems (except Australia—MSA) are based on carcass assessment. The principal components of these beef carcass grading systems are comprised of various traits such as carcass weight, sex, age or maturity, marbling, meat color, fat color, firmness, texture, lean maturity, fat thickness, and finally saleable meat yield (Polkinghorne and Thompson 2010).

The European grading system primarily focuses on yield estimation, which is based on describing carcass conformation (E—extremely muscled, U, R, O, and P—very poorly muscled) and external fat level (from 1—very lean to 5—very fat) (Polkinghorne and Thompson 2010). However, the European grading system does not provide any quality/palatability-related information. Marbling (intramuscular fat [IMF] or the dispersion of fat within the lean) has been identified as an important determinant of palatability because of its contribution to juiciness and flavor. Thus, marbling has been widely used (particularly in North American and Asian countries) within their beef carcass grading systems as a major factor in predicting eating

TABLE 15.2
Examples of Beef Quality Assurance Programs in Selected Countries around the World

Country	Program (Brand)	Quality Trait Requirements
Japan	Kobe beef	Beef only from the Tajima-gyu strain of Wagyu cattle, raised in Hyogo Perfecture, yield grade (A and B), quality scores (4 or higher), marbling scores (6 or higher), carcass weight less than 470 kg, fine meat texture, and excellent firmness
New Zealand	Beef and lamb quality mark	NZ origin, no growth promotants, electrical stimulation, chilling, and postmortem product aging regime, ultimate pH < 5.8 (or aged for a supplementary period and shear force testing after aging)
USA	Nolan Ryan tender-aged beef	Time-on-feed, mild implant regimen, electrical stimulation, postmortem product aging, and BeefCAM (instrumental muscle color assessment)
	Cattlemen's Collection	Time-on-feed, mild implant regimen, electrical stimulation, 14 days of postmortem product aging, ≤2 in. hump height, and instrumental muscle color assessment
	Swift's Chain Of Tenderness	Source-verified genetics, time-on-feed, electrical stimulation, high-temperature carcass conditioning, postmortem product aging, hump height, and muscle color assessment
	Safeway Rancher's Reserve Angus	Only from Red Angus or Black Angus cattle, time-on-feed, mild implant regimen, electrical stimulation, postmortem product aging, hump height, tender-cut suspension, and routine Warner-Bratzler shear force testing
Australia	Meat Standards Australia (MSA)	Bos indicus%, hormone growth promotant implant, carcass weight, fat depth, whether milk-fed veal, sex, hanging method, marbling score, ossification score, meat color, hump height, ultimate pH, postmortem aging time, cut by cooking method

quality because higher IMF is considered to improve eating quality traits (Hocquette et al. 2010).

In the USDA beef quality grading system, there are eight quality grades (Prime, Choice, Select, Standard, Commercial, Utility, Cutter, and Canner) and carcasses are largely separated by marbling and maturity (Figure 15.2) (USDA 1997). Marbling is divided into nine degrees (often called scores) based on the amount of IMF for the quality grading standards (ascending order): practically devoid, traces, slight, small, modest, moderate, slightly abundant, moderately abundant, and abundant. Beef carcass maturity is divided into five groups: A (9–30 months), B (30–42 months), C (42–72 months), D (72–96 months), and E (>96 months) and assessed by size, shape, and ossification of the bones and cartilage, particularly determining the split chine bones, and the color and texture of the lean tissue (Smith et al. 2008b).

The accuracy and efficacy of the USDA quality grades to predict tenderness is influenced by the cut (Smith et al. 2008b). In general, major meat cuts (mostly from the middle section) from carcasses graded in the top four USDA quality regions (Prime, Choice, Select, and Standard) are well aligned to predict palatability, but are less useful predictors of tenderness of other meat cuts (mostly chuck and round areas) due to differing amounts of connective tissue (Smith et al. 2008b). Any system that is based on using one muscle to predict eating quality in the rest of the carcass as used in the USDA system will have limitations, as the *Longissimus lumborum* muscle is an imperfect predictor of the tenderness of other muscles in the carcass (Rhee et al. 2004). In the United States, audits of beef tenderness showed little improvement between 1991 and 1996 (George et al. 1999) indicating the reliance on a quality assessment system based on carcass traits only (Smith et al. 2008b).

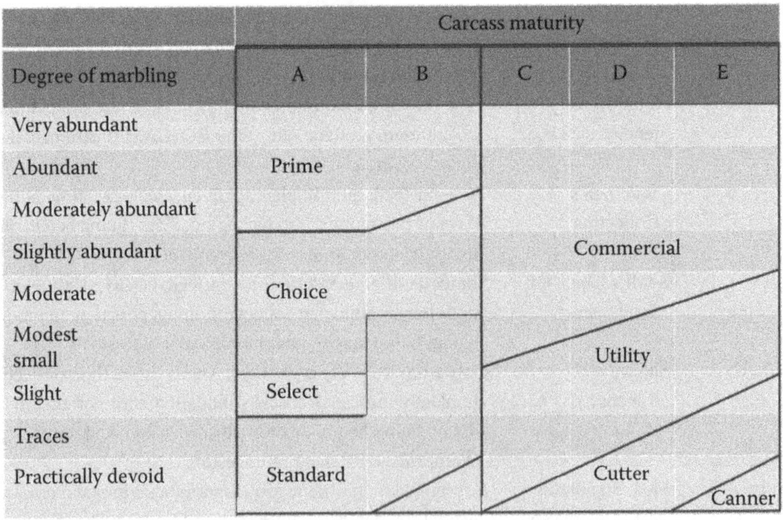

	Carcass maturity				
Degree of marbling	A	B	C	D	E
Very abundant					
Abundant	Prime				
Moderately abundant					
Slightly abundant				Commercial	
Moderate	Choice				
Modest small				Utility	
Slight	Select				
Traces					
Practically devoid	Standard			Cutter	Canner

FIGURE 15.2 USDA beef quality grade. (Reproduced by courtesy of the National Cattlemen's Beef Association.) Maturity increases from left to right (A through E).

In light of this limitation, and the variation in predicting eating quality attributes based on the current grading system, the use of principles from total quality management (TQM) was further developed as "palatability assurance critical control points (PACCP)" (Smith et al. 2008b). The principal concept of TQM (or PACCP) is to identify the factors affecting variability of the products throughout the whole value chain—from live animal production/management to postslaughter processing steps and work to improve the production process by providing continued measuring and monitoring of the identified critical variables (Smith et al. 2008b). Some of the examples of the TQM (or PACCP) type approach are given in Table 15.2.

In New Zealand, a Beef and Lamb Quality Mark program was introduced to domestic consumers in 1997 in response to the inconsistent tenderness of both beef and lamb, where a considerable proportion of beef (50%) and lamb (25%) was below acceptable tenderness (a predetermined shear force threshold) (Frazer 1997). After the establishment of the mandatory standards for the Quality Mark, the shear force of beef declined by 22% between 1997 and 1999 (Bickerstaffe et al. 2001). The fundamental component of the beef Quality Mark standard was to ensure the tenderness of the meat product—for 95% of the samples to have a shear force less than 80 N (based on conversion of MIRINZ Tenderometer values [11 kgF] using the model of Hopkins et al. 2011) through the requirement for meat to have an ultimate pH below 5.8 and/or an extended aging period. Regular shear force assessments were required to verify the efficacy and accuracy of the electrical stimulation, chilling, and aging regimes implemented by processors to meet the Quality Mark tenderness standard.

One of the most comprehensive and well-established PACCP-like systems for improving meat palatability is Australia's MSA grading system (Polkinghorne and Thompson 2010). The basic principle of the MSA grading system was to develop a prediction model for the eating quality of beef (Thompson 2002) based on statistical models, where anatomical cut descriptions were replaced with determination of the eating quality of muscles. The MSA prediction model for palatability assessment calculates four meat quality variables (tenderness, juiciness, flavor, and overall liking) for each cut by cooking method combination (Watson et al. 2008) based on sensory evaluation done by untrained consumers. From a weighting of the four traits, a MQ4 score is derived. Three ratings are given on the basis of estimated MQ4 scores (<46 as unsatisfactory, 47–63 graded 3 star—good everyday, 64–76 graded 4 star—better than everyday, and >76 graded 5 star—premium). The final recommended formula for the calculation of MQ4 scores to predict the final grade in Australia was 0.4 (tenderness) + 0.1 (juiciness) + 0.2 (flavor) + 0.3 (overall liking) (Watson et al. 2008). The major components of the MSA beef grading system are given in Table 15.1. Meat color is used as a threshold and if the score is above 3 (based on color chips from AUS-MEAT; Anonymous 2005), then the carcass is not graded. Later work has created categorizations based on the selling method of the cattle via either saleyard or direct to the abattoir. An example of the output from the model is given (Figure 15.3) for cuts/muscles cooked different ways taken from a steer carcass not treated with HGP, weighing 240 kg, with 75 mm hump height and a rib fat depth of 7 mm (Anonymous 2012c). The carcass was hung by the Achilles tendon, had an ossification score of 150, a marbling score of 270, an ultimate pH of 5.55, and the cuts were aged for 5 days. This shows the reduction in eating quality when hindleg cuts are

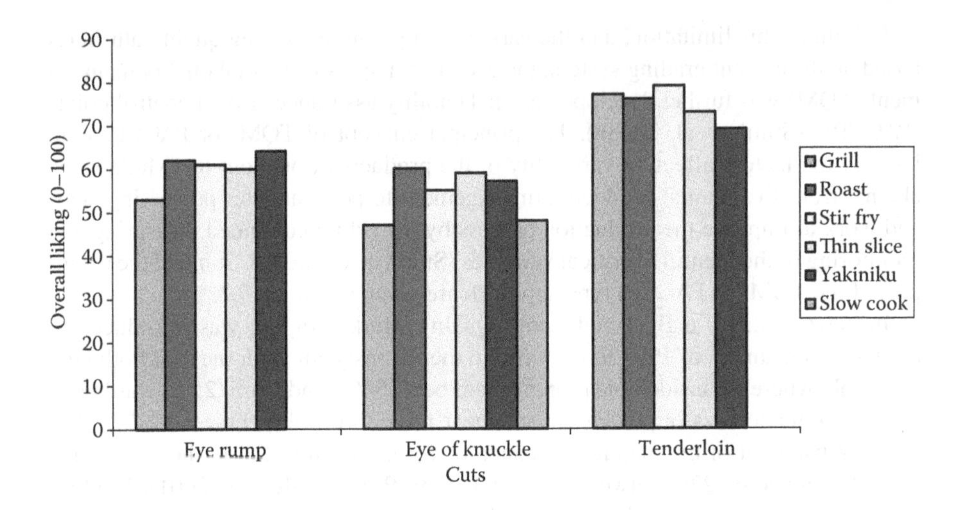

FIGURE 15.3 Effect of cut and cooking method on the eating quality (overall liking) of beef. (Adapted from Anonymous. 2012c. *Meat Standards Australia Beef Information Kit.* Meat & Livestock Australia Ltd, Sydney, Australia. p. 33. http://www.mla.com.au/files/0e6b8058-df65-4d30-a60f-9dbc009760bb/msa-beef-info-kit.pdf.)

grilled with roasting or cooking of the meat in thin slices producing a better result. The low content of connective tissue in the tenderloin is reflected by the higher eating quality score, irrespective of the cooking method.

The Australian developed MSA system of grading beef has highlighted the real value differences between carcasses that are of similar type and is designed to improve the "eating quality guarantee" which could be given to any piece of meat (Hopkins 2014). For example, as discussed by Polkinghorne (2006) rump steak is comprised of different muscles that have various levels of eating quality that will also interact with the cooking method so the true value of the cut will shift depending on which muscles and cooking method are used (see Figure 15.3). As pointed out above, this approach challenged the concept that single muscle or carcass measures can be used to predict eating quality and thus true value. The model has the capacity to provide eating quality scores for multiple muscles cooked by different methods (Polkinghorne 2006) and more recently has been able to provide predicted eating quality scores for 135 "cut by cooking method" combinations (Polkinghorne et al. 2008b).

From the MSA program it has been demonstrated at the retail level that products can be marketed according to predicted eating quality within cooking guidelines, without any mention of cuts (Polkinghorne 2006). This concept led to the development of new value added products, more appropriate use of particular cuts such as thin sliced topsides and seam boning (e.g., MSA 2010) with returns per carcass shown to increase by adopting this approach (Polkinghorne et al. 2008a). Extensive work in the United States has characterized the meat qualities of muscles in lower value cuts, including the chuck (Von Seggern et al. 2005) so that alternative uses of the cut could be made, applying the seam boning approach. This work has led to a comprehensive profile of the beef carcass for a range of meat quality traits (see Jones et al. 2005), but these traits did not include eating quality. The MSA system has been tested with Australian consumers (n = >67,900) over 15 years and has also been evaluated with consumers in overseas countries (Polkinghorne and Thompson 2010). Overall, these studies suggest that the MSA system can provide very consistent estimations of beef palatability in line with consumers' response in various countries. This indicates the potential application of the MSA system as a global standard, with some minor adjustments in each country (Polkinghorne and Thompson 2010).

15.2.3 Lamb Meat Grading and Quality Assurance Programs

The classification/grading systems for lamb carcasses are not as widely established as beef carcasses around the world. The EU system has two classification schemes for lamb carcasses: one for carcasses weighing above 13 kg, which then can be further evaluated based on carcass conformation based on the EUROP classification, and another for carcasses weighing less than 13 kg, which are not further considered for classification due to their poor morphology, low subcutaneous/internal fat ratio and/or light weight (Sañudo et al. 2000). In the Mediterranean area, where light lamb carcasses (<13 kg) are very common, carcasses are divided into three different weight categories (A, B, C as <7.0 kg, 7.1–10.0 kg, or 10.1–13.0 kg, respectively), and each weight category is subdivided to two quality classes based on meat color and fatness scores (Russo et al. 2003). However, the EU Mediterranean classification system is not associated with palatability and meat quality attributes (water-holding capacity, pH, cook losses, and color) (Sañudo et al. 2000; Russo et al. 2003). In the United States, lamb carcasses are evaluated for a quality grade, which is divided further into three classification sections as lamb, yearling mutton, and mutton depending upon the level of maturity and yield grade (1–5, with 1 representing the highest meat yield based on the amount of external fat present) (USDA 1997). There are four subjective quality grades (Prime, Choice, Good, and Utility) for each ovine classification based on the quality characteristics of the lean (indirectly evaluated by the quantity of fat streakings within and upon the inside flank muscles, and degree of firmness of lean and external fat), and the conformation of the carcass (Figure 15.4; USDA 1997).

Although the USDA quality grades of lamb carcasses provide general "palatability-indicating characteristics" of the resultant cuts, the accuracy and/or efficacy of the system as a prediction tool for assuring consumers satisfaction level has been criticized (Carpenter and King 1965a,b; Crouse et al. 1982; Jeremiah 1998). In fact, the adoption of systems to ensure consumers guaranteed eating quality for sheep meat has been limited around the world. In New Zealand, the "Beef and Lamb Quality Mark" program (as discussed in the previous section) was launched in response to survey studies which showed an unacceptable proportion of lamb exceeded the shear force threshold set as part of the accelerated conditioning and aging (AC&A) program (Frazer 1997). The objective was to ensure processors complied with the AC&A guidelines which included measurement of muscle samples for shear force at least twice a year to ensure compliance. Chilling guidelines were also set and this program reduced the proportion of tough lamb at the retail level (Anonymous 2000). This program was not, however, based on using consumers as the arbitrators of quality, which was the approach when the Australian sheep-meat eating quality (SMEQ)

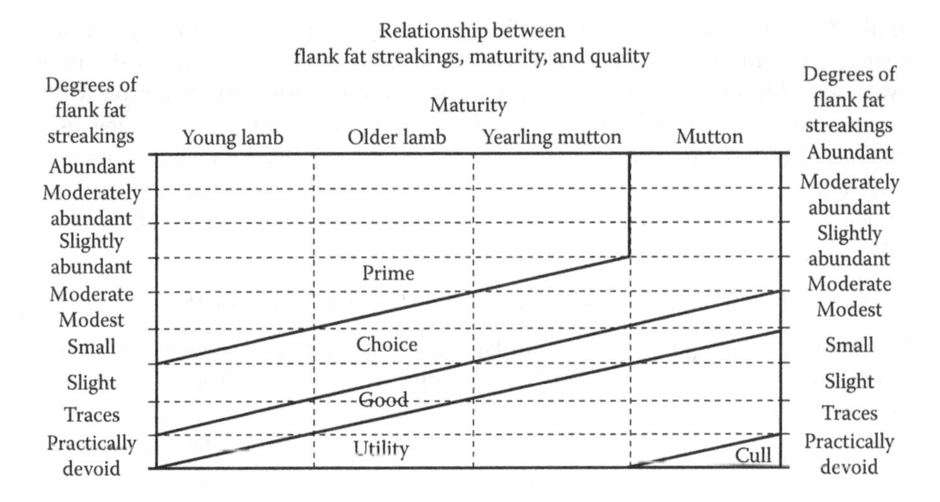

FIGURE 15.4 USDA lamb carcass quality grade. (Adapted from USDA. 1997. *Official United States Standards for Grades of Carcass Beef.* Agriculutural Marketing Service, United States Department of Agriculture, Washington, DC.)

program was developed. This program was initiated following the development of the MSA program in Australia for beef.

The eating quality of sheep-meat can be considered as a function of the production, processing, value-adding, and cooking methods used to prepare the product for consumption by the consumer (Thompson et al. 2005). As part of the SMEQ program, critical control points were identified and the impact of these on eating quality was quantified. These critical points were covered in a technical guide (Anonymous 2006). At the farm level, sheep/lambs cannot be sent for slaughter within 2 weeks of shearing; they must have a wool length of 5 mm; they must be on the property of consignment for a minimum of 2 weeks; water must be available during on farm curfew and lairage, and the total time off feed should not be greater than 48 h before slaughter (Anonymous 2013). Further to this, lambs must weigh more than 18 kg carcass weight, have a fat score of 2 (GR > 5 mm: GR is the total tissue depth 11 cm from the spine at the 12th rib) or higher (Anonymous 2013) and if they are Merino lambs, it is recommended that they grow at 150 g/day or more for 2 weeks prior to slaughter and if crossbred lambs at 100 g/day or more (Anonymous 2012a). Postdeath, the only current requirement is that carcasses are to have an "ideal" rate of pH fall in the *Longissimus lumborum* muscle with a target of 18–25°C at pH 6.0 as this was shown to result in superior eating quality compared to slower or faster rates of pH fall for the domestic market (Thompson et al. 2005). This outcome was generally consistent with the early studies of Locker and Hagyard (1963) who showed minimal shortening at close to 15°C, but the temperature range was subsequently increased from 18–25°C to 18–35°C (Anonymous 2007; van de Ven et al. 2013). Not every carcass is measured for pH decline and a sampling regime is applied to verify that abattoir processes allow the target to be achieved. A summary of the requirements for processing are given in Table 15.3. The development of MSA sheep meat lead to a significant

TABLE 15.3
Processing and Aging Conditions for Optimum Eating Quality in Different Markets

	Domestic Chilled Trade		Domestic or Export Chilled Trade
Hanging method	**Tenderstretch**	**Achilles**	**Achilles**
Electrical stimulation needed	No	Yes	No
Enter rigor (pH 6) at	8–35°C	18–35°C	8–18°C
Minimum aging period	5 days	5 days	10 days

Source: Adapted from Anonymous. 2012b. MSA, The MSA sheep meat information kit, p. 11. http://www.mla.com.au/Marketing-beef-and-lamb/Meat-Standards-Australia/MSA-sheepmeat.

Note: This is an outline only. Processes need to be tuned to match abattoir facilities and specific market needs.

increase in the use of new electrical stimulation technology, with more than 70% of the throughput of sheep and lambs on a tonnage basis per year in Australia in 2008 subjected to stimulation (Hopkins et al. 2008). There are now few sheep abattoirs in Australia that do not have this technology installed (Hopkins 2011).

The ranking of different cuts for eating quality based on a roast cooking method according to the age of the sheep from which the cut was taken is shown in Figure 15.5. This clearly shows that the cuts from the middle of the carcass (rack and short

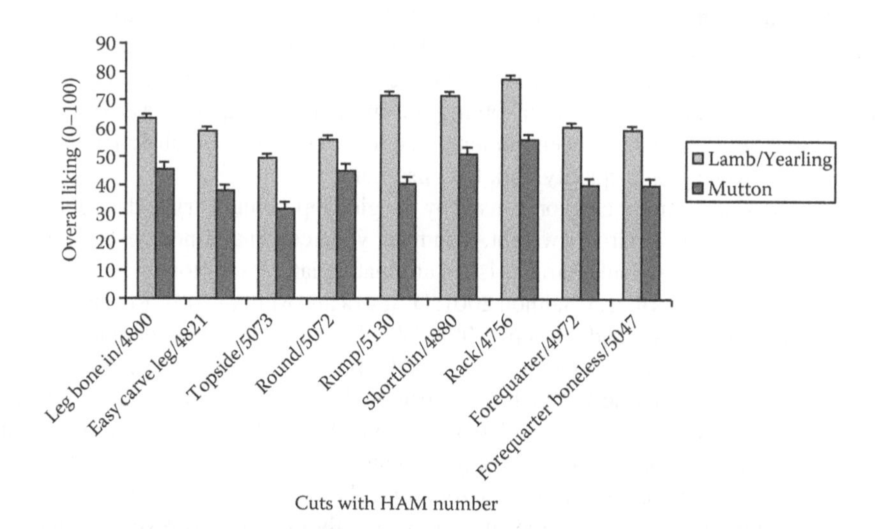

FIGURE 15.5 Effect of cut (with HAM number; Anonymous 1998) and sheep class on the eating quality (overall liking) of sheep meat. (Adapted from Pethick, D.W. et al. 2006. *Proc. NZ Soc. Anim. Prod.* 66:363–367.)

loin) provide the best eating quality where overall liking is a composite of tenderness, flavor, and juiciness. Of the cuts evaluated, the topside had the lowest eating quality and, based on these results, was not recommended for roasting or indeed for grilling when taken from the younger sheep and is not even given a grade when taken from older sheep. The data also clearly shows the lower eating quality of meat from older sheep, but does illustrate that some cuts from older sheep can reach acceptable levels for eating quality (e.g., rack and shortloin). For lamb, a target of 5% IMF has been suggested to help ensure a "good every day" (grade 3 out of a 5 level system) score, despite the literature suggesting that marbling is only one component of eating quality (Hopkins et al. 2006). This target level may be slightly less based on more recent work, with a value of 4% suggested by Pannier et al. (2014). In the future, it is anticipated that IMF could be measured as part of an upgrade of the system and that a cut by cook model will be developed, mirroring the beef MSA system, but at a lower level of complexity.

15.3 PRODUCING PORK WITH HIGH EATING QUALITY

Quality assurance systems, based on the principles of Hazard Analysis and Critical Control Point (HACCP), have been implemented in many countries to manage the risks associated with the production and transport of pigs and the processing and retailing of pork. The adoption of quality assurance principles allows companies to demonstrate to their customers that documented procedures have been followed, and standards have been met, in the delivery of safe and wholesome pork and pork products. On-farm standards, such as the Australian Pork Industry Quality Assurance program (APIQ✓®), typically cover management, animal welfare, food safety, traceability, and biosecurity to enable producers to follow good agricultural practices (GAP). Although guidelines and checklists have been developed for eating quality, these have largely been used to support individual company brands. Despite eating quality being the final outcome, few examples of industry systems (in contrast to individual company programs), involving the integration of quality assurance systems already in place, have been implemented to meet a common objective for consistently delivering high-quality pork to consumers.

Current grades and/or categories used by the global pork industry to describe pork quality, including sex, carcass weight, lean meat yield content, fatness level, and evidence of secondary sexual characteristics are inadequate descriptors of eating quality. Published reviews (e.g., Channon 2001; D'Souza and Mullan 2001; Rosenvold and Andersen 2003; Ngapo and Gariepy 2008; Wood et al. 2008; Channon and Warner 2011) have documented key factors, from production to consumption, that influence pork eating quality. These key factors are briefly described here, with further detail also provided elsewhere in this book (e.g., Chapter 12). Although extensive research has been conducted, it is currently not possible to accurately determine the effect of different combinations of pathway factors, and the interactions between them, on pork eating quality as these studies have largely only investigated several factors at a time (with data for interactions not necessarily published if found to be nonsignificant). This issue is a complex one and careful consideration of the approach to be used is required in order to generate both support, and system adoption, by the whole pork supply chain to drive value and result in increased demand for pork and pork products by consumers.

15.3.1 Optimizing Pork Quality

Flavor, tenderness, and juiciness have been identified as the key pork eating quality attributes that influence overall acceptability of pork (Wood et al. 1986; Cameron et al. 1990; Channon et al. 2004, 2013a, b, 2014). These three attributes are highly correlated to each other as well as to overall acceptability of different pork cuts. Ultimate pH, IMF content, and drip loss can all drive consumer purchasing decisions at the retail level, due to their influence on meat color and product appearance, with their relative influence on consumer purchasing decisions varying between countries (Ngapo et al. 2007). The sensory traits (flavor, tenderness, and juiciness) can be influenced by ultimate pH, IMF content, and drip loss, however the effects of these technological traits on eating quality are not necessarily consistent (Bidner et al. 2004). Particular focus is therefore needed to determine those pathway factors that can significantly influence sensory traits through their impact on muscle pH, IMF content, and muscle growth. In order to satisfy consumer requirements for tasty, tender, and juicy pork, the key issues that need to be addressed include the following:

15.3.1.1 Boar Taint

"Boar taint" refers to the undesirable, often intense, fecal, urine-like odor, and/or flavor of pork and has negative impacts on consumer acceptability of pork. Managing boar taint risks by using carcass weight strategies for entire males has its limitations (D'Souza et al. 2011), with weak correlations reported for both androstenone and skatole compared with hot carcass weight (ranging from 60 to 80 kg). The incidence of boar taint can be virtually eliminated by immunocastrating entire male pigs during the finisher phase of production (Dunshea et al. 2001; Lealiifano et al. 2011). Immunocastration also has animal welfare benefits on-farm, due to a reduction in aggressive behavior of entire males following the administration of the second vaccine, reducing the incidence and severity of carcass lesions (Lealiifano et al. 2011). In many countries, surgical castration of males at 2–5 days of age, without analgesia, has been a common practice to minimize risks of boar taint and reduce aggression. Increasing consumer concerns about animal welfare have led to the EU pork industry moving to voluntarily end the practice of surgically castrating pigs by January 1, 2018. Pork from immunocastrated males has been shown to be of similar eating quality to that from females and surgical castrates (D'Souza and Mullan 2002; Hennessy et al. 2006) and more acceptable than from entire males, even when entire males have low levels of androstenone and skatole (Font i Furnols et al. 2008). The noninclusion of entire males in any eating quality pathway system is a necessary consideration and can be scientifically justified.

15.3.1.2 Intramuscular Fat

Pigs of lean, fat growing genotypes tend to produce pork with lower IMF contents than slower growing genotypes. The fatty acid composition of IMF may also become more unsaturated in leaner animals (Cameron et al. 1990) and negative correlations between polyunsaturated fatty acids and eating quality traits of pork have been reported (Cameron and Enser 1991). Although positive effects on sensory attributes of pork of IMF content in excess of 2% have been reported (Bejerholm and Barton-Gade 1986; Touraille et al. 1989; Fernandez et al. 1999), Channon et al. (2014) reported very weak correlations between IMF content and sensory traits which may reflect the narrow range in IMF content of pork loin steaks in their study.

Increased feed intake of immunocastrated males following the administration of the second vaccine can result in increased fat deposition, increasing IMF levels, and lowering shear force compared with entire males (Batorek et al. 2012). Similar observations were made by Channon et al. (2013b) where immunocastrated males were found to have higher IMF levels of loin, silverside, and bolar blade muscles compared with entire males. Whilst considerable work has evaluated the effect of immunocastration on growth performance, feed efficiency, and boar taint compounds compared with other genders, it has been identified that few of these studies have also evaluated meat and eating quality (Channon et al. 2011; Batorek et al. 2012).

15.3.1.3 Rate and Extent of Muscle pH Decline Postslaughter

Management of factors that can influence ultimate pH is necessary as part of a pathway approach to optimize pork eating quality, due to the impact of pH (both the rate and extent of muscle pH decline) on muscle color, drip loss, purge, as well as tenderness, juiciness, and flavor (Huff Lonergan et al. 2010). Muscle pH can influence meat color when ultimate pH is close to the isoelectric point between 5.2 and 5.5 (Boler et al. 2010). Unlike beef and lamb, neither pH/muscle temperature windows have been established nor ultimate pH cut-offs developed to optimize the eating quality of pork, despite much work in this area.

Although a large number of studies have been conducted to determine the effect of transport time, transport distance, time off feed, and resting in lairage on technological quality attributes (including rate of pH and temperature decline postslaughter, ultimate pH, drip loss, and meat color), there is little published data on the effects on pork eating quality. Welfare standards for livestock transporters, as well as for meat processing establishments, are now in place in many countries, including Australia. These aim to ensure that animals are well handled and managed according to welfare requirements as well as enable businesses to demonstrate that they comply with industry standards and guidelines. These systems are supported by legislation and may be independently audited. In addition, the adoption of on-farm quality assurance programs (e.g., APIQ✓® by Australian pig producers; Pork Quality Assurance® (PQA) Plus in the United States; Red Tractor Assurance in the UK), which include management, animal welfare and food safety standards, stockperson training on-farm and at abattoirs (e.g., ProHand—http://www.animalwelfare.net.au), as well as animal welfare programs in place at processing establishments, play a role in reducing the incidence of low pH pork.

15.3.1.4 Impact of Ultimate pH on Pork Eating Quality

Acidic/sour flavor notes and juiciness may be influenced by pH and/or muscle fiber types (Aaslyng et al. 2007; Myers et al. 2009). More recently, Jose et al. (2013) confirmed the importance of ultimate pH on both meat and sensory attributes of pork and demonstrated the need to implement pathway parameters, both pre- and postslaughter, to achieve consistent muscle pH decline postslaughter to attain a normal pHu.

Chilling can influence pork quality as it influences the rate of muscle pH and temperature decline of muscles. In contrast to beef and lamb, pH decline is more rapid in pork and the muscles are therefore more likely to experience elevated temperatures during the onset phase of rigor. As pork has a higher proportion of white, type IIB muscle fibers present which have a higher glycogen content and are more likely to exhibit fast rates of pH fall, rigor onset may commence very early postslaughter and be completed at about 6 h postslaughter (Savell et al. 2004); which may significantly affect pork eating quality.

15.3.2 Pathway Approaches for Consistent Delivery of Pork Eating Quality

Tonsor and Schroeder (2013), in a study conducted for the US National Pork Board, noted that an economically relevant portion of US fresh pork does not meet retail pork quality targets. This position is not assisted by the inability of industry to adequately measure pork quality attributes in ways that reflect end user value, in real time and at low cost. In Australia, pork quality measures including ultimate pH, meat color, shear force, and IMF are not being routinely measured by processors. There are also no industry systems in place generating a "pull-through" from customers in relation to delivering consistent eating quality. Consequently, producers cannot obtain market signals from processors on pork quality.

The use of objective measurements to predict the sensory quality of pork is problematic, demonstrated by low correlations between key objective and subjective measures that have been reported. Several countries have implemented pork quality targets to assist with meeting quality objectives. In the United States, the National Pork Producers Council (NPPC 2002) Pork Quality Solutions Team provided the following targets for the *Longissimus* muscle: subjective color—NPPC scores 3.0–5.0; ultimate pH—5.6–5.9; tenderness (shear force)—<31.4 N; flavor—robust pork flavor; IMF content—2–4%; and drip loss—not to exceed 2.5%. These targets, however, are not part of an industry-wide program such as MSA for beef and lamb in Australia (Polkinghorne et al. 2008a,b). The 2012 National Retail Benchmarking Study, funded by Pork Checkoff, was conducted to determine the quality attributes of pork available at the retail level (Klinkner 2013). Average ultimate pH of both center cut loin chops and sirloin chops were within the target range (5.80 and 5.89, respectively), indicating that the US industry is meeting their objective to produce pork with pH > 5.7. Overall, it was concluded that US consumers favored pork with lower shear force, higher pH, and high IMF levels, as previously reported by Moeller et al. (2010). However, sensory data for collected samples was not presented. It was also identified that consistent cooking temperature recommendations need to be provided to consumers.

The British Pork Executive (BPEX 2010) developed Pork Quality Targets for use by industry to support their initiatives to supply pork of improved tenderness, juiciness, and flavor with less drip and better color to consumers. These targets based on findings of published studies are shown in Table 15.4.

TABLE 15.4
Pork Quality Targets Developed by the British Pork Industry

On-farm factors	*Ad libitum* feeding from 30 kg to slaughter
	Dietary exclusion of fish meal from finisher rations
	Maximum of 1.6% by weight of linoleic acid in finishing ration
	Maximum inclusion of 2% by weight of polyunsaturated fatty acids in finisher rations
	Minimum inclusion of 100iu/kg vitamin E
	Include dietary components, where appropriate, to reduce skatole concentrations
	Minimize risks associated with skatole through maintaining good shed cleanliness, ventilation, and shed design
	Optimize growth rates to minimize age at slaughter
Transport conditions to the abattoir	Optimal time off feed period of 8–12 h prior to slaughter
	Avoid >18 h time off feed
	Ensure space allowance requirements during transport are met
	Avoid ramps at loading
	Maintain pigs in stable social groups
	Minimize transit time to the abattoir
	Avoid use of electric goads
	Consider the use of toys during transport
Lairage	Avoid ramps at unloading
	Ensure pigs are promptly unloaded
	Maintain pigs in their social group in lairage and up to stunning
	Use water sprinklers in lairage in hot weather
	Ensure pen and race designs encourage pigs to move forward to the point of stunning without using goads
	Consider the use of toys during lairage
Processing	Effective high voltage electrical stimulation (HVES) at 20 min postslaughter followed a minimum aging period of 4 days to optimize loin quality
	To optimize leg quality, hang sides from the aitchbone within 1 h of stunning for a minimum of 12 h followed by aging for 4 days
	Aitchbone hanging can improve leg quality in both HVES and unstimulated carcasses
Chilling	Prevent muscle temperature of <10°C within the first 3 h after stunning unless the carcass has been stimulated
	Avoid overloading chillers
	Maintain storage temperature through the chain from carcasses to prepared cuts at 0–4°C (for maturation in vacuum packs hold at <3°C).
Maturation	For optimum tenderness, aim for an aging period from slaughter to retail of: 4 d for legs, 7 d for loin bone-in; 12 d loin boneless, 4 d for HVES loins
Carcass/cut selection	Ensure a minimum P2 fat depth of 8 mm.
	Ensure good visual appeal—avoid PSE, DFD, or blood splash
Cooking	70–80°C internal temperature of grilled and roasted cuts

Source: Adapted from BPEX. 2010. *Target Pork Quality: Opportunities for Improving the Quality of Pork.* British Pig Executive, a division of the Agricultural and Horticulture Development Board, Kenilworth, Warwickshire, UK.

These targets are available for use by industry as part of individual company procedures, but do not support an industry-wide eating quality assurance program. Standard 6.3 of the British Quality Assured Pork Standards (2013) states that pork primals, food service, and retail pork packs are subject to organoleptic sampling (including tenderness, succulence, flavor, and visual appearance) to ensure that a consistently acceptable standard is delivered, but this is at a company level and no further detail on actual targets are included in the Standard.

Taverner (2001) provided risk ratings of key factors that may impact pork eating quality along the value chain, based on research outcomes from Australia and overseas at that time (Table 15.5). Other sections of this book have covered the key pathway factors that influence the eating quality of pork (see Chapter 12). Taverner (2001) highlighted that those factors presenting the greatest risk to pork eating quality were those closest to the consumer, including cooking preparation (encapsulating cooking method, cut type, and final internal temperature and their interactions).

TABLE 15.5
Collation and Risk Rating of Key Factors Influencing Pork Eating Quality along the Value Chain

Critical Management Point	Rating	Issue
Consumer preparation—cooking method, temperature, cut type	******	Overcooking (>75°C) can negatively affect eating quality of all cuts of pork
	Product Preparation	
Packaging	**	Tenderness issues associated with retail ready MAP cuts
Moisture infusion/ enhancement	*****	Major benefits on juiciness and tenderness
Aging	*****	Benefits of aging on tenderness (>7 days postslaughter)
	Carcass Processing	
Chilling	****	Effect on shrinkage
Hanging	****	Aitchbone hanging benefits on leg and loin quality
Electrical stimulation	**	Low-voltage electrical stimulation 5 min postslaughter
Transport/lairage/preslaughter handling	**	Time in lairage >2 h; maximum of 24 h time off feed; minimize mixing of unfamiliar pigs
On-farm handling	*	Good agricultural practice to optimize animal welfare
Housing	*	Pen cleanliness to minimize risk of skatole taint
Nutrition	**	Ad libitum feeding from 30 kg to slaughter Taint from certain ingredients (fish meal/oil) Use of metabolic modifiers
Sex	***	Immunocastration of entire male pigs to improve animal welfare, reduce risk of boar taint and increase marbling
Breed	**	Use of the Duroc which is known to influence intramuscular fat content

Source: Based on Taverner, M.R. 2001. *Manipulating Pig Production VIII.* Australasian Pig Science Association: Werribee, Adelaide, Australia.

Note: Impact on eating quality—low* to high ******

Unfortunately, this is a critical factor over which the global pork industry as a whole has very little control, except through marketing efforts aimed at changing cooking practices used for pork, in accordance with cut type. More recently, Channon et al. (2013a,b) has shown that the cooking method and temperature of different cut types had a greater influence on the pork eating quality attributes of tenderness, juiciness, flavor, and overall liking than either gender or aging period. Considerable, and ongoing, efforts are needed to communicate optimal cooking methodologies for different pork cuts to result in practice change by consumers.

Taverner (2001) also noted that few of these pathway factors are independent. It is difficult to quantify the size and effect of these pathway interactions on pork eating quality attributes as published studies with multifactorial designs are not numerous and those studies that have investigated several factors have not reported interaction data when treatment effects were nonsignificant (Channon et al. 2011).

To date, very few studies have reported pork eating quality outcomes from commercial companies who have implemented a pathway approach to improve pork quality consistency. The study reported by D'Souza et al. (2012) is one of these. This study involved the implementation of eating quality interventions at the producer and processor level in two stages. The Stage 1 pathway stipulated; (i) halothane free pigs (to minimize risks of PSE and rapid rate of muscle pH decline postslaughter), (ii) pigs with a minimum of 50% Duroc sire lines (to optimize IMF content), and (iii) no entire males (to minimize risks of boar taint—immunocastrates, surgical castrates, and females only) and Stage 2 involved moisture infusion/enhancement (to optimize tenderness and juiciness) of fresh pork from pigs produced according to the Stage 1 specifications. Higher consumer scores for flavor, juiciness, tenderness, overall acceptability, and quality grade of pork loin (*Longissimus muscle*) were found for moisture-infused pork (Stage 1 and 2) compared with Stage 1 (Table 15.6), when cooked to a 75°C final internal temperature. Pork produced according to the Stage

TABLE 15.6
Pathway Effects on Consumer Sensory Quality of the *M. longissimus* Muscle

Brand	Generic Pork	Stage 1	Stage 1 and 2	LSD	Significance
Aroma[a]	55	63	57	6.54	0.002
Flavor[a]	54	66	76	6.11	<0.001
Juiciness[a]	43	58	75	6.85	<0.001
Tenderness[a]	41	59	75	7.40	<0.001
Overall acceptability[a]	48	64	76	6.67	<0.001
Quality grade[b]	2.9	3.5	4.0	0.28	<0.001

Source: Adapted from D'Souza, D.N. et al. 2012. Enhancing pork product quality and consistency: A pathway approach. *Proceedings of 58th International Congress of Meat Science and Technology*. August 12–17, 2012, Montreal, Canada (pp. PRODUCTP 83).

Note: LSD, least-significant difference.

[a] Acceptability score (line scale); 0 = dislike extremely and 100 = like extremely.

[b] Quality grade; 1 = unsatisfactory, 2 = below average, 3 = average, 4 = above average, 5 = premium.

1 pathway was also more acceptable than generic pork. Overall, the eating quality "fail rate" (determined by the percentage of consumers who scored pork as being unsatisfactory or below average for quality grade) was highest for generic pork (30%) compared with 15% for Stage 1 and 3% for Stages 1 and 2, respectively.

The quantification of the impact of key critical control points that influence eating quality attributes of fresh pork has been the focus of Australian research (Channon and Warner 2011) to support industry efforts to deliver consistently high-quality pork to customers based on a pathway approach. Consumer sensory studies are being used to validate pathway interventions to inform the development of a nonprescriptive, cost-effective industry system.

15.3.3 Interactions between Different Pathway Factors on Pork Eating Quality

An extensive database of previous research that reported the effects on pork eating quality has been compiled (Channon et al. 2011). The majority of studies were obtained from peer-reviewed journals and several unpublished final reports from previous Australian pork quality research. This work highlighted many data gaps including, but not limited to, the need to better quantify the size and extent of interactions between different pathway factors and whether other pathway interventions (e.g., extended aging, aitchbone hanging, moisture infusion, lower cooking temperatures) could be implemented to effectively manage any impacts on pork eating quality arising from earlier pathway factors (e.g., restricted feeding, use of metabolic modifiers; low IMF levels of pigs in crossbred, commercial genotypes). Although Moore et al. (2012) reported no interaction between the use of ractopamine and/or pST and moisture infusion postslaughter on sensory pork quality, the interaction data were not presented.

More specifically, whilst differences in eating quality (together with growth performance and boar taint compound levels) between surgical castrates, immunocastrates, and entire males have been investigated, few publications have compared eating quality attributes of immunocastrated males and females. Whilst interactions between gender and nutritional management (including the use of metabolic modifiers) on eating quality were found, the extent of these interactions are difficult to estimate as few studies have investigated effects on eating quality in addition to growth performance, carcass composition, and fatty acid composition effects. Interactions between multiple factors (i.e., cooking method, cooking temperature, final internal temperature, cut type and muscle type, as well as raw meat quality traits) on pork eating quality could not be fully elucidated from the literature, due to variations in methodologies used as well as many studies that have primarily included the loin muscle, prepared as a steak, roast, or chop.

Different methods used to cook meat, including roasting, grilling, frying, and casseroling/stewing, can also result in differences in surface temperature of the meat, the temperature profile through the meat, and the method of heat transfer (Bejerholm and Aaslyng 2003) with associated impacts on flavor, juiciness, and tenderness. Further information to fill these knowledge gaps is vital, particularly as cooking skills of consumers are extremely variable. A number of multifactorial studies, whilst including the loin for which much published data is available, have evaluated pathway effects on other major primal cuts, including leg and shoulders when prepared as roasts and stir fry cuts (Channon et al. 2013a,b). In addition, pork may be overcooked by many consumers in response to unfounded food safety concerns associated with cooking pork to a rare, medium rare, or medium degree of doneness. This is also exacerbated by the variable cooking skills of consumers. For rind-on pork products, the desire of consumers to achieve a perfect crackling may result in overcooking of pork due to the high oven temperatures used to optimize crackle quality. All these relationships, as complex as they are, need to be better understood and quantified in order to enable an eating quality system to be cuts-based.

This leads to which interventions may be implemented at the processor level to support a whole supply chain approach to pork eating quality. Hanging carcasses from the aitchbone has been shown to restrain or stretch the *Longissimus, Semimembranosus, Biceps femoris,* and *Gluteus medius* muscles that are otherwise free to shorten on the sides of carcasses hung from the Achilles tendon (Harris and Shorthose 1988). Aitchbone hanging has been included in the MSA system as a pathway intervention for both beef and lamb carcasses. The recommended aging time to reach acceptable quality was reduced in pork with aitchbone hanging (Channon et al. 2014).

Electrical stimulation has been shown to improve pork tenderness and reduce the aging period required to improve pork quality using constant voltage (Gigiel and James 1984; Taylor and Tantikov 1992; Taylor and Martoccia 1995; Taylor et al. 1995a,b; Maribo et al. 1999) and constant current systems (Channon et al. 2003). Taylor et al. (1995b) using a constant voltage stimulation system, found that although both hanging method and electrical stimulation influenced sensory tenderness and juiciness, hanging carcasses from the aitchbone had a greater effect than electrical stimulation.

Moisture infusion/enhancement of fresh pork cuts is an effective means available to pork processors to improve pork eating quality consistency, particularly tenderness and juiciness when neutral flavored brines are used (Cannon et al. 1993). As moisture-infused pork products are generally branded, quality standards and specifications for raw meat used in the manufacture of moisture-infused pork may be in place by individual processors in order to optimize product quality and reduce the incidence and costs associated with excessive purge.

These outcomes suggest that the inclusion of moisture infusion, hanging method, electrical stimulation, and aging period into a pathway model to improve pork sensory quality could each be additive factors. The size of additional benefits arising from the inclusion of multiple pathway factors is being further investigated in Australia as part of eating quality studies involving a number of major pork supply chains.

15.4 CONCLUSION

Economic implications for all participants of the meat marketing chain, including consumers, will have to be monitored. Success in improving fresh meat consumption by providing meat of consistently high quality will require cooperation and quality management throughout the entire value chain process. Each sector will need to deliver product that is ideal for the next sector of the chain to work with. If one link fails to deliver products with uniform quality levels, it is expected that negative impacts on consumer acceptability will be experienced. There is evidence that consumers will pay for eating quality and grading approaches, as seen for MSA beef in Australia, which is a good example of a whole of chain approach leading to greater consumer satisfaction.

REFERENCES

Aaslyng, M.D., Oksama, M., Olsen, E.V., Bejerholm, C., Baltzer, M., Andersen, G., Bredie, W.L.P., Byrne, D.V., and Gabrielsen, G. 2007. The impact of sensory quality of pork on consumer preference. *Meat Sci.* 76:61–73.

Anonymous. 2000. R&D brief number 65 How tender is the beef and lamb from NZ supermarkets and butcher shops? In *Meat & Wool NZ.* Wellington, New Zealand.

Anonymous. 2005. *Handbook of Australian Meat*, 7th edition. Authority for Uniform Specification Meat and Livestock, Brisbane, Australia.

Anonymous. 2006. *Improving Lamb and Sheepmeat Eating Quality. A Technical Guide for the Australian Sheepmeat Supply Chain.* Meat & Livestock Australia, Sydney, Australia, p. 59.

Anonymous. 2007. Producing quality sheep meat. Meat Technology Update, 6/07, Food Science Australia.

Anonymous. 2012a. MSA, Buy and sell MSA sheep & lambs. http://www.mla.com.au/Marketing-beef-and-lamb/Meat-Standards-Australia/MSA-sheepmeat

Anonymous. 2012b. MSA, The MSA sheep meat information kit, p. 11. http://www.mla.com.au/Marketing-beef-and-lamb/Meat-Standards-Australia/MSA-sheepmeat

Anonymous. 2012c. *Meat Standards Australia Beef Information Kit.* Meat & Livestock Australia Ltd, Sydney, Australia. p. 33. http://www.mla.com.au/files/0e6b8058-df65-4d30-a60f-9dbc009760bb/msa-beef-info-kit.pdf.

Anonymous. 2013. MSA, Standards manual, Section 5 Livestock supply. Meat & Livestock Australia, Sydney, Australia. http://www.mla.com.au/Marketing-beef-and-lamb/Meat-Standards-Australia/MSA-Standards.

Batorek, N., Candek-Potokar, M., Bonneau, M., and Van Milgen, J. 2012. Meta-analysis of the effect of immunocastration on production performance, reproductive organs and boar taint compounds in pigs. *Animal* 6:1330–1338.

Bejerholm, C., and Aaslyng, M.D. 2003. The influence of cooking technique and core temperature on results of a sensory analysis of pork—Depending on the raw meat quality. *Food Qual. Prefer.* 15:19–30.

Bejerholm, C., and Barton-Gade, P. 1986. *Effect of Intramuscular Fat Level on the Eating Quality of Pig Meat.* Manuscript No 720E. Danish Meat Research Institute, Roskilde, Denmark.

Bendall, J.R. 1978. Variability in rates of pH fall and of lactate production in the muscles on cooling beef carcasses. *Meat Sci.* 2:91–104.

Bickerstaffe, R., Bekhit, A.E.D., Robertson, L.J., Roberts, N., and Geesink, G.H. 2001 Impact of introductory specifications on the tenderness of retail meat. *Meat Sci.* 59:303–315.

Bidner, B.S., Ellis, M., Brewer, M.S., Campion, D.R., Wilson, E.R., and McKeith, F.K. 2004. Effect of ultimate pH on the quality characteristics of pork. *J. Muscle Foods* 15:139–154.

Boler, D.D., Dilger, A.C., Bidner, B.S., Carr, S.N., Eggert, J.M., Day, J.W., Ellis, M., McKeith, F.K., and Killefer, J. 2010. Ultimate pH explains variation in pork quality traits. *J. Muscle Foods* 21:119–130.

Bouton, P.E., Harris, P.V., Shorthose, W.R., and Baxter, R.I. 1973. A comparison of the effects of ageing, conditioning and skeletal restraint on the tenderness of mutton. *J. Food Sci.* 38:932–937.

BPEX. 2010. *Target Pork Quality: Opportunities for Improving the Quality of Pork*. British Pig Executive, a division of the Agricultural and Horticulture Development Board, Kenilworth, Warwickshire, UK.

Bratcher, C.L., Johnson, D.D., Littell, R.C., and Gwartney, B.L. 2005. The effects of quality grade, aging, and location within muscle on Warner–Bratzler shear force in beef muscles of locomotion. *Meat Sci.* 70(2):279–284.

British Quality Assured Pork Standards. 2013. http://www.bmpa.uk.com/Content/standards.aspx. Issue date 10 May 2013, Revision 5. pp. 1–13. (accessed on April 13, 2014).

Cameron, N., and Enser, M. 1991. Fatty acid composition of lipid in Longissimus Dorsi muscle of Duroc and British Landrace pigs and its relationship with eating quality. *Meat Sci.* 29:295–307.

Cameron, N.D., Warriss, P.D., Porter, S.J., and Enser, M.B. 1990. Comparison of Duroc and British Landrace pigs for meat and eating quality. *Meat Sci.* 27227–27247.

Cannon, J.E., McKeith, F.K., Martin, S.E., Novakofski, J., and Carr, T.R. 1993. Acceptability and shelf-life of marinated fresh and precooked pork. *J. Food Sci.* 58:1249–1253.

Carpenter, Z.L., and King, G.T. 1965a. Cooking method, marbling, and color as related to tenderness of lamb. *J. Animal Sci.* 24:291.

Carpenter, Z.L., and King, G.T. 1965b. Factors influencing quality of lamb carcasses. *J. Animal Sci.* 24:861.

Channon, H.A. 2001. Managing the eating quality of pork—What the processor can do. In Cranwell, P.D. (ed.), *Manipulating Pig Production VIII*. Australasian Pig Science Association, Adelaide, South Australia.

Channon, H.A., Baud, S.R., Kerr, M.G., and Walker, P.J. 2003. Effect of low voltage electrical stimulation of pig carcasses and ageing on sensory attributes of fresh pork. *Meat Sci.* 65:1315–1324.

Channon, H.A., D'Souza, D.N., Hamilton, A.J., and Dunshea, F.R. 2013a. Eating quality of pork shoulder roast and stir fry out performs cuts from the loin and silverside in male pigs. In Pluske, J.R. and Pluske, J. (eds.), *Manipulating Pig Production XIV*. Australasian Pig Science Association, Werribee, Victoria, Australia.

Channon, H.A., D'Souza, D.N., Hamilton, A.J., and Dunshea, F.R. 2013b. Gender, cut type, cooking method and endpoint temperature influence eating quality of different pork cuts. In Pluske, J.R. and Pluske, J. (eds.), *Manipulating Pig Production XIV*. Australasian Pig Science Association, Werribee, Victoria, Australia.

Channon, H.A., Hamilton, A.J., D'Souza, D.N., and Dunshea, F.R. 2011. Development of an eating quality system for the Australian pork industry. Paper read at Proceedings 57th International Congress of Meat Science and Technology, 7–12 August 2011, at Ghent, Belgium.

Channon, H.A., Kerr, M.G., and Walker, P.J. 2004. Effect of Duroc content, sex and ageing period on meat and eating quality attributes of pork loin. *Meat Sci.* 66:881–888.

Channon, H.A., Taverner, M.R., D'Souza, D.N., and Warner, R.D. 2014. Aitchbone hanging and ageing period are additive factors influencing pork eating quality. *Meat Sci.* 96:581–590.

Channon, H.A., and Warner, R.D. 2011. Delivering consistent quality Australian pork to consumers—A systems approach. In van Barneveld, R.J. (ed.), *Manipulating Pig Production XIII*. Australasian Pig Science Association, Werribee.

Cho, S.H., Kim, J., Park, B.Y., Seong, P.N., Kang, G.H., Kim, J.H., Jung, S.G., Im, S.K., and Kim, D.H. 2010. Assessment of meat quality properties and development of a palatability prediction model for Korean Hanwoo steer beef. *Meat Sci.* 86:236–242.

Crouse, J.D., Ferrell, C.L., Field, R.A., Busboom, J.R., and Miller, G.J. 1982. The relationship of fatty acid composition and carcass characteristics to meat flavour in lamb. *J. Food Quality* 5:203–214.

Davis, G.W., Smith, G.C., Carpenter, Z.L., Dutson, T.R., and Cross, H.R. 1979. Tenderness Variations among Beef Steaks from Carcasses of the Same USDA Quality Grade. *J. Animal Sci.* 49:103–114.

Dransfield, E., Etherington, D.J., and Taylor, M.A.J. 1992. Modelling post-mortem tenderisation—II: Enzyme changes during storage of electrically stimulated and non-stimulated beef. *Meat Sci.* 31:75–84.

D'Souza, D.N., Dunshea, F.R., Hewitt, R.J.E., Luxford, B.G., Meaney, D., Schwenke, F., Smits, R.J., and van Barneveld, R.J. 2011. High boar taint risk in entire male carcases. In van Barneveld, R.J. (ed.), *Manipulating Pig Production XIII*. Australasian Pig Science Association, Werribee, Australia, p. 259.

D'Souza, D.N., and Mullan, B.P. 2001. Managing the eating quality of pork—What the producer can do. In Cranwell, P.D. (ed.), *Manipulating Pig Production VIII*. Australasian Pig Science Association, Werribee, Victoria.

D'Souza, D.N., and Mullan, B.P. 2002. The effect of genotype, sex and management strategy on the eating quality of pork. *Meat Sci.* 60:95–101.

D'Souza, D.N., Trezona Murray, M., Dunshea, F.R., and Mullan, B.P. 2012. Enhancing pork product quality and consistency: A pathway approach. *Proceedings of 58th International Congress of Meat Science and Technology*, August 12–17, 2012, Montreal, Canada, pp. PRODUCTP 83.

Dunshea, F.R., Colantoni, C., Howard, K. et al. 2001. Vaccination of boars with a GnRH vaccine (Improvac) eliminates boar taint and increases growth performance. *J. Animal Sci.* 79:2524–2535.

Fernandez, X., Monin, G., Talmant, A., Mourot, J., and Lebret, B. 1999. Influence of intramuscular fat content on the quality of pig meat—2. Consumer acceptability of m. longissimus lumborum. *Meat Sci.* 53:67–72.

Font i Furnols, M., Gispert, M., Guerrero, L. et al. 2008. Consumers' sensory acceptability of pork from immunocastrated male pigs. *Meat Sci.* 80(4):1013–1018.

Frazer, A.E. 1997. New Zealand tenderness and local meat quality mark programmes In *Proceedings Meat Quality and Technology Transfer Workshops*. Auckland, New Zealand, pp. 37–47.

George, M.H., Tatum, J.D., Belk, K.E., and Smith, G.C. 1999. An audit of retail beef loin steak tenderness conducted in eight U.S. cities. *J. Animal Sci.* 77:1735–1741.

Gigiel, A.J., and James, S.J. 1984. Electrical stimulation and ultra-rapid chilling of pork. *Meat Sci.* 11:1–12.

Gruber, S.L., Tatum, J.D., Scanga, J.A., Chapman, P.L., Smith, G.C., and Belk, K.E. 2006. Effects of postmortem aging and USDA quality grade on Warner–Bratzler shear force values of seventeen individual beef muscles. *J. Animal Sci.* 84:3387–3396.

Harris, P.V., and Shorthose, W.R. 1988. Meat Texture. In Lawrie, R.A. (ed.), *Developments in Meat Science–4*. London Elsevier, London.

Hennessy, D.J., Singayan-Fajardo, J., and Quizon, M. 2006. Eating quality and acceptability of pork from improvac immunized boars. In *Proceedings of 18th International Pig Veterinary Society, Copenhagen, Denmark,* pp. 291.

Herring, H.K., Cassens, R.G., and Rriskey, E.J. 1965. Further studies on bovine muscle tenderness as influenced by carcass position, sarcomere length, and fiber diameter. *J. Food Sci.* 30:1049–1054.

Hertzman, C., Olsson, U., and Tornberg, E. 1993. The influence of high temperature, type of muscle and electrical stimulation on the course of rigor, ageing and tenderness of beef muscles. *Meat Sci.* 35:119–141.

Hocquette, J.F., Legrand, I., Jurie, C., Pethick, D.W., and Micol, D. 2010. Perception in France of the Australian system for the prediction of beef quality (Meat Standards Australia) with perspectives for the European beef sector. *Animal Prod. Sci.* 51:30–36.

Hopkins, D.L. 2011. Processing technology changes in the Australian sheep meat industry: An overview. *Animal Prod. Sci.* 51:399–405.

Hopkins, D.L. 2014. Beef processing and carcass and meat quality. In Cottle, D.J. and Kahn, L.P. (eds.), *Beef Cattle Production and Trade*. CSIRO Publishing, Melbourne, pp. 17–46.

Hopkins, D.L., Hegarty, R.S., Walker, P.J., and Pethick, D.W. 2006. Relationship between animal age, intramuscular fat, cooking loss, pH, shear force and eating quality of aged meat from young sheep. *Aust. J. Exp. Agric.* 46:978–984.

Hopkins, D.L., Toohey, E.S., Kerr, M.J., and Van de Ven, R. 2011. Comparison of the G2 Tenderometer and the Lloyd Texture Analyser for measuring shear force in sheep and beef meats. *Animal Prod. Sci.* 51:71–76.

Hopkins, D.L., Toohey, E.S., Pearce, K.L., and Richards, I. 2008. Some important changes in the Australian sheep meat processing industry. *Aust. J. Exp. Agric.* 48:752–756.

Hostetler, R.L., Landmann, W.A., Link, B.A., and Fitzhugh, H.A. 1970. Influence of carcass position during rigor mortis on tenderness of beef muscles: Comparison of two treatments. *J. Animal Sci.* 31:47–50.

Hostetler, R.L., Link, B.A., Landmann, W.A., and Fitzhugh, H.A. 1973. Effect of carcass suspension method on sensory panel scores for some major bovine muscles. *J. Food Sci.* 38:264–267.

Huff-Lonergan, E., and Lonergan, S.M. 2005. Mechanisms of water-holding capacity of meat: The role of postmortem biochemical and structural changes. *Meat Sci.* 71:194–204.

Huff Lonergan, E., Zhang, W., and Lonergan, S.M. 2010. Biochemistry of postmortem muscle—Lessons on mechanisms of meat tenderization. *Meat Sci.* 86:184–195.

Hwang, I.H., Devine, C.E., and Hopkins, D.L. 2003. The biochemical and physical effects of electrical stimulation on beef and sheep meat tenderness. *Meat Sci.* 65(2):677–691.

Hwang, I.H., and Thompson, J.M. 2001. The interaction between pH and temperature decline early postmortem on the calpain system and objective tenderness in electrically stimulated beef longissimus dorsi muscle. *Meat Sci.* 58:167–174.

Jeremiah, L.E. 1998. Development of a quality classification system for lamb carcasses. *Meat Sci.* 48:211–223.

Jones, S.J., Calkins, C.R., Johnson, D.D., and Gwartney, B.L. 2005. *Bovine Mycology.* University of Nebraska, Lincoln, NE, http://bovine.unl.edu.

Jose, C.G., Trezona Murray, M., Channon, H.A., and D'Souza, D.N. 2013. Eating quality linked to ultimate pH and tenderness in Australian fresh pork. In Pluske, J.R. and Pluske, J. (eds.), *Manipulating Pig Production XIV.* Australasian Pig Science Association, Werribee, Victoria, Australia, p. 34.

Kim, Y.H.B., Stuart, A., Nygaard, G., and Rosenvold, K. 2012. High pre rigor temperature limits the ageing potential of beef that is not completely overcome by electrical stimulation and muscle restraining. *Meat Sci.* 91:62–68.

Kim, Y.H.B., Warner, R.D., and Rosenvold, K. 2014. Influence of high pre-rigor temperature and fast pH fall on muscle proteins and meat quality: A review. *Animal Prod. Sci.* 54:375–395.

Klinkner, B.T. 2013. *National Retail Pork Benchmarking Study: Characterizing Pork Quality Attributes of Multiple Cuts in the Self-Serve Meat Case.* M.S. Thesis. North Dakota State University, Fargo, North Dakota, pp. 1–58.

Koohmaraie, M., Kennick, W.H., Elgasim, E.A., and Anglemier, A.F. 1984. Effects of postmortem storage on muscle protein degradation: Analysis by SDS-polyacrylamide gel electrophoresis. *J. Food Sci.* 49:292–293.

Lealiifano, A.K., Pluske, J.R., Nicholls, R.R., Dunshea, F.R., Campbell, R.G., Hennessy, D.P., Miller, D. W., Hansen, C.F. and Mullan, B.P. 2011. Reducing the length of time between harvest and the secondary gonadotropin-releasing factor immunization improves growth performance and clears boar taint compounds in male finishing pigs. *J. Animal Sci.* 89:2782–2792.

Ledward, D.A. 1985. Post-slaughter influences on the formation of metmyoglobin in beef muscles. *Meat Sci.* 15:149–171.

Lee, Y.B., and Ashmore, C.R. 1985. Effect of early postmortem temperature on beef tenderness. *J. Animal Sci.* 60:1588–1596.

Locker, R.H., and Hagyard, C.J. 1963. A cold shortening effect in beef muscles. *J. Sci. Food Agric.* 14:787–793.

Maribo, H., Ertbjerg, P., Andersson, M., Barton-Gade, P., and Møller, A.J. 1999. Electrical stimulation of pigs—Effect on pH fall, meat quality and Cathepsin B + L activity. *Meat Sci.* 52:179–187.

Marsh, B.B. 1954. Rigor mortis in beef. *J. Sci. Food Agric.* 5:70–75.

Marsh, B.B., Lochner, J.V., Takahashi, G., and Kragness, D.D. 1981. Effects of early postmortem pH and temperature on beef tenderness. *Meat Sci.* 5:479–483.

Marsh, B.B., Ringkob, T.P., Russell, R.L., Swartz, D.R., and Pagel, L.A. 1987. Effects of early-postmortem glycolytic rate on beef tenderness. *Meat Sci.* 21:241–248.

Melody, J.L., Lonergan, S.M., Rowe, L.J., Huiatt, T.W., Mayes, M.S., and Huff-Lonergan, E. 2004. Early postmortem biochemical factors influence tenderness and water-holding capacity of three porcine muscles. *J. Animal Sci.* 82:1195–1205.

Miller, M.F., Carr, M.A., Ramsey, C.B., Crockett, K.L., and Hoover, L.C. 2001 Consumer thresholds for establishing the value of beef tenderness. *J. Animal Sci.* 79:3062–3068.

Moeller, S.J., Miller, R.K., Edwards, K.K., Zerby, H.N., Logan, K.E., Aldredge, T.L., Stahl, C.A., Boggess, M., and Box-Steffensmeier, J.M. 2010. Consumer perceptions of pork eating quality as affected by pork quality attributes and end-point cooked temperature. *Meat Sci.* 84:14–22.

Moore, K.L., Mullan, B.P., and D'Souza, D.N. 2012. The interaction between ractopamine supplementation, porcine somatotropin and moisture infusion on pork quality. *Meat Sci.* 92:125–131.

MSA. 2010. *What next for MSA?* http://www.mla.com.au/files/00674356-2089-430e-bfdc-9e33010516ca/mla-producer-forum-2010-michellegorman.pdf and *MSA Beef brochure* http://www.mla.com.au/files/1ecd5a95-a823-4af4-b14b-9d5f007f9d59/msa-beef-brochure.pdf.

Myers, A.J., Scramlin, S.M., Dilger, A.C., Souza, C.M., McKeith, F.K., and Killefer, J. 2009. Contribution of lean, fat, muscle color and degree of doneness to pork and beef species flavor. *Meat Sci.* 82:59–63.

NPPC. 2002. Pork Quality Targets—Fact sheet. Sourced online at: http://www.pork.org/filelibrary/Factsheets/PIGFactsheets/NEWfactSheets/12-04-02g.pdf. (accessed on April 12, 2014).

Ngapo, T.M., and Gariepy, C. 2008. Factors affecting the eating quality of pork. *Crit. Rev. Food Sci. Nutr.* 48:599–633.

Ngapo, T.M., Martin, J.F., and Dransfield, E. 2007. International preferences for pork appearance: I. Consumer choices. *Food Qual. Pref.* 18:26–36.

Pannier, L., Gardner, G.E., Pearce, K.L., McDonagh, M., Ball, A.J., Jacob, R.H, and Pethick, D.W. 2014. Associations of sire estimated breeding values and objective meat quality measurements with sensory scores in Australian lamb. *Meat Sci.* 96:1076–1087.

Pen, S., Kim, Y.H.B., Luc, G., and Young, O.A. 2012. Effect of pre rigor stretching on beef tenderness development. *Meat Sci.* 92:681–686.

Pethick, D.W., Pleasants, A.B., Gee, A.M., Hopkins, D.L., and Ross, I.R. 2006. Eating quality of commercial meat cuts from Australian lambs and sheep. *Proc. NZ Soc. Anim. Prod.* 66:363–367.

Pike, M.M., Ringkob, T.P., Beekman, D.D., Koh, Y.O., and Gerthoffer, W.T. 1993. Quadratic relationship between early-post-mortem glycolytic rate and beef tenderness. *Meat Sci.* 34:13–26.

Polkinghorne, R., Philpott, J., Gee, A., Doljanin, A., and Innes, J. 2008a. Development of a commercial system to apply the Meat Standards Australia grading model to optimise the return on eating quality in a beef supply chain. *Aust. J. Exp. Agric.* 48:1451–1458.

Polkinghorne, R., Thompson, J.M., Watson, R., Gee, A., and Porter, M. 2008b. Evolution of the Meat Standards Australia (MSA) beef grading system. *Aust. J. Exp. Agric.* 48:1351–1359.

Polkinghorne, R.J. 2006. Implementing palatability assured critical control point (PACCP) approach to satisfy consumer demands. *Meat Sci.* 74:180–187.

Polkinghorne, R.J., and Thompson, J.M. 2010. Meat standards and grading: A world view. *Meat Sci.* 86:227–235.

Rhee, M.S., Wheeler, T.L., Shackelford, T.L., and Koohmaraie, M. 2004. Variation in palatability and biochemical traits within and among eleven beef muscles. *J. Animal Sci.* 82:534–550.

Roeber, D.L., Cannell, R.C., Belk, K.E., Tatum, J.D., and Smith, G.C. 2000. Effects of a unique application of electrical stimulation on tenderness, color, and quality attributes of the beef longissimus muscle. *J. Animal Sci.* 78:1504–1509.

Rosenvold, K., and Andersen, H.J. 2003. Factors of significance for pork quality—A review. *Meat Sci.* 64:219–237.

Russo, C., Preziuso, G., and Verità, P. 2003. EU carcass classification system: Carcass and meat quality in light lambs. *Meat Sci.* 64:411–416.

Sammel, L.M., Hunt, M.C., Kropf, D.H., Hachmeister, K.A., Kastner, C.L., and Johnson, D.E. 2002. Influence of chemical characteristics of beef inside and outside semimembranosus on color traits. *J. Food Sci.* 67:1323–1330.

Sañudo, C., Alfonso, M., Sánchez, A., Delfa, R., and Teixeira, A. 2000. Carcass and meat quality in light lambs from different fat classes in the EU carcass classification system. *Meat Sci.* 56:89–94.

Savell, J.W., Mueller, S.L., and Baird, B.E.. 2004. The chilling of carcases. Paper read at 50th International Congress of Meat Science and Technology, at Helsinki, Finland.

Savell, J.W., Mueller, S.L., and Baird, B.E. 2005. The chilling of carcasses. *Meat Sci.* 70:449–459.

Shackelford, S.D, Wheeler, T.L., Meade, M.K., Reagan, J.O., Byrnes, B.L., and Koohmaraie, M. 2001. Consumer impressions of Tender Select beef. *J. Animal Sci.* 79:2605–2614.

Smith, G.C., Arango, T.C., and Carpenter, Z.L. 1971. Effects of physical and mechanical treatments on the tenderness of the beef longissimus. *J. Food Sci.* 36:445–449.

Smith, G.C., Dutson, T.R., Hostetler, R.L., and Carpenter, Z.L. 1976. Fatness, rate of chilling and tenderness of lamb. *J. Food Sci.* 41:748–756.

Smith, G.C., Tatum, J.D., and Belk, K.E. 2008a International perspective: Characterization of United States department of agriculture and meat standards Australia systems for assessing beef quality. *Aust. J. Exp. Agric.* 48:1465–1480.

Smith, G.C., Tatum, J.D., Belk, K.E., and Scanga, J.A. 2008b. *Post-Harvest Practices for Enhancing Beef Tenderness.* National Cattlemen's Beef Association, Centennial, CO, USA.

Taverner, M.R. 2001. Managing the eating quality of pork: Conclusions. In Cranwell, P.D. (ed.), *Manipulating Pig Production VIII.* Australasian Pig Science Association: Werribee, Adelaide, Australia.

Taylor, A.A., and Martoccia, L. 1995. The effect of low voltage and high voltage electrical stimulation on pork quality. *Meat Sci.* 39:319–326.

Taylor, A.A., Nute, G.R., and Warkup, C.C. 1995a. The effect of chilling, electrical stimulation and conditioning on pork eating quality. *Meat Sci.* 39:339–347.

Taylor, A.A., Perry, A.M., and Warkup, C.C. 1995b. Improving pork quality by electrical stimulation or pelvic suspension of carcasses. *Meat Sci.* 39:327–337.

Taylor, A.A., and Tantikov, M.Z. 1992. Effect of different electrical stimulation and chilling treatments on pork quality. *Meat Sci.* 31:381–395.

Taylor, J.M., Toohey, E.S., van de Ven, R., and Hopkins, D.L. 2012. SmartStretch™ technology. IV. The impact on the meat quality of hot-boned beef rostbiff (*m gluteus medius*). *Meat Sci.* 91:527–532.

Thompson, J.M. 2002. Managing meat tenderness. *Meat Sci.* 62:295–308.

Thompson, J.M., Hopkins, D.L., D'Sousa, D., Walker, P.J., Baud, S.R., and Pethick, D.W. 2005. The impact of processing on sensory and objective measurements of sheep meat eating quality. *Aust. J. Exp. Agric.* 45:561–573.

Thompson, J.M., Perry, D., Daly, B., Gardner, G.E., Johnston, D.J., and Pethick, D.W. 2006. Genetic and environmental effects on the muscle structure response post-mortem. *Meat Sci.* 74:59–65.

Tonsor, G.T., and Schroeder, T.C. 2013. Economic needs assessment: Pork quality grading system. Final report prepared for National Pork Board, USA.

Toohey, E.S., van de Ven, R., Thompson, J.M., Geesink, G.H., and Hopkins, D.L. 2012a. SmartStretch™ Technology: V. The impact of SmartStretch™ technology on beef topsides (m. semimembranosus) meat quality traits under commercial processing conditions. *Meat Sci.* 92:24–29.

Toohey, E.S., van de Ven, R., Thompson, J.M., Geesink, G.H., and Hopkins, D.L. 2012b. SmartStretch™ Technology. I. Improving the tenderness of sheep topsides (m. semimembranosus) using a meat stretching device. *Meat Sci.* 91:142–147.

Toohey, E.S., van de Ven, R., Thompson, J.M., Geesink, G.H., and Hopkins, D.L. 2012c. SmartStretch™ Technology. II. Improving the tenderness of leg meat from sheep using a meat stretching device. *Meat Sci.* 91:125–130.

Touraille, C., Monin, G., and Legault, C. 1989. Eating quality of meat from European × Chinese crossbred pigs. *Meat Sci.* 25:177–186.

USDA. 1997. *Official United States Standards for Grades of Carcass Beef.* Agriculutural Marketing Service, United States Department of Agriculture, Washington, DC.

Van de Ven, R.J., Pearce, K.L., and Hopkins, D.L. 2013. Modelling the decline of pH in muscles of lamb carcases. *Meat Sci.* 93:138–143.

Von Seggern, D.D., Calkins, C.R., Johnson, D.D., Brickler, J.E., and Gwartney, B.L. 2005. Muscle profiling: Characterizing the muscles of the beef chuck and round. *Meat Sci.* 71:39–51.

Wang, H., Claus, J.R., and Marriott, N.G. 1994. Selected skeletal alterations to improve tenderness of beef round muscles. *Journal of Muscle Foods* 5:137–147.

Warner, R.D., Dunshea, F.R., Gutzke, D., Lau, J., and Kearney, G. 2014. Factors influencing the incidence of high rigor temperature in beef carcasses in Australia. *Animal Prod. Sci.* 54:363–374.

Watson, R., Polkinghorne, R., and Thompson, J.M. 2008 Development of the meat standards Australia (MSA) prediction model for beef palatability. *Aust. J. Exp. Agric.* 48:1368–1379.

Wood, J.D., Enser, M., Fisher, A.V., Nute, G.R., Sheard, P.R., Richardson, R.I., Hughes, S.I., and Whittington, F.M. 2008. Fat deposition, fatty acid composition and meat quality: A review. *Meat Sci.* 78:343–358.

Wood, J.D., Jones, R.C.D., Francombe, M.A., and Whelehan, O.P. 1986. The effects of fat thickness and sex on pig meat quality with special reference to the problems associated with overleanness. 2. Laboratory and trained taste panel results. *Animal Prod.* 43:535–544.

Index

C

N

Q

R